THE GEOGRAPHICAL ANALYSIS OF POPULATION

With Applications to Planning and Business

DAVID A. PLANE
Professor and Head
Department of Geography and Regional Development
University of Arizona

PETER A. ROGERSON
Professor and Chair
Department of Geography
State University of New York at Buffalo

JOHN WILEY & SONS, INC.
New York   Chichester   Brisbane   Toronto   Singapore

ACQUISITIONS EDITOR Chris Rogers/Marian Provenzano/Barry Harmon
MARKETING MANAGER Catherine Faduska
SENIOR PRODUCTION EDITOR Sandra Russell
DESIGNER Karin Gerdes Kincheloe
MANUFACTURING MANAGER Andrea Price
ILLUSTRATION COORDINATOR Rosa Bryant

This book was set in 10/12 Times Roman by V&M Graphics.

Recognizing the importance of preserving what has been written, it is a policy of John Wiley & Sons, Inc. to have books of enduring value published in the United States printed on acid-free paper, and we exert our best efforts to that end.

Copyright © 1994, by John Wiley & Sons, Inc.

All rights reserved. Published simultaneously in Canada.

Reproduction or translation of any part of this work beyond that permitted by Sections 107 and 108 of the 1976 United States Copyright Act without the permission of the copyright owner is unlawful. Requests for permission or further information should be addressed to the Permissions Department, John Wiley & Sons, Inc.

Library of Congress Cataloging in Publications Data:
Plane, David.
   The geographical analysis of population with applications to planning and business / David A. Plane, Peter A. Rogerson.
     p. cm.
   Includes index.

   1. Population geography.   2. Population—Economic aspects.
3. Migration, Internal.   4. Emigration and immigration.
I. Rogerson, Peter. II. Title.
HB1951.P58 1994   93-48671
304.6—dc20   CIP

Printed in Singapore

10 9 8 7 6 5 4 3 2 1

*For our baby-boom-echo children,
Emily, Ellen, Bethany, and Christopher,
and their baby-boomer moms,
Kathy and Laura.*

PREFACE

An understanding of population distribution, composition, and change is essential for making decisions in both the public and private sectors. Future school enrollments depend critically upon the expected size and age composition of future populations. The size and location of new parks, the design of sewage treatment facilities, and the planning of transportation networks all require a thorough knowledge of how populations are expected to change. In the private sector, the success of advertising campaigns is often linked directly with an ability to identify and target specific demographic groups or segments of the market. The location of new businesses also relies critically upon accurate consideration of demographic factors.

Improvements in the power and ease of computing, as well as the development of specific technologies and databases (such as geographic information systems and the Census Bureau's TIGER cartographic database), have fueled the growth of these applications by enhancing the analyst's ability to carry them out.

A primary objective we had throughout the writing of this book was to convey to the reader the nature of planning and business applications that make use of information on population and population change. We pursued that objective by adopting a geographical perspective and by focusing primarily at a local scale of analysis.

The analysis and the implications of population change have long occupied researchers from a wide range of fields. Demographers, market researchers, geographers, economists, planners, mathematicians, and sociologists have all studied various aspects of population growth and distribution. Our emphasis upon the *geographical* analysis of population reflects a belief that a geographical perspective has much value to contribute to the continuing growth of demographic analysis in planning and business applications. Geographical concepts such as location, accessibility, distance, and interaction are therefore emphasized, as are the inherently geographical aspects of population analysis, such as the study of migration flows and the mapping of population distribution.

Population analysis may be carried out at many geographic scales. Most texts have focused upon topics such as the world population explosion, global hunger issues, and national birth control policies. We have chosen to focus upon smaller geographic scales. Hence we emphasize the illustration of techniques of small-area population estimation and projection, market-demand estimation, retail and public facility location, school enrollment projection, and transportation and recreation demand forecasting. During the past decade, there has been a tremendous increase in interest in such local-scale applications. Throughout, we suggest that a geographical perspective is useful in understanding and analyzing these local-area applications.

Chapters 1 and 2 introduce some of the fundamental geographic notions that provide a foundation for the geographical analysis of population. The next two chapters provide a description of the measurement and analysis of the components of population change. Chapter 3 contains material on fertility and mortality, while Chapter 4 focuses exclusively upon the migration component of change. The following three chapters focus upon demographic analysis. Chapters 5 and 6 treat the traditional demographic problems of population estimation and projection, and in Chapter 7 we devote special attention to the modeling and forecasting of migration flows. The second half of the book is concerned with the implications that population distribution, composition, and change have on applications in the public and private sectors. In Chapter 8, we use examples in transportation and recreation planning to illustrate problems in facility planning. Chapters 9, 10, and 11 also focus upon the effects of population on planning and business, with a progression from the organization of people into households and the use of the household as the unit of analysis (Chapter 9), to the description and analysis of how those households are arranged in space and may be segmented into separate markets (Chapter 10), to the use of geographic information systems in demographic analysis (Chapter 11). In the final chapter we discuss more general issues, such as geographic scale and long-term perspectives on population change, that have a bearing on much of the material presented earlier in the book.

We have deliberately kept the level of mathematics to that of high school algebra. There are one or two exceptions to this (specifically, the use of elementary matrix algebra and the discussion of regression equations), and where they occur, we have made a concerted effort to provide the essentials, so that all students should be able to follow the exposition. In addition, in the Appendices we have added a short review of logarithms for those students who may require it, and a brief overview of the basic concepts of matrix algebra.

In all cases we have attempted to lay out the steps in particular methods in as straightforward a manner as possible. The reader should therefore be less likely to get "bogged down" in the details of a particular technique, and be more likely to appreciate the rest of the discussion on the use of the method, its assumptions, and its limitations.

At the conclusion of each of chapters 2–11 we have provided a set of exercises. These give the reader an opportunity to "try out" the methods described in the chapter. Solutions to most of the exercises are found in the back of the book.

Like many interstate migrants, this book was started on the east coast, and the west coast was its final destination. We first began to put together an outline on the Maine coast, while sanding the sailboat *Ruthless*. The official end of our joint efforts came in July of 1993 on Pescadero State Beach in California, where a coin flip determined the order of authorship.

In addition to the inspiration provided by places, many people have assisted us, either directly, or in shaping our perspectives. The number of individuals who assist somewhere along the way during the process of writing a book increases at least proportionally with the time it takes to write the book, and hence we are grateful to quite a large number of people. Colleagues from the academic community who contributed their time and ideas to help us with the project included Brigitte Waldorf, Richard Morrill, Arthur Getis, Jim Huff, Patricia Gober, Thomas Kontuly, Gordon Mulligan, Roberty Nunley, Andrew Isserman, Paul Densham, Richard Greene, David Sawicki, Kevin McHugh, Brian Sommers, Ge Lin, David Stocking, and Larry Abrams. We are also grateful for the personal assistance provided by the extremely professional staff at Wiley—Chris Rogers, Marian Provenzano, Cathy Faduska, Sandra Russell, Karin Kincheloe, and Rosa Bryant—as well as to the many others who work behind the scenes. The assistance of NSF Grant SES-9022192 to the Center for Advanced Study in the Behavioral Sciences, where Peter Rogerson was a Fellow in 1992–93, is gratefully acknowledged. And finally we would like to thank Barry Harmon, former Wiley geography editor, who worked with us in developing the project. Without Barry's enthusiasm for the venture over the past half decade we might never have completed the book.

David A. Plane
Peter A. Rogerson

TABLE OF CONTENTS

CHAPTER 1   INTRODUCTION   1

1.1   Why Study Population?   1
Local-Scale Applications Require Broad System-Scale Perspectives   2

1.2   The World Scale: Potential Consequences of Population Change   2
Fibonnacci's Rabbits   2
Current and Past Rates of World Population Growth   4
Implications of Falling Birth Rates   7

1.3   The National Scale: Age Composition and Migration Pattern Shifts   8
Impact of the Baby Boom   8
Broad-Scale Migration Pattern Changes   10
The Repercussions of Age Composition Changes on Migration Patterns   12

1.4   The Local Scale: A School Reassignment Case Study   13
Case Study: Catalina Foothills School District Elementary School Reassignment Problem   14
Reasons for Our Focus on the Local Scale   18

1.5   The Interdisciplinary Organization of Population Research   19
Demography and Population Studies   19
Population Geography   19
Population Analysis in Planning, Marketing, and Regional Science   20

1.6   Sources of Population Data   21
U.S. Census   21
Private-Sector Sources   22

CHAPTER 2   POPULATION DISTRIBUTION AND COMPOSITION   23

2.1   Introduction   23
2.2   Population Distribution   23
Population Dot Maps   24

Population Density 25
The Index of Concentration 28
The Lorenz Curve 29
Centers of Population 31

2.3 Measuring Accessibility 37
The Threshold Accessibility Index 37
The Aggregate Accessibility Index 38

2.4 Population Composition 41
Median Age 42
The Dependency Ratio 42
Economic Participation Measures 43
The Sex Ratio 44
Population Pyramids 46
Location Quotients 48

2.5 Geographic Association 50
2.6 Summary 53

CHAPTER 3 DESCRIBING DEMOGRAPHIC CHANGE 57

3.1 Introduction 57
3.2 Population Growth (and Decline) 53
Percentage Changes 58
Geometric Rates of Change 58
Exponential Rates of Change 59
Determining the Geometric and Exponential Rates of Change 61
Logistic Growth 62

3.3 Components of Change and Demographic Accounting 64
Demographic Rates and Probabilities 64
Components of Change 65
Demographic Accounting 66
Illustrative Example: Minor Flows 68

3.4 Measures of Mortality 70
Standardized Death Rates 71
Measures of Infant Mortality 74

3.5 Life Tables: Construction and Interpretation 75
Interpretation of a Life Table as a Cohort of Individuals 75
Alternative Interpretation as a Stationary Population 77
Life Table Construction 77
Survival Ratios 78

3.6 Measures of Fertility 81
Period versus Cohort Rates 86

3.7 Summary 86

CHAPTER 4   MIGRATION: ANALYZING THE GEOGRAPHIC PATTERNS 91

4.1 Introduction 91
4.2 Migration Definition and Measurement 92
Importance of Areal Unit Definition 93
Fixed Period versus Registry-type Data 94

4.3 Measures of Migration 95
Transition Probabilities 95
Gross Migration Rates 97
Net Migration Rates 98
Demographic Effectiveness 98

4.4 The Relationship Between In- and Out-Migration 100
The Intuitive Perspective 100
The Positive Correlation Perspective 100
Lowry's "Asymmetrical" Relationship of In- and Out-Migration 102
Beale's Findings 102
Kriesberg and Vining's Measure of In- and Out-Migration Contributions 103
The Nonlinearity of the Relationship between Out-Migration and Net Migration 104
The Cross-Regional Variance of In-Migration and Out-Migration 105
The Entropy of In- and Out-Migration 105

4.5 Disaggregating by Migrant Characteristics 106
Migration Age Schedules 107
Migration Expectancies and Migraproduction Rates 108
Migrant Selectivity 111
Beaten Paths and the Role of Information 112
Return Migration 113
Duration of Residence Effects and Cumulative Inertia 114

4.6 Methods for Analyzing Geographic Patterns of Migration 115
Principal Component Analysis 115
The Extremal Tendencies of Sonis 118
Spatial Shift-Share Analysis 121

4.7 Summary 124

CHAPTER 5   POPULATION ESTIMATION 127

5.1 Introduction 127
The Parlance of Estimates 128
The Principal Types of Estimation Techniques 130

5.2 Simple Interpolation and Extrapolation 130
Linear Interpolation 130
Alternative Extrapolation Techniques 132

5.3 The Regression or Ratio-Correlation Method 134
A Real World Ratio-Correlation Equation 136
Problems with Ratio-Correlation Estimates 136

5.4 Component Methods 137
Component Method II 137
The Administrative Records Method 140

5.5 Housing-Unit Methods 142
Monitoring Temporal Change in the Number of Housing Units 142
Measuring Occupancy Rates 144
The Persons-per-Housing-Unit Factor 145

5.6 Evaluating Estimates 146
Two Methods for "Testing" Estimation Procedures 146
Some Rules of Thumb to Improve Estimates 147
A Case Study in the Pitfalls of Unequal Weighting 150

5.7 Summary 151

CHAPTER 6 POPULATION PROJECTIONS 155

6.1 Introduction 155
6.2 Elementary Extrapolative Methods 156
Linear Extrapolation 157
Geometric and Exponential Extrapolation 158

6.3 The Single-Region Cohort Component Model 160
Matrix Form of the Cohort Component Model 164
Example 166

6.4 Cohort Component Models with Migration 169
Cohort Survival with Constant Net Migration Rates 169
The Markov Model for Population Redistribution 170
The Interregional Cohort Component Model 174
Accounting-Based Population Projections 176
Other Applications of the Cohort Concept 177

6.5 School Enrollment Projections 177
The Grade Progression Ratio Method 178
The Housing Unit Method 181

6.6 Projecting the Demographic Structure of Organizations 183
6.7 Summary 184

CHAPTER 7 MODELING AND FORECASTING MIGRATION 189

7.1 Introduction 189
7.2 What Migration Quantity to Model? 190
The Volatility of Net Migration 191
The Fallacy of Using Net Migration Rates for Forecasting 192

7.3 The Gravity Model 196
Distance Deterrence 197
Constrained and Unconstrained Gravity Models 199
Fitting an Unconstrained Model 200
Singly Constrained Gravity Models 203
Other Forms of the Gravity Model 204

7.4 The Intervening Opportunities Model 206
An Example: The Opportunity and Constrained Gravity Models Compared 208

7.5 Temporally Varying Transitional Probabilities 211
The Causative Matrix Model 211
Feeney's Model 215
The Destination Population Weighted (DPW) Model 218
Improving on the Basic DPW Model 218

7.6 Economic Gravity Models 219
Lowry's Model 220
The Role of Amenities 222
The Need for a Micro Perspective in Behavioral Migration Research 223
An Important Accounting Constraint: Milne's "Seemingly Unrelated Regression Approach" 224

7.7 Toward Economic–Demographic Models 225
Comprehensive Economic–Demographic Modeling Frameworks 227

7.8 Summary 227

CHAPTER 8 THE ROLE OF POPULATION IN INFRASTRUCTURE PLANNING 231

8.1 Introduction 231
8.2 Urban Travel Forecasting 232
The Role of Free Choice and Constraints in Urban Travel Demand Analysis 232
Steps in the Urban Travel Forecasting Process 233

8.3 Recreation Facilities Planning 245
Recreation Supply 248
Recreation Demand 248
Matching Forecasts of Demand with Supply 254

8.4 Site Location 255
8.5 Impediments to the Production of Improved Forecasts 259
8.6 Optimal Demand Assignment Methods 260
The Un-unified–Unified School District Problem 261
8.7 The Use of Forecasts in Decision Making 269
8.8 Summary 270

CHAPTER 9   HOUSEHOLD DEMOGRAPHY   275

9.1 Introduction 275
9.2 A Brief Look at Households, Location, and Relocation in the United States 276
9.3 Geographic Concepts Relevant to the Study of Households 279
Vacancy Chains 280
9.4 Demographic and Geographic Analysis of Households 284
Householders and Markers 285
Householder Rate Method 285
The Household Membership Rate Method 287
Microsimulation 291
Matching Housing Demand with Housing Supply 293
9.5 Summary 296

CHAPTER 10   DEMOGRAPHICS   299

10.1 Introduction 299
10.2 The Measurement of Diversity and Segregation 301
Diversity Measures 301
Segregation Measures 303
"Cracking and Packing": Diversity and Segregation in Political Redistricting 307
10.3 Factorial Ecology 308
Case Study: Postindustrial Demographic Change in Cleveland 309
10.4 Cluster Analysis for Defining Socially Homogeneous Areas 320
Similarity Measures 322
Agglomerative Clustering 323
Iterative Clustering 327
10.5 Life-Style Clustering for Market Segmentation and Targeting 328
Case Study: The Geography of Grit Magazine *Subscription* 329
10.6 Summary 334

CHAPTER 11 DEMOGRAPHIC AND GEOGRAPHIC INFORMATION SYSTEMS FOR POPULATION ANALYSIS 339

11.1 Introduction 339
11.2 A Brief Historical Account of Early Population Mapping 340
11.3 TIGER: A Digital Cartographic Database 341

The TIGER System 343

11.4 Geographic Information Systems 345
11.5 Demographic and Geographic Information Systems for Population Analysis 345
11.6 Some Capabilities of Demographic and Geographic Information Systems Relevant to Population Analysis 350

Areas of Polygons 350
Areal Interpolation 351
Dasymetric Mapping 352
Choropleth Maps and Geographic Scale 354

11.7 Limitations of GIS Used for Population Analysis 356
11.8 Prospects 357
11.9 Summary 358

CHAPTER 12 CONCLUSIONS 361

12.1 Importance of the Geographical Analysis of Population 361
12.2 Scale 361

Scale Dependence in Choice of Method 361
Different Questions for Different Scales 362
The Modifiable Areal Unit Problem 362
Positive Benefits from Studying Phenomena at Alternative Spatial Scales 363
Results at a Given Scale May Hide Significant Spatial Variation 364

12.3 Long-Term Perspectives on Population Change 366

The Interaction Between Rates and Composition 366

12.4 Geographic Perspectives on Future Population Change 369

APPENDIX A ADDRESSES FOR U.S. ORGANIZATIONS OF INTEREST 373
APPENDIX B GEOGRAPHICAL SUBUNITS AND HIERARCHICAL RELATIONSHIPS 374
APPENDIX C LOGARITHMS 379
APPENDIX D MATRIX ALGEBRA 380

REFERENCES 382
SOLUTIONS TO SELECTED EXERCISES 397
AUTHOR INDEX 408
SUBJECT INDEX 410

> **Imagine all the people....**
> —*John Lennon*

CHAPTER 1
INTRODUCTION

1.1 Why Study Population?
1.2 The World Scale: Potential Consequences of Population Change
1.3 The National Scale: Age Composition and Migration Pattern Shifts
1.4 The Local Scale: A School Reassignment Case Study
1.5 The Interdisciplinary Organization of Population Research
1.6 Sources of Population Data

1.1 WHY STUDY POPULATION?

Population growth and decline, the distribution of population across areas, and the geographic patterns of population characteristics—both economic and social—all exert profound influences on our everyday lives. In this book we present a variety of useful methods for understanding the population dimension that underlies many public policy and business decisions.

Why study population at all? An immense literature on various aspects of population has been published by researchers in demography, geography, planning, economics, sociology, mathematics, and ecology. An examination of the many disciplines contributing to the field makes it clear that there are many possible motivations.

Perhaps the most basic motivation is our simple quest to understand the world in which we live. Why do some areas grow while others decline? Is growth fueled by natural increase—the excess of births over deaths—or by net in-migration from other regions? Why do people move? Do different types of people move for different reasons? Why do fertility rates vary over space and time? What are the consequences of prevailing rates of fertility and mortality? What are the consequences of an aging population for the future structure of the economy? Are we in danger of a population explosion that tests the capacity of our resources? Are we likely to witness declining national population levels in postindustrial societies? Answers to each of these questions and many others are sought both to satisfy our innate curiosity and to help make sense of potentially significant "big picture" societal issues.

Another motivation for the study of population is the recognition that an understanding of population composition, distribution, and change is essential for making decisions in both the public and private sectors. For example, school enrollment levels depend critically upon the size and age composition of population at the neighborhood scale. The number of new parks, the design of sewage treatment facilities, and the planning of transportation networks all require a thorough knowledge of how local populations are likely to change. In the private sector, the success of advertising campaigns is often directly linked with the ability to identify and target specific demographic groups or segments of the market down to the ZIP code level or even smaller geographic units. Choosing the location of new businesses also relies critically on accurate considerations of local demographic factors.

This second group of motivations—closer to home and of more immediate concern to many decisionmakers, practicing planners, and business people—are largely the ones that the methods set forth in this book are intended to address. Our focus is particularly on the local scale of analysis and on practical techniques for finding solutions to real public policy, planning, and business problems.

Local-Scale Applications Require Broad System-Scale Perspectives

In this introductory chapter we seek to emphasize that local-scale demographic trends are often reflections of larger-scale national and global trends. The well-equipped geographical analyst of population must bring to any planning or business problem not only specific demographic technical know-how, but also a broad-based understanding of population trends and issues. It is necessary to understand the context of the questions being asked so as to be able to evaluate the relevancy and significance of the results provided by the various methods and models of population geography. The local-scale planning and business problems that the geographical analyst of population is asked to address are played out against a backdrop of the ever-changing dynamics of world- and national-scale population systems. Although this book is organized around *practical techniques*, we begin our presentation of the geographical analysis of population by discussing in some detail the demographic context within which local-scale public and private-sector decisions are made in the United States.

The next three sections of this introductory chapter are about population issues at three different geographic scales: global, national, and local. We examine in Section 1.2 the extraordinarily rapid numerical growth of our species that has characterized the most recent period of human occupancy of the earth. We next zoom down our geographic focus to consider, in Section 1.3, some current U.S. national-scale demographic trends. Then in Section 1.4 we reach the local scale, presenting a case study of an actual school reassignment problem to illustrate a number of the concerns addressed by the methods found in the remaining chapters of the book. Subsequently, Section 1.5 discusses the nature of the scientific field of population geography—as distinct from demography—and of the new applied field of demographics. We conclude the chapter by overviewing, in Section 1.6, some basic sources of population information that may be consulted by any reader who seeks to undertake a real-world geographical analysis of population.

1.2 THE WORLD SCALE: POTENTIAL CONSEQUENCES OF POPULATION CHANGE

Fibonnacci's Rabbits

According to Asimov (1977, pp. 47–49), one of the first persons to study population change in an analytical way was the great mathematician of the Middle Ages, Leonardo

Fibonnacci. Living around the year 1200 in the major commercial city of Pisa, Fibonnacci was dedicated to trying to convince his local area merchants to convert from Roman to Arabic numerals. Like the ongoing effort in the United States to adopt the metric system of weights and measures, this number system conversion would prove to take several centuries to complete. The population problem posed by Fibonnacci is much more quickly solved!

Suppose you start out the year with a single pair of newborn rabbits. How many pairs of rabbits will you have by year's end? Make the following three assumptions:

1. Rabbits take two months to become mature and start bearing young.
2. Every pair of mature rabbits breeds each month and produces another pair: one male and one female.
3. Rabbits don't mind incest.

The solution to Fibonnacci's problem—144 pairs—is illustrated in Table 1.1. During January and February the initial breeding pair is still maturing. Not until March 1 is the first new pair of rabbits born. On April 1 a second pair of siblings is issued, while the elder brother and sister are still in puberty. A third new pair is brought into the world on May 1 by the original parents, but in addition the first third-generation rabbits are born to the March hares. On June 1 the January, March, and April rabbits are all having litters. Similar logic, carried forward further, results in the final answer of 144 pairs of rabbits by year's end.

As you may recall from your younger school days, there is an alternative, less hairy method for finding the solution than keeping track of all the rabbits' genealogies. This simple population problem illustrates the so-called Fibonnacci series in which each number is the sum of the previous two. Thus it is easily computed that by the end of the 13th month 233 rabbit pairs would exist, after 14 months there would be 377, and so forth. The rabbit population according to this very primitive growth model

TABLE 1.1 Fibonnacci Series Showing the Number of Pairs of Rabbits after Successive Months of Breeding

Month	Pairs of Rabbits
1	1
2	1
3	2
4	3
5	5
6	8
7	13
8	21
9	34
10	55
11	89
12	144

Source: Solution to the problem described in Asimov (1977, pp. 47–49).

zooms upward at an ever faster and faster rate. Asimov notes that the 55th Fibonnacci number is over a trillion, meaning that such a breeding frenzy carried out over just a little more than four and a half years produces more than 2 trillion rabbits, and that in 30 years there would be more rabbits than the estimated number of subatomic particles of matter in the entire universe!

Of course, this simple example is not wholly realistic—rabbits, after all, die. It is the relationship between births and deaths that keeps a population's growth in check. And yet, the actual history of recent human population growth is nearly as extraordinary.

Current and Past Rates of World Population Growth

In mid-1993 the world's population stood at about 5.5 billion. The annual rate of growth (or of what a demographer would term *natural increase*—the excess of births over deaths) was 1.6 percent (Population Reference Bureau, 1993). The consequences of these two very basic descriptors of our planet can be illustrated quite simply and dramatically. Had the human population increased by only 1 percent per year over the past 7,000 years, we would now be facing a standing-room-only situation; in fact, we would need a planet with a radius 14 times the distance from the Sun to Neptune just to have sufficient space for everyone to stand back to back (Durand, 1970)!

Clearly the world's population has not always grown at its present rate, nor can it continue to grow at this rate. Westing (1981), refining a somewhat higher earlier estimate by Keyfitz (1966), estimates that only about 50 billion persons have ever lived. Thus about 1 in 10 of all people ever born are walking the planet today.

As illustrated by Figure 1.1, it wasn't until about 1650 that the current, faster-growth regime began. According to the *International Encyclopedia of Population* (1982, article on World Population, pp. 679–81), only after hunter/gatherer societies gave way to agrarian-based ones was there an extremely slow growth in the world's human population. From 8000 to 6000 B.C., when the first urban civilizations were

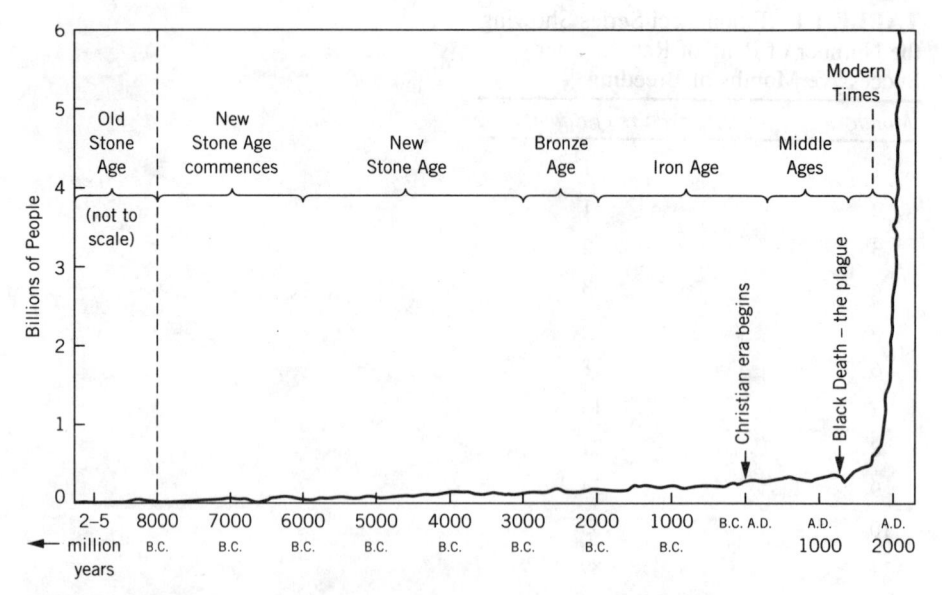

FIGURE 1.1 Historical Growth of the World's Human Population

Source: Population Reference Bureau. (Reprinted by permission.)

being formed (first in Mesopotamia and thence in Egypt, Crete, India, China, and Peru), global population went from perhaps 5 to 10 million. At the time of Christ's birth around A.D. 1 there were probably no more than 300 million people alive, with some estimates (such as Westing's of 200 million) being still lower. The conventionally accepted population figure for 1650—when population growth began to accelerate dramatically—is 500 million, but the course of growth up until that point had been anything but smooth. The most notable decline likely to have ever occurred in the world's population was associated with the bubonic plague or "Black Death" epidemic around 1348, which caused Europe to lose almost a quarter of its people. Among the hardest hit areas was Sweden, whose population was cut in half.

Since around 1650, medical advances have lowered mortality and increased life spans. Whereas during hunter/gatherer times the average person could expect to live only 18 to 20 years, and while throughout the early agrarian period 25 to 30 years was the norm, today life expectancies at birth average 61 years for males and 64 years for females in developing countries. The average life spans in more developed countries are 71 years for males and 78 years for females; for the United States, the numbers are 72 and 79 years, respectively (Population Reference Bureau, 1993).

The period of very rapid population growth that occurs during the period of history when mortality rates decrease from high to low levels and before fertility rates drop commensurately is called the *demographic transition* (Notestein, 1945). As illustrated by the time series of Danish rates shown in Figure 1.2, such transitions in vital rates have characterized the population histories of developed countries during the industrialization era. These kinds of demographic transitions have not been completed, however, in much of the lesser developed parts of the world. About 77 percent of the

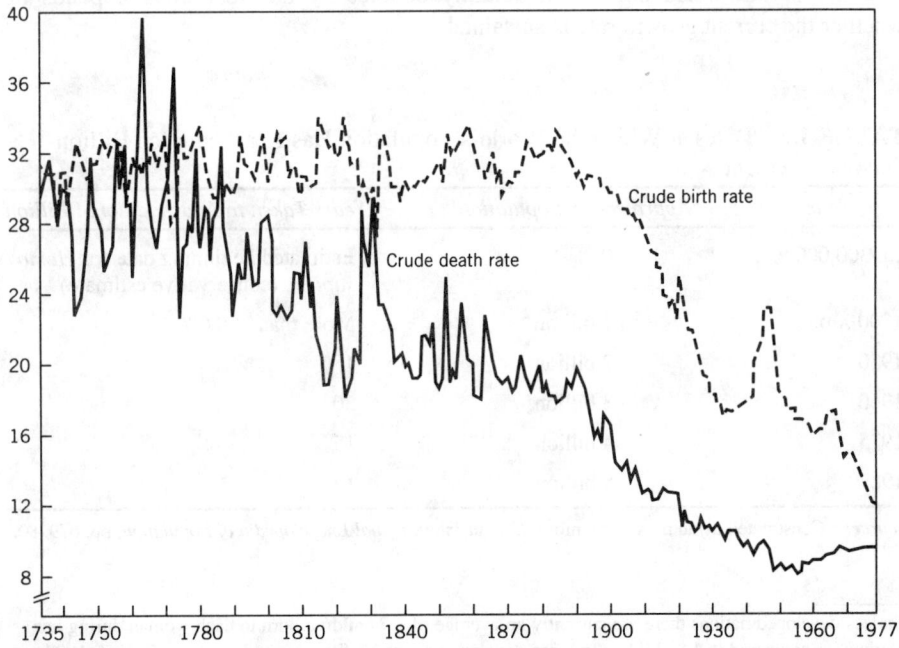

FIGURE 1.2 The Demographic Transition in Birth and Death Rates—Historical Data for Denmark

Source: Jones, 1981, p. 166. (Reprinted by permission.)

world's total population lives in less developed countries (4.2 out of the 5.5 billion total people) where birth rates, though generally dropping, are still on average twice as high as in the developed world.[1] There is thus considerable momentum for future global population growth.

A very effective visual presentation of the magnitude of recent world population growth is a 6½ minute videotape distributed by the Washington, D.C., organization Zero Population Growth. On the screen a world map is shown with numbers in the corner ticking off the years 1 A.D. to 2020. Each time another million people are added to the human population over this period, a small dot in an appropriate part of the world appears. As one watches the tape, surprisingly little change occurs until the final moments—when the dots mushroom rapidly over much of the map.

Table 1.2 lists the approximate dates at which the human population passed each of the five successive billion marks. Though probably not as visually striking as the videotape, the rapidly shrinking periods of time it has taken to add each billion people are a poignant illustration of the seriousness of our ongoing population explosion.

It is often useful to associate annual growth rates with the time it would take for a population to double, if that rate stayed constant into the future. In Chapter 3 we shall delve into detail about methods for representing the growth of a population, but for now a simple approximation commonly used for computing doubling times will serve our purposes. It turns out that a fairly good estimate of the number of years required for a population to become twice as large is found by dividing 0.7 by the annual percentage growth rate expressed as a decimal.[2] For example, a population growing at a constant annual rate of 1 percent doubles in size about every 70 years (0.7 / 0.01); a growth rate of 2 percent implies a doubling every 35 years; and so forth. Table 1.3 shows how the world's hypothetical population doubling times have decreased over the past few centuries, all the way down to today's very short 42 years. Whether or not world population *actually* doubles by the year 2035 depends on whether the current growth rate is sustained.

TABLE 1.2 Dates at Which the World's Population Has Reached Each Billion-Person Increment

Year	Approximate Population	Years Taken to Add Additional Billion
ca. 300,000 B.C.	0	Estimated beginning date for *Homo sapiens* (conservative estimate)
1800 A.D.	1 billion	More than 300,000
1930	2 billion	130
1960	3 billion	30
1975	4 billion	15
1987	5 billion	12

Source: Constructed by authors from information in *International Encyclopedia of Population*, pp. 679–80.

---

[1]In less developed nations there are currently an average of 3.7 children born to each woman during her lifetime, as compared to 1.8 children in more developed countries (Population Reference Bureau, 1993, statistics on total fertility rates).

[2]The actual formula, assuming a continuous growth rate exponential model, is the natural logarithm of 2 divided by the growth rate; the natural logarithm of 2 is approximately 0.6931471—or roughly 0.7!

TABLE 1.3 Historical Evolution of World's Population Doubling Times

Year	Doubling Time at Growth Rate for that Year
1600	260 years
1900	90 years
1950	75 years
1993	42 years

Source: Compiled by authors from data in *International Encyclopedia of Population* (1982) and Population Reference Bureau (1993).

Implications of Falling Birth Rates

We do not intend to be alarmists here. It was fashionable to present examples like these in the late 1960s and early 1970s, a period of increased awareness and concern over rapid acceleration in the use of nonrenewable resources. Indeed, some people are now making arguments about dangers inherent with zero or negative population growth rates, such as those now in evidence in certain developed countries and in the formerly communist nations of Eastern Europe. Examination of a table of growth statistics for all the nations of the world (Population Reference Bureau, 1993) reveals that the current doubling time of 42 years is the weighted average of a doubling time of just 35 years in less developed countries, but a lengthy 162 years in more developed nations.

Some long-term projections made recently by the U.S. Bureau of the Census show the population of the United States leveling off at about 350 million around the year 2050, before beginning a decline in succeeding decades. In 1993 a few countries were already experiencing low rates of natural *decrease*. Recording fewer births than deaths were Germany, Hungary, Bulgaria, the Ukraine, Latvia, and Estonia—while Europe, as a whole, had a rate of natural increase of only 0.2 percent. It is one of the ironies of our times that in many nations of the developing world a major public policy issue is the promotion of birth control, while in, for instance, the nations of Eastern Europe national population policies have been aimed at encouraging higher fertility levels because of the fears of impending labor shortages.

Although no one at present is very much concerned about extinction of the human race through a failure to procreate in sufficient numbers, eventually negative growth rates might lead us to be as concerned about perpetuation of our species as we now should be about the exhaustion of natural resources. The principal point to note here is that strikingly different consequences can be associated with a seemingly small change in demographic rates. To illustrate this point, an average of three children born to every female would lead to an *acceleration* of our current population explosion; but if every female were to give birth to just two children, the species would eventually become extinct (because not all female children would live to the age when they, too, can "replace" their parents).

Throughout this discussion of world-scale demographic issues the reader may have noted that we describe world rates as *weighted averages* of the rates for different parts of the globe. It is virtually impossible to understand any worldwide phenomenon without examining the geographic distribution that underlies it! We focus now on population issues of particular concern in the United States. Here again we shall find the need to zoom down to an even more fine scale level of geographic resolution to fully understand national trends.

1.3 THE NATIONAL SCALE: AGE COMPOSITION AND MIGRATION PATTERN SHIFTS

Powerful demographic forces have been at work throughout the course of U.S. history, fueling national and local policy debates and helping shape economic fortunes and individual psyches. In recent decades two of the more significant trends have been (1) a major shift in age composition, as the post–World War II baby boom generation has passed through the years of labor force entry and into middle age, and (2) a set of sweeping changes in the patterns of inter- and intraregional migration.

Impact of the Baby Boom

The baby boom generation is conventionally defined to include persons born during a period of high fertility rates lasting from 1946 to 1964. The former date coincides with the end of World War II, and the latter is actually the year in which total births in the United States reached a maximum. As shown in the *population pyramid* in Figure 1.3, in 1980 the large baby boom generation ranged in age from 16 to 34. During the 1970s and early 1980s the baby boomers contributed to a tremendous increase in housing demand and labor supply. Increased competition for housing and for entry-level jobs contributed to the stress on families and households. The sheer size of this generation has exerted an important influence on such societal concerns as the crime rate and the divorce rate, as well as leading to significant effects on the economy.

As traced out by Easterlin (1980), the heightened labor force supply pressure induced when a boom generation comes of age acts to depress wage growth, and members of that large generation—especially its youngest members—face relatively more difficult economic prospects and chances for advancement than did their parents. One result of this is likely to be a decline in fertility rates. The decision to have children is, to a certain degree, an expression of optimism about the future, so it is not surprising that fertility levels are low when young adults hold pessimistic views about their own future chances to achieve prosperity.

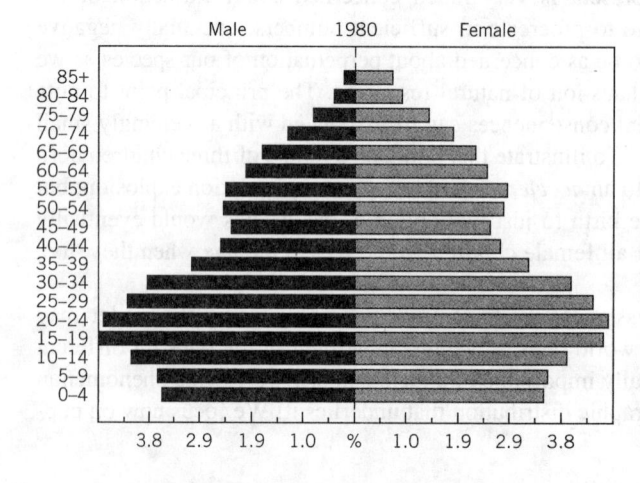

FIGURE 1.3 Population Pyramid Showing Age Composition of the U.S. Resident Population on April 1, 1980

Source: Constructed by the authors from 1980 decennial census data.

During the 1970s U.S. fertility rates were, in fact, at very low levels. Many in the baby boom generation decided to forgo, or at least to postpone, having children. The decline in fertility rates and the emergence of the smaller "baby bust" cohorts shown at the base of the 1980 national population pyramid in Figure 1.3 impacted several areas. For instance, within many school districts, pitched battles were fought over the selection and timing of elementary school closings. In the business sector, many companies needed to rethink product lines that formerly had been geared to youth-oriented markets.

Before the smoke had long settled from the school closing battles, and just as businesses were adjusting to the realities of the new demographics of product demand, an interesting phenomenon began to take place: both the number of births and the birth rate began to increase! Now, during the 1990s, kindergarten and elementary grades are once again swollen as members of the baby boom *echo* generation begin their schooling; and product marketing oriented toward parents of young children is once again in vogue. In the next section of this chapter we shall examine a case study of the interplay of geography and demography as one school district in a rapidly growing region of the country tries to accommodate new elementary grade enrollments.

The upturn in births began in 1979, and birth rates turned upward as well. These increases are due primarily to the larger numbers of persons of childbearing age in the baby boom generation. Although the large cohorts born from 1946 to 1964 began to enter the prime reproductive age groups during the late 1970s and early 1980s, because many couples elected to postpone childbirth, it was not until the mid-1980s that a substantial new "mini" baby boom was fully underway. In turn, the children of the next demographic phenomenon, the baby bust echo, will soon be following on the heels of those of the baby boom echo in causing problems for school planners.

The number of women of childbearing ages has continued to go up during the past decade as has the number of births, even though the nation's overall pregnancy *rate* has begun a downward trend. Data from the National Center for Health Statistics (cited in Kalish, 1993) suggest that the drop in pregnancy rates from 1980 to 1988 can be explained by the baby boomers entering middle age. Back in 1980 members of this generation were 16 to 34 years old, with the median age falling squarely on the highest fertility ages of the early and mid-twenties:

> But by 1988, their age span had shifted to 24 to 42. Because women in different age groups have distinctive patterns of conception and pregnancy outcomes . . . overall birth rates are dropping as this giant generation moves farther and farther from its prime childbearing years. (Kalish, 1993)

Significant changes in the composition of households have also been associated with the aging of the baby boom generation. In particular, the average size of households has declined, while the rate of growth in the number of households has exceeded the rate of population growth.

Household size has declined as a result of higher divorce rates, postponed or forgone childbearing, and relatively low fertility rates. The number of households has grown because of high divorce rates, a slight increase in age at first marriage, and, simply, the large number of baby boom individuals who left home for the first time.

Demographers have traditionally used the individual as their basic unit of analysis. For many practical applications, however, the household is the more appropriate concept. We emphasize the demography of the household throughout the book, especially in Chapter 9. Product marketing and planning for future housing needs are just two examples where the study of household change is more relevant than the study of population change.

The continued aging of the baby boom generation and of the new, more diversified household types in which they live, will continue to have a substantial impact well into the future. The median age of the U.S. population is now over 30 and increasing. Beginning in the year 2011 and continuing through 2029 baby boomers will be celebrating their 65th birthdays. During this period very pressing demands are likely to be placed on our publicly funded social security and health care systems. As shown in Figure 1.4, as currently projected the fractions of the overall population in the traditional working age groups—the groups paying into federal entitlement programs—will be approximately equal to those in the elderly groups who will be receiving payouts. This is a very different situation than has existed to date, when the number of persons paying into these systems has greatly exceeded the number drawing monies out. (Cf. Figures 1.4 and 1.3.)

Broad-Scale Migration Pattern Changes

In many parts of the country shifting mobility and migration patterns have been at least as significant as the broad-scale age composition changes just described. In terms of the overall propensity of the population to move, for three decades after World War II there was remarkably little change. Approximately 20 percent of Americans moved from one housing unit to another in any given year. Then in the mid- to late-1970s a decline began to be noticed. By 1983 only 16.1 percent of the population could be counted as movers. This decline was due, in part, to the increase in female labor force participation. Two-worker households face more restrictions on mobility than single-worker ones (Mincer, 1978). From 1975–76 to 1982–83 there was a marked decline in the mobility of married couples, whereas the movement rate among the never-married population actually increased slightly. Significant impacts on metropolitan housing markets occur when mobility changes for specific demographic segments.

In terms of the geographic patterns of U.S. population movement, some very remarkable trends have characterized the most recent three decades. Since the 1960s, and particularly during the 1970s, there were extremely strong currents of Frostbelt-to-Sunbelt migration. As examined by Vining and Pallone (1982), U.S. trends were consistent with net migration patterns in many other developed nations in which the traditional national "core" region has lost population to the traditional "peripheral" regions. Trends in net population losses by the Northeast and Midwest regions to the South and West from 1935–40 to 1975–80 are shown in Table 1.4.

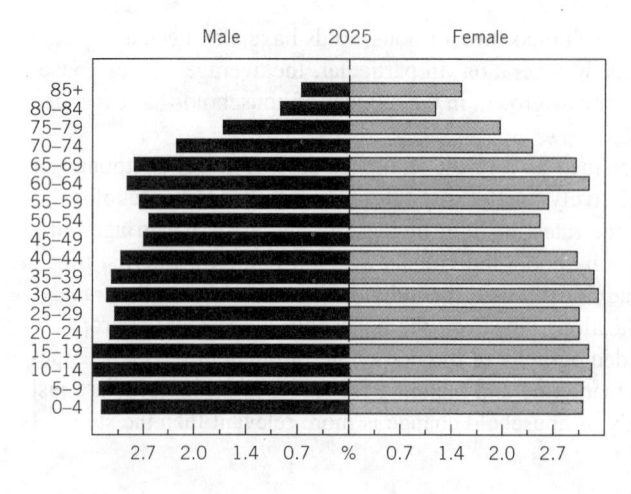

FIGURE 1.4 Population Pyramid Showing the Projected Age Composition of the U.S. Resident Population on July 1, 2025

Source: Constructed by the authors based on U.S. Bureau of the Census projections issued in 1992 (middle series).

TABLE 1.4 Average Annual Population Losses from the Northeast and Midwest Regions to the South and West Because of Net Out-Migration

Period	Annual Net Loss from the Northeast	Annual Net Loss from the Midwest
1935–40	28,000	95,000
1949–50	110,000	79,000
1955–60	120,000	176,000
1965–70	132,000	138,000
1975–80	336,000	297,000
1985–90	226,000	167,000

Note: Data are rounded to the nearest thousand.

Source: Computed by the authors from data in reports of the U.S. decennial censuses of 1940, 1950, 1960, 1970, and 1980 and from unpublished data in a press release for the 1990 census.

Although Frostbelt-to-Sunbelt migration continued in the 1980s, there was a reduction in overall volume. For instance, net migration flows from the Northeast to the South and West declined from 633,000 during 1978–79 to 480,000 during 1982–83. And population movement patterns became more regionally complex than in the 1970s. Figure 1.5 shows states having population gains and losses from net internal migration during the period 1985–90. By the second half of the 1980s all Northeastern states except New York were experiencing positive net in-migration as were four Midwestern ones; on the other hand, four Southern and six Western states were losing

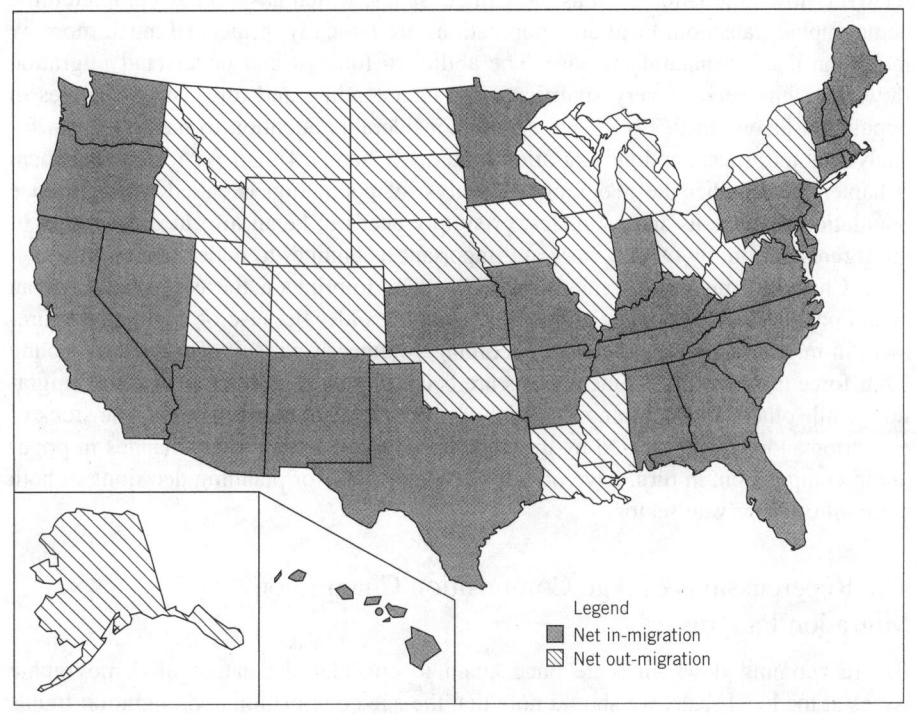

FIGURE 1.5 Direction of Net Interstate Population Exchange for U.S. States, 1985–90

Source: Mapped by the authors from 1990 U.S. decennial census data.

population in their migration interchanges with the rest of the nation. A major factor underlying the most recent trends seems to have been the recent oil glut. Maps such as that shown in Figure 1.5 suggest a 1980s counterbalancing of the major in-movement to states with energy-dependent economies that followed the Arab oil embargo of 1973 and the subsequent drive for energy independence. (See, for example, Plane and Rogerson, 1989; McHugh and Gober, 1992.)

Reversals of trends at a different geographic scale have generated heated academic debate and received considerable popular attention. The 1980 census revealed that for the first time since nationhood, nonmetropolitan areas were growing faster than metropolitan areas. While total U.S. population grew by 11.4 percent during the 1970s, nonmetropolitan counties experienced a growth rate of 15.1 percent. In contrast, the growth rate for metropolitan counties over the same period was only 10.2 percent.

There have been many reasons suggested for this phenomenon of the 1970s. Among the leading ones are increased retirement migration (in part due to a larger cohort of elderly persons reaching early retirement age), migration to amenity-rich areas, a mid-decade recession that disproportionately hit urban areas specialized in manufacturing, and baby boomers going to state universities located in small towns. Interestingly, since 1980 the so-called nonmetropolitan turnaround reversed itself, and metropolitan areas are once again growing faster than the nonmetropolitan portions of the nation.

Changing migration patterns and trends are extremely important for decision makers, planners, and businesspeople. One important consequence of the trends we have just described is obviously the effect on the level of population change experienced in any given place. An area's population may change either through natural increase (the difference between births and deaths) or through net migration (the difference between in- and out-movement). Because both fertility and mortality rates are relatively low in a nation such as the United States, which has already completed the demographic transition, local-area populations are typically influenced much more by migration than by natural increase. The ability to forecast and understand migration flows has thus become very significant in understanding and forecasting changes in population levels. In this book we devote considerable attention to methods both for analyzing and understanding migration patterns (Chapter 4) and for forecasting them (Chapter 7). Whether the geographical analyst of population is called on to produce population projections and estimates, or simply needs to be an informed consumer of the figures put out by others, these are important techniques to understand.

Changing migration trends also leave their imprint on the composition of an area's population. Some populations are becoming older because of significant retirement in-migration, while others are becoming younger as a result of significant young labor force in-migration. Some regions are becoming more affluent because of migration, while others are becoming less so. Some regions gain *human capital* (the stock of education and skills possessed by the population) while others lose. Changes in population composition, in turn, have obvious consequences for planning decisions in both the public and private sectors.

The Repercussions of Age Composition Changes on Migration Patterns

Before zooming down in scale once again to consider the nature of demographic issues at the local scale, we should note that the age composition and migration trends we have talked about at the national scale are themselves likely to be interconnected. As set forth in Plane and Rogerson (1991), age composition change appears to be an important regulator of the timing and spatial patterns of United States migration. We believe it was anything but accidental that the massive speedup in core-to-periphery migration occurred during the same period of time as the passage of the baby boom generation through the years of labor force entry. And the slowdown that we have been witnessing more recently is likely tied in with the continued aging of the baby boomers into their less mobile, midcareer years.

In a nutshell, we believe that two primary effects take place during a period when a generation larger than the previous one comes of age and finds a very tight labor market in their region of birth: (1) overall mobility rates are depressed as the number of opportunities per capita are reduced, and (2) longer-distance migration becomes a more salient alternative. On the other hand, when a smaller cohort follows a larger one, mobility rates should be higher. We could thus posit a long-term cycle in mobility rates similar to the graph in Figure 1.6. (For further details, see Plane and Rogerson, 1991; Plane, 1992, 1993b.)

We also believe that there may be some evidence of a delayed mobility effect among members of the baby boom generation, just as there was the much noted delayed fertility effect. Members of a large generation that experienced reduced mobility during their labor market entry years may at some point later in their careers realize unsatisfied, pent-up migration demand. (See Plane and Rogerson, 1991.)

1.4 THE LOCAL SCALE: A SCHOOL REASSIGNMENT CASE STUDY

On any given day, in any of tens of thousands of localities across the nation, public policy and business decision makers debate a broad spectrum of problems and issues directly or indirectly related to population distribution and demographic trends. Many of these problems and issues strain the ability of even the most sophisticated methods to provide simple answers. In the end, many public- and private-sector deci-

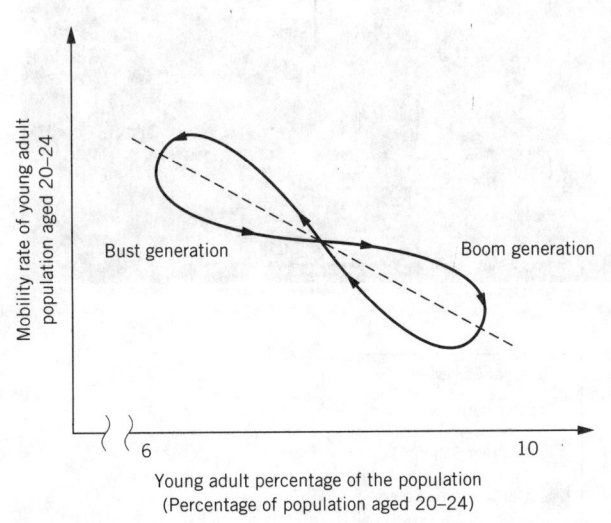

FIGURE 1.6 Hypothesized Intergenerational Cycle in the Geographic Mobility of Young Adults

Source: Adapted from Plane 1993b. (Reprinted by permission.)

sions reflect a trade-off between various criteria—all of which may not be possible to satisfy simultaneously. Our goal in writing this book is to equip the geographical analyst of population to provide the best possible information about such problems, recognizing that in many cases no amount of sophistication with respect to techniques employed can substitute, in the end, for sound decision making.

In this section we focus on the *local scale* of geographical analysis because it is within the friendly confines of many of our localities that population issues most often arise, calling for the input of the population specialist. Subsequent chapters of the book present a variety of examples of different types of local-area planning problems. Here we choose to describe a single real issue facing one of thousands of school boards around the nation. This case study was chosen to illustrate a number of common aspects of problems the analyst may face and, in particular, show how the geographical dimension at the local scale often lends considerable additional complexity to national-scale demographic trends.

Case Study: Catalina Foothills School District Elementary School Reassignment Problem

The map in Figure 1.7 shows Arizona school district number 16 located in the foothills of the Catalina Mountains of metropolitan Tucson. An area of semirugged terrain and lush desert vegetation, the district encompasses a variety of single- and multifamily suburban residential subdivisions.

The area has been experiencing rapid growth during the past 25 years. The oldest neighborhoods, built beginning in the 1930s at very low five-acres-per-housing-unit densities, are found in the southwest corner of the district. Many of the other subdivisions in the western half of the district are now almost fully developed at an average density of one acre per lot. The fastest-growing areas today are located on the eastern side, where some 2,000 building sites currently platted could eventually be built upon. Apartment and townhouse developments are scattered throughout the area,

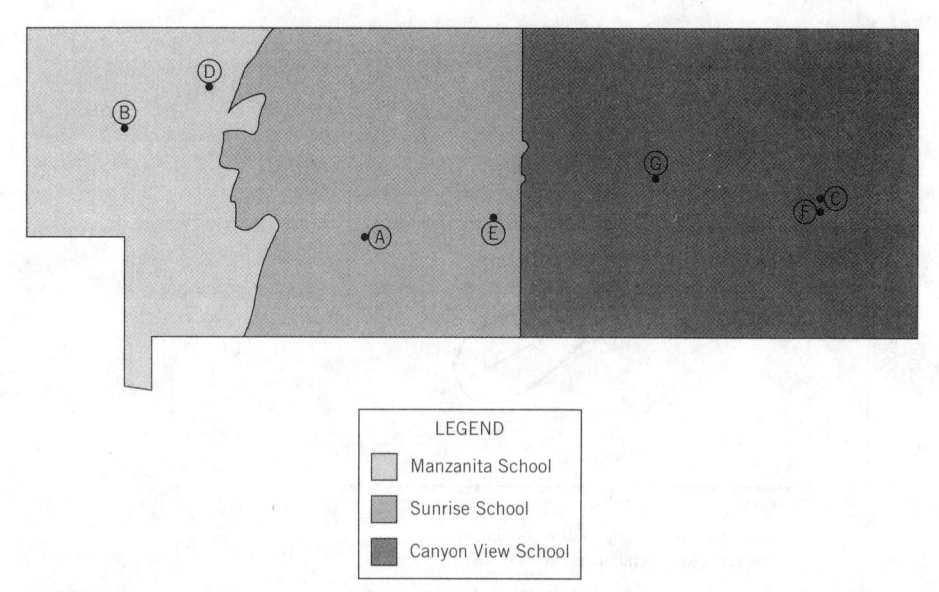

FIGURE 1.7 Location of Schools and 1993 Elementary Catchment Area Boundaries for the Tucson, Arizona, Catalina Foothills School District

with the highest density of such multiple-unit residential land use found on the far eastern edge. The two most significant employment centers in the district are major destination resorts, and the only real commercial developments are several shopping centers in the approximate center of the area.

Internally the racial and ethnic composition of the district does not vary greatly; the entire district has a somewhat lower representation of minority subpopulations than does the greater Tucson metropolitan area as a whole. In this regard, this case study could be considered somewhat simpler than those in many other sections of metropolitan areas where racial balance among schools is also of prime concern.

In fall 1993 the district opened its first high school near the community's commercial center at point A in Figure 1.7. (Prior to 1992 all 9th- through 12th-grade students were bused to other districts' high schools, with District 16 paying their tuition.) Two middle schools, located at points B and C, accommodate grades 6, 7, and 8.

As a result of the national baby boom echo cohort now in the primary school ages, and because of the rapid rate of new home construction in the area (many home buyers being young adults with younger children), the district's three elementary schools are currently overcrowded. With the addition of temporary classrooms each is housing considerably more students than its original 550-pupil design capacity. Enrollments in February 1993 were 660 pupils at Manzanita, the westernmost of the three (at point D), 586 at Sunrise Drive (point E), and 725 at Canyon View (point F). The catchment areas for the three schools as defined are shown in Figure 1.7.

The district recently voted to authorize the sale of bonds to construct a fourth elementary school. Because these bonds will exhaust the district's bond-indebtedness capacity, a fifth school for the primary grades probably cannot be built until after the year 2000. The district's board has purchased an eastside site (point G) and announced its intention to open the new fourth school there at the beginning of the 1994–95 school year. A citizen advisory committee was appointed in early 1993 to draw up boundaries for the catchment (or attendance) areas of elementary schools after the new facility opens. The committee was charged with assigning pupils to the four elementary schools so as to balance six criteria:

1. Set corresponding elementary and middle school boundaries such that attendance areas for Manzanita and Sunrise Drive (elementary) Schools would constitute that for Orange Grove Middle School and attendance zones for Canyon View and the new elementary school would be the zone for Espero Canyon Middle School.

2. Minimize disruption to present school boundaries wherever possible.

3. Set the boundaries such that they will remain valid for the longest time frame possible.

4. Keep neighborhoods intact wherever possible.

5. Provide an equitable distribution of students among the schools (at both elementary and middle school levels), recognizing that the desired capacity for elementary schools is 550.

6. Set the boundaries so that the student transportation system (bus, bike, or walking) to schools is to the maximum benefit of the district.

While few would object to any of these criteria, it quickly became apparent that it would be impossible to fully satisfy all of them. After several meetings the citizen's committee voted to distribute for comment a map of proposed new boundary lines.

16   Chapter 1 • Introduction

Figure 1.8 shows the committee's four future catchment areas. Current elementary school pupils brought home the map along with notification of a public hearing to be held later that same week. The hearing brought out in force parents of the many children who would be displaced under the proposal from their current schools. Parents gave voice to many cogent points regarding what they perceived to be desirable alterations to the proposal.

Close examination of the map in Figure 1.8 can lead us to infer how the committee approached a number of the thorny problems inherent to their assignment. With the exception, perhaps, of the easternmost line between the existing Canyon View school and the new school, the boundaries were not drawn to follow the "natural" catchment areas that a geographer might likely identify. A geographer's ideal functional attendance zones would be drawn so as to send, for the most part, all children to the closest school (via street distance), but modified slightly so as not to divide up neighborhoods. Such ideal lines would take into account the natural boundaries afforded by major "washes" (arroyos, or dry creek beds) that generally demarcate one neighborhood or subdivision from the next, or they would traverse nonresidential land use zones. Such boundaries would recognize the very real social barriers represented by major high-speed thoroughfares that children are forbidden to cross on bike or foot.

The lines proposed by the committee deviate significantly from child-based neighborhood territories. For example, under the committee's plan, children in areas near the numbers I, II, III, and IV on Figure 1.8 would find themselves bused past the closest school to attend one considerably more distant. The reasons for the committee members' choices are (1) they began by assigning all walkers to their present schools so as to cut down on busing costs (hence the "island"-type zone around Sunrise Drive school) and (2) they tried to leave sufficient pupil capacity in each school to accommodate further enrollment growth expected in the next few years. As may be seen in

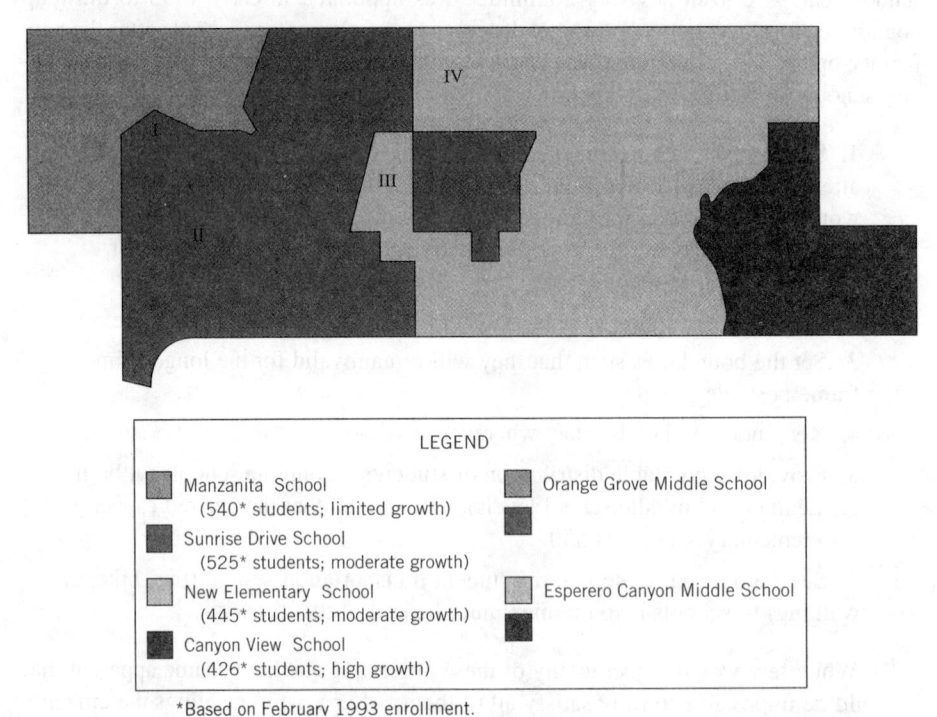

LEGEND

- Manzanita School (540* students; limited growth)
- Sunrise Drive School (525* students; moderate growth)
- New Elementary School (445* students; moderate growth)
- Canyon View School (426* students; high growth)
- Orange Grove Middle School
- Esperero Canyon Middle School

*Based on February 1993 enrollment.

FIGURE 1.8 New Elementary and Middle School Catchment Area Boundaries Proposed by the Advisory Committee to the Catalina Foothills District School Board

the legend of the map, based on current-year enrollments the most space for new growth was reserved at the two eastside schools, where future new home construction is expected to be focused, whereas the limited-growth Manzanita school attendance area already has 540 children in the primary grades.

The dilemma faced by the committee in large part comes about by virtue of the fact that when one seeks to add a fourth facility, after three facilities have already been built, it is not possible to have all four be in optimum locations. The new school site is a good one to address the current, and even larger pending, deficit of classrooms on the eastside. The current overenrollment problem at Manzanita school on the westside, however, is quite suboptimally "solved" by proposing costly busing of children lengthy distances to the east—taking them away from their current friends and neighbors. For the most part, the attendance zone for Sunrise Drive school (located close to the center of the district) would become geographically detached from the immediate vicinity of the school itself.

In attacking the proposal, concerned parents of children to be affected focused on the benefits of neighborhood-based schools, arguing that in suburban areas such as this, the school often serves as the only true community focal point. They vehemently presented the case for preserving as neighborhoods units that sometimes transcended the subdivision boundaries by which the area had originally been developed.

These parents also expressed their concern about the apparent short-term nature of the solution proposed. With such high growth rates forecast on the eastside of the district—where entire new subdivisions are yet to be built—traditional school enrollment forecasting methods such as the grade-progression method (which we shall detail in Chapter 6) are wholly inadequate.

Broader-scale demographic trends such as the aging of the baby boom echo cohort may be easily overwhelmed by migration of new households into the area. In rapid-growth regions such as Arizona, almost wholly uncertain is the type of development that might occur 10 or more years in the future. The implications for school planning are radically different if the approximately 2,000 unbuilt-upon lots in the district are marketed by their developers to elderly, retirement in-migrants rather than to young, labor-force-age in-movers. In this way, the demographically driven decisions in the privately-controlled real estate sector impose massive uncertainty on the demographically driven decisions of public officials!

The concerned parents also argued that short-term flexibility might be desirable in exceeding the 550-pupil optimum size for a school if, in the longer term, the neighborhoods in that school's attendance zone could be expected to "age." In particular, parents in the area designated "I" made the case that the subdivisions feeding their current neighborhood elementary school (Manzanita) have by now been mostly built out. National-scale demographic trends would, in this one part of the district, seem to be quite relevant. Before long it will be the baby bust echo children rather than the baby boom echo children who will use the elementary grade classrooms. Why, therefore, rip asunder natural attendance zones to achieve a more uniform short-term distribution of pupils among schools if, in the long term, any such equality may prove a very elusive goal to achieve? In essence, these parents argued that the committee had placed undue weight on criterion 5 (treating total school size more like a strict constraint) and had given insufficient attention to meeting the goals expressed in criteria 2, 3, and 4.

Of course, underlying the very tricky technical issues associated with a problem such as this are even more fundamental philosophical ones of equity, fairness, and how to plan for an uncertain future. Whereas the methods touted in this book can produce outputs that may become quite helpful inputs into the decision-making process, this case study has hopefully illustrated that such tools are not a panacea for all the world's ills.

A computerized, optimal school assignment algorithm based on solving the Transportation Problem of Linear Programming (a method we will present in Chapter 8) might very well have drawn the exact same boundaries as those recommended by the Catalina Foothills citizen advisory committee! However, to replicate such boundaries, the programmer of the algorithm would have to specify a mathematical objective function and a set of constraints that reflect the same relative implicit weighting of criteria that the committee members used more intuitively. Computerized methods generate radically different proposals under different sets of alternative weightings of criteria.

After generating solutions based on a given set of assumptions, a further stage of analysis is almost always called for. In this *evaluatory* stage, the geographical analyst of population should examine the outcome of the method used to decide whether all the right assumptions were made to start with. In the case of the school reassignment problem just presented, the map produced by the committee is an understandable one—given how members of the committee chose to weight the criteria—but the outcome may or may not really be the best thing for the children of the school district. One of the two authors of this book has a seven-year-old daughter who would be among those children most affected by the committee's proposal. He felt compelled to testify loudly and vehemently that all his years of geographic and demographic training suggested that the proposed map should be torn up!

Reasons for Our Focus on the Local Scale

It is at the local scale of analysis that the expertise of the geographical analyst of population is often the most severely tried. It is here that problems of limited data are most serious, here where the least certain of the three components of population change—migration—exerts its strongest influence, and here where the impacts of decisions hit home the quickest. Not coincidentally, however, it is at this scale of geographical analysis where students of population geography are most likely to find future employment and where their knowledge can most frequently be put to good purposes. For these reasons, and in line with the current slogan to "think globally and act locally," in this textbook we set forth a new framework for the subdiscipline of population geography.

In addition to the increased emphasis on the *local scale of analysis*, our perspective is one that focuses more heavily than current population geography texts on *practical techniques*. We bring in some aspects of geographical analysis more often found in texts for subfields other than population geography. At the same time, we purposefully exclude items that are commonly found in population geography texts but that we feel to be less intrinsically geographical or less useful for real, local-scale public- or private-sector planning applications.

Because our applied approach is an action-oriented one, we shall not dwell overly long here in the introductory chapter on the nature of the academic disciplines of demography, of population geography, and of population studies more generally. Some basic concept of each of these, however, may help the beginning student as well as the practitioner already in the field to know where to seek out further information and data pertinent to a course assignment or a "real-world" problem. We thus conclude this introductory chapter by giving an overview of the organization of research on population in the United States, as well some tips on where to find population literature and demographic data.

1.5 THE INTERDISCIPLINARY ORGANIZATION OF POPULATION RESEARCH

Demography and Population Studies

Population research is highly interdisciplinary. *Demography* is the statistical study of population, whereas *population studies* is a broader term that incorporates many approaches other than the purely mathematical. Demography began through the numerical analysis of mortality statistics and later came to embrace analyses of fertility, and, to a lesser extent, migration and population distribution and composition.

For the most part there are not academic departments of demography. Universities have historically tended to house demography programs in their sociology departments. Also common are independent, interdisciplinary population studies programs or population institutes.

The primary U.S. professional association for academic population researchers as well as for federal, state, and local demographers is the Population Association of America (PAA). Although the largest group of PAA members received their academic training in sociology, also very numerous in the group are population and labor economists. Sizable cohorts of statisticians, geographers, and planners also participate in PAA affairs.

The journal of the PAA is *Demography*. It is highly oriented toward methods and population theory. Reflecting emphases within many sociology and demography programs, coverage of fertility research is quite extensive. Although topics of particular interest to the *geographical* analyst of population—such as distribution and migration—are treated, the research literature on these subjects is spread throughout journals in many different fields.

A very useful source of information on where to find articles on any particular population topic is the *Population Index*. Published quarterly by the Office of Population Research of Princeton University, it is provided to persons belonging to the PAA as a membership benefit. Containing abstracts of papers published in an extraordinarily broad set of sources from around the world, the *Population Index* is easily used. Topical, author, and geographical indexes as well as an annual cumulative index make it easy to research population topics.

Within the PAA a subgroup of researchers and practitioners interested in *applied demography* has been formed. A newsletter sent to members paying a small additional fee contains useful information about practical planning and business-related demographic problems and methods.

Population Geography

Within the broad and eclectic academic discipline of geography, population studies have long assumed an important role. For the geographer it is the *locational* dimension that assumes principal importance. Whereas spatial studies of fertility and mortality patterns can be found in the geographic literature, by far the largest body of population research by geographers focuses on the migration component of population change. Geographic mobility, in general, has been of central importance within human geography; especially significant within the subfield of urban geography have been studies of the spatial distribution of demographic characteristics.

In the United States the geographers' national academic organization is the Association of American Geographers (AAG); an umbrella international group called the

International Geographical Union embraces scholars from many nations. The two journals of the AAG are the more scholarly *Annals of the AAG* and the more applied *Professional Geographer*.

In an often cited article published in the *Annals of the AAG*, Trewartha (1953) set forth the case for a separate subdiscipline of population geography. He presented the view that the field of inquiry for geography should encompass three subcomponents: the physical earth, the people who inhabit the physical earth, and the works of the people of the physical earth. The first of these branches is conventionally called physical geography, while the third was then termed cultural geography. Trewartha argued that the second branch, population geography, should have equal standing with the other two. Today a two-part division of the discipline into physical and human geography is more commonly used, with population geography being a well-recognized subpart of human geography.

Some years ago the AAG created specialty groups to rally together researchers who studied similar topics. One of these, the Population Specialty Group (PSG), carries the flag for geographers in the United States who are interested in spatial demographic phenomena. A chapter of the book *Geography in America* (Gaile and Willmott, eds., 1989) authored by members of the PSG is a good, relatively current "snapshot" of research in geography on population topics. The Population Commission of the International Geographical Union puts out a newsletter and sponsors periodic meetings of population geographers from around the world.

Population Analysis in Planning, Marketing, and Regional Science

Although generally rather practical and applied in nature, geography does tend to be more "academic" in its purview than the field of planning. Planning education seeks largely to provide professional training for practitioners who will go to work largely in city and regional planning agencies. Training in local-scale demographic analysis is a key part of an accredited planning school curriculum. The American Planning Association (APA) publishes the *Journal of the American Planning Association* with articles on applied methods of population analysis.

Within the business college environment, marketing departments tend to be those most oriented toward demographic analysis. Market segmentation analysis and target marketing of products are techniques of central importance for which the demographic dimension is of primary concern. The American Marketing Association (AMA) is the principal scholarly organization. Its publications, the *Journal of Marketing Research* and the *Journal of Marketing* are the most prestigious publication outlets for U.S. researchers.

One final academic field deserves note here. Regional science is an international multidisciplinary field focused on the mathematical analysis of regions. Drawing together in particular regional economists and quantitatively oriented economic geographers (as well as planners, urban engineers, and operations researchers) regional scientists publish on population topics, especially on migration analysis. The Regional Science Association International (RSAI) is the umbrella worldwide organization and publishes *Papers in Regional Science: The Journal of the RSAI*. Among other regional science journals, *The Journal of Regional Science* is the best known.

Appendix A provides addresses of the academic professional associations that we have just described as well as several private-sector organizations distributing materials of interest to the geographical analyst of population.

1.6 SOURCES OF POPULATION DATA

U.S. Census

"Raw" population data needed for actual demographic research comes from various sources. In the United States, the primary source is the decennial census carried out by the U.S. Bureau of the Census. Completed in each of the years ending in a zero since 1790, this census has grown from a simple tally of the names of heads of households (with columns for such counts as the numbers of women, children, and slaves) to encompass a broad range of economic and social characteristics of individuals. In addition, the decennial census of housing is carried out in conjunction with the census of population, thereby giving information on the characteristics of housing units inhabited by people with different socioeconomic characteristics. Table 1.5 lists the categories of information about which the American people were surveyed in 1990. Appendix B describes the geographic units by which 1990 census data have been tabulated.

A useful source of practical advice on how to use U.S. census data is the book *Analysis with local census data: portraits of change* by Myers (1992). For more per-

TABLE 1.5 Items Included in the 1990 Decennial Censuses of Population and Housing

	Population Characteristics	Housing Characteristics
100 percent data:	Household relationship	Number of units in structure
	Sex	Number of rooms in unit
	Race	Tenure (owned or rented)
	Age	Value of home or monthly rent
	Marital status	Congregate housing (meals included in rent)
	Hispanic origin	Vacancy characteristics
Sample data:	Place of birth	Year moved into residence
	Citzenship and year of entry to the United States	Number of bedrooms
		Plumbing and kitchen facilities
	Education (enrollment and attainment)	Telephone in unit
		Vehicles available
	Ancestry	Heating fuel
	Migration (residence in 1985)	Source of water and method of sewage disposal
	Language spoken at home	
	Veteran status	Year structure built
	Disability	Condominium status
	Fertility	Farm residence
	Labor force status	Shelter costs, including utilities
	Place of work and means of journey to work	
	Year last worked	
	Occupation, industry, and class of worker	
	Work experience in 1989	
	Income in 1989	

Source: 1990 U.S. census forms.

sonal assistance, many states have a State Data Center with staff to answer questions or provide help. A nationwide address list of agencies regularly preparing population and housing estimates is available (U.S. Department of Commerce, 1990).

The Statistical Abstract of the United States, a compendium of summary demographic and economic data, is issued each year by the federal government. Most other nations publish demographic yearbooks. The United Nations produces good international comparative data. Besides its annual *Demographic Yearbook,* it provides summary data on a more timely basis in *Population and Statistical Reports.*

Private-Sector Sources

Two private-sector sources are of particular interest to the student or practicing geographical analyst of population. The first of these is the Population Reference Bureau (PRB). The most famous of its many publications is its annual *World Population Data Sheet*—a handy compilation of comparable, current demographic data and rates for all nations. More recently PRB has issued a similar *U.S. Population Data Sheet* containing state-level data. Also distributed to PRB members is an informative newsletter, *The Population Bulletin*, as well as various special reports and teaching materials on particular demographic topics.

A Dow Jones publication, *American Demographics*, has carved out a niche of providing quick-reference demographic information to businesses, as well as being the outlet for advertising by the ever-expanding U.S. demographic information industry. As we shall detail in Chapter 10, the noun *demographics*—which came into use around the mid-1960s—has come to refer to a specialized branch of applied demography, especially in terms of using information on small areas to identify markets for products.

> **Location matters.**
> —*The geographer's motto*

CHAPTER 2
POPULATION DISTRIBUTION AND COMPOSITION

2.1 Introduction
2.2 Population Distribution
2.3 Measuring Accessibility
2.4 Population Composition
2.5 Geographic Association
2.6 Summary

2.1 INTRODUCTION

Before presenting the more applied aspects of the geographical analysis of population beginning with Chapter 5, we cover in Chapters 2, 3, and 4 some important foundational concepts about how human populations are distributed and how these distributions change over time. The methods we present in these early chapters are low-level ones in the sense that they may be used to analyze a wide variety of questions that the analyst may encounter. We attempt to motivate many of the techniques presented, however, by posing them in the context of an actual public planning or business problem.

This chapter begins our examination of the geographical analysis of population by introducing two important, related concepts: *population distribution* and *population composition*. We discuss methods for analyzing a population distribution—the geographic pattern of the location of people or of a subgroup of people—and for examining the composition of a population—the characteristics of people in an area. Two further geographic concepts that are extremely useful for practical problems involving distribution and composition are presented. *Accessibility* is a measure of the proximity of a location to a geographically distributed set of people. *Geographic association* has to do with how the characteristics of people covary across an area.

2.2 POPULATION DISTRIBUTION

Intrinsic to the definition of *population* is the notion of a *set of people*, which has traditionally required the specification of a geographic *area*. At the global scale the concept of population is simple: it is the membership of the species *Homo sapiens*. At the more local scale, however, our increased propensity to travel and move has made the one-to-one assignment of persons to geographic points more problematic and less conceptually appealing than in the past.

Population counts have traditionally been made on the basis of place of inhabitance, as in the case of the U.S. decennial censuses in which persons are enumerated at their "usual place of residence." Increasingly, however, such definitions are proving inadequate to the needs of population analysts. Seasonal movements of population (for example, of transients and the homeless) and the diurnal movements of population within Daily Urban Systems (metropolitan areas) mean that the numbers of persons in various places at various times that are relevant for business and planning purposes require definitions not based on characterizing individuals by fixed geographic coordinates. (See, e.g., McHugh, 1990; McHugh and Mings, 1991; Clark, 1980, 1983; Clark and Burt, 1980; Plane, 1981.)

The school of geographic research known as *time geography* (see, e.g., Hägerstrand, 1970; Dicken and Lloyd, 1981, pp. 20–29) has been influential in creating awareness of the temporal dimension inherent in human geographic distribution. Time geographers have proposed such ideas as the *daily activity bundle* and the *life space prism* to illustrate how each of our lives is expressed geographically as a set of pathways extending in interesting patterns across the surface of the Earth. (See Figure 2.1.)

Population Dot Maps

Having noted the difficulties with the one-to-one assignment of persons to geographic points (issues to which we shall return in Chapter 5, "Population Estimation"), we note that to study population distribution usually requires exactly such an assignment! Perhaps the most direct way of describing a distribution is with a *population dot map*. According to Monmonier and Schnell (1988), population dot-mapping and, indeed, the symbolic representation of population originated with an 1830 map of the distribution of people in France, on which each dot represented 10,000 people. We shall return to examine population mapping in more detail in Chapter 11 when we describe the burgeoning use of geographic information systems (GIS) for population analysis.

The representation of population distribution on a dot map is straightforward. In principle a mark is made to indicate the location of each individual, most commonly

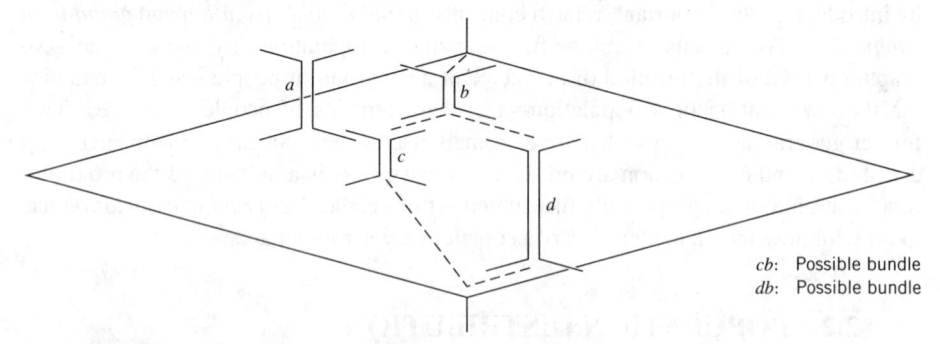

FIGURE 2.1 Daily Activity Bundles and the Life Space Prism

Source: Hägerstrand (1970, p.16). (Reprinted by permission of the Regional Science Association International.)

the location of his or her housing unit (though depending on the geographic scale of the map, aggregate marks to represent the location of many individuals may be required). Geopositioning systems (GPS) now exist that permit analysts in the field to read off the precise geographic coordinates of housing units. Several national censuses have already assigned grid coordinates to basic data. Although currently housing units in the United States are not individually *geocoded,* beginning with 1990, U.S. census data are all assignable to geocoded "blocks" defined for the entire national territory (a block being, as in everyday usage, the area enclosed by a set of connecting streets or roads). Air photos are commonly used in small-area population dot-mapping to pinpoint locations of residents within a census block.

As we have noted, at larger geographic scales, the dots on a population dot map may represent more than a single person. As detailed in standard cartography texts (see, e.g., Dent, 1990; Robinson, et al. 1984), design issues include choosing the appropriate size and number of symbols to represent various numbers of persons and how best to place the symbols to faithfully render the distribution of persons on the land. Monmonier and Schnell (1988, pp. 160–200) provide a nice discussion of all types of maps useful for portraying population distributions.

A population dot map that has captured considerable popular attention and acclaim is the white-on-black "nighttime" map of the distribution of the U.S. population produced by the U.S. Bureau of the Census, first based on the 1970 counts and reissued in updated form after the 1980 census (Figure 2.2).

Population Density

A population dot map such as the Bureau of the Census's provides a conceptually pleasing picture of a population distribution, but for many purposes it is desirable to compute summary measures to understand the attributes of a distribution. Such summary measures generally require the division of the study area into a set of smaller geographic units. Probably the most familiar of these is *population density.* According to Clarke (1972, p. 28), the measure was originally used on maps prepared in 1837 for planning Irish railroads.

The densities of a set of geographic subdivisions, $j = 1, 2, \ldots, r$, may be computed as

$$D_j = P_j / A_j$$

where P_j is the population of subdivision j, and A_j is its area (preferably its land area rather than its total area). The densities, D_j, for each of the r subdivisions are thus expressed in terms of persons per unit of area (such as persons per square mile, persons per square kilometer, persons per acre, or persons per hectare).

The world's density is currently approaching 100 persons per square mile, since the population is approximately 5.5 billion (Population Reference Bureau, 1993) and the land area is 57,821,000 square miles. This is about the same density as that of the states of Kentucky and Louisiana. Note that density inherently smooths out the distribution of population within the areas for which it is measured. The world includes vast unpopulated areas such as Greenland (with a density of only 0.07 persons per square mile) as well as extremely dense urban areas such as Hong Kong (with a density of 247,000).

A number of ways have been suggested to refine the density concept. The standard measure that we have been discussing is sometimes termed *crude* or *arithmetic density.*

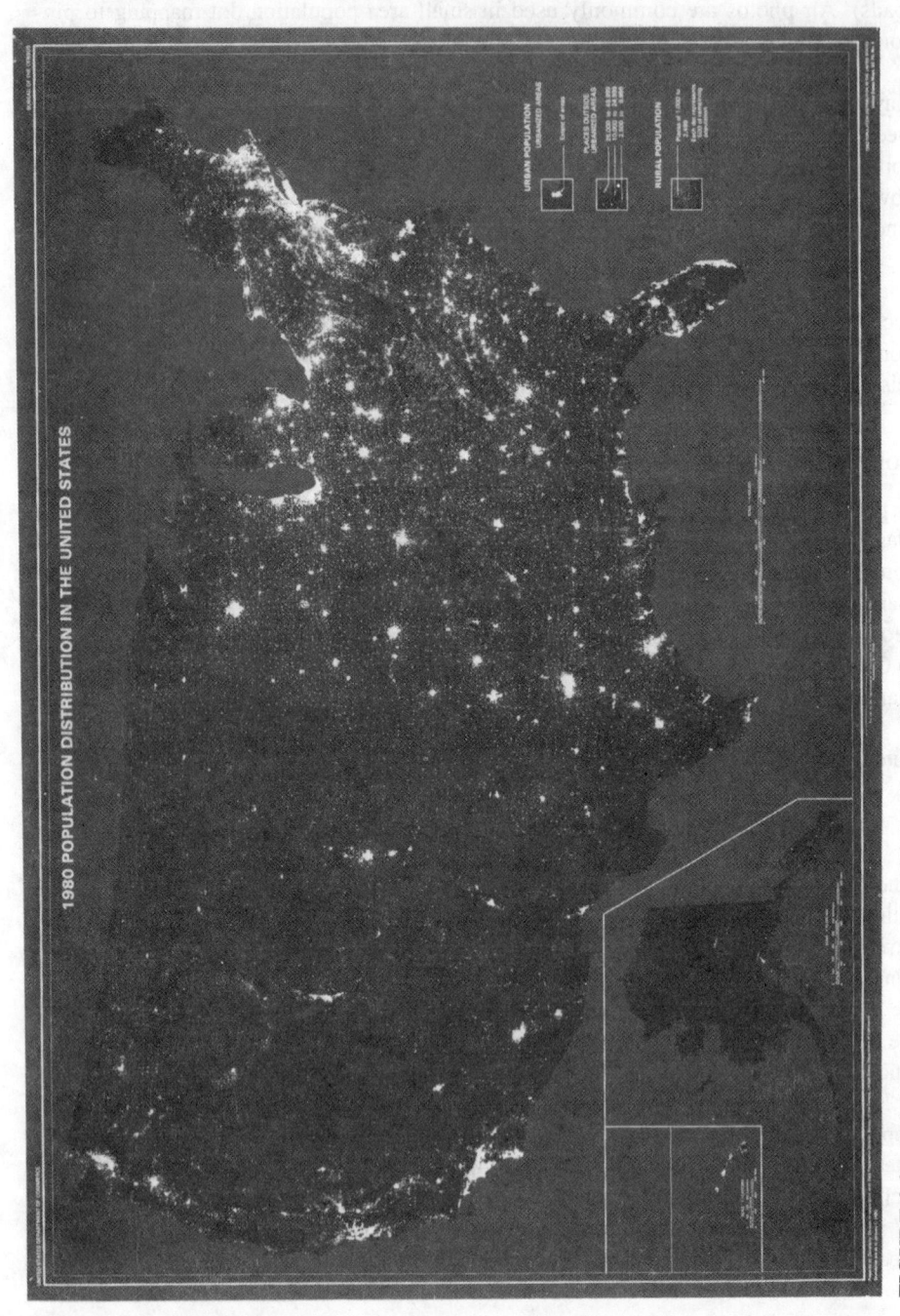

FIGURE 2.2 U.S. Bureau of the Census 1980 Population Dot Map

For national-scale applications the *physiological* or *nutritional density,* computed as population divided by arable land, may provide a useful comparative index of a nation's self-sustainability in terms of food. A variant is the *agricultural density,* which has the same denominator, but whose numerator includes only the farm population. The agricultural density provides perspective on the labor-to-land intensity of farming. It is influenced both by the level of technology characteristic of the society and by the mix of crops grown. We return to practical aspects of density computation in Chapter 11, where we present the concept of dasymetric mapping.

Density is inherently an *areal-specific* concept, meaning that its patterns are strongly affected by the definition of the subunits employed in its computation. *Choropleth* maps are thus more widely used than *isoline* (contour) maps for portraying its patterns. There has, however, been considerable interest in urban geography in trying to understand the continuous patterns of density characteristic of metropolitan areas—that is, to estimate the value of density as we use a finer and finer mesh of areal subunits and to understand how such limiting values are spatially distributed.

A seminal paper by Colin Clark (1951) posited and tested a *negative exponential* model of population distribution (Figure 2.3). The mathematical function plotted has the form

$$D(x) = D_0 e^{-bx}$$

Here $D(x)$ is the density of the metropolitan area at any point x miles away from the "center" (for instance, the peak land value intersection in the Central Business District), D_0 is a parameter representing the hypothetical "central density" at this point, e is the base of the natural logarithms (approximately 2.71828), and b is a second parameter that controls how rapidly densities fall off with increasing distance from the center. The b parameter is the slope of the straight-line relationship that the model posits between the *natural logarithm* of population density and distance from the center of the urban area. (For a short refresher on logarithms, see Appendix C.)

As pointed out in an important modification of this model proposed by Newling (1969), the negative exponential representation of urban density patterns is a better simplification of the daytime than the nighttime distribution of people. Because of the substitution of commercial for residential land use in the Central Business District

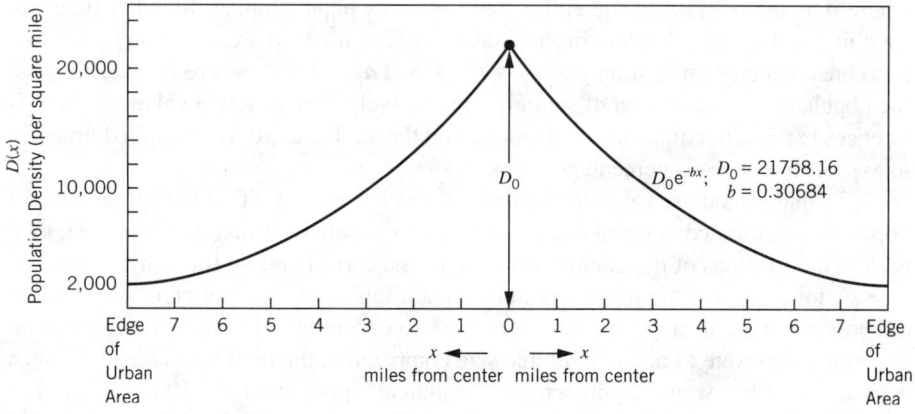

FIGURE 2.3 Negative Exponential Distribution of Population for an Urban Area of 1,000,000

(CBD), residential densities typically are highest along a rim some distance away from the location $x = 0$, with the CBD itself appearing as a "central crater." Newling's modified model proposes a theoretical urban density distribution that resembles a volcano if graphed in three dimensions.

Except in the special case of the limiting values used in these urban density gradient models, density provides a single number for each of a set of areal subdivisions. Two useful devices for *summarizing* how population is distributed across entire sets of subdivisions are (1) the index of concentration and (2) the Lorenz curve.

The Index of Concentration

Considerable attention in recent decades has been given to questions about the tendency of the American population to disperse itself ever more widely. At the local scale, planners are frequently concerned with the effects of urban sprawl as the suburbanization that began during the past century has dramatically intensified since the automobile became the primary means of intraurban transport. At a larger geographic scale, during the 1970s a *migration turnaround* occurred that saw the longstanding metropolitanization trend reversed, as nonmetropolitan counties grew more rapidly than metropolitan ones. (See, e.g., Long, 1981.) This "back to the land" movement, however, proved to be relatively short-lived; beginning again in the 1980s, growth in metro areas outstripped that in nonmetro areas. At the broadest, interregional scale, population deconcentration has been an ongoing phenomenon. During the entire post–World War II period there has been net migration out from the more densely settled Northeast and Midwest regions to the South and West. In the 1970s this trend accelerated dramatically, due, in part, to the coming of age of the baby boom generation (Plane, 1992). This Snowbelt-to-Sunbelt population redistribution has captured considerable popular attention.

How can a population analyst measure the level of population concentration represented by any given distribution? An intuitively pleasing measure is the *index of concentration*. First set forth by Edgar Hoover (1941) to examine interstate population redistribution, it is now in such applications commonly referred to as the *Hoover index*. It is calculated as

$$H = 50 \sum_{i=1}^{r} |p_i - a_i|$$

where p_i is the *fraction* of the entire study region's population, P, found at time t in subunit i, and a_i is the *fraction* of the total study region's land area, A, in i. These fractions are computed quite simply as $p_i = P_i / P$ and $a_i = A_i / A$, where P_i and A_i denote the population and land area of subunit i, respectively. The absolute values of the differences in these fractions are summed up, and the total quantity is multiplied times 50 to express the index in percentage terms.

To understand the values taken on by the index, consider first the special case of population distributed at equal density across all r subareas. If that were the case, then each of the fractions of population would equal the corresponding fraction of area (i.e., $p_i = a_i$, for all i), and the index would be equal to its lower limit of zero. At the other extreme, suppose all population, P, were to be concentrated in a single subunit, and this unit's area were to be of negligible size compared to the total land area, A. In such a case the index would approach its theoretical upper limit of 100 percent. For instance, if the states of the United States (plus the District of Columbia) were used as subunits $i = 1, 2, \ldots, 51$, then the index would have a maximum value of 99.9981 percent, corresponding to the unlikely event that all Americans tried to crowd into the 69 square miles of the nation's capital (quite in contrast to recent out-movement patterns from D.C.!). To compute this limiting value, simply take the difference between 1.0 (the hypothetical fraction of population resident in D.C.) and D.C.'s actual fraction of the nation's land area, 0.000019. It turns out that the sum of all the other absolute differences (between 0, the fraction of the population, and each of the other states' fractions of the land area) will equal an identical amount; this is why the formula says to multiply by 50 rather than 100 to convert the index into percentage terms.

The value of the index based on the population of states recorded in the 1990 census is intermediate between the limiting values: 46.03 percent. The interpretation of an index value such as this is simple. It tells us the percentage of the total population that would have to relocate in order to achieve equal densities in each state (or, equivalently, it tells what fraction of the total land area of the country would have to be moved to other states—but this is a bit harder to visualize!).

One of the principal uses of the index is, in fact, to examine whether actual population growth trends exhibit tendencies toward more uniform distribution, or whether they serve to focus population in a more concentrated pattern. For example, we may compare the 1990 index of 46.03 to a value of 48.04 percent computed based on 1950 census data. The lowering of the index over time tells us that the long-term trend has been toward a more uniform, or less concentrated, pattern of population distribution at the state level. Note that deconcentration, as defined in this fashion, can take place despite a significant increase in overall population densities. During the 40-year post–World War II period from 1950 to 1990, the density of the U.S. population rose from 43 to 70 persons per square mile.

Since Hoover's seminal (1941) paper, several influential studies of U.S. population redistribution have made use of the index of concentration. Duncan, Cuzzort, and Duncan (1961) and Vining and Strauss (1977) show the time trend in the index for a variety of subunits of the nation representing different scales of geographic aggregation. A technical point about the index that such a comparison exposes is that its value is strongly influenced by the number of subunits, r, that are used. The fewer the units, the lower the index.

What is particularly interesting, however, is that different trends may be picked up at different scales of spatial resolution. Vining and Strauss's study disclosed that in the 1970s population was deconcentrating regardless of which of the five sets of units from the earlier Duncan, Cuzzort, and Duncan study were used. Particularly revealing was the reversal during the 1970s of a 50- or 60-year trend toward increasing concentration indicated by the index values computed for counties, state economic areas, or economic subareas. These results were influential in establishing the case that the 1970s migration turnaround represented a clean break from past trends. At the same time, the analysis showed that the broad-scale pattern of largely westward deconcentration continued throughout the century, as picked up by the continuously decreasing index values when computed for the nine census divisions of the nation. Trends were found to be the most complex at the state level, with deconcentration indicated from 1900 to 1940, and then increasing concentration until 1960, when the increasing trend in the index flattens out. Beginning in 1970, deconcentration at the state level reemerges. At this intermediate scale we are probably observing a combination of the continuing deconcentration at the broad interregional scale and the less steady course of metropolitanization at the more localized one.

The Lorenz Curve

The *Lorenz curve* is a useful graphical technique for examining the extent to which a population is concentrated or uniformly dispersed across a set of geographic units. Figure 2.4 shows two Lorenz curves: one for the 1990 distribution of population across the counties of Vermont and one for the distribution in Arizona. Table 2.1 (on pages 32 and 33) illustrates the computations made prior to drawing the curves. The first step was to compute the densities of each county and rank those for each state in order from least to most dense. Then, the percentages of the state's total land area and population found in each county were calculated. Finally, each of these percentages were added up to produce the cumulative numbers needed to plot the curves on the graph.

The Figure 2.4 curves clearly reveal the different settlement patterns characteristic of Vermont and Arizona. Vermont's curve hugs closer to the 45-degree line than does Arizona's, which bows out more. If all counties of a state had the same population density, then the Lorenz curve would lie atop the 45-degree line, whereas in the hypothetical case of a state's entire population being concentrated at a single point the curve would have a right-angled shape, tracing the horizontal axis and the right edge of the graph. Vermont's settlement pattern is for the most part one of uniform distribution, with much of the population living in small villages or fairly evenly dispersed across the rural landscape. Almost all of Arizona's population, on the other hand, is found in either the Phoenix or Tucson metropolitan counties (Maricopa and Pima), with vast stretches of unpopulated lands elsewhere.

The index of concentration that we have previously studied is related to the Lorenz curve. Perhaps the easiest way to compute the index is to find the largest difference between the cumulative area and cumulative population percentages, as shown in the column headed |Cum A − Cum P| in Table 2.1. As illustrated by the final column of the table, however, the same value may be obtained by adding all the *uncumulated* differences and dividing the sum by 2. It turns out that this is a formula that we have already discussed! Look back at the formula for H, the Hoover index (index of concentration).

The Hoover index has a graphical interpretation with respect to the Lorenz curve. The largest absolute difference of the cumulative percentages is geometrically

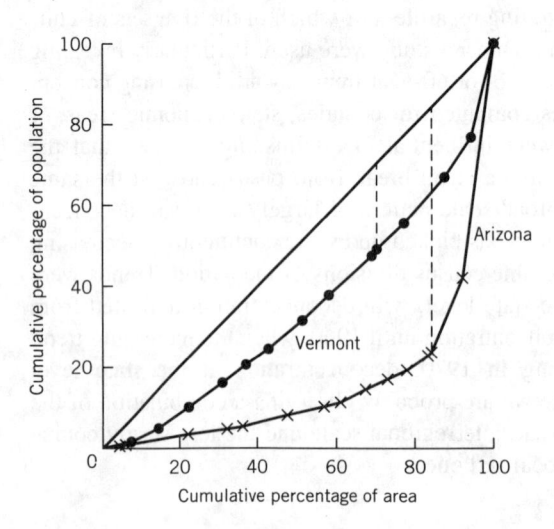

FIGURE 2.4 Lorenz Curves for Vermont and Arizona Counties, 1990

the greatest vertical distance between the curve and the 45-degree line. Vermont's and Arizona's 1990 Hoover indexes are indicated on Figure 2.4 by the dashed vertical lines. Thus, the index is a measure of the bowing out of the Lorenz curve. As we've already discussed algebraically in the previous section, its value varies from 0 percent, indicative of completely uniform distribution across the areal units, to an upper limit of 100 percent. In the former case, there would be no deviation of the curve away from the 45-degree line; in the latter, the maximum deviation would be found along the extreme righthand side of the graph and the area above the curve would encompass the entire triangle under the 45-degree line—corresponding to the case of all areal units except the last having no population, and this last unit having no land area!

Lorenz curves are also commonly used by economists in a nongeographic way to study, for instance, how income is spread among households. Sometimes the index of concentration as we have discussed it here is used in conjunction, but also commonly employed is the *Gini index*. The Gini index is the ratio of the area lying between the Lorenz curve and the 45-degree line to the entire triangular area lying below the 45-degree line. It thus varies between 0 and 1.0. As employed, for instance, in Mulligan (1991), the Gini index may also be computed without any "reference distribution." Lorenz curves and Hoover and Gini indexes are also commonly used to study the spatial segregation of ethnic minority groups. In this latter application the horizontal axis is used for the cumulative percentages of *total* population found in, say, the census tracts of an urban area, and the vertical axis for the cumulative percentages of the *minority group's* population. In Chapter 10 on demographics we reconsider these measures for such purposes, and we give the full formula for computing the Gini index. One of this chapter's exercises foreshadows that discussion and illustrates this usage of the Lorenz curve and index of concentration.

Centers of Population

Another geographical technique to characterize a population distribution is to locate its center. Most people are familiar with the concept of, for example, the center of population for the United States, but can you explain how such a location is actually determined?

Population Centroid

The most common method is to find the *population centroid*, also called the *mean center*, the *mean point*, the *center of gravity*, or sometimes simply the *center of population*. Conceptually, if the mythological Atlas were to hold up the entire area for which a center is being computed—let's say the United States—and assuming that people were the only objects contributing to the weight (and also assuming everyone weighs the same!), the point where he would have to stand to balance the country would be the centroid.

Actual calculation of a population centroid involves weighting each of the n populated geographic coordinates by the number of people residing at that coordinate:

$$\bar{x} = \sum_{i=1}^{n} P_i x_i \bigg/ \sum_{i=1}^{n} P_i, \quad \bar{y} = \sum_{i=1}^{n} P_i y_i \bigg/ \sum_{i=1}^{n} P_i,$$

where $\bar{x}$ and $\bar{y}$ are the coordinates of the centroid, P_i is the population located at point i, and x_i and y_i are the coordinates of i. This formula is readily usable for small areas

TABLE 2.1 Computations Needed to Plot Lorenz Curves for Vermont and Arizona Populations, 1990

Vermont County	Population	Land Area	Density	Percentage of State Total — Population	Percentage of State Total — Area	Cumulative Percentage of Total — Population	Cumulative Percentage of Total — Area	|Cum A − Cum P|	|Pct A − Pct P|
Essex	6,405	666	9.62	1.1	7.2	1.1	7.2	6.0	6.0
Orleans	24,053	697	34.51	4.3	7.5	5.4	14.7	9.3	3.2
Orange	26,149	690	37.90	4.6	7.4	10.1	22.1	12.1	2.8
Addison	32,953	773	42.63	5.9	8.3	15.9	30.5	14.6	2.5
Caledonia	27,846	651	42.77	4.9	7.0	20.9	37.5	16.6	2.1
Lamoille	19,735	461	42.81	3.5	5.0	24.4	42.5	18.1	1.5
Windham	41,588	787	52.84	7.4	8.5	31.8	51.0	19.2	1.1
Bennington	35,845	677	52.95	6.4	7.3	38.1	58.3	20.1	0.9
Windsor	54,055	972	55.61	9.6	10.5	47.7	68.7	21.0	0.9
Grand Isle	5,318	89	59.75	0.9	1.0	48.7	69.7	21.0[a]	0.0
Franklin	39,980	649	61.60	7.1	7.0	55.8	76.7	20.9	0.1
Rutland	62,142	932	66.68	11.0	10.1	66.8	86.7	19.9	1.0
Washington	54,928	690	79.61	9.8	7.4	76.6	94.2	17.6	2.3
Chittenden	131,761	540	244.00	23.4	5.8	100.0	100.0	0.0	17.6
State	562,758	9,273	60.69	100.0	100.0				21.0[b]

2.2 Population Distribution

Arizona County	Population	Land Area	Density	Percentage of State Total Population	Percentage of State Total Area	Cumulative Percentage of Total Population	Cumulative Percentage of Total Area	|Cum A − Cum P|	|Pct A − Pct P|
La Paz	13,844	4,430	3.13	0.4	3.9	0.4	3.9	3.5	3.5
Greenlee	8,008	1,837	4.36	0.2	1.6	0.6	5.5	4.9	1.4
Coconino	96,591	18,608	5.19	2.6	16.4	3.2	21.9	18.7	13.8
Apache	61,591	11,211	5.49	1.7	9.9	4.9	31.8	26.9	8.2
Graham	26,554	4,630	5.74	0.7	4.1	5.6	35.9	30.2	3.4
Mohave	93,497	13,285	7.04	2.6	11.7	8.2	47.6	39.4	9.2
Navajo	77,658	9,955	7.80	2.1	8.8	10.3	56.3	46.0	6.7
Gila	40,216	4,752	8.46	1.1	4.2	11.4	60.5	49.1	3.1
Yavapai	107,714	8,123	13.26	2.9	7.2	14.3	67.7	53.3	4.2
Cochise	97,624	6,218	15.70	2.7	5.5	17.0	73.2	56.2	2.8
Yuma	106,895	5,564	19.21	2.9	4.9	19.9	78.1	58.1	2.0
Pinal	116,379	5,343	21.78	3.2	4.7	23.1	82.8	59.7	1.5
Santa Cruz	29,676	1,238	23.97	0.8	1.1	23.9	83.9	**60.0**[a]	0.3
Pima	666,880	9,187	72.59	18.2	8.1	42.1	92.0	49.9	10.1
Maricopa	2,122,101	9,127	232.51	57.9	8.0	100.0	100.0	0.0	49.9
State	3,665,228	113,508	32.29	100.0	100.0				**60.0**[b]

[a] The Hoover index is equal to the largest absolute difference between the cumulated areas and cumulated populations.
[b] The Hoover index is equal to one-half the sum of the absolute uncumulated differences.

in which simple rectangular (Euclidean) geometry closely approximates locations on the curved surface of the earth. Figure 2.5 gives an example. Other procedures for finding population centroids make use of the fact that the centroid is the point that minimizes the sum of all the squared distances to the dispersed population. When computing a center for a large area such as the United States, x_i and y_i would be the longitudes and latitudes of the centers of the smallest units for which population data could be obtained, and spherical (great circle) distances would be used.

One of the uses of population centers is to characterize how a distribution shifts geographically over time. The drift of the U.S. population over the past two centuries has of course been primarily westward; Figure 2.6 shows that in the past few decades it has also taken a southward turn. The 1980 census revealed the center of population had for the first time crossed the Mississippi River.

Another very important use of the population centroid is for computing more detailed measures of distribution, such as the accessibility index that we will discuss shortly, for inferring such characteristics as the average distance people move from a set of migration data, or for use in demand analysis models such as we will describe in Chapter 8. In these cases it is necessary to *aggregate* a dispersed population at a "representative" location.

Very commonly in actual practice, centroids are estimated rather than precisely computed. There may be no data available at a finer scale of aggregation, or time and cost considerations may suggest a "quick and dirty" determination of centroid locations. In such circumstances the analyst would typically examine a street map or air photo to eyeball the approximate centroid, taking into account the apparent densities of population within the borders of each subunit for which a centroid is needed. The hope is that such a procedure comes closer to picking out the true population centers than would using the geographic centers of the subunits. In metropolitan-scale analy-

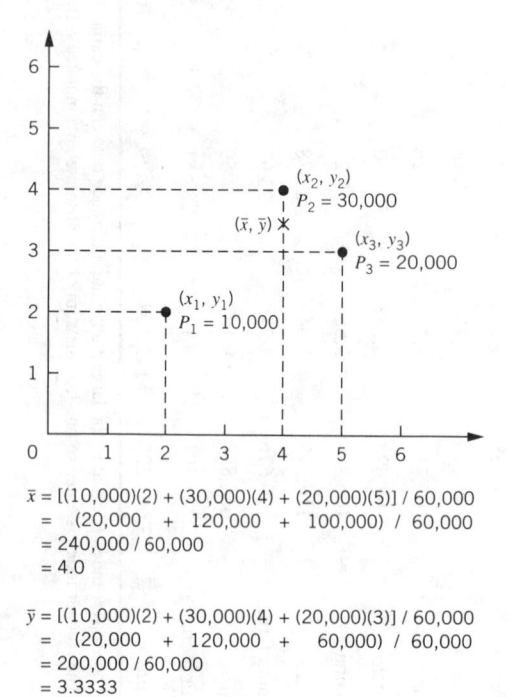

$\bar{x} = [(10,000)(2) + (30,000)(4) + (20,000)(5)] / 60,000$
$\phantom{\bar{x}} = (20,000 + 120,000 + 100,000) / 60,000$
$\phantom{\bar{x}} = 240,000 / 60,000$
$\phantom{\bar{x}} = 4.0$

$\bar{y} = [(10,000)(2) + (30,000)(4) + (20,000)(3)] / 60,000$
$\phantom{\bar{y}} = (20,000 + 120,000 + 60,000) / 60,000$
$\phantom{\bar{y}} = 200,000 / 60,000$
$\phantom{\bar{y}} = 3.3333$

FIGURE 2.5 Example of the Computation of a Population Centroid

FIGURE 2.6 Westward Drift of the U.S. Population Centroid, 1790–1990

ses, using *block group*[1] rather than census tract data minimizes errors from incorrectly located centroids.

Median Center

Centers of population other than the centroid may be defined. (See Clarke, 1972.) The most useful of these is the *median center.* The median center is the *point of minimum aggregate travel.*

Unfortunately it is not quite as simple to find the median center as it is to find the mean center. Interest has run high in simple computational procedures for locating median centers because of the obvious applicability of the concept for optimally locating such facilities as schools, hospitals, and retail outlets. Consider, for example, the problem of choosing the best location for a new regional hospital to serve the three cities indicated in Figure 2.5. One measure of optimality would be to minimize the total distance needed to be traveled if everyone in the region were to travel to the site.

One method to solve such problems is set forth in Griffith and Amrhein (1991). Again, we are considering the case of small areas in which we can ignore the curvature of the earth. The method is called an *iterative algorithm* because it involves finding successively better approximations to the true median center. The median center of a population is the point (X, Y) that minimizes the formula:

$$D = \sum_{i=1}^{n} P_i \sqrt{(x_i - X)^2 + (y_i - Y)^2} = \sum_{i=1}^{n} P_i d_i \qquad (2.1)$$

This expression defines the aggregate sum of travel distances, D, to a point (X, Y) because the population, P_i, at each location defined by coordinates (x_i, y_i) is being multiplied by the Euclidean distance, d_i, between that point and (X, Y). The equation uses

---

[1] A block group is, as the name suggests, simply a group of city blocks. This unit is intermediate between the census tract and the block. See Appendix B for a detailed description of Bureau of the Census statistical units. Although only very restricted amounts of information are disclosed for blocks themselves, in principle the whole range of information available for census tracts may be found for block groups. Although better for many analyses than tract data, block group data have found limited use in the past because they have only been available on microfiche or magnetic tape, whereas tract data have been published in book form.

the Pythagorean formula which states that the sum of the squares of the lengths of the two shorter sides of a right triangle is equal to the square of the hypotenuse. Figure 2.7 illustrates graphically the calculation of the distances, d_i, required for Equation 2.1.

To find the values for X and Y that result in the smallest possible value of total travel distance (which we will call D^*), begin by taking a first approximation (X_0, Y_0) set equal to the coordinates of the mean center of population, (x, y). Then, successively update the approximations X_τ and Y_τ obtained at each step τ by using the formula

$$X_\tau = \left(\sum_{i=1}^{n} P_i x_i / \sqrt{(x_i - X_{\tau-1})^2 + (y_i - Y_{\tau-1})^2} \right) \bigg/ \left(\sum_{i=1}^{n} P_i / \sqrt{(x_i - X_{\tau-1})^2 + (y_i - Y_{\tau-1})^2} \right)$$

$$= \left(\sum_{i=1}^{n} P_i x_i / d_{i,\tau-1} \right) \bigg/ \left(\sum_{i=1}^{n} P_i / d_{i,\tau-1} \right)$$

$$Y_\tau = \left(\sum_{i=1}^{n} P_i y_i / \sqrt{(x_i - X_{\tau-1})^2 + (y_i - Y_{\tau-1})^2} \right) \bigg/ \left(\sum_{i=1}^{n} P_i / \sqrt{(x_i - X_{\tau-1})^2 + (y_i - Y_{\tau-1})^2} \right)$$

$$= \left(\sum_{i=1}^{n} P_i y_i / d_{i,\tau-1} \right) \bigg/ \left(\sum_{i=1}^{n} P_i / d_{i,\tau-1} \right)$$

The distances $d_{i,\tau-1}$ found from the previous estimate of the median center are used to compute new estimates until there is negligible change in the locations from one step to the next.

The calculations to find a median center for the three-city example in Figure 2.5 are shown in Table 2.2. The iterations were halted when each coordinate changed by less than 0.000001 miles. This is called the *convergence criterion*. The minimum sum of travel distances for our new hospital site is $D^* = 56{,}569$ miles for the 60,000 persons in the study area. Note that the median center is at city 2. This is because city 2's population *is equal to the sum of the others*. This illustrates a property of the median center that is a well-known principle in regional science where it has been much studied in the context of industrial location analysis. The analogous concept is that of a "predominant ideal weight." (See, e.g., Hoover and Giarratani, 1984.)

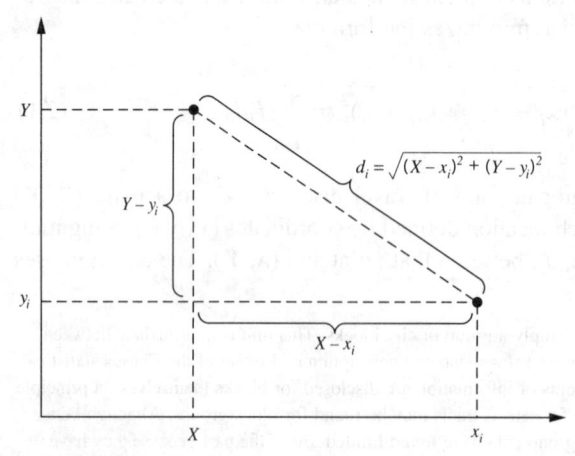

FIGURE 2.7 Illustration of How to Calculate Distances Using the Pythagorean Formula

TABLE 2.2 Calculation of the Median Center for the Three Cities Shown in Figure 2.5

Iteration (τ)	X_τ	Y_τ
0	4.000000	3.333333
1	4.156356	3.599404
2	4.127957	3.711517
3	4.089674	3.784589
4	4.062907	3.839305
5	4.044796	3.880354
.	.	.
.	.	.
.	.	.
38	4.000002	3.999993
39	4.000002	3.999995
40	4.000001	3.999996

Convergence criterion: .000001

Median center $(X,Y) = (4.00000, 4.00000)$

The different location of the median center for this example from the mean center (or centroid) marked on Figure 2.5 illustrates the fundamental difference between the mean and median center concepts. For the mean center the weighting of population at different points is by the square of the distance, so the mean center assigns greater importance to shortening the longest distances than does the median center. The median center treats an extra mile traveled by any person equivalently to any mile traveled by anyone else.

2.3 MEASURING ACCESSIBILITY

The median center has introduced the notion of "optimal locations" with respect to a dispersed population. In many real-world public policy and business site-location applications there is much interest in choosing, if not the *most* convenient, then at least a *quite* convenient site to serve the surrounding population.

The concept of *accessibility* provides an index of the relative desirability of different sites. An accessibility measure gives the analyst a comparative perspective on how close various sites are to the entire dispersed population of the study area. Indexes may be computed for a set of preselected sites (for example, parcels of available land) or a map can be produced showing a continuous *accessibility surface*. From an accessibility map the value of the index at any location may be found.

The Threshold Accessibility Index

One of the more common ways to measure accessibility is simply to report the number of persons living within a circular area of radius R miles (or other distance units) from the site. This is a rather gross measure, and the results obtained may be sensitive to the choice of R, which is called the *threshold radius*.

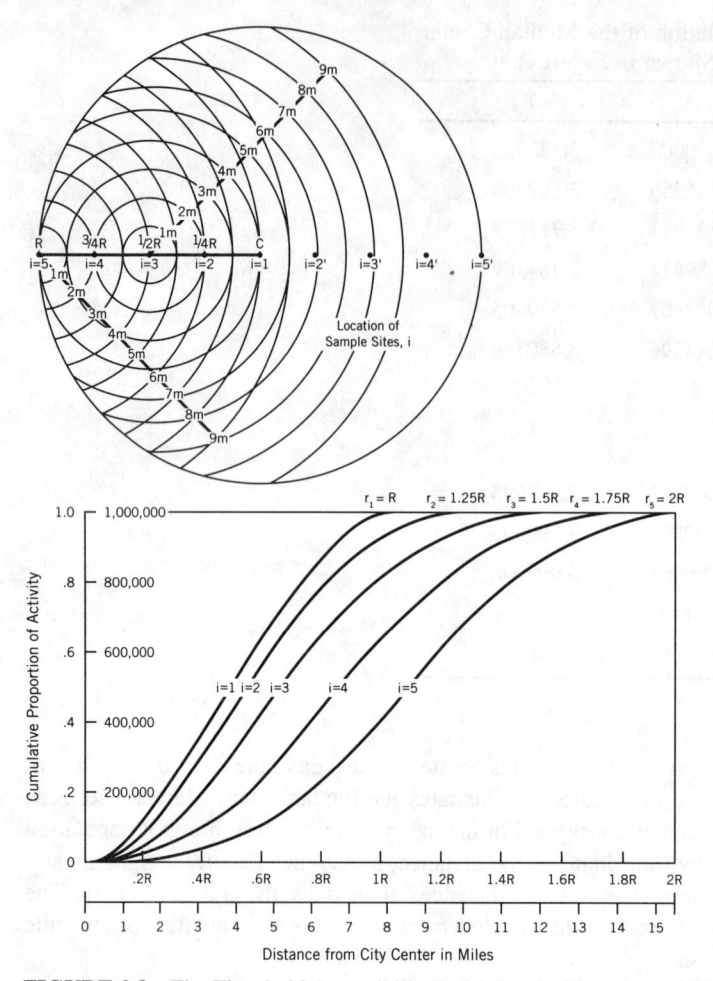

FIGURE 2.8 The Threshold Accessibility Index for the Hypothetical Urban Area of 1,000,000 Shown in Figure 2.3

Figure 2.8 shows pictorially and graphically the *threshold accessibility index* values for two locations in a hypothetical urban area of one million population that has a perfectly symmetrical, negative exponential population distribution (i.e., a distribution that follows exactly the model of Colin Clark discussed in Section 2.2). All possible threshold radii, *R,* are considered. The bottom panel of the figure shows that smooth, S-shaped functions are obtained for each of the sample sites indicated in the top panel. In such a hypothetical urban area, the more central the site, the more accessible it is. In a real urban area, however, with less smooth variation in population densities, the analogous curves will quite likely cross one another—meaning that the relative ranking of the accessibility of different sites will depend on the value of *R* chosen.

The Aggregate Accessibility Index

A common alternative to a threshold accessibility measure is provided by the following aggregate accessibility index:

$$V_j = \sum_{i=1}^{r} (P_i / d_{ij})$$

The accessibility index value, V_j, at any site j is found by first taking the population, P_i, of every subarea i in the study area and dividing by the distance d_{ij} from its centroid to the site. Then these ratios are summed up for all r subareas, $i = 1, 2, \ldots, r$.

This kind of accessibility index is sometimes called *population potential*. (See, e.g., the discussion in Coffey, 1981.) The word *potential* derives from the usage of the term in physics, the field from which the geographer's accessibility concept was borrowed. The name, however, may cause needless confusion because the most accessible locations need not exhibit the greatest potential for population growth! The index is also sometimes referred to as *Hansen accessibility* after the author of a classic (1959) paper in the *Journal of the American Planning Association* which, in fact, explained the relationship between accessibility and the probability of vacant, urban land being developed.

To illustrate the computation and use of the aggregate accessibility index, suppose you have just been elected to the city council of a hypothetical town. As illustrated in Figure 2.9, your district encompasses four precincts. These have populations of

$P_1 = 1,000$
$P_2 = 2,000$
$P_3 = 3,000$
$P_4 = 4,000$

Their population centroids are indicated by large dots on the map. You wish to choose a site in the district for an office where citizens can conveniently come to present their concerns to you—and you can conveniently dispense patronage!

In such a real world application, the problem would likely be broken down into two phases: (1) site identification and (2) site selection. The results of the first phase would be a limited set of available, feasible office sites from which a choice is to be made in the second phase—with accessibility being one of the primary decision variables. Let's suppose that phase (1) has identified three feasible sites ($j = 1, 2,$ and 3) identified by the X's in Figure 2.9.

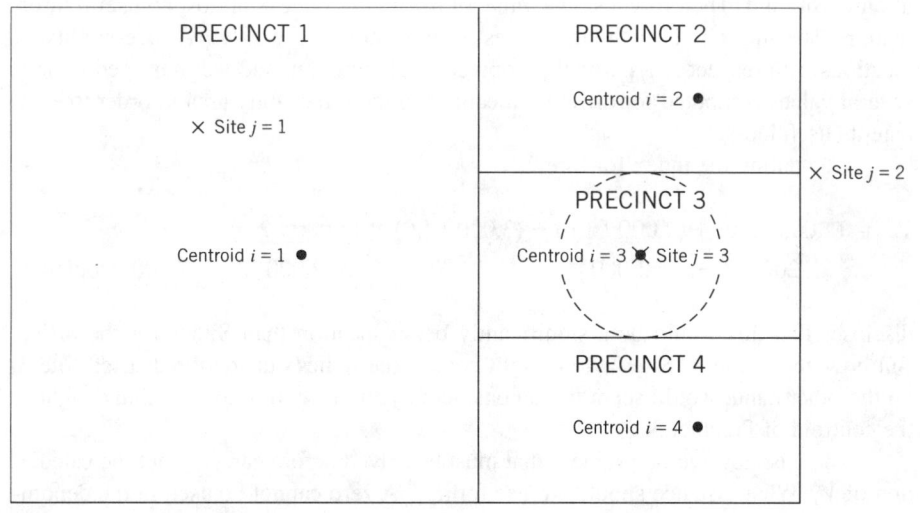

FIGURE 2.9 Location of Precincts, Their Population Centroids, and Potential Office Sites in a Hypothetical City Council District

TABLE 2.3 Distances (in Kilometers) between Population Centroids and Office Sites for the City Council District Accessibility Index Example

	To site:	$j = 1$	$j = 2$	$j = 3$
From precinct:	$i = 1$	2	5	4
	$i = 2$	4	1	1
	$i = 3$	6	1	—
	$i = 4$	8	2	1

Table 2.3 shows the distances, in kilometers, between the four precinct centroids and the three sites.

Using the population and distance figures, the accessibility index value for Site 1 may be computed as

$$V_1 = (1{,}000 / 2.0) + (2{,}000 / 4.0) + (3{,}000 / 6.0) + (4{,}000 / 8.0)$$
$$= 500 + 500 + 500 + 500 = 2{,}000 \text{ people / km}$$

Note that each of the fractions contributes the same amount to the overall value of the index. In this made-up example, the larger the population of the precinct, the farther away it is from this site. This is hardly the optimal situation to maximize accessibility!

Note, also, the unit of the index: people/kilometers. At first consideration this may seem odd; we are more used to, for instance, the units of density—persons per square kilometer. It is, however, simply a result of the accessibility concept. The closer the people of the district are, in aggregate, to a site, the smaller will be the distances in the denominators of the fractions, and thus the greater will be the value of accessibility.

We have suggested that the distances in the fractions of the index for Site 1 are less than ideal. So, is 2,000 people/km a *low* value for accessibility? It turns out that the magnitudes of the aggregate accessibility index vary depending on the population included in the overall study area, as well as on how dispersed the population distribution is within it. Therefore a single value of the index conveys almost no useful information. The index is best thought of as an indicator of the *relative* accessibility of locations with respect to a particular population distribution, and we thus need to have several values computed for different locations in the same study area in order to evaluate its usefulness.

Computing the index for Site 2,

$$V_2 = (1{,}000 / 5.0) + (2{,}000 / 1.0) + (3{,}000 / 1.0) + (4{,}000 / 2.0)$$
$$= 200 + 2{,}000 + 3{,}000 + 2{,}000 = 7{,}200 \text{ people/km}$$

discloses that this would be a significantly better location than Site 1 for the office. But how about Site 3? Site 2, you might note, actually lies outside the district. Site 3, on the other hand, would seem to be considerably more central, as it is found right at the centroid of Precinct 3.

There is, however, a problem that must be solved before carrying out the calculation of V_3. What distance should we use for d_{33}? A zero cannot be used in the denominator of one of the terms of the index, or it will "blow up." And no one, in fact, is likely to be residing in the vacant office space that we are calling Site 3. Furthermore, even if someone were, most of the population of Precinct 3 still lives some distance away from the centroid. There are several ways to handle the computation of this part of the accessibility measure (sometimes called *self-potential!*). One of the simplest is to use half the *nearest neighbor distance*. In other words, find the closest of the other population centroids and take half the distance between that one and the site. Examining the distances in Table 2.3, we see that the centroids of Precincts 2 and 4 are both 1.0 km from that of Precinct 3. We would therefore assume that the typical resident of Precinct 3 lives 0.5 km from Site 3. As illustrated by the dashed circle of radius 0.5 km in Figure 2.9, this seems like a quite plausible assumption; some residents of the precinct actually live farther away, and some live even closer. Thus

$$V_3 = (1{,}000 / 4.0) + (2{,}000 / 1.0) + (3{,}000 / 0.5) + (4{,}000 / 1.0)$$
$$= \quad 250 \quad + \quad 2{,}000 \quad + \quad 6{,}000 \quad + \quad 4{,}000 \quad = 12{,}250 \text{ people / km}$$

Site 3, from the accessibility standpoint, is the clear winner!

Figure 2.10 shows a complete accessibility surface for the study area hypothetical town used in the city council district example. Each line connects up points having the indicated accessibility index values. It is not possible to do such a detailed map from only the three site-specific calculations that we showed. To draw such a surface a computer program would be used to calculate the values at a large number of points and linear interpolation would be used to place each contour line. The analyst would also need finer-scaled population data (say, at the block group or block level) to do a very accurate map.

2.4 POPULATION COMPOSITION

The concept of the *composition* of a population is intrinsically bound together with that of population distribution. Unless we are concerned with the characteristics of the entire world's population, the aggregate characteristics that typify a set of people are strongly affected by how the set is geographically defined. This is certainly

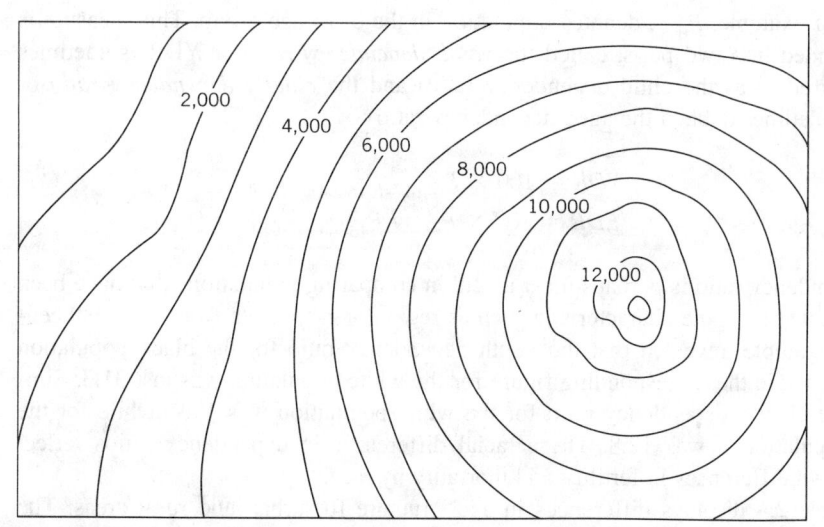

FIGURE 2.10 The Population Accessibility Surface for the Hypothetical City Shown in Figure 2.9

well recognized, and has been much studied, in the context of political redistricting—a topic that we take up in Chapter 10. In fact, in the vast majority of the practical applications we talk about in this book, the concern is with the *distribution* of population characteristics.

In this section we present a selection of the many measures commonly used for analyzing population composition.

Median Age

The ages of the members of a population have broad-reaching implications for demographic processes and for public planning and business applications. Most demographers, if told they could have data with detail on only a single population characteristic, would, without much hesitation, choose age.

There are numerous aggregate measures employed to summarize the age composition of a population. As a measure of central tendency, the *median age* is usually preferable to the mean age. Either, however, may mask important characteristics of the age composition. For example, despite the well-known large flows of elderly migrants to Arizona, that state's population has a lower median age than does the nation as a whole. The reason is the relatively large number of children due to the ethnic composition of the state (with its large Spanish-origin and American Indian subpopulations having higher than average fertility). On the other hand, there are smaller percentages of Arizona's population in the middle, working ages than in the nation as a whole. The next measure we consider is aimed at determining this aspect of an area's age composition.

The Dependency Ratio

A useful measure often employed by demographers and planners compares the youth and elderly populations to the population of working age. The *dependency ratio* is usually defined as

$$DR = 100 \times (P_{0-14} + P_{65+}) / P_{15-64}$$

where, for example, P_{0-14} denotes population in the 0–14 age group. This measure is often divided into two parts, called the *youth dependency ratio* or YDR (sometimes also referred to as the child dependency ratio) and the *elderly dependency ratio* or EDR (sometimes dubbed the aged dependency ratio):

$$YDR = 100 \times P_{0-14} / P_{15-64}$$
$$EDR = 100 \times P_{65+} / P_{15-64}$$

The dependency ratio is perhaps most useful in comparing populations that have been disaggregated by some characteristic such as region or race. Data from the 1980 census, for example, revealed that the youth dependency ratio for the black population was 42.0, while the corresponding figure for the white population was just 31.3. Similarly, the elderly dependency ratio for the white population was 19.9, while for the black population it was 12.8. These racial differences in dependency ratios reflect well-known differences in fertility and mortality by race.

Table 2.4 displays differences in age structure for urban and rural areas. The youth dependency ratio is clearly higher in rural areas than in urban areas, reflecting,

TABLE 2.4 Dependency Ratios for U.S. Regions

	Northeast	Midwest	South	West
Elderly (EDR)				
Urban	19.2	17.1	17.1	14.8
Rural	16.2	18.0	17.3	14.8
Youth (YDR)				
Urban	30.5	33.6	33.6	32.9
Rural	36.7	39.6	38.8	39.2

Source: U.S. Bureau of the Census, *1980 Census of Population,* United States Summary, PC(1), Table 54.

in part, the higher fertility rates characteristic of rural areas. The low youth dependency ratio in the Northeast likely reflects the substantial out-migration of young couples and their families prior to 1980. Note also that the elderly dependency ratio is lowest in the West. We may conclude that the influx of young adult migrants and their families has had more effect on the age structure of the West than has in-migration by the elderly. The most interesting of the urban–rural differences is that of the Northeast, where the high elderly dependency ratio may be attributed in part to outflows of the nonelderly to other regions.

Economic Participation Measures

Whereas the dependency ratio measures the extent to which a population has a large or small percentage of persons of economically active age, the *labor force participation rate* (LFPR) measures the extent to which persons in the traditional working years actually are working or trying to find work. The rate is most commonly defined as a percentage:

$$\text{LFPR} = (\text{LF} / P_{15-64}) \times 100 = [(E + U) / P_{15-64}] \times 100$$

where LF is the labor force—made up of the total number of employed persons, E, and unemployed persons who are actively seeking work, U. The labor force excludes all persons not actively seeking employment, including so-called discouraged workers (the hard-core unemployed who have given up on seeking employment). The denominator is total population 15 to 64 years of age, as in the dependency ratio.

Clarke (1972, p. 85) presented some interesting data on male economic activity (labor force participation) rates by age for developed and lesser developed nations in 1950. For all ages combined, developed countries had a rate of 61.5 percent, whereas for countries in the least developed category it was 58.5 percent. Interestingly, however, the age-specific rates in the least developed countries were higher than (or equal to) the comparable rates for developed nations. Especially notable were the differences for those age 10 to 14 (30.8 percent economically active in the least developed nations versus only 4.9 percent in the most developed category) and for those age 15 to 19 (81.8 percent versus 68.9 percent). This is an example of *Simpson's paradox.* Because of the relatively higher number of children in lesser developed nations than developed ones, and because labor force participation rates for children are below the overall male average, the aggregate rate turns out to be lower for lesser developed countries—despite the much greater prevalence of child labor!

Two related measures are the *unemployment rate*:

$$UR = (U/P_{15-64}) \times 100$$

and the *employment-to-population ratio:*

$$EPR = E/P_{15-64}$$

The definition of what constitutes labor force participation is rather problematic—especially in terms of how to determine who among the jobless are actively seeking employment. Though less commonly used than the labor force participation rate, many econometricians prefer the EPR because of the difficulty of interpreting unemployment data and trends.

Most of the variation in U.S. labor force participation rates over time has been due to the rising number of women working in the "money-wage" sector. (Traditional economic data do not include home work as employment.) Geographically there is also considerable variation in female labor force participation, although exact statistics are hard to find. Worldwide, differing definitions (when statistics are available) make comparisons hazardous. Table 2.5, however, classifies rates in general terms. Note that it matters a great deal whether we talk about a country's rural or urban areas.

The Sex Ratio

> Females are capable of producing about 18 babies during their lifespan,
> half of which are usually one or the other sex.
>
> —Quoted from the *Rockland* (N.Y.) *Review* in
> *The New Yorker,* January 28, 1991

As foreshadowed by our discussion of economic participation measures, sex (now alternatively termed gender) is also an extremely important dimension for disaggregating population data. The *sex ratio* of a population is defined as the number of males per 100 females. A number greater than 100, therefore, implies there are more males than females, whereas a sex ratio of less than 100 indicates the greater prevalence of females than males.

Although the sex ratio varies from country to country, in most cases it is below 100. Clarke (1972) describes three major factors that influence this ratio and that cause it to deviate from 100. One such factor is the preponderance of male births, which is a

TABLE 2.5 Female Labor Force Participation Rates for Selected World Regions

	Rural Areas	*Urban Areas*
Arab nations	Low	High
Latin America	Low	High
Africa	High	Low
India	High	Low
Southeast Asia	High	High

biological characteristic not only of human populations, but of most mammals. The U.S. sex ratio at birth is about 105, and the trend over time has been upward. Negative relationships between the sex ratio at birth and both the age of parents and birth order have been noted; older parents are more likely to have girls, and younger siblings are more likely to be female. In addition, blacks tend to have lower sex ratios at birth than whites. Reasons for these biological differences are not very well understood.

A second factor influencing the sex ratio is the differential mortality of males and females. In general, male mortality rates are significantly higher than those of females. This of course leads to a tendency for sex ratios to be less than 100. Differential mortality has a much more significant influence on the overall sex ratio for a country than does the excess of male births; hence national sex ratios tend to be less than 100.

The underlying cause for the higher mortality rates of males has not been definitively established. Because male mortality is higher at virtually all ages (indeed, even beginning at conception!), biological causes are undoubtedly of some significance. Figure 2.11 shows the inverse relationship between the sex ratio and age for the United States; the ratio declines from over 100 at birth to well under 50 for older age groups. There is also substantial geographic variation in the degree to which mortality varies by sex. Figure 2.12 gives a sampling of life expectancies by sex for various nations. Notice that the more developed countries generally exhibit the greatest variation between male and female mortality.

Finally, migration may also exert a sizable influence on regional and national sex ratios, because many migration streams are sex-selective. For example, when the western United States was settled during the 19th century, migration from the East was predominantly by men. Consequently, sex ratios in the West increased while

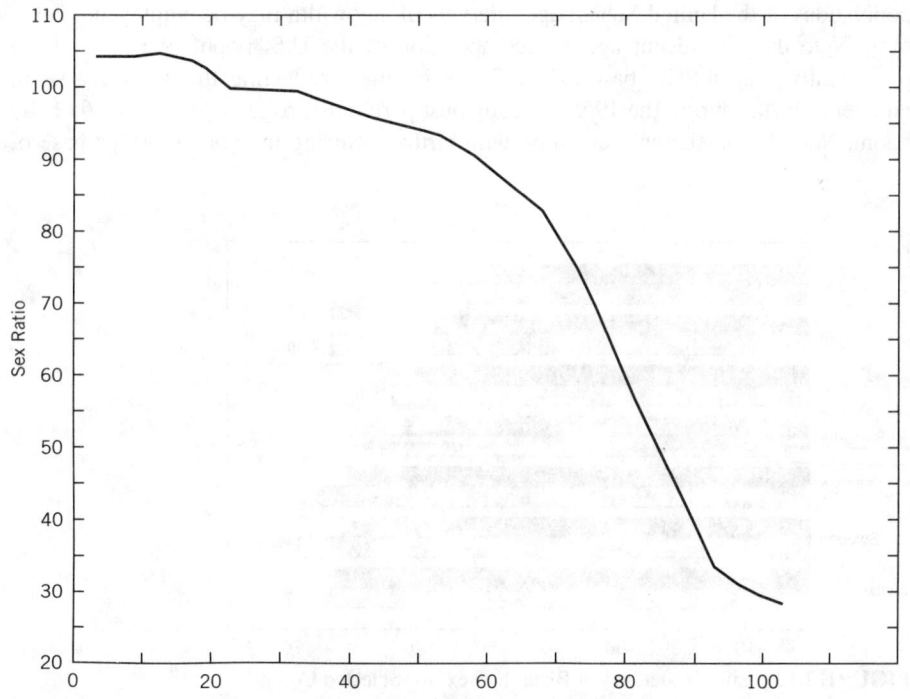

FIGURE 2.11 Inverse Relationship between U.S. Sex Ratio and Age

Source: 1988 Population Estimates, U.S. Bureau of the Census.

46   Chapter 2 • *Population Distribution and Composition*

those in the East declined. Likewise, the vast waves of immigration into the United States during the late 19th century and early 20th centuries were predominantly male, causing the sex ratio to rise in those states receiving the bulk of the immigrants. Because U.S. immigration and internal migration is not as sex-selective as it once was, these effects are relatively minor today. Table 2.6, however, shows that the sex ratio in the western United States is still higher than it is in any other region of the country. It is also interesting to observe that sex ratios are higher in rural than urban regions; females have historically had higher rural-to-urban migration rates than males, reflective perhaps of the dearth of jobs predominantly oriented toward females in rural areas. Table 2.7 shows that the 19th-century decline of sex ratios in the East that we noted has continued into the 20th century in New Jersey. The reasons, however, might differ. Whereas earlier out-migration to the western frontier (with high sex ratios) may have been a leading cause, today the aging of these eastern states' populations is probably the major part of the story. Note that by 1990 the ratio had turned upwards, perhaps as a result of decelerating male-female mortality differences, as well as, perhaps, changes in migration.

Population Pyramids

Population pyramids, such as those we showed in Chapter 1 for the U.S. population in 1980 and 2025, provide a convenient graphical means for describing the combined age and sex structure of a population. Population pyramids are a special form of a horizontal bar graph, with age groups on the vertical axis and the numbers of males and females in these groups on the horizontal axis. By longstanding convention, men are shown on the left and women on the right. Although it is common to portray pyramids with five-year age groups (as in Figures 1.3 and 1.4 for the 1980 and projected future populations of the United States) age intervals of any width may be employed.

Note that the recent age–sex composition of the U.S. population is far from pyramidally shaped. The base of the figure is small, reflecting the relatively small number of births during the 1965–79 baby bust period that followed the 1946–64 baby boom. Note, however, that beginning with births occurring in about 1980 the base of

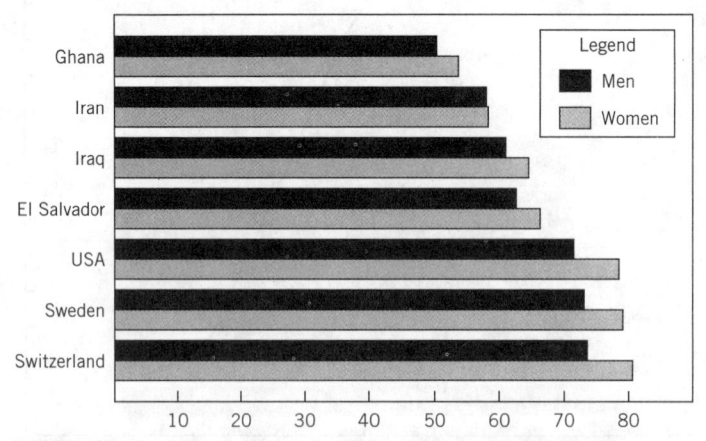

FIGURE 2.12   Life Expectancy at Birth, by Sex, for Selected Countries

Source: Graphed from data in *1990 World Almanac.*

TABLE 2.6 Sex Ratios by Region and Urban/Rural Residence, United States (1980)

	Northeast	Midwest	South	West	Total
Urban	89.7	92.0	92.3	96.7	92.5
Rural	98.8	101.0	98.6	105.4	100.1
Total	91.5	94.6	94.3	98.0	94.5

the pyramid begins to turn outward, reflecting the increase in births attributable to the large number of baby boomers that were then coming into childbearing age. Children in these youngest age groups are members of the baby boom *echo* generation. The large midriff of the pyramid is formed by their by now middle-aged baby boomer parents. The top of the pyramid also deviates from a pyramidal form, failing to collapse inward very rapidly until the oldest age groups are reached. This reflects the low mortality that has been reached in the United States.

Other, more subtle features of the age–sex composition may be noted from population pyramids. The base of the pyramid is generally wider on the left than the right, reflecting the higher sex ratio at birth, whereas the top of the figure is weighted more heavily on the right, consistent with the higher mortality rates of males than females. For some nations, such as the former Soviet Union, the effects of wars would be particularly notable in terms of the jagged appearance of especially the left edge of the graph. Single-year-of-age pyramids have also been used to graphically demonstrate the degree to which people lie about their ages! Shryock and Siegel (1976) note that in many societies there seems to be a preference for reporting ages that end in 0 or 5. The consequent overestimation of the number of individuals in particular age groups is referred to as *age heaping*. One way to avoid this problem in survey work is to ask for a respondent's year of birth rather than age.

On the more local scale, age–sex structure may diverge widely from the characteristic shapes of national population pyramids. As already noted in passing in our discussion of the sex ratio, many patterns of migration are highly selective by age and

TABLE 2.7 New Jersey's Sex Ratio over Time

Year	Ratio
1910	102.8
1920	101.5
1930	101.0
1940	98.9
1950	97.1
1960	96.0
1970	93.7
1980	92.2
1990	93.5

Source: Computed by the authors from U.S. decennial census data.

sex. (We detail such empirical regularities in Chapter 4.) The result is that local-area pyramids are often quite unusual. For example, Figures 2.13 and 2.14 show those for Green Valley, Arizona, in 1990 and Grafton County, New Hampshire, in 1970. Green Valley is a retirement community developed during the 1970s; note the predominance of females at the oldest ages. Grafton County includes Hanover, where Dartmouth College admitted its first females in 1972—making it the last of the Ivy League schools to become coeducational.

Location Quotients

As we mentioned at the beginning of this section on population composition, considerable interest focuses on the geographic distribution of persons of different types in different locations. Raw counts of specific types of persons can often be difficult to compare because of the differing total numbers of persons in the geographic subunits for which data are made available.

Location quotients are useful to compare the concentration of persons in a specific subgroup in each of a set of geographic subunits to the concentration of such persons in the entire, broader study area. Consider, for example, the distribution of members of an ethnic group in a city. Table 2.8 gives the raw numbers of persons "of Spanish origin" and of total persons in selected neighborhoods of Tucson, Arizona. To compute the location quotients for these neighborhoods, the standard formula is

$$Q_i = \frac{S_i / S^*}{P_i / P^*}$$

where Q is the location quotient, i references the specific neighborhood, * indicates the city as a whole, and S and P denote Spanish-origin and total population, respectively. Thus the location quotient is a ratio of ratios.

A more intuitive understanding may be gained by rearranging the formula as

$$Q_i = \frac{S_i / P_i}{S^* / P^*}$$

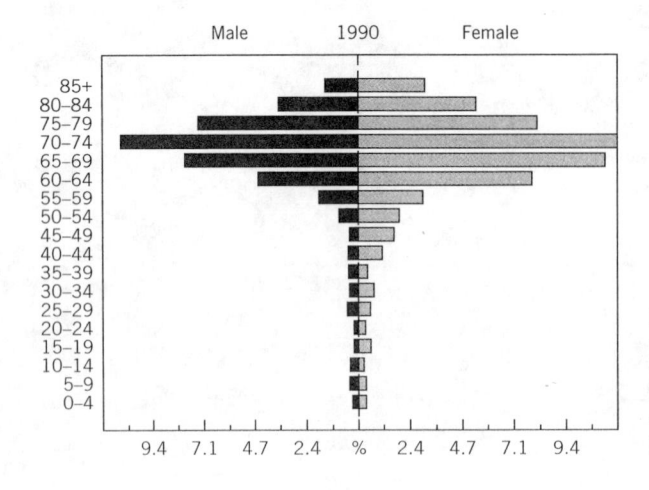

FIGURE 2.13 Population Pyramid for Green Valley, Arizona, 1990

Source: Graphed from 1990 U.S. Decennial Census Data.

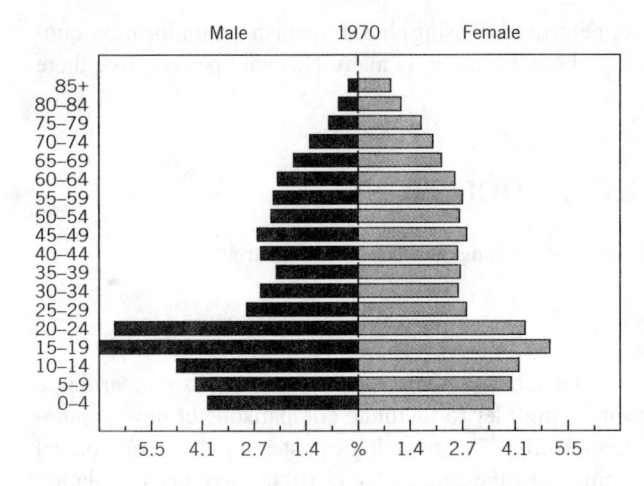

FIGURE 2.14 Population Pyramid for Grafton County, New Hampshire, 1970

Source: Graphed from 1970 U.S. Decennial Census Data.

Were the numerator, S_i / P_i, to be multiplied by 100 we would simply have the percentage of the population in neighborhood i that is of Spanish origin. Similarly, multiplying the denominator by 100 would give the citywide Hispanic percentage. As may be seen easily from this version of the formula, a location quotient of 1.0 would be found if a neighborhood has exactly the same Spanish-origin percentage as the city as a whole.

The first neighborhood listed in Table 2.8 comes very close to having the same Hispanic percentage as the city of Tucson as a whole. The location quotient for the Rincon Heights neighborhood is (903 / 3,632) / (82,609 / 336,465) = 1.01. For the Armory Park neighborhood, the location quotient is 1.90. There is almost twice the concentration of persons of Spanish origin in this neighborhood as in the city as a whole. The Manzo neighborhood illustrates the fact that the upper bound of the location quotient is fixed by the percentage of the subgroup population found for the "reference region" (in this case, the city of Tucson). Virtually its entire population is Hispanic, and its Q value is 3.80, meaning almost four times the city average. With a citywide percentage of 24.6, the maximum location quotient possible for this variable is 1 / 0.246 = 4.07. Finally, the Sam Hughes neighborhood (dubbed Barrio Volvo

TABLE 2.8 Spanish-Origin Population of Selected Tucson, Arizona, Neighborhoods for Use in Sample Computation of Location Quotients

	Spanish-Origin Population	Total Population	Percentage of Spanish Origin
Rincon Heights	903	3,632	24.9
Armory Park	1,002	2,149	46.6
Manzo	2,861	3,070	93.2
Sam Hughes	258	4,391	5.9
Total city of Tucson	82,609	336,465	24.6

Source: Special neighborhood tabulations, City of Tucson Planning Department from 1980 U.S. Census of Population data.

because of its yuppie-occupied, gentrified housing) has a Spanish-origin location quotient of only 0.24, meaning only about a quarter as many Hispanic persons live there as is typical for Tucson.

2.5 GEOGRAPHIC ASSOCIATION

Some circumstantial evidence is very strong, as when you find a trout in the milk.

—Henry David Thoreau

Our example of the use of location quotients shows one way to standardize a measure of population composition in order to facilitate comparisons of how population characteristics vary geographically. The next logical step in the geographical analysis of population is to examine how the geographic distribution of one population characteristic corresponds to that of other variables. We call this concept *geographic association*. We will take up this topic in detail in Chapter 10 when we discuss demographics and market segmentation analysis. We conclude this chapter, however, by discussing a simple, bivariate geographic association technique to illustrate the intertwining of the concepts of population distribution and population composition.

Perhaps the most elementary method for analyzing geographic association is by *cross-tabulating* two or more variables in a *contingency table*. Many population data are cross-tabulated quantities to begin with, such as numbers of males and females broken down by age, or breakdowns of persons by both race and years of schooling. Our focus here, however, is on how aggregate population characteristics of spatial subunits correspond to one another across the entire set of such units defined within a broader study area.

As an example, suppose we wished to probe further the matter of the Hispanic population distribution in Tucson. We might suspect that this ethnic group has been a disadvantaged one and is more likely to be found in poorer rather than wealthier parts of the city. Our hypothesis, then, is that there is a negative or *inverse* geographic association between Hispanic-origin ethnicity and income level. Such an association is one in which areas having high values of one of the variables tend to have low values of the other, and vice versa. The other type of geographic association is a *direct* one. (That is, areas having high values on one variable also tend to have high values on the other, whereas other areas tend to have low values of both.) The third and final possibility is that there is *no* statistically detectable geographic association between the variables.

A number of steps were involved in performing the contingency table analysis of the association between Hispanic ethnicity and income illustrated in Tables 2.9 and 2.10. First, the location quotients for all 75 Tucson neighborhoods were computed and 1980 census data on the median income of families in the neighborhoods were found. Next, both of these variables were broken down into three categories representing low, medium, and high values. Sometimes variables are inherently categorical, but in instances such as here, where a continuous range of values is possible, it is necessary to carefully specify the *break points*. This is part art and part science. On the one hand, it is nice to choose round values to define the categories (such as the 0.5 and 1.5 critical Q values and the $15,000 and $20,000 income level cutoffs), but there is also a statistical requirement that we shall discuss shortly.

TABLE 2.9 Observed Frequencies (O_{ij}) of Neighborhoods for Tucson, Arizona, Cross-tabulated by Population of Spanish Origin (Location Quotients) and Median Family Income

	Income (citywide median = $17,413)			Total
	$0–$14,999	$15,000–$19,999	$20,000 and Over	(row proportion, R_i)
Spanish origin				
$Q < 0.5$	10	9	*17*	36
				(0.48000)
$0.5 \leq Q < 1.5$	8	*8*	3	19
				(0.25333)
$Q \geq 1.5$	*14*	6	0	20
				(0.26667)
Total	32	23	20	75
(column proportion, C_j)	(0.42667)	(0.30667)	(0.26667)	(1.00000)

Source: Special neighborhood tabulations, City of Tucson Planning Department from 1980 U.S. Census of Population data.

The third step was to examine the list of observed values for the two variables and cross-classify each of the 75 neighborhoods as belonging in one of the nine boxes or cells in Table 2.9. Following these assignments, tallying up the numbers of neighborhoods in each cell gave the *observed frequencies*, $\{O_{ij}\}$, that are shown (for example, $O_{11} = 10$, meaning that 10 neighborhoods had both low location quotients and low income levels). Then the rows and columns were summed, and the *row and column proportions, R_i and C_j*, respectively, were calculated. For instance, $R_1 = .48000$ means that 48.000 percent (36 out of 75) of the neighborhoods had location quotients less than 0.5.

Although it is interesting to examine the observed frequencies recorded in Table 2.9, much more may be learned by comparing these to the frequencies that we would expect to find if there were no geographic association between the variables. Table 2.10 shows such *expected frequencies*. The formula to find the frequencies assuming *statistical independence* is

$$E_{ij} = R_i \times C_j \times N$$

where E_{ij} is the expected value for the cell in row i, column j, and N is the total number of observations (i.e., 75). For example, $E_{13} = R_1 \times C_3 \times N = .48000 \times .26667 \times 75 = 9.60$. An important statistical requirement alluded to earlier is that these expected frequencies cannot be too low if valid conclusions are to be drawn from the analysis. A common, rough rule of thumb is that each should exceed 5.0. If after computing the E_{ij}s one or more is found to be less than 5.0, then the break points of the variables should be changed or the number of categories reduced. (For example, a 2-by-2, 2-by-3, or 3-by-2 table could be used rather than the 3-by-3 table shown here).

Once all the expected frequencies have been computed (and found to exceed 5.0!), the observed values may be compared to them. In Table 2.9 each of the observed values that is larger than the corresponding expected frequency has been shown in bold italics. There are three of them—in the cells corresponding to low location quotients and high incomes (O_{13}), medium location quotients and medium incomes (O_{22}), and high location quotients and low incomes (O_{31}). This pattern is, in fact, characteristic of an inverse association. The higher the Hispanic location quotient, the more likely it is that the neighborhood has low family income.

So at this point we are beginning to suspect that our hypothesis is supported by the neighborhood data. A remaining question, however, is how *significant* is this inverse association that has been uncovered? The *chi-square statistic* (χ^2) is used to check the level of confidence with which we may make assertions about the association. Chi-square is simply a measure of the divergence between the observed and expected cell values. It is computed as

$$\chi^2 = \sum_{i=1}^{m}\sum_{j=1}^{n}\left[\left(O_{ij} - E_{ij}\right)^2 / E_{ij}\right]$$

Table 2.11 shows the computation of the value for our example. We then simply consult a standard table of χ^2, reading across the row for the appropriate *degrees of freedom* (df) given the size of the contingency table for which the value was computed.[2] In our case the value of 18.44 with four degrees of freedom exceeds the critical value for the .01 probability level (13.277) but is slightly smaller than the value for the .001 level (18.465). This indicates that there is less than a 1 percent chance, but slightly greater than a 0.1 percent chance, that the association between the variables occurred due to random variation. A probability level of .05 or .01 is conventionally used as the critical one, so we can confidently state that there is an inverse geographical association between population of Spanish origin and family income for Tucson's neighborhoods.

TABLE 2.10 Expected Frequencies (E_{ij}) of Neighborhoods for Tucson, Arizona, Cross-tabulated by Population of Spanish Origin (Location Quotients) and Median Family Income

	\$0–\$14,999	\$15,000–\$19,999	\$20,000 and Over	Total (row proportion, R_i)
Spanish Origin				
$Q < 0.5$	15.36	11.04	9.60	36 (0.48000)
$0.5 \leq Q < 1.5$	8.11	5.83	5.07	19 (0.25333)
$Q \geq 1.5$	8.53	6.13	5.33	20 (0.26667)
Total (column proportion, C_j)	32 (0.42667)	23 (0.30667)	20 (0.26667)	75 (1.00000)

Income (citywide median = \$17,413)

---

[2] Because ours is a 3-by-3 table, df = 4. The general formula is df = $(r-1) \times (c-1)$, where r and c are the numbers of rows and columns, respectively. One way to think about the degrees of freedom is that they equal the number of entries that could be made in the body of the table before all the other values are determined—given that row and column totals are previously known.

TABLE 2.11 Calculation of the χ^2 Statistic for the Tucson Neighborhood Contingency Table Analysis

i	j	$(O_{ij} - E_{ij})$	$(O_{ij} - E_{ij})^2$	$(O_{ij} - E_{ij})^2/E_{ij}$
1	1	−5.36	28.73	1.87
1	2	−2.04	4.16	0.38
1	3	7.40	54.76	5.70
2	1	−0.11	0.01	0.00
2	2	2.17	4.71	0.81
2	3	−2.07	4.28	0.84
3	1	5.47	29.92	3.51
3	2	−0.13	0.02	0.00
3	3	−5.33	28.41	5.33

$$\chi^2 = \Sigma_i \Sigma_j [(O_{ij} - E_{ij})^2 / E_{ij}] = 18.44$$

Note that the kind of association that we have uncovered is a property of the neighborhoods rather than of individuals as enumerated in the census itself. We must be careful about asserting that we have shown that persons of Spanish origin have lower incomes than the population as a whole. It could be persons other than Hispanics in the neighborhoods that actually have the strongest effect on the median neighborhood income levels. This possible booby trap is known as the *ecological fallacy*. To make definitive statements about income levels of families of different ethnicities requires a cross-tabulation of the individual-level census data.

The contingency table and chi-square type of analysis we have been discussing is just one of many ways that may be used to analyze the geographic association between demographic variables. For instance, *regression analysis* may be used if the original variables—like those in this example—are continuous ones rather than categorical ones. For instance, we might have found the best-fitting regression line for the 75 neighborhoods' Spanish-origin location quotients and median family incomes. The line would be downward-sloping, indicating a negative association. The R^2 statistic rather than χ^2 statistic would indicate the strength of the relationship.

2.6 SUMMARY

In this chapter we have emphasized how population distribution and population composition are interrelated. The relative prevalence of a demographic characteristic is highly dependent on the geographic units used in its measurement—and thus on the geographic distribution of the characteristic. We have explored a variety of measures for analyzing population distribution and composition, concluding with a discussion of how the two concepts may be brought together through the concept of geographic association. The techniques presented were all relatively low-level ones, meaning that they may form parts of much more sophisticated analyses of real population problems.

In the next chapter we turn from methods and concepts for analyzing the *static* patterns of population distribution and composition to techniques and ideas focusing on the *dynamics* of a population system.

EXERCISES

1. Compute the mean center (centroid) of population for four cities with the following (x, y) coordinates and populations:

City (i)	x_i	y_i	P_i
1	1	2	1,000
2	2	1	2,000
3	3	3	2,000
4	5	4	7,000

2. Suppose you were asked to advise on the optimal location for a new regional fire station to serve the four cities of Exercise 1. What are the coordinates of the point that minimizes total travel distance to the 12,000 people? (*Hint:* This can be solved conceptually. Lengthy computations and formulas are not required!) What is the minimum total travel distance? Why, for a fire station, might a point other than that which minimizes total travel distance be optimal?

3. Suppose the population of each of the four cities in Exercise 1 is 3,000. Compute the new centroid and median center as well as the minimum total travel distance. Use a convergence criterion of .000001 in finding the median center. (*Note:* If you take on the task of calculating the median centers you would be well advised, if you have the expertise, to write a simple computer program. The hand calculations can become quite tedious!)

4. The 1990 total populations and Irish populations of the five census tracts of the West End District of the imaginary town of Yenralb were:

Tract Number	Total Population	Irish Population
7.07	7,016	714
7.08	7,777	865
8	8,080	6,506
13.01	5,223	3,788
13.02	3,904	1,127

Draw a Lorenz curve showing the concentration of Irish in the West End of Yenralb. Note that for this exercise you are not comparing the distribution of population versus the distribution of land area, but rather the distribution of one subpopulation versus the distribution of total population.

5. Find the value of the Hoover index (index of concentration) for the data in Exercise 4.

6. A team of urban geographers has estimated that the density (in persons per square mile) of the Planeville Urbanized Area is well approximated by Colin Clark's negative exponential model when a D_0 parameter of 2,000 and a b parameter of 0.08 are used.

On the other hand, densities in the Rogersonburgh Urbanized Area are better predicted with values of $D_0 = 5,000$ and $b = 0.40$. Calculate the models' expected densities at successive numbers of miles away from the center of each urban area. (Plug in values of $x = 0, 1, 2$, etc.) Stop when density dips below 1,000 persons/sq mi (one of the defining attributes for land to be included in an urbanized area). Which urbanized area exhibits greater urban sprawl? Which do you think has a larger total population? Explain why.

7. A supermarket chain is considering two sites for opening a new store in the West End of the city of Yenralb. The distances (in kilometers) between each site and the population centroids of the West End's five census tracts are:

Tract:	7.07	7.08	8	13.01	13.02
Site A	5.0	2.5	3.4	1.1	2.9
Site B	3.8	4.2	1.9	2.2	7.3

Using the 1990 total populations for these tracts shown in Exercise 4, calculate the accessibility index values for each site. Which site would you advise the chain to pick?

8. Construct a population pyramid for New York State in 1980. Use the following data on population by sex for five-year age groups:

Age Group	Males	Females
0–4	580,389	555,536
5–9	506,546	579,333
10–14	717,004	689,056
15–19	803,639	794,932
20–24	737,556	782,390
25–29	690,997	737,379
30–34	649,344	702,495
35–39	526,286	580,621
40–44	446,856	489,456
45–49	433,243	478,268
50–54	469,110	527,492
55–59	456,018	523,352
60–64	382,502	458,505
65–69	312,502	412,020
70–74	229,349	338,557
75–79	153,100	260,457
80–84	88,035	173,764
85 and over	57,946	135,037
Total	8,339,422	9,218,650

9. Calculate the overall sex ratio for New York's 1980 population using the data in Exercise 8. Then calculate the sex ratio for each age group and graph the sex ratio as a function of age. Compare this graph to Figure 2.11 which shows the relationship by age for the entire United States.

10. Calculate the youth and elderly dependency ratios for the combined male and female populations of New York State in 1980, again using the data given for Exercise 8.

11. If the Irish make up 40.0 percent of the overall population of Yenralb, what are the location quotients for Irish concentration in the five census tracts in Exercise 4?

12. The following table shows the observed frequencies of census tract location quotients for the Irish and Portuguese ethnic groups for the entire city of Yenralb. Construct a table of expected frequencies.

	Irish L.Q.		
Portuguese L.Q.	Low	High	Total
Low	16	5	21
High	7	12	19
Total	23	17	40

13. Compute the value of the χ^2 statistic for the observed and expected frequencies of the contingency table of Exercise 12. Is there a statistically significant geographic association between the locations of Irish and Portuguese in Yenralb? Is it a direct or inverse relationship? Explain why.

> **What I want is Facts. Teach these boys and girls nothing but Facts. Facts alone are wanted in life. Plant nothing else, and root out everything else.**
>
> *From Hard Times, by Charles Dickens*

CHAPTER 3
DESCRIBING DEMOGRAPHIC CHANGE

3.1 Introduction
3.2 Population Growth (and Decline)
3.3 Components of Change and Demographic Accounting
3.4 Measures of Mortality
3.5 Life Tables: Construction and Interpretation
3.6 Measures of Fertility
3.7 Summary

3.1 INTRODUCTION

Geographers, demographers, planners, and those interested in marketing demographics are all concerned with past population change as well as change that might occur in the future. This chapter and the following one provide a foundation for studying such changes by describing demographic change and its components.

This chapter begins with a general treatment of elementary models of population growth and decline. Following this, we turn to *components* of population change: fertility, mortality, and migration. In this chapter we address the fertility and mortality components of change; Chapter 4 is devoted entirely to the migration component.

The components of population change together determine the rate of population growth or decline. Understanding the evolution of these components is essential for comprehending the dynamics of population change.

3.2 POPULATION GROWTH (AND DECLINE)

In this section we examine the temporal evolution of populations that are subject to constant rates of population growth. There are many reasons to study the dynamics of populations experiencing constant growth rates. For example, it is of interest to compare the demographic fate of regions with differing rates of population growth. In addition, historical rates of population growth often prove useful in making estimates of current population, as we shall see in Chapter 5. Finally, historical rates of population growth may also be used to construct elementary projections of future levels of population. (See Chapter 6).

Percentage Changes

We begin by reviewing how simple percentage rates of change are calculated. Genesee County in New York State had a population of 59,336 in 1980 and a census tally of 60,060 in 1990. The percentage change in population is derived by (a) dividing the most recent population by the earlier population, (b) subtracting 1 from the result, and (c) multiplying by 100 percent to convert the answer to a percentage. Thus the percentage population change in Genesee County during the 1980s was

$$[(60,060 / 59,336) - 1] \times 100\% = 1.22\%$$

The same procedure is applied when there is a decline in population. For example, the city of Buffalo, New York, had a population of 357,002 in 1980 and a census count of 328,123 in 1990. Its percentage change in population during the 1980s was

$$[328,123 / 357,002) - 1] \times 100\% = -8.1\%$$

Geometric Rates of Change

To understand how we may use growth rates to understand the consequences for future population growth (or decline), let us first consider what the population of Genesee County would be in the year 2000, should the 1.22 percent growth rate that held during the 1980s persist during the 1990s. Its population in the year 2000 would equal its 1990 population (60,060) plus an increment of 1.22 percent of 60,060. To calculate the increment, we (a) convert the percentage to a decimal by dividing the percentage by 100 and (b) multiply the result by the initial population (in this case, 60,060). The population in the year 2000 under a scenario of constant growth rates is therefore

$$60,060 + (1.22 / 100) \times 60,060 = 60,793$$

Alternatively, we could write the preceding equation as

$$60,060[1 + (1.22 / 100)] = 60,793 \qquad (3.1)$$

The population expected under the constant decadal growth rate scenario in the year 2010 is similarly found from

$$60,793[1 + (1.22 / 100)] = 61,535 \qquad (3.2)$$

Thus a general expression for finding the population one decade in the future is

$$P_{t+1} = P_t (1 + r) \qquad (3.3)$$

where P_t is the current population, P_{t+1} is the population one decade in the future, and r is the decadal growth rate, expressed as a decimal. There is of course no reason to restrict this method to using time periods of 10 years; in fact any time period length may be used, and it is common to use Equation 3.3 to find the population one year from the present, using a growth rate, r, defined as an annual rate of growth.

Note that we have derived the 2010 population from the 2000 population, and the 2000 population from the 1990 population. By inserting (3.1) into (3.2), we may express the 2010 population in terms of the 1990 population:

$$61{,}535 = 60{,}793(1 + .0122) = 60{,}060\,(1 + .0122)(1 + .0122)$$

The population two time periods in the future may therefore be written as

$$P_{t+2} = P_{t+1}\,(1 + r)(1 + r) = P_t\,(1 + r)^2 \qquad (3.4)$$

assuming that it grows at a constant rate of r during each period.

A useful analogy may be drawn between populations growing at constant rates and money that accumulates interest at a constant rate in a bank. How we calculate the population at some future point in time is precisely the same as how we calculate the amount of money accumulated in the bank at a future point in time. In both cases, we need the starting amount (in terms of either money or people) and a rate of growth (either an interest rate or a rate of population growth). When the observed rate of growth is applied periodically as we just did, growth will occur geometrically. By generalizing the argument in (3.4), the population at a future point in time, P_t, may be determined from

$$P_t = P_0\,(1 + r)^t \qquad (3.5)$$

where P_0 is the initial population, r is the per period rate of population growth, and t is the number of time periods that future point is away from the initial point in time. For example, if a population of 200,000 grew at an annual rate of 1 percent per year, in 20 years it would grow to

$$200{,}000\,(1.01)^{20} = 244{,}038$$

Exponential Rates of Change

Just as interest on savings is often compounded continuously, it is also often assumed that the population growth rate is an instantaneous one that may be applied continuously. (After all, births and deaths do not all occur at the same time on a single day of the year!) This leads to exponential population growth; the population at a future point is determined in this case from

$$P_t = P_0\,e^{rt} \qquad (3.6)$$

where e is the number 2.71828... and r represents the continuous rate of change. If a population of 200,000 were to grow continuously at a rate of 1 percent for 20 years, it would grow to

$$200{,}000\,e^{.01\,(20)} = 244{,}281$$

60 Chapter 3 • Describing Demographic Change

Note that population, like money, grows faster when it is compounded continuously, rather than periodically.

The amount of actual, absolute growth associated with exponential growth appears to start out slowly, but over time the amount of growth may become mind-boggling. (This helps to explain why financial advisors are keen to illustrate the benefits of compounding—you can become a millionaire quite easily by investing a small amount of money now; what they often don't explain is that it takes quite a while before this happens!)

Particularly alarming to those concerned with the rapid pace of world population growth is the added concern that the growth rate itself has been increasing. Approximate annual world population growth rates were 0.3 percent in 1600, 0.8 percent in 1900, and 0.9 percent in 1950. The annual world population growth rate is now in the neighborhood of 1.7 percent. If the world's population had grown at a constant annual rate of 1.7 percent since 1000 B.C., and if the world had started with just two people at that time, world population would reach 1.84×10^{22} by the year 2000! A population this large would completely occupy a sphere (assuming that each person occupies only 2 sq ft) that had a radius approximately equal to 10 million miles. This in turn would imply that the earth would have to have a radius that extended from its core to almost halfway to the planet Venus!

The consequence of continued geometric growth of the world's population is indicated in Figure 3.1; the graph is based upon a July 1992 population of 5.42 billion (Population Reference Bureau, 1992) and a 1.7 percent annual rate of growth. Within 190 years the population would increase roughly 24-fold to 129 billion!

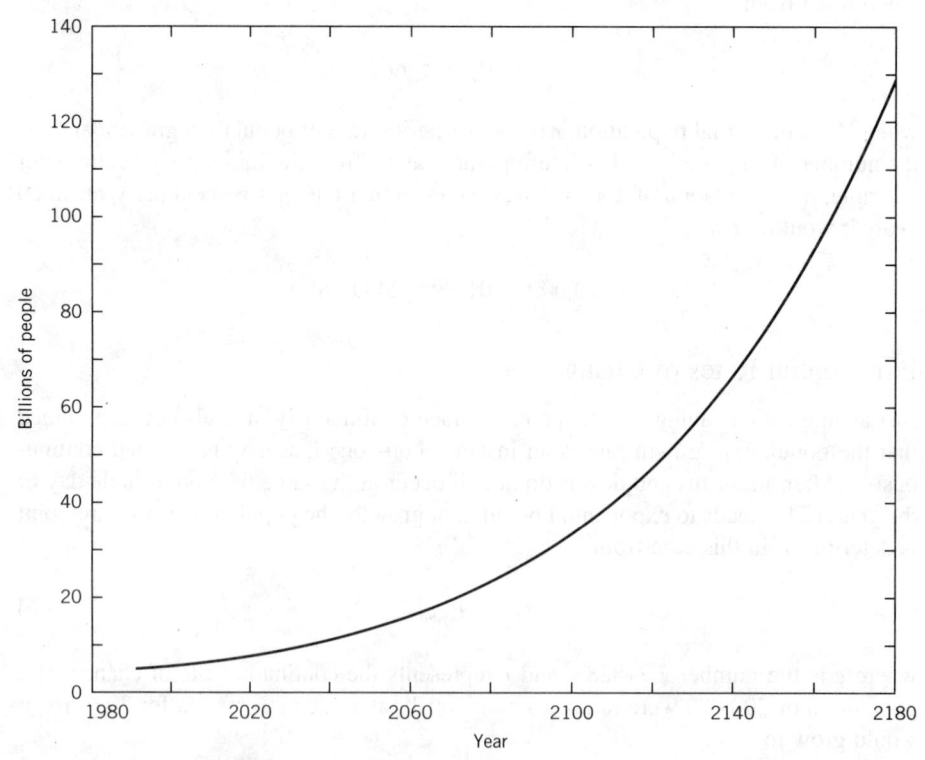

FIGURE 3.1 The Effect of an Annual Growth Rate of 1.7 Percent on World Population Growth

It should be clear that caution must be given to the use and interpretation of constant population growth rates. It is simply unrealistic to use any growth rate (with the exception of $r = 0$) for too long a period of time—either the population will grow to an absurdly high number, or it will shrink to zero. Population growth rates prove most useful for comparative purposes, for describing the short-run growth experience of regions, and for elementary short-run extrapolations.

Determining the Geometric and Exponential Rates of Change

We have now seen how to trace the effects of a given rate of growth on the evolution of population size. But how are geometric or exponential growth rates determined? A population's rate of growth over a time period may be found directly from either of the models given above (Equations 3.5 and 3.6).

With data on population for two points in time, Equations 3.5 and 3.6 are solved for r. Thus to solve for the annual rate of growth in Equation 3.5, divide both sides by P_0, and raise both sides to the power $(1/t)$:

$$(P_t / P_0)^{1/t} = 1 + r$$

Then subtract 1 from both sides to solve for the annual growth rate, r:

$$r = (P_t / P_0)^{1/t} - 1$$

Note that this corresponds to the procedure used earlier to calculate the decadal growth rate for Genesee County (with $t = 1$ decade). To solve for r in Equation 3.6, first divide both sides by P_0:

$$P_t / P_0 = e^{rt}$$

Then take the natural logarithms of both sides (for a review of logarithms, see Appendix C):

$$\ln(P_t / P_0) = rt$$

Finally, divide both sides by t to obtain the growth rate, r:

$$r = [\ln(P_t / P_0)] / t$$

For example, the population of the United States rose from 226,545,805 in 1980 to 248,709,873 in 1990. This implies an annual rate of increase of

$$r = (248{,}709{,}873 / 226{,}545{,}805)^{1/10} - 1 = .00938$$

which is equivalent to an annual increase of 0.938 percent. The population increase implies a *continuous* rate of increase of

$$\ln(248{,}709{,}873 / 226{,}545{,}805) / 10 = .00933$$

or 0.933 percent.

The Time Required for a Given Change in Population

We may also solve for the time it will take for a current population, P_0, to grow (or decline) to some predetermined population level, P_t, assuming a constant rate of growth. For the geometric model, we find t by dividing both sides of Equation 3.5 by P_0, taking the natural logarithm of both sides, and dividing by the natural logarithm of $(1 + r)$:

$$t = \frac{\ln(P_t / P_0)}{\ln(1 + r)}$$

For the exponential model, dividing both sides of Equation 3.6 by P_0, taking the natural logarithms of both sides, and dividing by r yields

$$t = \frac{\ln(P_t / P_0)}{r} \tag{3.7}$$

It is common, for example, to compute the *doubling times* associated with particular population growth rates. A doubling time is defined as the length of time it would take for a population to double in size, assuming that its current growth rate stayed constant. By substituting $P_t = 2P_0$ in Equation 3.7, we have an expression for the doubling time under exponential growth:

$$\text{Doubling time} = \frac{\ln 2}{r} = \frac{.6931}{r}$$

Thus the doubling time for the world's population (given its current 1.7 percent growth rate) is approximately $.6931/.017 = 40.8$ years.

Logistic Growth

The *logistic* curve has often been adopted as a more realistic long-run description of population growth. Under logistic growth, a population starts out with growth rates that are close to exponential, but as the population grows, the rate of growth slows, and eventually an upper limit is reached on the size of the population. While exponential growth has been used to describe the unchecked growth of species not subject to resource constraints, the upper limit that characterizes logistic growth is consistent with the more usual case of populations that do have resource constraints (as well as other conditions) that ultimately limit their growth.

The S-shaped curve of population growth is consistent with the *demographic transition* discussed in Chapter 1. The demographic transition often describes the trajectories of birth and death rates in a country over time. In the first (high stationary) phase, birth and death rates are both high, and the rate of population growth is low. In the second (early expanding) stage, death rates decline with advances in health care, and population grows at an increasing rate. In the third (late expanding) stage, there is a decline in the rate of decrease in mortality rates, and birth rates begin to fall as industrialization and urbanization take place. The population is still increasing, but at a decreasing rate. In the fourth (low stationary) phase, birth and death rates are both low, and the level of population is relatively stable. Figure 3.2 sketches the demographic transition and the logistic growth in population that accompanies it.

The curve describing logistic growth is

$$P_t = \frac{k}{1 + ae^{bt}}$$

where P_t is the population at time t, e is the number 2.718..., and k, a, and b are parameters to be estimated using available data. The parameter k may be interpreted as the upper limit that the population approaches, a is a parameter that determines the level of population at time $t = 0$, and b determines the shape of the logistic curve (i.e., the rate at which the curve steepens and levels off over time).

Pearl and Reed (1920) popularized the use of the logistic curve in population analysis. In 1940, Pearl, Reed, and Kish published an article in *Science* demonstrating the remarkably good fit of the logistic curve to the size of the U.S. population from 1790 to 1910. The curve is given by

$$P_t = \frac{197.27}{1 + 67.32e^{-.0313t}}$$

where P_t is the population in millions and t is time as measured in years since 1790. Note that the upper limit on population implied by this curve is 197 million.

Furthermore, the authors showed that had that curve been used to predict U.S. population from 1920 to 1940, the error would have been exceptionally small (less than 2 percent in each of the three censuses from 1920 to 1940). However, the pitfalls associated with extrapolative projection methods of this type are readily seen in Figure 3.3. With the onset of the baby boom in 1946, the size of the population accelerated rapidly. The upper limit of 197 million suggested by the logistic was surpassed before

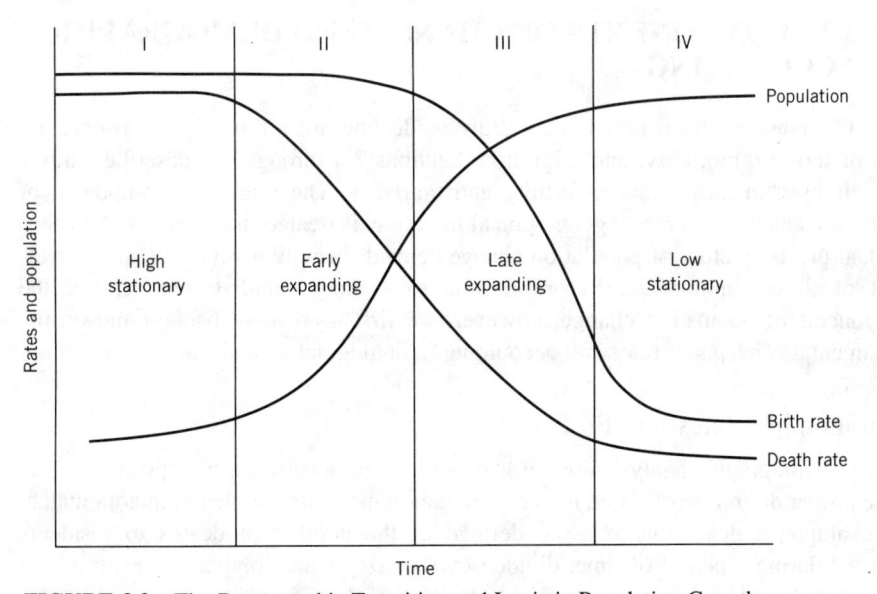

FIGURE 3.2 The Demographic Transition and Logistic Population Growth

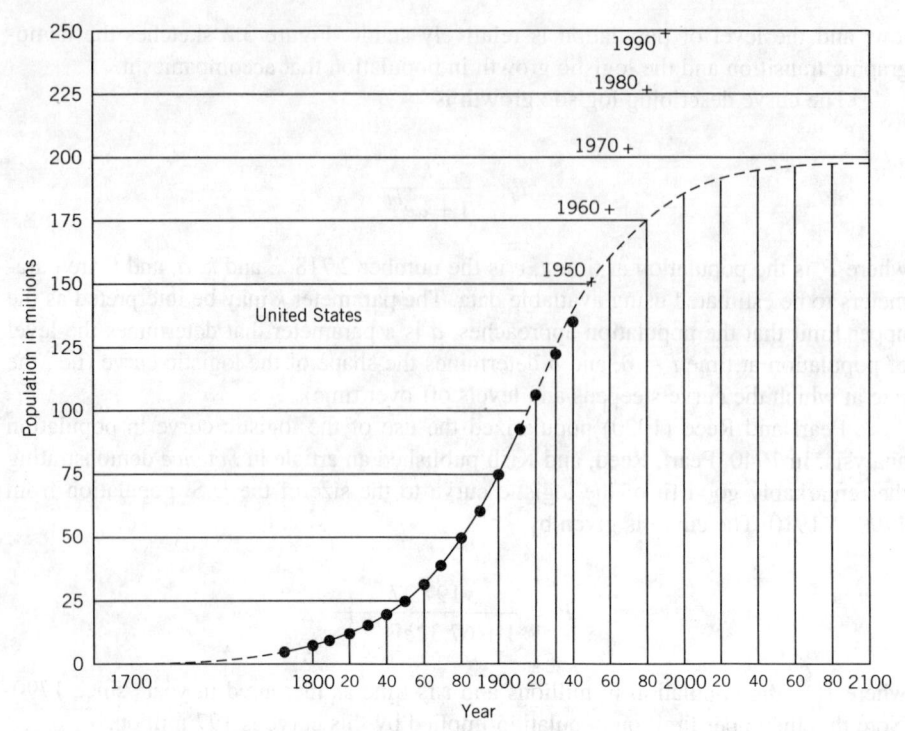

FIGURE 3.3 Logistic Population Growth in the United States

Source: Pearl, Reed, and Kish (1940). Reprinted by permission.

1970. It was clear that U.S. population size was not simply following some mechanistic curve, and that more attention had to be given to understanding the components of population change.

3.3 COMPONENTS OF CHANGE AND DEMOGRAPHIC ACCOUNTING

The rates at which populations grow or decline are ultimately determined by rates of fertility, mortality, and migration. Sections 3.4 through 3.6 describe various methods used in measuring of fertility and mortality. The migration component of population change, inherently geographical in nature, is treated in Chapter 4. An accurate and precise picture of population change depends heavily upon an adequate treatment of these components. Before turning to a more detailed treatment of the components of population change, however, we first need more background on the fundamental concepts of rates and accounting in demographic analysis.

Demographic Rates and Probabilities

Rates in demographic analysis are defined as ratios of occurrence to exposure. A key to the proper definition of a rate is to ensure that numerators and denominators match. For example, a death rate may be defined as the number of deaths to residents recorded during a period of time, divided by the size of the population "exposed" to the risk of death. Each person exposed to the risk of death is assigned a weight equal to the proportion of the time period for which the individual was exposed.

The most expedient method for getting a rough idea of the size of the "at-risk" population for use in the denominator of a rate is to use a midperiod population. Thus if 200 deaths to residents are recorded during a year, and the starting and ending populations are 18,000 and 22,000, respectively, the death rate is calculated as

$$d = \frac{200}{(18{,}000 + 22{,}000)/2} = \frac{200}{20{,}000} = .01$$

Birth and death rates are often expressed per 1,000 people at risk. Thus, in the preceding example, the death rate may also be expressed as $0.01 \times 1{,}000 = 10$ deaths per thousand at risk of dying. This form of expression is in contrast to that used for growth rates, which as we have seen are typically multiplied by 100 to be expressed in percentage terms.

The equation for the death rate implicitly assumes that the original 18,000 were exposed to the risk of dying for the entire period, and that the additional 4,000 people were exposed for half of the period; that is, $18{,}000 + (1/2)\,4{,}000 = 20{,}000$. However, the calculation of the proper at-risk population is more complicated than this. Migrants moving into a region from other regions are exposed to the risk of death for, on average, only half of the period (assuming a flow of people into the region that is uniform over time). Babies who are born during the period in other regions and migrate into the region are exposed to the risk of death for an average of only one fourth of the time period (since we may assume that they are on average born midway through the period, and that they migrate, on average, midway through the remaining half of the period).

Probabilities are distinguished from rates in the definition of both the numerator and the denominator. For probabilities, the denominator may generally be interpreted as an initial, rather than midyear, population. The numerator contains the number of events occurring to that initial population. Thus a probability of death is calculated by dividing the number of deaths occurring to an initial population by the size of that initial population. In the preceding example, assuming that 190 of the 200 deaths occur to people in the starting population, the probability of death is

$$\frac{190}{18{,}000} = .01056$$

Note that to define this probability properly, the data ideally should be detailed enough to provide the number of deaths occurring to the initial population. Since this is rarely the case, estimates of this quantity need to be made. Making such estimates requires knowledge of demographic accounting, and we will address this topic after first discussing the general concept of demographic components of change.

Components of Change

The relation between population change and its components is one of the most basic of all relations in demography:

$$P_{t+1} = P_t + B(t, t+1) - D(t, t+1) + I(t, t+1) - O(t, t+1) \quad (3.8)$$

where P_t denotes the population at time t. B, D, I, and O refer to births, deaths, in-migration, and out-migration, respectively. The notation $(t, t+1)$ reflects the fact that the events have occurred between exact time t and exact time $t + 1$. The excess of births over deaths $[B(t, t + 1) - D(t, t + 1)]$ is referred to as *natural increase*, and the difference between in- and out-migration $[I(t, t + 1) - O(t, t + 1)]$ is termed *net migration*. An important use of this demographic accounting equation is in migration estimation. We discuss this use in Chapter 4.

Demographic Accounting

Equation 3.8 is sometimes called the *demographic accounting equation*. Figure 3.4 provides an example of this simple form of demographic accounting. The city of Buffalo's population in 1980 is obtained from information on 1970 population and from information on births, deaths, in-migration, and out-migration during the decade thusly:

$$357{,}000 = 463{,}000 + 60{,}400 - 56{,}100 + 60{,}000 - 170{,}300$$

The decline in Buffalo's population during the 1970s is clearly attributable to net out-migration $(60{,}000 - 170{,}300 = -110{,}300)$; natural increase is small, yet positive $(60{,}400 - 56{,}100 = 4{,}300)$.

The accounting of population change requires consideration of what are termed *stock* variables and *flow* variables. A *stock* variable refers to an estimate or count of a quantity at a given point in time, while a *flow* variable refers to a quantity that is measured over a time period. In the preceding example, the 1970 population and the 1980 population are stock variables, and the components of change (births, deaths, in-migration, and out-migration) are flow variables. The division of terms into stock and flow variables provides a convenient means for understanding the process of demographic change.

Of course we can add more specific detail to the accounting equation. We might, for example, wish to describe the population change for a particular age group. It is also desirable to account for the peculiarities associated with the way that data are often collected. For example, simply subtracting recorded deaths from the initial population in the preceding equation is not entirely correct, since some of those recorded deaths are to individuals who have moved into the region during the time period. Such individuals do not get counted in the in-migration figures since they are not surviving in-migrants and because they are not there at the end of the period to respond to surveys about prior region of residence! However, they are counted in the tally of deaths to residents to the region. Proper accounting of population change calls for this group of individuals to be subtracted from the death statistics, since they do not in fact constitute decrements to the original population.

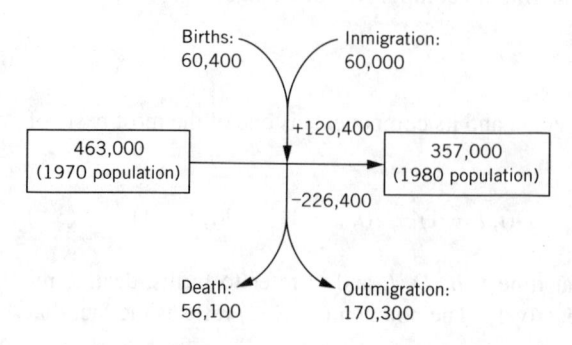

FIGURE 3.4 Components of Change in the City of Buffalo, 1970 to 1980

Source: Rogerson and Stack (1987). Reprinted by permission.

Likewise some additional deaths *should* be counted as decrements to the initial population, but are not counted because of out-migration that occurred before time $t + 1$. These individuals are not counted as out-migrant decrements to the original population because they are not surviving out-migrants and are not alive to be counted as out-migrants at the end of the period. But they do in fact constitute decrements to the original population, and hence the total number of deaths needs to be augmented by the number of such individuals.

Including these and similar factors leads to an accounting of population change in region i that is more precise than Equation 3.8:

$$P_{*i} = P_{i*} + I_i - O_i + (B_{i*} + \sum_{j \neq 1} B_{ji} - \sum_{j \neq 1} B_{ij} - \sum_{j \neq 1} B_{i\delta(j)} - B_{i\delta(i)})$$

$$- (D_{*i} + \sum_{j \neq 1} D_{ji} - \sum_{j \neq 1} D_{ij} - \sum_{j \neq 1} B_{j\delta(i)} - B_{i\delta(i)})$$

where

P_{i*} is the initial population, say at time t.

P_{*i} is the population at the end of the period, say time $t + 1$.

I_i and O_i represent surviving in- and out-migrants for region i during the period $(t, t + 1)$.

B_{i*} is the number of recorded births in region i during $(t, t + 1)$.

B_{ji} is the number of infants born in some other region j during $(t, t + 1)$ who survive and reside in region i at time $t + 1$.

B_{ij} is the number of infants born in region i during $(t, t + 1)$ who survive and reside in some other region j at time $t + 1$.

$B_{i\delta(j)}$ is the number of infants born in region i who move to some other region j, and die before time $t + 1$.

$B_{i\delta(i)}$ is the number of non-migrating infants born in region i who do not survive until time $t + 1$.

$B_{j\delta(i)}$ is the number of deaths occurring in region i before $t + 1$ to babies born in some other region j.

D_{*i} is the number of deaths to residents of region i during $(t, t + 1)$.

D_{ji} is the number of deaths to migrants from j to some other region i during $(t, t + 1)$.

D_{ij} is the number of deaths to migrants from i to some other region j during $(t, t + 1)$.

The term $B_{i\delta(i)}$ appears twice in the equation, and the terms cancel. We have included them in the equation to remind the reader that infants born during the period that fail to survive until the end of the period do not constitute either increments or decrements to the initial population. They must therefore be subtracted from the total number of births to residents to avoid counting them as increments, and they must be subtracted from the tally of deaths to residents to avoid counting them as decrements.

Figure 3.5 displays an account of regional population change in a system with N regions. The upper left quadrant of the matrix represents individuals who are alive at both the beginning and end of the time period. The upper right quadrant consists of deaths to individuals who were alive at the beginning of the period. The lower left

$$\begin{bmatrix} P_{11} & P_{12} & \cdots & P_{1N} & D_{11} & D_{12} & \cdots & D_{1N} \\ P_{21} & P_{22} & \cdots & P_{2N} & D_{21} & D_{22} & \cdots & D_{2N} \\ \vdots & & & P_{NN} & \vdots & & & \\ P_{N1} & P_{N2} & \cdots & P_{NN} & D_{N1} & D_{N2} & \cdots & D_{NN} \\ \hline B_{11} & B_{12} & \cdots & B_{1N} & B_{1\delta(1)} & B_{1\delta(2)} & \cdots & B_{1\delta(N)} \\ B_{21} & B_{22} & \cdots & B_{2N} & B_{2\delta(1)} & B_{2\delta(2)} & \cdots & B_{2\delta(N)} \\ \vdots & & & & \vdots & & & \\ B_{N1} & B_{N2} & \cdots & B_{NN} & B_{N\delta(1)} & B_{N\delta(2)} & \cdots & B_{N\delta(N)} \end{bmatrix}$$

FIGURE 3.5 The Rees–Wilson Accounting Matrix

quadrant represents infants born during the period who survive to the end of the period, and the lower right quadrant consists of infants born during the period who die during the period.

Row and column sums of the matrix have intuitive interpretations. The row sums in the upper half of the matrix are the initial region-specific population totals. The column sums for the left-hand side of the matrix are the region-specific populations at the end of the period. The row sums for the bottom half of the matrix represent region-specific total births, and the column sums for the right-hand side of the matrix are the region-specific death totals.

Illustrative Example: Minor Flows

We now illustrate some of these concepts through an accounting of population change for the elderly male cohort in Florida during the period 1975 to 1980. Since our example focuses on the elderly cohort, the reader should note that (a) the death rates are age-specific and (b) a more complete accounting of population change for all age groups would also include consideration of births, as represented by the terms in the lower half of Figure 3.5.

The male population age 60 and over was approximately 758,200 on July 1, 1975. The number of resident deaths that occurred for this cohort between July 1, 1975, and July 1, 1980, was 184,970. The number of in-migrants to Florida in this cohort during the period 1975–80 was 126,286, and the corresponding number of out-migrants was 24,165. Using the elementary accounting scheme, we would estimate the number of males age 65 and over in Florida on July 1, 1980, by subtracting deaths from the initial population, and by adding net migration to the result:

$$758{,}200 - 184{,}970 + 126{,}286 - 24{,}165 = 675{,}351$$

The problem with this approach is that it does not account for the fact that some of the 184,970 deaths occurred to in-migrants, and hence should not be subtracted from the initial population. We need to estimate the number of deaths that occurred to in-migrants. Likewise, we need to estimate the number of deaths that occurred to out-migrants, so that those deaths can be subtracted from the initial population.

To estimate the number of individuals who migrate and do not survive, Rees and Wilson (1977) equate the death rate associated with this group of individuals with the death rate at their destination region, j. (Note that an argument could also be made to use the death rate at the *origin*, but we shall follow common practice here.) Using d_j to denote the death rate at the destination,

$$d_j = \frac{D_{ij}}{0.5P_{ij} + 0.25D_{ij}} \tag{3.9}$$

The death rate for this group of individuals is expressed by the right-hand side of (3.9) and equals the number of deaths in the group, D_{ij}, divided by the population at risk of migrating and dying. The appropriate at-risk population is equal to the population that migrates and survives (weighted by one-half, since they are in the destination region for an average of one-half of the time period) plus the population that migrates and dies (weighted by one-fourth, since this group is in the destination region for an average of one-fourth of the period before dying). This is equated to the overall death rate for region j, d_j. The quantity d_j is found by dividing the total number of deaths in region j, D_{*j}, by the population at risk of death. Rees and Wilson note that in general this at-risk population is approximately

$$P_{j*} - 0.5\ D_{*j} + 0.5\ B_{i*} + 0.5\sum_{i \neq j}(P_{ij} - P_{ji})$$

ignoring the minor flows. We shall use this approximation in our example of elderly males in Florida (ignoring the birth term, since no births reach the elderly cohort that quickly!) since it simplifies the presentation of our illustrative example. Following the example, we return to this point to describe how the true at-risk population could be used. In our example, the mortality rate for elderly males in Florida is

$$d_{FL} = \frac{184{,}970}{758{,}200 - (184{,}970/2) + .5 \times (126{,}286 - 24{,}165)} = .2581$$

The corresponding mortality rate for elderly males in the United States was similarly computed and found to be $d_{US} = .3058$.

Now, solving (3.9) for D_{ij} in terms of d_j and P_{ij} yields

$$D_{ij} = \frac{0.5\,d_j}{1 - (0.25\,d_j)}\,P_{ij}$$

Therefore, after rounding to the nearest integer, we find that

$$D_{US,FL} = \frac{(0.5)(.2581)(126{,}286)}{[1 - (0.25)(.2581)]} = 17{,}421$$

$$D_{FL,US} = \frac{(0.5)(.3058)(24{,}165)}{[1 - (0.25)(.3058)]} = 4{,}001$$

With these estimates in hand, we now estimate the male population age 65 and over in Florida on July 1, 1980 as

$$758{,}200 - (184{,}970 - 17{,}421 + 4{,}001) + 126{,}286 - 24{,}165 = 688{,}771$$

which is approximately 2 percent higher than the naive estimate of 675,351.

Other quantities associated with Florida's elderly male population may now be found via accounting identities. For example, the number of deaths to nonmigrants who were alive in 1975 (the $D_{FL,FL}$ term) is equal to the number of deaths to residents of Florida, minus the number of deaths to in-migrants:

$$184{,}970 - 17{,}421 = 167{,}549$$

The number of nonmigrating survivors among those alive at the beginning of the period (i.e., the $P_{FL,FL}$ term) equals the initial population minus the number of deaths to nonmigrants, minus the number of deaths to out-migrants, minus the number of surviving out-migrants:

$$758{,}200 - 167{,}549 - 4{,}001 - 24{,}165 = 562{,}485$$

Although in some cases the minor flows are so minor that this extra accounting detail is hardly worth the trouble, in situations such as those just outlined, it is important to account for the minor flows. Popular destinations for elderly migrants, such as Florida and Arizona, have a significant number of migrants who do not survive until the end of the period, and significant errors may otherwise result for particular age groups. Use of the accounting scheme is most valuable in these circumstances as well as in fostering clear definitions of rates and probabilities.

In the calculations just described, an approximation was used for the at-risk population used in the denominator of d_j. The reason for this is that the true at-risk population also includes minor flows that are unknown at the time of the calculation. The population at risk of death in region j is

$$P_{jj} + 0.5P_{ji} + 0.5D_{jj} + 0.25D_{ji} + 0.5\,P_{ij} + 0.25D_{ij}$$
$$+\ 0.5B_{jj} + 0.25B_{j\delta(j)} + 0.25B_{ji} + 0.125B_{j\delta(i)} + 0.25B_{ij} + 0.125B_{i\delta(j)}$$

Once the minor flows have been determined, d_j can be recalculated using this at-risk population. For our example, the mortality rate for elderly males in Florida, using this more precise definition of at-risk population (where again the B terms representing birth equal zero) and using the minor flows just estimated previously, is $d_{FL} = .2447$. The entire analysis is then repeated, and new, updated minor flows are derived by using the more accurate value of d_{FL}. In the present example, the minor flows would be changed only slightly, since the new value of d_{FL} is quite close to the original. After the updated minor flows have been calculated, yet another revised set of d_j values may be derived. The iterative process continues in this manner until there are negligible changes in the values of the minor flows.

For purposes of comparison, we can also use the results of our example to define the probabilities of death. There are $167{,}549 + 4{,}001$ deaths that occur to the initial population of 758,200. Hence the probability of death for this cohort is equal to $171{,}550/758{,}200 = 0.2263$.

3.4 MEASURES OF MORTALITY

> Either that wallpaper goes, or I go.
> —Oscar Wilde's last words

Perhaps the simplest measure used to describe the rate at which deaths occur is the *crude death rate* (CDR):

$$\text{CDR} = 1{,}000\left(\frac{D}{P}\right)$$

Here D is the number of deaths during the year, and P is the population at risk of dying during the year (often approximated by the midyear population). A distinct advantage of the crude death rate is its ease of calculation. However, it does not account for the age and sex structure of the population. Countries may therefore exhibit similar crude death rates despite substantial differences in the underlying force of mortality. Thus West Germany, the United Kingdom, and Austria all had the same crude death rate (12) in 1987 as did India, despite the fact that the life expectancy at birth in the former countries was 74 years, while for India it was only 55 years. The crude death rate was high in Austria, West Germany, and the United Kingdom simply because approximately 15 per cent of their populations were over 65 years old; only 4 percent of the population of India was over 65 (Population Reference Bureau, 1987), so the latter's crude death rate may be attributed more to a strong underlying force of mortality.

Despite the fact that the crude death rate is dependent upon age composition, it does provide a simple summary measure that is useful in describing mortality's contribution to population change. Two regions with similarly high CDRs will have their natural increase diminished by a similar amount (relative to their population sizes), whether it is achieved by high levels of mortality or by an age structure with a high proportion of elderly. *Age-specific mortality rates* ($_hM_x$) facilitate the comparison of regional mortality rates by focusing upon specific age cohorts:

$$_hM_x = \frac{_hD_x}{_hP_x}$$

where $_hD_x$ is the number of annual deaths to individuals age x to $x + h$ and $_hP_x$ is the midyear population age x to $x + h$. The age-specific mortality rate is sometimes multiplied by 1,000 to express the number of deaths per 1,000 people in a particular age group.

The notation that we adopt here to define age groups is common in demographic analysis. The subscript on the left refers to the width of the age group (in years, unless otherwise stated). The subscript on the right refers to the youngest age in the age group. Thus $_5P_{50}$, for example, would refer to the size of the population in the five-year age group beginning with age 50. It would include all people who had reached their 50th birthday, but had not yet reached their 55th birthday. A common way to express this age group in tables is "50-54." When we use the words "age x to $x + h$" we are referring to all individuals at least age x that have an age up to, but not including, age $x + h$. The age group width in most demographic analyses is $h = 5$, and values of $h = 1$ and $h = 10$ are also common.

Standardized Death Rates

A region's *standardized death rate* (SDR) is a crude death rate that has been adjusted for the differences in age composition between that region and some standard population (which is most often taken to be the country as a whole). Such standardized death rates are useful in many circumstances. For example, assessing the effect of environmental factors on death rates in a region can only be done by properly accounting for the particular age structure of the region. There are two commonly used methods of adjustment. The method of *direct standardization* involves calculating a weighted average of the region's age-specific mortality rates, where the weights represent the age-specific sizes of the standard population. Thus

$$\text{SDR}_1 = 1{,}000 \frac{\sum_a M_a P_{a,s}}{P_s} \qquad (3.10)$$

where SDR_1 is the standardized death rate, $P_{a,s}$ is the number of people in age group a in the standard population, and P_s is the total population of the standard. Table 3.1 depicts the age-specific mortality rates for New York and Arizona, as well as the age structure of the U. S. population in 1988. Employing Equation 3.10, the standardized death rate for New York is

$$\frac{1{,}000\,[.00272(18{,}456) + .00066(72{,}053) + \ldots]}{245{,}808} = 9.283$$

where 245,808 is the total U.S. population in thousands. For Arizona, the standardized death rate is

$$\frac{1{,}000\,[.00269(18{,}456) + .00076(72{,}053) + \ldots]}{245{,}808} = 7.947$$

These rates may be interpreted as the crude death rates that would have been obtained if the individual states had the age structure of the country as a whole. New York's 1988 crude death rate was 9.84 (or $1{,}000 \times 176{,}289 / 17{,}909{,}000$). Since this is higher than the standardized rate, we may infer that New York had a relatively greater proportion of its population in age groups with high mortality rates, in comparison with the age composition of the country. Similarly, the actual crude death rate in Arizona was 7.92 (or $1{,}000 \times 27{,}646 / 3{,}489{,}000$) in 1988, which is close to the standardized rate. This indicates that any difference between the age structure of Arizona and the age structure of the United States had little effect on the adjustment of death rates. Indeed, Table 3.2 shows that New York had a relatively small fraction

TABLE 3.1 Data Required for Direct Standardization: U.S. Population and State Mortality Rates

Age	Population	New York	Arizona
0–4	18,456	.00272	.00269
5–24	72,053	.00066	.00076
25–34	43,675	.00192	.00146
35–44	35,264	.00321	.00224
45–54	24,163	.00540	.00484
55–64	21,830	.01212	.01102
65–74	17,897	.02793	.02408
75+	12,470	.08753	.07359

Note: Population in thousands.

Source: U.S. Bureau of the Census (1989), U.S. Department of Health and Human Services, (1990).

TABLE 3.2 Age Composition, 1988

Age	New York	Arizona	United States
0–4	.0712	.0857	.0751
5–24	.2805	.2946	.2931
25–34	.1685	.1760	.1777
35–44	.1436	.1373	.1435
45–54	.1066	.0903	.0983
55–64	.0996	.0880	.0888
65–74	.0752	.0794	.0728
75+	.0548	.0487	.0507

Note: Columns sum to one.

of people under age 34 (which is likely a result of the age-selective nature of recent out-migration; see Chapter 4) as well as a relatively high percentage of people age 45 and over. The table also shows that while Arizona had a relatively high proportion of its population in one of the high mortality rate age groups (65–74), it also had a relatively high fraction of its population in one of the low mortality rate age groups (<5). In addition, it had a relatively low fraction in the oldest age category (75+). The net effect is that Arizona's crude death rate would not change much were its age structure identical to that of the United States.

The method of *indirect* standardization is used in those situations where less data are available—in particular, where regional age-specific mortality rates may not be easily obtained. Age-specific mortality rates in the standard population are used to derive the expected number of deaths in the regional population. The ratio of actual to expected deaths is then multiplied by the crude death rate in the standard population to derive the standardized rate. Thus

$$\text{SDR}_2 = \frac{d_r}{\sum_a M_a P_{a,r}} \text{CDR}_s$$

where d_r is the number of deaths in region r, M_a is the age-specific mortality rate in the standard population, $P_{a,r}$ is the size of age group a in the regional population, and CDR_s is the crude death rate in the standard population. Using the data in Table 3.3, the standardized death rate for New York, using indirect standardization, is

$$\frac{176,289}{[.00251(1,275,000) + .00065(5,023,000) + \ldots]} \times 8.817 = 9.262$$

The corresponding standardized rate for Arizona is

$$\frac{27,646}{[.00251(299,000) + .00065(1,028,000) + \ldots]} \times 8.817 = 7.947$$

As the reader will note, the two methods lead to quite similar results.

TABLE 3.3 Data Required for Indirect Standardization

Age	U.S. Death Rate	Population (thousands) New York	Arizona
0–4	.00251	1,275	299
5–24	.00065	5,023	1,028
25–34	.00135	3,018	614
35–44	.00220	2,572	479
45–54	.00486	1,909	315
55–64	.01236	1,784	307
65–74	.02730	1,347	277
75+	.08513	981	170

Total deaths in New York: 176,289.

Total deaths in Arizona: 27,646.

U.S. CDR: 8.818.

Sources: U.S. Bureau of the Census, (1989). U.S. Department of Health and Human Services, (1990b).

Measures of Infant Mortality

The *infant mortality rate* ($_1M_0$) is defined as the number of deaths to infants less than age one, per 1,000 live births:

$$_1M_0 = 1,000 \left(\frac{_1D_0}{B} \right)$$

where $_1D_0$ is the number of annual deaths to infants less than one year old, and B is the number of live births. In 1992, the rate for the world was 70—implying 70 deaths in the first year of life, for every 1,000 live births. There is considerable geographical variability in infant mortality rates throughout the world. In 1992, Africa had an infant mortality rate of 94, and Afghanistan's rate was 168. The United States had a low infant mortality rate (8.6), but not nearly as low as in many other developed countries. Japan's rate was among the lowest (4.4), as were rates in Finland (5.8), Sweden (6.2), and Canada (6.8). All of Western Europe had an infant mortality rate of 7.0.

Mortality during the first year is heavily concentrated in the first few weeks of life, and for comparative purposes it is useful to define a *neonatal mortality rate* ($_{28\ days}M_0$) using deaths to infants less than a month old:

$$_{28\ days}M_0 = 1,000 \left(\frac{_{28\ days}D_0}{B} \right)$$

Data from the National Center for Health Statistics show that the neonatal mortality rate in the United States has declined from 20.5 neonatal deaths per thousand live births in 1950 to 6.2 in 1989. The neonatal mortality rate was slightly higher in metropolitan areas (6.3) than in nonmetropolitan areas (5.8), reflecting in part differences by race. The nonwhite population is disproportionately located in metropolitan areas and has a higher neonatal mortality rate (10.3) than does the white population (5.1).

An alternative means for describing the mortality experience of a population is the life table. The next section deals with the construction and use of life tables.

3.5 LIFE TABLES: CONSTRUCTION AND INTERPRETATION

A useful way to summarize the mortality experience of a population is by means of a *life table*. Conveyed by the table is such information as the probability of surviving from one age to another, and the expected number of years of life remaining for an individual who has attained a given age x.

In addition to summarizing the mortality experience of a population, life tables have many other applications. For example, they are used by life insurance companies to set insurance premiums. One of the primary demographic uses of life tables lies in the construction of survival ratios used in other demographic applications such as population projections. (See Chapter 6.) As we shall see in the next subsection, they also are used in defining fertility rates that use the likelihood of surviving from one childbearing age group to the next.

The life table methodology is a general one — that is, there is no reason to confine the application to mortality. Although the life table describes the attrition of individuals from a population through their deaths at various ages, many other situations also call for an analysis of the successive decrements to some starting population. Thus the rate at which the single population gets married, the rate at which young adults enter the labor force, and the rate at which patients leave a hospital all may be subject to this type of analysis. Increments may be included as well. A marital status life table, for example, includes not only decrements to the state of marriage through death and divorce, but also increments to the state of marriage from the single population.

Interpretation of a Life Table as a Cohort of Individuals

A typical life table is illustrated in Table 3.4. One interpretation of the life table is as a summary of the mortality experience of a single, hypothetical group (i.e., cohort) of individuals, all born at the same time. The initial size of this hypothetical cohort is given by l_0, (in this example equal to its usual value of 100,000) and is termed the *radix*. The individual columns of the table may be interpreted as follows:

$_hM_x$—the observed age-specific mortality rate for individuals age x to $x + h$.

$_hq_x$—the probability that an individual age x will die before attaining age $x + h$.

l_x—the number of individuals in the cohort surviving to exact age x.

$_hd_x$—the number of individuals in the cohort dying between ages x and $x + h$.

$_hL_x$—the number of person-years lived by the l_x individuals between ages x and $x + h$.

T_x—the cumulative number of person-years lived by the cohort beyond age x.

e_x—the expectation of life, in years, conditional upon survival to exact age x.

TABLE 3.4 Life Table for Evanston, Illinois, Males, 1980

Age	D	P	$_hM_x$	$_hq_x$	l_x	$_hd_x$	$_hL_x$	T_x	e_x
0–1	6	393	.01527	.01515	100,000	1515	98,939	7,116,208	71.16
1–4	1	1,367	.00073	.00293	98,485	288	393,363	7,017,269	71.25
5–9	0	1,944	.00000	.00000	98,197	0	490,984	6,623,905	67.46
10–14	1	2,320	.00043	.00215	98,197	211	490,455	6,132,921	62.46
15–19	2	3,713	.00054	.00269	97,985	264	489,268	5,642,466	57.59
20–24	2	4,669	.00043	.00214	97,722	209	488,086	5,153,198	52.73
25–29	2	3,879	.00052	.00257	97,513	251	486,936	4,665,112	47.84
30–34	4	3,008	.00133	.00663	97,262	645	484,697	4,178,175	42.96
35–39	5	2,017	.00248	.01232	96,617	1190	480,110	3,693,478	38.23
40–44	4	1,862	.00215	.01068	95,427	1020	474,586	3,213,368	33.67
45–49	12	1,614	.00743	.03650	94,407	3446	463,424	2,738,782	29.01
50–54	9	1,489	.00604	.02977	90,962	2708	448,039	2,275,358	25.01
55–59	24	1,476	.01626	.07813	88,254	6895	424,032	1,827,319	20.71
60–64	31	1,241	.02498	.11756	81,359	9564	382,884	1,403,287	17.25
65–69	31	1,001	.03097	.14,372	71,795	10,318	333,178	1,020,403	14.21
70–74	42	888	.04730	.21148	61,476	13,001	274,880	687,225	11.18
75–79	57	620	.09194	.37377	48,475	18,119	197,080	412,345	8.51
80–84	45	529	.08507	.35074	30,357	10,647	125,165	215,265	7.09
85+	77	352	.21875	1.00000	19,709	19,709	90,100	90,100	4.57

Source: Data courtesy the State of Illinois. Calculations by the authors.

The $_hD_x$ and $_hP_x$ columns represent observed age-specific deaths and populations, respectively, and are the only information that is needed to construct the life table. They are not usually included in the life table itself; they are presented in Table 3.4 merely for expositional purposes. The data for observed deaths represent deaths over a one-year period to individuals age x to $x + h$, and the population figures represent the size of age group x to $x + h$ at the midpoint of this one-year time period. In practice, the average annual number of deaths over a 3-year period is often used instead of an annual figure, to smooth out year-to-year fluctuations that may occur in the number of deaths. This is particularly important for regions where populations are small; without smoothing, unreasonable age-specific mortality rates resulting from an unusually large or small number of deaths would sometimes be derived.

The hypothetical cohort is subjected to the set of age-specific mortality rates ($_hM_x$), beginning at birth. Within each age group, a given number of individuals ($_hd_x$) in the cohort die, implying that the number of individuals reaching age x, given by l_x, declines with age (as expected).

Note that the life table captures the mortality experience of a population that dies according to a particular set of age-specific mortality rates. Implicit are the assumptions that (1) as the cohort ages, members will die according to those rates and (2) that the rates will not change throughout the lifetime of all members of the cohort.

Alternative Interpretation as a Stationary Population

An alternative interpretation of the life table is made possible by assuming that l_0 births are added to the population each year. Individuals are again assumed to be subject to a schedule of age-specific mortality rates (the $_hM_x$). An unchanging or stationary population is achieved since there are l_0 births (by assumption) and l_0 deaths each year. To see why there are l_0 deaths per year, first recognize that there must be d_x deaths each year to individuals in age group x, $x + h$. Since the sum of the d_x column equals l_0, there are l_0 deaths each year. With this perspective we have the following interpretations for the life table quantities:

l_x—the number of individuals in the stationary population reaching age x each year.

$_hd_x$—the number of deaths each year to individuals in the stationary population age x to $x + h$.

$_hL_x$—the size of the stationary population in age group x to $x + h$.

T_x—the size of the stationary population age x or greater.

Since the $_hL_x$ values denote the size of the stationary population, the L column may be interpreted as the age composition of the stationary population (keeping in mind the assumptions, namely that 100,000 births are added each year, and all individuals are subject to the schedule of age-specific mortality rates). Thus the total size of the stationary population is T_0, the sum of the L column. Note that since there are l_0 births per year, the crude birth rate in the stationary population is simply equal to 1,000 (l_0 / T_0), or 1,000 / e_0.

Life Table Construction

As previously indicated, the only information needed to construct a life table is a set of age-specific annual mortality rates ($_hM_x$). These are derived by simply dividing observed annual deaths ($_hD_x$) by the midyear population ($_hP_x$). For example, from Table 3.4, the annual mortality rate for males age 55 to 59 in Evanston, Illinois, was 24 / 1,476 = .01626 in 1980. Note from the table that the mortality rates follow the typical J-shaped mortality curve, with rates rising steadily with age, with the exception of small declines in mortality after age 1.

Life table construction proceeds by converting the annual age-specific mortality rates, $_hM_x$, into probabilities of dying, $_hq_x$. Conceptually the probability that an individual age x does not survive until the beginning of the next age group is equal to the number of deaths to the cohort between ages x and $x + h$, divided by the size of the cohort alive at exact age x. Since the *annual* observed number of deaths may be written as the age-specific mortality rate multiplied by the midyear population (i.e., $_hD_x = {_hM_x} {_hP_x}$), the number of deaths expected to the population age x to $x + h$ over an h-year period may be represented by $h \, _hM_x \, _hP_x$. For the denominator, we need the size of the population at the *beginning* of the time period. This is equivalent to the midperiod population plus half of all deaths occurring during the period, $_hP_x + (h/2) \, _hM_x \, _hP_x$. The implicit assumption is that deaths are uniformly distributed over the period.

Putting all of this together, we have

$$_hq_x = \frac{h \, _hM_x \, _hP_x}{_hP_x + (h/2) \, _hM_x \, _hP_x}$$

This may be simplified to

$$_hq_x = \frac{2h \, _hM_x}{(2 + h \, _hM_x)}$$

Thus the probability that an Evanston male age 35 does not survive to age 40 is equal to $(2 \times 5 \times .00248) / [2 + (5 \times .00248)] = .01232$.[1]

For each age group, the number of deaths, $_hd_x$, is determined by multiplying the number of individuals attaining age x, l_x, by the probability $_hq_x$ that they do not survive until the beginning of the next age group:

$$_hd_x = l_x \, _hq_x$$

The number of individuals who survive to the beginning of the next age group is simply equal to l_x minus the number of deaths occurring to individuals age x to $x + h$:

$$l_{x+h} = l_x - {_hd_x}$$

For example, in Table 3.4 the number of deaths occurring to the cohort between ages 60 and 65 equals $.11756 \times 81,359 = 9,564$. Since 81,359 members of the cohort survive to age 60, the number surviving to age 65 equals $81,359 - 9,564 = 71,795$.

The values of q, d, and l are next computed for all of the remaining age groups. For each age group, the value of q is first derived from M, and then d is determined. From this, the value of l for the next age group is found.

---

[1] It is usual to report the M and q values to an accuracy of about five decimal places. It is therefore best to retain several extra decimal places (one or, preferably, two) in the value of M so that q may be computed accurately to the desired number of decimal places. The values of M displayed in Table 3.4 are rounded off to an accuracy of five decimal places, though seven decimal places were actually used to calculate the reported values of q.

After q, d, and l have been found for all age groups, the values of L, T, and e may be derived. The number of person-years lived by the cohort during a period of h years is determined by multiplying an estimate of the number of persons alive at the midpoint of the age group, $(l_x + l_{x+h}) / 2$, by the number of years in the cohort, h:

$$L_x = \frac{h(l_x + l_{x+h})}{2}$$

This calculation of L assumes that deaths to members of the cohort that occur between ages x and $x + h$ are distributed uniformly over the age group.

Returning to our example, the number of person-years lived by the Evanston cohort between ages 40 and 45 is equal to $5 \times (95,427 + 94,407) / 2 = 474,585$. Table 3.4 shows this value of L as 474,586 because extra decimal places have also been carried along in the calculations (though they are not reported in the table) for the quantities q and l. Note that the maximum number of person-years possible for any cohort over a 5-year period is 500,000; this would result only if the entire cohort of 100,000 survived for the entire period.

After the values of L have been derived for all age groups, the total number of person-years lived by individuals age x and over (T_x) is found by cumulating $_hL_x$ from x to the last group in the life table:

$$_hT_x = \sum_{i=x}^{z} {_hL_i}$$

where z is used to denote the oldest cohort in the life table. In practice, once the L column has been determined, it proves easiest to find the T column by starting with $T_z = L_z$. Then it is a simple matter of working up the T column, each time adding the appropriate value of L:

$$_hT_x = {_hT_{x+h}} + {_hL_x}$$

For example, the number of person-years remaining to be lived by the Evanston cohort beyond age 75 equals $90,100 + 125,165 + 197,080 = 412,345$.

The final column of the life table is calculated by dividing the number of person-years lived beyond age x by the number of persons reaching age x; this yields the average remaining life expectancy for those currently age x:

$$e_x = \frac{_hT_x}{l_x}$$

Thus the expectation of remaining life for Evanston males that attain age 25 is equal to $4,665,112 / 97,513 = 47.84$ years.

The first and last age groups present minor complications in the derivation of life table functions. Since deaths occurring to infants less than a year old are more likely in the first half of that year than in the second, those less than 1 year old are often tabulated separately, irrespective of the value of h. It has been found empirically that for those less than 1 year old, the estimate

$$L_0 = .3l_0 + .7l_1 \tag{3.11}$$

is more accurate than that which results from the standard method just described:

$$L_0 = \frac{l_0 + l_1}{2} = .5l_0 + .5l_1 \qquad (3.12)$$

The reader should note that Equation 3.11 results in a smaller number of person-years than the standard method (Equation 3.12) since it places a relatively higher weight on the smaller number (l_1) and a relatively lower weight on the larger number (l_0).

We then need to remember that in the case where the age group width is say, 5, $h = 4$ (instead of $h = 5$) should be applied in the calculation of L for the age group 1 to 4, since age group 0 to 1 has been treated separately.

The last age group is of course open-ended, and this merits special attention as well. Rather than calculating q using the value of M, as is done for other rows of the table, q is simply set equal to 1.00 since everyone reaching this age group must necessarily die within it. This leads to

$$_\infty d_z = l_z$$

where z again denotes the last age group. Finally, the number of person-years lived by the last age group requires special consideration. Let $m_x = d_x / L_x$ be the annual age-specific mortality rate for individuals in the stationary life table population. Assuming that the age-specific mortality rates in the observed and stationary populations are equal (i.e. $M_z = m_z$), we have

$$M_z = m_z = \frac{d_z}{L_z}; \text{ that is, } L_z = \frac{d_z}{M_z}$$

Readers interested in further details pertaining to life tables may wish to consult Shryock and Siegel (1976) or Namboodiri and Suchindran (1987); they give detailed accounts of the construction, interpretation, and use of life tables.

Survival Ratios

As indicated at the beginning of this section, one of the chief uses of the life table is to derive various forms of survival ratios. Questions regarding survival between exact ages are answered by using the l values from the life table. Thus the probability that a male age 10 lives to his 60th birthday is, using Table 3.4, $l_{60} / l_{10} = 81,359 / 98,197$ or .8285. Similarly, the probability that a newborn lives to age 20 is given by $l_{20} / l_0 = 97,722 / 100,000 = .97722$.

Questions of survival pertaining to age groups are answered using the L values. The likelihood of an individual age 15 to 19 surviving 5 years to be counted in age group 20 to 24 is $L_{20} / L_{15} = 488,086 / 489,268 = .9976$. This result seems fairly intuitive in light of the stationary population interpretation of the life table; of the 489,268 males in the stationary population age 15 to 19, 488,086 will survive 5 years to be counted in the stationary population of age 20 to 24.

Some questions regarding survival concern the last, open-ended age group. The T values from the life table may be used in these instances. For example, the proportion of males 65 and over that live another 10 years is $T_{75} / T_{65} = 412,345 / 1,020,403$ or .8285. The proportion of males over age 65 in the stationary population is given by $T_{65} / T_0 = 1,020,403 / 7,116,208 = .1434$.

Other questions regarding survival mix the concepts of exact ages and age groups, necessitating the use of both l and L in the survival ratio calculation. For example, the proportion of 60-to-64–year-old males who reach their 65th birthday is derived as

$$\frac{5l_{65}}{L_{60}} = \frac{5(71,795)}{382,884} = .9376$$

The proportion of male infants born over a 5-year period surviving 20 years to be counted in age group 20 to 24 is

$$\frac{L_{20-24}}{5l_0} = \frac{488,086}{500,000} = .97617$$

Finally, questions of joint survival are often of interest. For brothers age 50 and 45, what is the probability that the younger of the two survives for 15 years, and the older of the two does not? Assuming that the survival of one is independent of the survival of the other, the answer is the product of the appropriate survival probabilities:

$$[1 - (L_{65} / L_{50})](L_{60} / L_{45}) = [1 - (333,178 / 448,039)](382,884 / 463,424) = .2118$$

Probabilities of widowhood may be calculated in a similar fashion when life tables for both males and females are available.

We shall return to the use of survival rates both in the next section of this chapter and in the treatment of cohort survival models of population projection in Chapter 6.

3.6 MEASURES OF FERTILITY

> The turtle lives 'twixt plated decks
> which practically conceal its sex.
> I think it is clever of the turtle
> in such a fix, to be so fertile.
>
> —Ogden Nash

Many of the measures of mortality may be used to define analogous measures of fertility. For example, the *crude birth rate* (CBR) is found by dividing the number of annual births by the midyear population and by multiplying by 1,000:

$$\text{CBR} = 1,000 \left(\frac{B}{P} \right)$$

where B is the number of annual births and P denotes the population at risk of giving birth (often represented by the midyear population). The crude birth rate in the United States in 1988 may be derived from the total number of births (3,909,510) and the total population (245,881,000):

$$1,000 \left(\frac{3,909,510}{245,881,000} \right) = 15.9$$

The CBR has a simple interpretation—in this example, it implies that during the year 15.9 births occurred for every 1,000 people in the country. Like the crude death rate, a desirable feature of the crude birth rate is its ease of calculation. Also like the crude death rate, the crude birth rate does not account for the age and sex structure of the population. Hence women in regions exhibiting similar crude birth rates may have substantially different propensities to have children. For example, in 1987 the Dominican Republic and South Africa had identical crude birth rates (33) despite the fact that on average women in South Africa had 4.6 children, while those in the Dominican Republic had just 4.0 children. The Dominican Republic had a crude birth rate as high as South Africa's despite lower rates of childbearing because of an age composition effect—the Dominican Republic had a relatively higher proportion of its women in the childbearing age groups.

Likewise, in a comparison of rural and urban fertility, urban areas can have higher crude birth rates because of the greater proportion of women in childbearing age groups. This could occur despite the fact that the average number of children ever born to women is typically higher in rural areas.

Age-standardized birth rates may be derived using either direct or indirect standardization, though this practice does not seem to be as common as it is for mortality. This is likely due to the array of alternative measures of fertility that we shall now discuss.

The *general fertility rate* (GFR) is defined as the number of live births per 1,000 women of childbearing age:

$$\text{GFR} = 1,000 \left(\frac{B}{P^f_{15-44}} \right)$$

where B is the number of births during a year, and P^f_{15-44} is the midperiod female population in age group 15 to 44. The age group 15 to 44 is most often used to represent women of childbearing age, though occasionally either or both of the age groups 10 to 14 and 45 to 49 are also included. Note that once again, the midyear population is used in the denominator. For the United States in 1988, the general fertility rate was

$$1,000 \left(\frac{3,897,495}{58,187,000} \right) = 67.0$$

since there were 3,897,495 births to females age 15 to 44, with 58,187,000 females in that age category.

An even finer analysis of age-specific fertility patterns may be carried out by using the *age-specific fertility rate* (ASFR):

$$_hF_x = \left(\frac{_hB_x}{_hP^f_x} \right)$$

where $_hF_x$ is the age-specific fertility rate among women age x to $x + h$, $_hB_x$ is the number of live births to women age x to $x + h$ during the year, $_hP^f_x$ is the midyear female population in age group x to $x + h$, and h is the width of the age cohort. Like age-specific mortality rates, the values of F are also often multiplied by 1,000. The final column of Table 3.5 depicts age-specific fertility rates for the United States in 1988.

The decision to use either the crude birth rate or more specific measures such as the GFR or age-specific fertility rates depends upon the primary interest of the analyst. If a comparison of two or more regions is being undertaken, or if added detail is required in an analysis of fertility, rates that account for the age–sex structure of the

TABLE 3.5 Births, Population, and Age-Specific Birth Rates, 1988

Age	Births	Population	Rate[a]
10–14	10,588	8,144,000	1.3
15–19	478,353	8,924,000	53.6
20–24	1,067,472	9,574,000	111.5
25–29	1,239,256	10,928,000	113.4
30–34	803,547	10,903,000	73.7
35–39	269,518	9,660,000	27.9
40–44	39,349	8,198,000	4.8
45–49	1,427	7,135,000	0.2

Source: U.S. Department of Health and Human Services (1990a).
[a]Rate is given as births per thousand population.

population should be used. If, however, the interest is solely in the contribution of fertility to recent population change, then it is perfectly reasonable to use the CBR. In this latter case, the net effect on population change may be the same whether many women are having few children or few women are having many children.

A particularly interesting way to portray age-specific fertility rates is through a contour map with age on one axis and time on the other. Figure 3.6 depicts age-specific fertility rates for the United States. There are three ways to look at this figure. Taking a horizontal cross section yields a picture of fertility for a given age over time. The baby boom is revealed as the top of the mountain (i.e., at the middle of the concentric rings centered on age group 20 to 24 in 1957). Taking a vertical cross section reveals how fertility varies with age in a given year. Finally, following the diagonal lines running from the lower left corner of the diagram to the upper right corner results in the portrayal of fertility for a given cohort. The delayed and reduced childbearing of cohorts recently passing through their childbearing age groups is discernable. Of interest is whether these recent cohorts will catch up at all to previous cohorts in their completed childbearing through higher age-specific fertility rates at older ages. This will only be revealed as additional years of data are added to the right-hand side of the figure.

The *total fertility rate* (TFR) may be interpreted as the expected total number of children that a woman will have. The TFR assumes that women will survive at least until the end of the childbearing age groups, and that they will bear children according to the prevailing age-specific rates as they progress through their childbearing years. This measure of fertility is particularly easy to comprehend and is widely used to compare the rates at which women in different regions are bearing children.

The TFR is calculated by summing the age-specific fertility rates over all childbearing age groups, and multiplying the result by the width of the age group used:

$$\text{TFR} = h \sum_x {}_h F_x$$

Using the data in Table 3.5, the total fertility rate in the United States in 1988 was

$$5 \times (1.3 + 53.6 + 111.5 + \ldots + 0.2) / 1{,}000 = 1.932$$

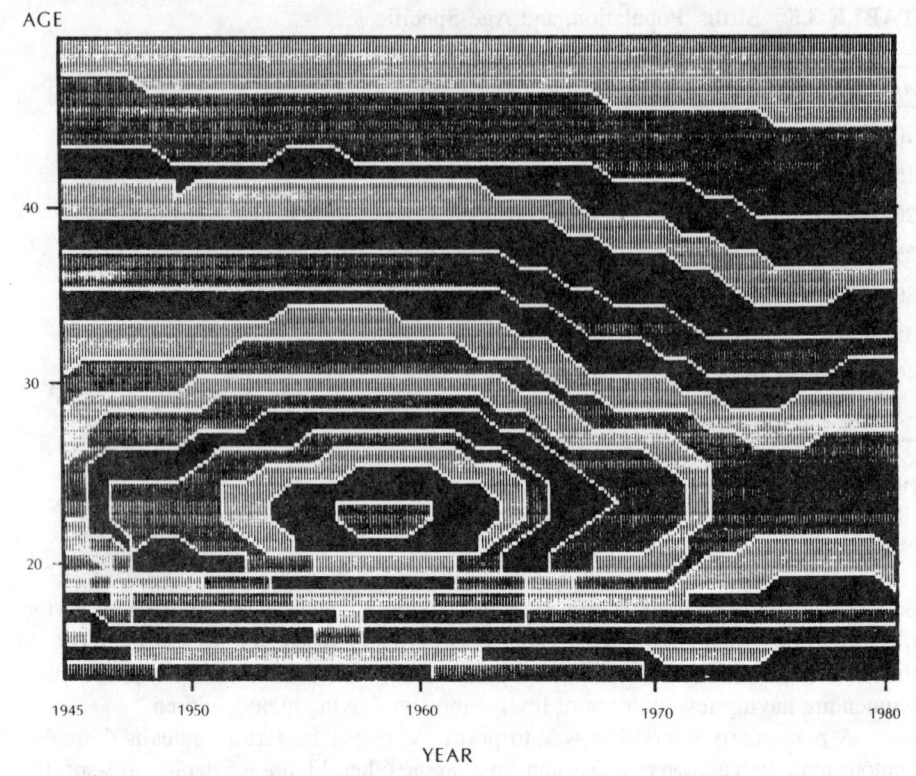

This LEXIS map shows the age-specific contours of U.S. cohort fertility from 1945 to 1980. The map is three-dimensional; age groups are on the Y-axis, while fertility bulges outward. The contours are logarithmic, building to a "mountain peak" of baby boom era fertility—261 live births per 1,000 women in the 20 to 24 age group in 1957.

FIGURE 3.6 U.S. Age-Specific Fertility Rates.

Source: Population Reference Bureau, *Population Today*, vol. 15 (October, 1987), p. 2.

It is often of interest to determine whether a population is growing fast enough to replace itself. The answer to this question is related to the mean number of female offspring that females are producing. This leads to the definition of the gross *reproduction rate* (GRR) as the expected total number of *female* children a woman will have, again assuming that women survive through the childbearing age groups and have children according to prevailing age-specific rates. It is calculated by simply multiplying the total fertility rate by the percentage of births that are female. Using the fact that in the United States roughly 105 males are born for every 100 females, the gross reproduction rate in 1988 was $1.932 \times (100 / 205) = .9424$. Gross reproduction rates in the neighborhood of 1 are consistent with zero rates of population growth in the long run; females will be approximately replacing themselves with one female in this instance. It should be clear that total fertility rates in the neighborhood of 2 will also be consistent with zero population growth in the long run since approximately half of all births are male and half are female.

To determine precisely whether a population will grow or decline in the long run, we need to account for the possibility that some of the female babies born may not themselves survive to the childbearing age groups. That is, we need to know the mean number of females surviving to childbearing age that would be produced by females subjected to age-specific fertility rates. This leads to the definition of the *net reproduction rate* (NRR):

$$\text{NRR} = wh \sum_x {}_hF_x \frac{{}_hL_x}{h\,l_0} = \frac{w}{l_0} \sum_x {}_hF_x \, {}_hL_x$$

where w is the proportion of births that are female, and again h is the width of the age interval. The probability that women survive to a particular childbearing age group (x to $x + h$) is equal to ${}_hL_x / h\,l_0$. Using life table data from the U.S. Department of Health and Human Services (1991), L values are used with the age-specific fertility rates in Table 3.5 to find the net reproduction rate for the United States in 1988:

$[(100 / 205) / 100{,}000] \times [.0013(493{,}954) + .0536(493{,}108)$
$\quad + .1151(491{,}802) + .1134(490{,}324) + .0737(488{,}539)$
$\quad + .0028(486{,}163) + .0048(482{,}754) + .0002(477{,}562)] = .9244$

which is slightly less than the gross reproduction rate, as expected.

A net reproduction rate greater than 1 leads to a population growing without limit in the long run, while a NRR less than 1 will ultimately lead to extinction. There is therefore a fine line between long-run growth and decline, since two children per female will lead to extinction while three per female will lead to exponential growth without limit.

Because populations do not have stable population growth rates over very long periods of time, the NRR is not intended as a serious predictor of the ultimate fate of the population. Rather, the attractive feature of the NRR is its ability to provide an indication of where a population is currently headed with respect to long-term population growth or decline.

It is important to recognize that a NRR equal to 1 leads to zero population growth only in the long run. Implied is the notion that should current rates of fertility continue to prevail, zero population growth will eventually be achieved. There is a certain amount of "momentum" associated with population change, so that an immediate reduction in age-specific fertility rates to rates consistent with zero population growth does not achieve zero population growth immediately. For example, if relatively higher fertility rates had prevailed in the past, population growth would continue for some time; the large number of children born during the period of high fertility would in turn produce a relatively large absolute number of offspring even if the total fertility rate had been reduced to replacement level.

Keyfitz (1985) notes that one definition of momentum is the amount by which a population would grow if its rate of childbearing immediately dropped to a net reproduction rate equal to 1 (consistent with zero population growth in the long run). He finds that momentum defined in this way may be approximated by the ratio of ultimate to current population:

$$\frac{be_0}{1{,}000\sqrt{\text{NRR}}}$$

where e_0 is the expectation of life at birth, NRR is the net reproduction rate before the drop in fertility rates, and b is the CBR. For example, a country with a life expectancy of 60, an NRR of 2, and a CBR of 30 would grow by an additional

$$\left[\frac{(30)(60)}{(1,000)(\sqrt{2})} - 1\right] \times 100\% = 27.3\%$$

before its population leveled off following an immediate drop to replacement-level fertility.

Period versus Cohort Rates

The discussion of measures of fertility has, to this point, assumed that data are collected across age groups for a given point in time. These rates are termed *period rates*, since they apply to data collected for a specified period. Period fertility rates are sometimes misleading because they can suggest age patterns of childbearing that may not be realistic. For instance, if fertility rates are low for all age groups in a given year, it may be incorrect to infer that the younger age groups will continue to bear children at the current low rates that typify the older age groups; these younger cohorts may simply be postponing their childbearing.

An alternative way of defining fertility rates is for a *cohort*; this overcomes the deficiencies inherent in the period rates since the childbearing behavior of a particular group of females is tracked over time. For example, completed total fertility rates are often reported for women who have passed through their childbearing ages. Calculation of cohort rates clearly requires more data, since fertility data must be collected over a series of years.

Similar comments apply to our earlier discussion of mortality rates and life tables. Cohort mortality rates and cohort life tables may be defined, but in practice are less common, due to the more demanding data requirements. The majority of practical work in population analysis uses period rates, and their limitations should therefore be kept in mind.

3.7 SUMMARY

On the surface, the measurement and accounting of population growth and change is a straightforward process. (See, e.g., Equation 3.8.) As we have seen in this chapter, however, the complexities involved are often subtle so careful attention is required to properly address questions of growth and change. This point is particularly relevant to the measurement and study of migration, the topic of the next chapter.

In subsequent chapters we build upon the concepts introduced here. In Chapter 5, we use rates of change to derive population estimates, and in Chapter 6, we use the rates and probabilities discussed here to describe the process of population projection.

EXERCISES

1. A town with a population of 13,000 grows at an annual rate of 1 percent for 20 years. What is its population at the end of the 20-year period?

2. A city of population 133,000 is declining at an annual rate of 2.1 percent. What will its population be in 10 years if the current rate of decline remains the same?

3. The population of a town is 13,000 in 1980 and 18,000 in 1990. What was its annual rate of increase? What was its continuous rate of increase?

4. How long will it take for a population of 15,000, growing at an annual rate of 1.5 percent, to reach 20,000?

5. How long would it take for the population in Exercise 4 to reach 20,000 if it were to grow at a continuous rate of 1.5 percent?

6. Find the expression for the tripling time of a population, that is, the time it takes for a population to triple in size when it grows continuously at a constant rate.

7. The estimated female population age 60 and over in Florida was 971,800 on July 1, 1975. The number of deaths to female residents in this cohort from July 1, 1975, to July 1, 1980, was 149,200. The number of in-migrants in this age group during this period was 153,800, and the number of out-migrants was 45,000. Assume that the mortality rate of females in this age group for the rest of the United States is 0.19. Use the standard demographic accounting equation (3.8) to find the estimated female population of Florida on July 1, 1980. Next, use the Rees-Wilson concept of at-risk population to account for deaths to in-migrants and out-migrants. Comment on the differences between the two answers. Comment also on the differences between your solution and the illustration for Florida males discussed in Section 3.3.

The following data should be used to answer Exercises 8 through 15:

New York State had the following age–sex structure in 1980:

Age	Male	Female
0–4	580,389	555,536
5–9	605,546	579,333
10–14	717,004	689,056
15–19	803,639	794,932
20–24	737,556	782,390
25–29	690,997	737,379
30–34	649,344	702,495
35–39	526,286	580,621
40–44	446,856	489,456
45–49	433,243	478,268
50–54	469,110	527,492
55–59	456,018	523,352
60–64	382,502	458,505
65–69	312,502	412,020
70–74	229,349	338,557
75–79	153,100	260,457
80–84	88,035	173,764
85+	57,946	135,037
Total	8,339,422	9,218,650

Source: U.S. Bureau of the Census, decennial census data.

Births to New York State females by sex of child in 1980 were:

Age	Male	Female
0–14	242	269
15–19	14,244	13,455
20–24	35,618	34,054
25–29	40,530	39,235
30–34	23,904	22,722
35–39	6,907	6,585
40–44	1,065	1,060
45+	51	36

Source: New York State Department of Health. 1981. *Vital Statistics of New York State, 1980.*

Deaths to New York State residents by age and sex in 1980 were:

Age	Male	Female
<1	1,673	1,321
<5	1,954	1,537
5–9	196	123
10–14	223	145
15–19	936	354
20–24	1,401	453
25–29	1,424	486
30–34	1,441	615
35–39	1,366	736
40–44	1,749	1,047
45–49	2,648	1,606
50–54	4,398	2,767
55–59	6,504	4,083
60–64	8,523	5,655
65–69	10,652	7,545
70–74	11,808	9,594
75+	33,473	46,303
Total	88,710	83,059

Source: New York State Department of Health, personal communication.

8. Find the crude birth and death rates for New York State in 1980.

9. Find the general fertility rate.

10. Find the age-specific fertility rates for each childbearing age group.

11. Find the total fertility rate.

12. Find the gross reproduction rate.

13. Construct a life table for females by assuming a hypothetical cohort or radix (i.e., l_0) of 100,000.

a. Use the data on deaths and population to compute age-specific mortality rates.

b. Use the age-specific mortality rates to derive age-specific probabilities of dying (i.e., the q values).

c. Use the l and q values for a specific age group to find the number of deaths (d) for each age group in the hypothetical cohort. Then find the value of l for the next age group by using d and the previous value of l.

d. Find the L values by using the information already obtained for l.

e. Find the T values by cumulating the L values from the bottom of the table.

f. Find the conditional expectation of remaining life (e) using the values of T and l.

14. Repeat Exercise 13 using data for males.

15. Using the completed life table for females, find the net reproduction rate.

> Bypasses are devices that allow some people to dash from point A to point B very fast. People living at point C, being a point directly in between, are often given to wonder what's so great about point A that so many people from point B are so keen to get there and what's so great about point B that so many people from point A are so keen to get there. They often wish that people would just once and for all work out where ... they wanted to be.
>
> —Douglas Adams, *The Hitchhiker's Guide to the Galaxy.*

CHAPTER 4
MIGRATION: ANALYZING THE GEOGRAPHIC PATTERNS

4.1 Introduction
4.2 Migration Definition and Measurement
4.3 Measures of Migration
4.4 The Relationship Between In- and Out-Migration
4.5 Disaggregating by Migrant Characteristics
4.6 Methods for Analyzing Geographic Patterns of Migration
4.7 Summary

4.1 INTRODUCTION

Of the components of population change—births, deaths, and migration—migration has been the most researched by geographers. Studies of human spatial mobility have, in fact, accounted for a significant percentage of all published papers in the entire eclectic discipline of geography. White (1980) found the percentage to be 8.5 percent in the four leading, broad-based journals of the field. Mobility and migration studies totally dominate the subfield of population geography. (See Gaile and Wilmott, 1989, pp. 258–89, for an assessment of the state of the art of population geography.) This penchant is in sharp contrast to the research carried out by demographers more generally, for whom research on fertility clearly predominates.

Migration is often the most difficult component of population change to accurately model and forecast. For applied demographic analysis such as population estimation and projection, the treatment of migration is likely to be the thorniest, yet most significant, issue the analyst will face.

For these reasons we give special emphasis to migration in this book. In this chapter we discuss fundamental geographic characteristics of migration and set forth a number of useful techniques that have been developed to *analyze* its spatial patterns. Following the presentation of techniques for making population estimates and projections in Chapters 5 and 6, we return to an explicit discussion of the migration component of population change in Chapter 7. There we examine methods to *model* and *forecast* migration.

Our treatment of migration processes here is necessarily broad-brush, emphasizing only some of the more important characteristics necessary for understanding migration patterns. Our emphasis is on *macro* properties of migration (characteristics of migration streams and migration considered as a form of interregional complementarity) as opposed to *micro* perspectives (the individual decision process, impacts of migration on people, and so forth). Clark (1986) provides a more detailed assessment of the overall geographic literature on migration, including some detailed treatment of behavioral research. Here, as throughout the book, we emphasize how to use geographical techniques for practical applications.

4.2 · MIGRATION DEFINITION AND MEASUREMENT

Unlike birth and death (which have fairly precise definitions, laying aside philosophical debates), rather different perspectives on migration result depending on how we define and measure it. Ask yourself a seemingly simple question: "How many times have I moved?" Most of you will wrestle with the question of what counts as a move. Does a move from one dormitory room to another in the same building count? How about a semester studying abroad (while maintaining a legal residence back home)? Or a move from one neighborhood to another, but in the same city?

To deal with these issues, different types of moves are typically given different names. An important distinction is that usually made between *international* and *internal* (within the same nation) migration, with the role of government regulations typically playing a larger role in the former case. For international migration, the people who come into a country to assume residence are generally called *immigrants,* and those giving up their abode to move to another nation are called *emigrants.* For internal migration, persons coming into a specific areal unit are most properly termed *in-migrants* while those moving away are known as *out-migrants.*

Spatial mobility (as opposed to *social mobility*—changing between social groups) is a term that covers a huge range of different kinds of movement, including (a) travel behavior, such as commuting between home and workplace, (b) *seasonal* or *temporary movement,* such as that undertaken by migrant workers, transients, nomads, and students (certain of these repetitive types of moves sometimes being referred to also as *circulation*), (c) intraurban *residential mobility,* which can be thought of as changing housing units without necessarily changing jobs or leaving the functional labor market area, (d) relocation to another area of one's region or to another region of the country, typically involving a change in job or in labor force status, and (e) international movement.

In this typology of spatial mobility only categories (d) and (e) are normally referred to as *migration*. How is the distinction between (c) and (d) made? The notion that internal migration (d) may involve a change in job while residential mobility

(c) does not is conceptually useful, but in practice some other rule is needed to provide a clear operational distinction. As an example, suppose in Kenmore, New York, the Jones family next door to the Smith family moves to Del Ray Beach, Florida. Mr. and Mrs. Jones, however, have been retired for several years prior to leaving Kenmore—so they are not experiencing any change in employment status. The Smiths then buy the former Jones residence and move next door. On moving day Mr. Smith lands a job on the assembly line of the local refrigerator assembly plant and quits his job as a butcher at the nearby supermarket; Mrs. Smith continues her employment as a plumber with the same firm she has worked for during the past 36 years. Mr. Smith's job change was incidental to his family's move, whereas the Joneses are likely to experience a significant change in life-style as a result of theirs. Because of instances like these we usually eschew an activity-based approach to defining migration.

Migration is almost always operationally defined to be a move across a political boundary of some sort. For U.S. internal population movements, the Bureau of the Census counts as migrants only those persons who change their usual place of residence from one county to another.

Importance of Areal Unit Definition

The usual definition of migration is based on the assumption that *interareal* moves will typically be longer than *intra-areal* moves, and they will thus more likely involve a major modification in individuals' daily activity spaces as well. But as Figure 4.1 shows, this need not always be the case.

The fraction of the population that gets measured as being migrants not only depends on individual propensities to move, but also is affected by the size, shape, and population distribution within the areal units used to define the phenomenon. (See Rogerson 1990a.) In general, however, the greater the areal aggregation, the lower the proportion of all moves that will be counted. Table 4.1 shows the relative frequency of intercounty and interstate migration compared to all residential moves in the United States for 1986–87.

Unlike the birth and death components of population change, migration affects population change in *two* geographic units. There is often a greater element of free choice associated with migration than with either birth or death. Unlike the two components of natural increase, migration need not be a once in a lifetime occurrence; people are counted as migrants at many different stages of the life cycle.

Just as spatial unit definition plays a major role in migration measurement, so too does the temporal frame of reference.

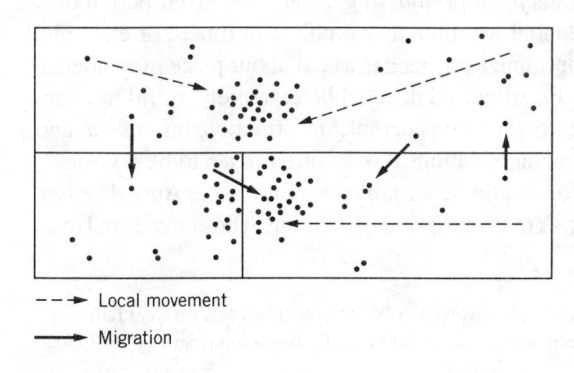

FIGURE 4.1 Example of Migration Flows That Are Shorter In Length than Local Movement When Migration Definition Involves Crossing an Areal Unit Boundary

TABLE 4.1 Relative Frequency of Movement Taking Place within Counties, States, and Regions of the United States, 1986–87

	Persons (in thousands)	Percentage
Total population	235,089	100.0%
Nonmovers (same house in U.S.)	191,396	81.4
Movers (different house in U.S.)	42,551	18.1
Movers from abroad	1,142	0.5

		% of movers	% of migrants
Movers in same county	27,196	63.9	—
Migrants (different county in U.S.)	15,355	36.1	100.0
Movers in same state	35,958	84.5	—
Same county	27,196	63.9	0
Migrants (different county in state)	8,762	20.6	57.1
Interstate migrants	6,593	15.5	42.9
Same region (different state)	3,055	7.2	19.9
Different region	3,538	8.3	23.0

Source: Computed by the authors from data in U.S. Bureau of the Census (1989c, pp. 16–17).

Fixed Period versus Registry-Type Data

Recall from Chapter 3 that migration is a flow rather than a stock variable, meaning that a time period must be given to report it. With respect to the time dimension, two major types of migration data may be collected. The first type is that recorded on, for instance, the population *registries* maintained by many nations in Europe and Asia (but not by the United States or Canada). Conceptually these are similar to records kept in almost all nations on vital statistics such as births and deaths. Whenever a person moves, a form is filed with the appropriate local officials (in some nations, the police). It would seem that with this type of data it would be a simple matter to tally up the number of moves made during, for example, one year to get an aggregate measure of migration. The results obtained, however, may not suit all the purposes for which migration data are used in population analysis. If the same person makes several moves during that one-year period, each of the moves will be counted.

The alternative to registry data is *fixed-period* migration data. Fixed-period data are based on a comparison of residential location at two points in time. For example, since 1940 a question on the U.S. decennial census has asked about place of residence five years prior to the census date.[1] Fixed-period data, while extremely useful in many types of analysis, suffer from the loss of considerable information on *repeat* and *return* movement—although for some applications this feature proves to be a virtue.

An example of repeat migration would be a Maine resident on the April 1, 1990, census who had been living in New York on April 1, 1985, but who moved to Texas

---

[1] In 1950, because of the large number of Americans involved in World War II in 1945, a 1-year rather than 5-year question was asked. Thus the decennial census statistics are for migration from 1949 to 1950.

in 1986 and subsequently to Maine in 1989. In the fixed-period 1985–90 migration data from the census, however, this person would be credited as a New York–to–Maine migrant, whereas two moves were actually made, neither from New York to Maine!

An example of a return migrant would be a person who moved in 1986 from Why, Arizona, to Anaheim, California, who got tired of battling the freeway traffic and moved back to Why in 1988. Because the county of residence at the time of the 1990 census and 5 years earlier were the same, this person is not considered a migrant at all even though two interstate moves were made during the 1985–90 period!

Obviously there is a relationship between the length of the fixed time interval over which migration is measured and the undercounting of actual moves that occurs as a result. Kitsul and Philipov (1981) and Rogerson (1990b) discuss problems associated with switching between 1-year and 5-year data (these being by far the two most common time intervals used to measure migration).

In addition to the 5-year decennial census data, other sources of migration information in the United States are 1-year estimates from the Bureau of the Census Current Population Survey (CPS) (see, e.g., U.S. Bureau of the Census 1992), and 1-year tabulations of state-to-state and county-to-county migration flow estimates created by the Bureau of the Census from matching the addresses listed on individual tax returns filed in consecutive years with the Internal Revenue Service. (Engels and Healy, 1981, describe these data.) Isserman, Plane, and McMillen (1982) provide a survey and assessment of U.S. migration data available from the federal government.

4.3 MEASURES OF MIGRATION

Once migration data are tabulated from the information recorded on registries, or from fixed-period survey questions (such as those on the decennial census or the CPS), or from matches of administrative records (such as the Census/IRS data), an important issue for the geographical analyst of migration is how to compare the amounts of migration taking place from and to different areas. In this section we first examine a variety of commonly used migration *rates* and then the lesser-used—but quite useful—concept of the *demographic effectiveness* or *efficiency* of migration.

Transition Probabilities

Let $\{m_{ij}\}$ be the flows (i.e., the numbers) of migrants in the migration streams from each origin region i to each destination region j during the time interval from exact time $t - 1$ to time t. Let P_i be the population of origin region i at time $t - 1$ that survives and lives somewhere in the migration system (in one of its n total regions) at time t. Then *transition probabilities* may be computed:

$$p_{ij} = m_{ij} / P_i; \qquad i = 1, 2, \ldots, n; \qquad j = 1, 2, \ldots, n; \qquad i \neq j \qquad (4.1)$$

which show the relative proportions of the population beginning the period in region i who end it in each of the other regions. Traditionally, additional terms $\{p_{ii}\}$ are computed representing the fraction of the population in each region i at $t - 1$ still there at time t. These p_{ii} terms thus include nonmovers as well as persons who move *within* region i itself.

The p_{ij} and p_{ii} fractions are traditionally arranged in the form of a table called a *transition probability matrix*, **P**. For migration between the four census regions of the United States from 1985 to 1990 the transition probability matrix computed from decennial census data is

$$\mathbf{P} = \begin{bmatrix} .9424 & .0076 & .0384 & .0115 \\ .0067 & .9429 & .0318 & .0185 \\ .0111 & .0163 & .9562 & .0164 \\ .0084 & .0156 & .0257 & .9504 \end{bmatrix} \qquad (4.2)$$

Each row corresponds to an origin region and each column to a destination. Here we have arranged the Bureau of the Census regions according to their typical sort order (1 = Northeast; 2 = Midwest; 3 = South; 4 = West), so, for example, the transition probability $p_{13} = 0.0384$ indicates that 3.84 percent of the population who were in the Northeast in 1985 and who survived to be counted in one of the four census regions 5 years later lived in the South in 1990. And $p_{33} = .9562$ indicates that 95.62 percent of the 1985 residents of the South who survived to be counted in a region 5 years later were still somewhere in the South region in 1990. Note that the largest of the diagonal elements $\{p_{ii}\}$ is this one for the South, meaning that in terms of interregional migration in 1985–90, southerners were the least likely to have left their region of residence.

An important property of the transition probability matrix is that if we add up all the elements across any of the rows of a transition probability matrix, we should get 1.0000 (neglecting any minor rounding errors). This is guaranteed because of the way transition probabilities were defined in Equation 4.1. Try summing the rows of the matrix in Equation 4.2 to verify this important *stochastic* property of transition probability matrices. This, in fact, is why they are called probabilities! Now try adding down each of the columns. You will find that, while the sums come close to 1.0000, they do vary significantly. There is no useful interpretation of these column sum quantities.

Because of the stochastic property of transition matrices, analysts often give the probabilities pseudo individual-level interpretations. You might read, for instance, that a Midwesterner had a 3.18 percent probability of choosing to move to the South in 1985–90. Remember, however, that this is simply based on the average of all actual interregional migration and nonmigration decisions. Every individual actually considers and responds to very different sets of influences so we should be cautious in considering a transition probability as a measure of individual proclivity.

One good descriptive use of transition probabilities is to interpret them, as suggested by Rogers (1992), as a form of speedometer. By comparing them over time the analyst can see the changing average tendencies of people to migrate from any particular region to any other. They could thus be thought of as a measure of the speed or velocity of flow in the migration stream from i to j. For example, migration to the South actually became more common for Northeasterners in the 1980s than it had been in the 1970s. Using decennial census data, $p_{13} = .0370$ for 1975–80 as compared to the 1985–90 transition probability reported earlier of .0384.

Just as some rivers are sluggish, while others run rapidly, so too with migration streams. It might seem logical to use transition probabilities as a gauge to compare the flows in all the different migration streams of the system. Considerable caution, however, should be used in making such interpretations. The populations and areal sizes of the regions of a system may vary considerably—factors that pollute pure interstream comparisons. For purposes of comparing the "velocity" of movement between regions we prefer to focus on the population redistributive role of migration and to use the demographic effectiveness measure that we will discuss later on in this section. To obtain a true picture of the relative levels of gross interchange in different migration streams we believe it is necessary to compare the actual volumes of flow to the estimated volumes that would be obtained from some sort of gravity model—see our discussion of "migration space" in Chapter 7.

A common use of transition probabilities is for making population projections. We return to examine their modeling uses in detail in both Chapters 6 and 7.

Gross Migration Rates

The *gross out-migration* from any specific region k may be found by adding up all the flows out of k to other regions:

$$\text{OM}_k = \sum_{j \neq k} m_{kj}$$

Similarly, *gross in-migration* to region k is found by adding up all the flows in its in-migration streams:

$$\text{IM}_k = \sum_{i \neq k} m_{ik}$$

A common means to standardize these quantities to facilitate intertemporal and interareal comparisons is to express them as the *out-migration* and *in-migration rates* for region k by dividing by the population of k:

$$\text{OMR}_k = (\text{OM}_k / P_k) \times 1{,}000$$
$$\text{IMR}_k = (\text{IM}_k / P_k) \times 1{,}000$$

As we discussed in Chapter 3, the best measure of the at-risk population is to use the population of k at the middle of the time interval. Quite commonly, however, these migration rates are computed using beginning-of-period population. The gross out-migration rate is a conceptually sound one, since the denominator does include those persons in region k who are "at risk" of becoming out-migrants. The in-migration rate, however, is not a properly constructed demographic rate. Its denominator, in fact, contains exactly those people in the system who cannot become in-migrants—persons already in region k! A more proper gross in-migration rate would be calculated using the population of all the regions in the system other than k:

$$\text{IMR}_k = \left[\text{IM}_k \Big/ \left(\sum_{i \neq k} P_i \right) \right] \times 1{,}000$$

This latter expression, however, is rarely used in actual migration analyses.

A typical use of gross migration rates is to compare the impacts of in- and out-migration on an areal unit, and to make comparisons across a set of areal units. Table 4.2, for example, shows in- and out-migration rates for the four census regions during 1985–90. Notice that the out-migration rates shown here differ slightly from those implied by summing the off-diagonal elements of the transition probability matrix.

TABLE 4.2 Gross and Net Migration Rates for U.S. Regions, 1985–90

Region	In-Migration Rate	Out-Migration Rate	Net Migration Rate
Northeast	34.4	58.3	−23.9
Midwest	42.1	57.5	−15.4
South	61.8	43.4	+18.4
West	60.9	49.3	+11.6

Source: Calculated by the authors from 1990 U.S. decennial census data.
Note: Rates are per thousand midperiod surviving population enumerated on the 1990 census as residing in one of the four regions.

This is because these rates were calculated using an average of the 1985 and 1990 populations covered in the flow table. Like with transition probabilities, some care should be given when talking about the meaning of out-migration rates because of the very different relative populations, land areas, and population concentrations near the borders of the various regional units. As with, for instance, population density measures, there is a tendency for small areas to exhibit the most extreme rates.

Net Migration Rates

The rightmost column of Table 4.2 shows an additional measure. *Net migration* (NM) to area k is simply the difference between gross in- and out-migration:

$$NM_k = IM_k - OM_k$$

and thus the net migration rate, NMR_k, is just the difference between the in- and out-migration rates. Note that unlike gross migration, net migration can take on either positive or negative values. A region with more in-migrants than out-migrants is said to have negative net migration or net out-migration.

Net migration rates are of interest as descriptors of the relative impact of migration on population change, but (as we detail in Chapter 7) their use for modeling and projection applications is problematic because of the differences in the true pools of "at-risk" population for in- and out-migration. As Morrison observed,

> There are no "net migrants"; there are, rather, people who are arriving at places or leaving them. (1977, p. 61)

We now examine a measure based on the relative proportions of people arriving and leaving.

Demographic Effectiveness

The *demographic effectiveness* of migration to region j is calculated as a percentage:

$$E_j = 100 \, (N_j / T_j)$$

where N_j is net migration (the difference between gross in- and out-migration) and T_j is total migration (the sum of in- and out-migration).

This statistic was first suggested in Thomas (1941), a study of long-term change in the Swedish migration system. Shryock (1959, 1964) extended the concept, which he referred to as demographic *efficiency*, to examine U.S. migration. For many purposes we would argue that the effectiveness measure is better than the more commonly used net migration "rate." E_j simply tells the percentage of "turnover" (to use Thomas's and Shryock's term for total migration) that results in population change. Effectiveness is solely a function of events taking place in the current period in a particular region and not of that region's population size. It does not share the small or large denominator problems inherent with the use of other sorts of growth rates for standardizing population change for variably sized geographic units.

The demographic effectiveness of migration in and out of each of the four census regions of the United States during 1985–90 was

Northeast	−25.8
Midwest	−15.4
South	+17.6
West	+10.5

Plane (1984) used a summary measure of *system effectiveness* equivalent to one suggested by Shryock:

$$E = 100 \sum_j |N_j| / \sum_j T_j$$

For the 1935–40 to 1965–70 periods the effectiveness of U.S. interstate migration declined from 21.1 percent to only 11.6 percent. Then for the 1975–80 period, system effectiveness rose to 16.8 percent. The system effectiveness of the most recent decennial census data (1985–90) declined from its 1975–80 level to the same low point registered in 1965–70, 11.6 percent.

Plane (1984) analyzed the web of changes in the 2,550 streams of movement between the 50 states (plus the District of Columbia) that gave rise to the very high overall demographic effectiveness of movement in 1975–80. *Stream effectiveness* may be calculated as

$$e_{ij} = 100 \, (n_{ij} / t_{ij})$$

where the net exchange, n_{ij}, equals the difference between the gross migration flow from state i into state j and that out from j to i,

$$n_{ij} = m_{ij} - m_{ji}$$

and t_{ij} equals their sum:

$$t_{ij} = m_{ij} + m_{ji}$$

McHugh and Gober (1992) and Plane (1994, forthcoming) used the stream effectiveness measure to examine short-term changes in U.S. interstate migration during the 1980s.

4.4 THE RELATIONSHIP BETWEEN IN- AND OUT-MIGRATION

If migration rates are computed in the usual manner, with the population of the region itself as the denominator, what will be the expected relationship between gross in- and out-migration across the regions of an internal migration system?

The Intuitive Perspective

Your best hunch would probably be that the relationship is as depicted in Figure 4.2. The regions with the highest rates of in-migration are shown as having the lowest rates of out-migration. It seems sensible to suppose that in-migrants will be attracted in large numbers to areas where economic conditions are good and where the environment for living is of high quality; in such places there should be little incentive for people to leave. Conversely, few in-migrants should be drawn to regions less economically well off and where living conditions are poorer; in such areas we would predict high propensities to out-migrate.

If a region's in- and out-migration rates fall anywhere along the dashed diagonal line in Figure 4.2 the net migration to this region is zero, because the number of in-movers is identical to the number of out-movers. Above and to the left of this line, out-migration rates exceed in-migration ones, so the dots in this section correspond to regions of net out-migration. Below and to the right of it are dots for regions with net in-migration. Our intuition, therefore, suggests an *inverse* or negative correlation between in- and out-migration rates across a set of regions. Such a commonsense-based perspective is embedded in a fair proportion of theoretical work on migration (e.g., the definition of the "master equation" used to control regional population growth dynamics in Weidlich and Haag, 1986).

The Positive Correlation Perspective

In many actual cases, however, just the opposite type of relationship has been found. For instance, for 1975–76, 1976–77, and 1978–79 U.S. interstate migration the correlation coefficients (r) between gross in- and out-migration rates were found to be .93, .91, and .84, respectively (Plane, 1981). The positive correlation viewpoint on the relationship is most often associated with the findings of Lowry (1966) and modeling work of Cordey-Hayes (e.g., 1975).

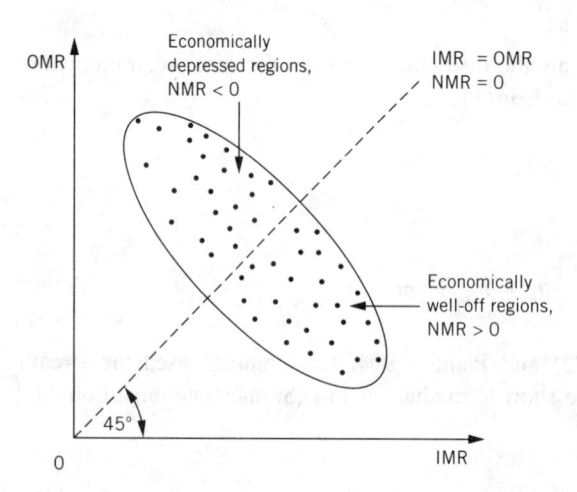

FIGURE 4.2 The Intuitive Relationship between In- and Out-Migration Rates

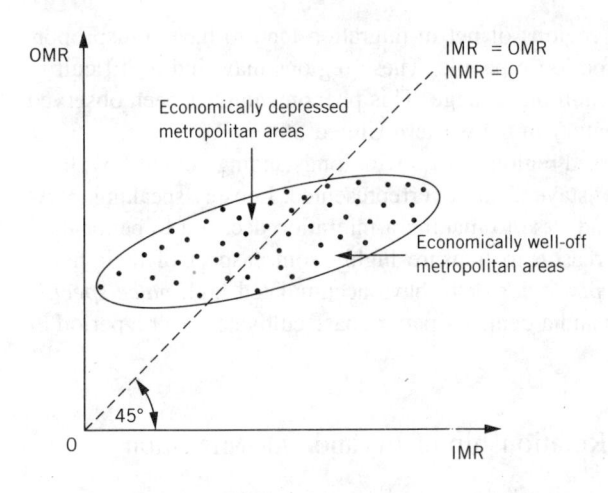

FIGURE 4.3 The Lowry and Cordey-Hayes Relationship between In- and Out-Migration Rates

Source: Drawn by the authors based on findings in Lowry (1966) and Cordey-Hayes (1975).

Gross migration rates for a set of regions exhibiting positive correlation between gross in- and out-migration would look schematically like those plotted in Figure 4.3. The regions of *net* in-migration are characterized not only by high rates of gross in-migration but also by higher than average rates of gross out-migration. Regions of net outflow, on the other hand, have lower than average rates of both in- and out-migration.

Three primary explanations (labor market turnover, age composition, and migrant stock) have been advanced to help explain the many situations where positive correlation has been found.

Labor Market Turnover

This explanation contests that people are generally more willing to change jobs and thus to consider migration when the economic conditions they are experiencing are relatively good. There are typically more total jobs "turning over" in relatively well-off places. The high numbers of job openings not only attract many in-migrants, they also encourage existing members of the local labor force to look to move up into better positions—including positions outside the area. Much of the total volume of migration in nations with highly developed economies now consists of interchange between pairs of the most economically advanced regions, rather than of the more demographically effective movement into these areas from less well-to-do places.

Age Composition

As we shall see shortly when we begin to disaggregate migration streams by the characteristics of the migrants, there is a strong relationship between age and movement propensities. The highest rates of movement for labor force migrants are for very young adults. Regions that have had high in-migration in recent times therefore have larger proportions of their total populations in the highly mobile early labor force ages, whereas regions of net out-movement have typically had their stock of such persons depleted. The regions of previous in-migration thus have populations in those age groups in which people are always more likely to consider out-migrating.

Migrant Stock

Closely connected to the second argument is the notion of *migrant stock*. It has been said that "migration is a lot like sinning—if you've done it once, you're more likely to consider doing it again." Thus regions of net in-migration tend to have a disproportionate share of hypermobile, footloose people. These regions may find it difficult to hold onto their population as conditions change. This phenomenon has been observed frequently over the past half-century in the western United States.

The populations of depressed regions, those with longstanding net outflow, have populations in which hard-core "stayers" are overrepresented. Loosely speaking, *stayers* are those who are strongly adverse to making a migration decision to better their lot in life. Persons who have reason to treasure highly some aspect of their home region are said to have strong *place ties* or to have accumulated high *place-specific capital*. Extended kinship webs and a dentist's patient base cultivated over a period of years are examples.

Lowry's "Asymmetrical" Relationship of In- and Out-Migration

Lowry's (1966) findings brought the relationship of in- and out-migration to the forefront of migration research. His results showed there to be an *asymmetrical* relationship between economic conditions in origin and in destination regions. When his "economic gravity model" (which we examine in Chapter 7) was fit to migration flows among 90 U.S. metropolitan areas, a significant correlation was found with wage rates and unemployment rates at the destination, but not at the origin of the streams. This triggered many followup studies (see, e.g., Stone, 1971; Miller, 1973; Feder, 1982) and gave occasion for a number of highly regarded persons in the regional economics establishment to assert that out-migration was little affected by economic conditions. Some even went so far as to adopt a fatalistic attitude about regional development policies aimed at halting the exodus of population from economically depressed regions. (See, e.g., Morrison, 1973.)

Beale's Findings

Beale (1969) in an often-cited but unpublished paper presented findings that led to our adoption of a more balanced and sophisticated perspective on the relationship between in- and out-migration rates. People knew that through much of the 20th century (except, perhaps, for the 1930s Depression and 1970s "turnaround" decade) *rural* areas in the United States had higher than average out-migration rates and lost a substantial portion of their population base through net out-migration.

Beale pointed up the importance of the spatial units one uses for a migration analysis. We should consider the fact that Lowry's results derived from intermetropolitan flows. And the positive correlation given earlier was for interstate flows. Most states encompass both urban and rural areas, with the population of the former often predominating. Beale's choice was to use an old set of statistical units called State Economic Areas (SEAs), which were aggregations of contiguous counties having similar sectoral economic structure. SEAs thus separated out metropolitan areas from sections of the country with rural or resource-based economies.

When Beale graphed in- and out-migration rates for SEAs, he found a relationship like the banana shape in Figure 4.4. Two groupings of areas with high *gross* out-migration rates are now evident. The first has high net *out*-migration (i.e., depressed, rural SEAs); the second has high net *in*-migration (i.e., the same well-off metropolitan areas present in Lowry's study). Beale's findings represent a combination of the "intuitive" and Lowry/Cordey-Hayes perspectives.

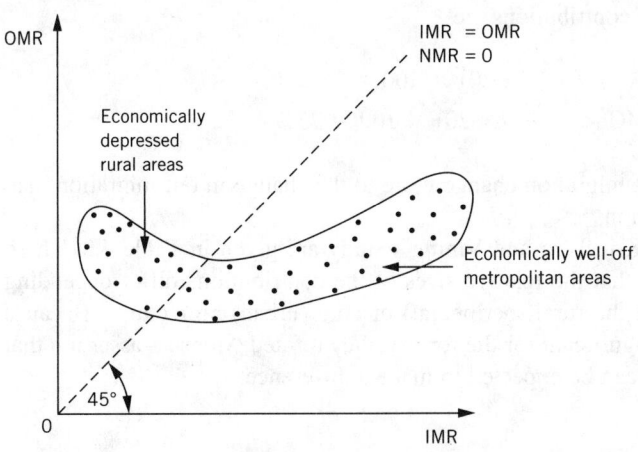

FIGURE 4.4 The Beale Relationship between In- and Out-Migration Rates

Source: Drawn by the authors based on findings in Beale (1969).

Kriesberg and Vining's Measure of In- and Out-Migration Contributions

Kriesberg and Vining (1978) later confirmed a *dynamic* relationship for Japanese inter-provincial flows similar to the cross-sectional one demonstrated by Beale. They showed that for peripheral, rural areas it was the changes in out-migration and not in-migration that contributed most substantially to net migration change. For the "core," urbanized prefectures the dynamics of net migration were more responsive to changes in gross in-migration. (For further analysis of Japanese migration patterns during the 1970s "turn-around" period, see Ishikawa, 1992.)

Kriesberg and Vining's study made use of quantities that the population analyst may find useful in general for the analysis of net migration shifts. Letting IMR, OMR, and NMR stand for the in-, out-, and net migration rates, respectively, and letting Δ indicate the change from one time period to the next (for example, ΔIMR = $IMR_{t+1} - IMR_t$), the *in-migration contribution* (IMC) and *out-migration contribution* (OMC) are given by

$$IMC = (\Delta IMR\ /\ \Delta NMR) \times 100$$
$$OMC = (-\ \Delta OMR\ /\ \Delta NMR) \times 100$$

Notice that these are in percentage terms and convince yourself that they should add up to 100! The minus sign is in the equation of the out-migration contribution because an *increase* in out-migration over time implies a *decrease* in net migration.

To see how these work, consider the following hypothetical case:

Year	IMR	OMR	NMR
1	5	25	− 20
2	10	10	0
Δ	+ 5	− 15	+ 20

The in- and out-migration contributions are

$$\text{IMC} = (+5 / +20) \times 100 = 25\%$$
$$\text{OMC} = (+15 / +20) \times 100 = 75\%$$

so we can attribute the net migration change more to the change in out-migration than to the change in in-migration.

Selected results of Kriesberg and Vining's study are given in Table 4.3. These illustrate their conclusion that the relative sizes of the contributions differ depending on whether one looked at the rural (peripheral) or core (urban) prefectures. Because out-migration was more significant for the former, they refuted Alonso's assertion that only in-migration change can be expected to make a difference.

The Nonlinearity of the Relationship between Out-Migration and Net Migration

The conclusion that we may draw from this case study of the examination of the cross-regional relationship of in- and out-migration is that there will, in fact, likely be some asymmetries in the responses of in- and out-migration to economic conditions. We now know a good bit more about this subject than when Lowry published his surprising results in 1966. There is a relationship between out-migration and net migration, but it isn't quite as simple as that between in-migration and net migration.

Figure 4.5 shows Beale's findings graphed in a somewhat different way than in Figure 4.4. This graph, in fact, is how Beale originally demonstrated the relationship. On the horizontal axis are net migration rates. He broke the entire range of the rates for SEAs into a fairly large number of numerical intervals and then computed and graphed the average gross in- and out-migration rates for those SEAs whose net migration rates fell into each interval.

We can observe that there is a monotonically increasing relationship between net and gross in-migration as plotted on the graph. The relationship between net migration and out-migration, however, is U-shaped and nonlinear, expressive of the trade-offs between several different influences that work in offsetting fashions.

A final important observation can be made about the various relationships we have been examining. Notice that in the case of both the Lowry/Cordey-Hayes and Beale relationships (cf. Figures 4.3 and 4.4), in-migration rates vary over a greater range than out-migration rates do.

TABLE 4.3 The Relative Contributions of Out-Migration Change and In-Migration Change to Net Migration Change for Japanese Prefectures During the Early 1970s

Year	Rural Prefectures OMC	Rural Prefectures IMC	Core Prefectures OMC	Core Prefectures IMC
1971–72	79%	21%	8%	92%
1972–73	61%	39%	11%	89%
1973–74	105%	−5%	−35%	135%

Source: Selected results contained in Kriesberg and Vining (1978).

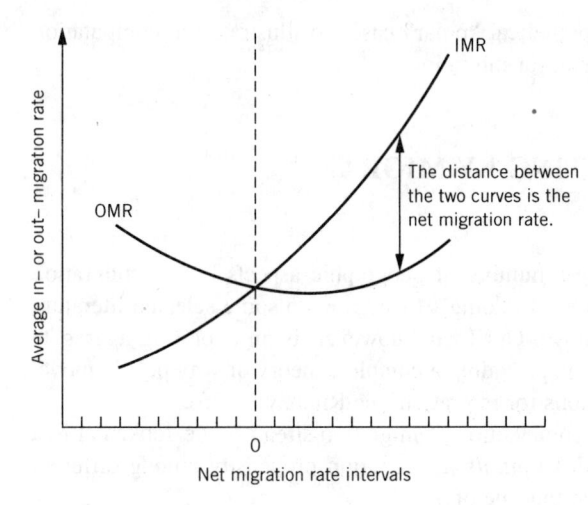

FIGURE 4.5 The Relationships of In- and Out-Migration Rates with Net Migration Rates According to Beale's Findings

Source: Drawn by the authors based on Beale (1969).

The Cross-Regional Variance of In-Migration and Out-Migration

There is, in fact, typically greater variation in in-migration rates than out-migration rates across a set of regions (although this is not true in all countries for all time periods). This may be attributable to the "competing influences" arguments for out-migration that we have just presented. A contributing factor is the structural difference in the pools of persons "at risk" of becoming in- versus out-migrants. As we noted when discussing the computation of in- and out-migration rates, only the residents of the region itself are eligible to elect to leave it, whereas there are many more people out in all the other regions of the system who may choose to become in-migrants to it. Plane, Rogerson, and Rosen (1984) reported that for U.S. states during the second half of the 1970s the ratios of in-migration rate variation to out-migration rate variation (V_{IR} / V_{OR}) ranged from 1.46 to 2.09. Their study examined in detail the theoretical repercussions of the different at-risk population pools and demonstrated that skewed distributions of interregional attractiveness—such as that which actually existed for labor market turnover variables at the time—tend to exacerbate the effect.

The Entropy of In- and Out-Migration

Another way to examine the asymmetries of in- and out-migration is through the entropy statistic (ENT), a measure of the uniformity of a distribution. Its general formula is

$$\text{ENT} = -\sum_i X_i \ln X_i$$

where X_i is one of the values of the variable and "ln" stands for the natural (base *e*) logarithm. When the entropy statistic is calculated for the proportion of gross out-migration in the whole system accounted for by each origin region, we can expect to find more often than not a higher entropy value—meaning greater uniformity—than when we calculate the statistic for the gross in-migration proportions for each destination. Higher row entropies than column entropies suggest that migration tends to be spatially focused on certain destinations, which is another way of thinking about the finding that the rates for out-migration have lower variance than those for in-migration.

Table 4.4 shows some hypothetical "polar" cases to illustrate the computation and meaning of the entropy statistic for migration.

4.5 DISAGGREGATING BY MIGRANT CHARACTERISTICS

In this section we examine a number of geographic aspects of how migration streams are differentially composed. In doing so we draw on some selected literature about theories for why people move. Our focus, however, is on exploring aggregate empirical regularities rather than on providing a complete theory of why people move. For a general review of explanations for migration, see Ritchey (1976).

We begin our study of the composition of migration streams—or, looked at in a complementary fashion, of the *differentials* in migration propensity among different groups of people—by focusing on the role of age.

TABLE 4.4 Measuring the "Spatial Focusing" of Migration with the Entropy Statistic—Hypothetical Polar Cases

Case A. Extreme Focusing

	Migration Matrix, M			OM_j
	⎡ 1	1	998 ⎤	1,000
	⎢ 1	1	998 ⎥	1,000
	⎣ 1	1	998 ⎦	1,000
IM_j	3	3	2,994	3,000

Row entropy:

ENT = $-(1{,}000 / 3{,}000) \ln (1{,}000 / 3{,}000) - (1{,}000 / 3{,}000) \ln (1{,}000 / 3{,}000)$
   $- (1{,}000 / 3{,}000) \ln (1{,}000 / 3{,}000)$
   = 1.098613

Column entropy:

ENT = $-(3 / 3{,}000) \ln (3 / 3{,}000) - (3 / 3{,}000) \ln (3 / 3{,}000)$
   $- (2{,}994 / 3{,}000) \ln (2{,}994 / 3{,}000)$
   = 0.011802

Case B. Extreme Dispersal

	Migration Matrix, M			OM_j
	⎡ 334	333	333 ⎤	1,000
	⎢ 333	334	333 ⎥	1,000
	⎣ 333	333	334 ⎦	1,000
IM_j	1,000	1,000	1,000	3,000

Row and column entropies are identical:

ENT = $-(1{,}000 / 3{,}000) \ln (1{,}000 / 3{,}000) - (1{,}000 / 3{,}000) \ln (1{,}000 / 3{,}000)$
   $- (1{,}000 / 3{,}000) \ln (1{,}000 / 3{,}000)$
   = 1.098613

Migration Age Schedules

Although many demographic and socioeconomic characteristics influence the likelihood that an individual will migrate, perhaps none is as important as age. Probabilities of movement vary throughout the life course, peaking in the young adult years that coincide with nest-leaving events such as marriage, seeking and obtaining employment, and going away to college. Figure 4.6 displays age-specific mobility rates in the United States in 1990–91.

With the exception of those under age 5, children have relatively low rates of mobility. This youngest group has a higher rate of mobility than children age 5 to 14 because (a) children migrate with their parents (usually), and the parents of those under 5 are themselves likely to be in an age group that experiences a higher degree of mobility, and (b) those age 5 to 14 are tied to particular school districts making their parents less inclined to move because of the disruption it would cause to their children's social ties and academic environment. After individuals complete their education and settle into jobs and marriages, the mobility rate declines steadily. Sometimes a much smaller, secondary peak that corresponds to retirement-related mobility is observed.

The universality (or at least high generalizability) of this age pattern has been commented upon by Rogers, Raquillet, and Castro (1978). They discuss the use of model migration schedules as a way to summarize the age pattern of migration. Model migration schedules collapse the information contained in a table of age-specific migration rates into a smaller number of parameters that describe how the migration rate varies with age. Figure 4.7, taken from Rogers and Castro (1986), illustrates how a model schedule can be derived from separate components representing migration during the labor force years, as well as migration during the pre- and post-labor force years.

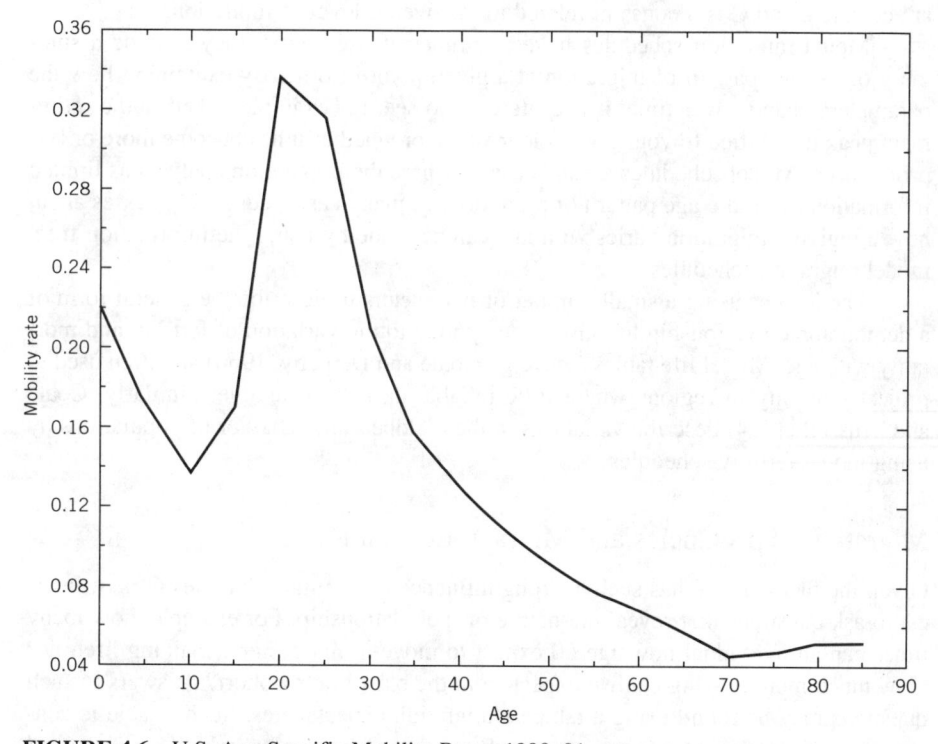

FIGURE 4.6 U.S. Age-Specific Mobility Rates, 1990–91

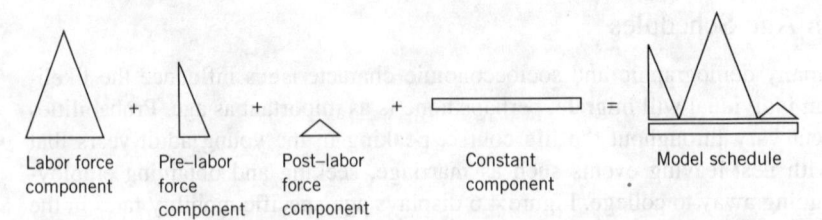

FIGURE 4.7 The Fundamental Components of a Model Migration Schedule
Source: Rogers and Castro (1986, p. 175). Reprinted by permission.

In practice, rather than represent these building blocks as triangles, Rogers and Castro suggest that they be represented by smoother exponential curves. Accordingly, the migration rate as a function of age, x, is described as the sum of four curves:

$$M_x = A_1 \exp(-\alpha_1 x)$$
$$+ A_2 \exp\{-\alpha_2(x - \mu_2) - \exp[-\lambda_2(x - \mu_2)]\}$$
$$+ A_3 \exp\{-\alpha_3(x - \mu_3) - \exp[-\lambda_3(x - \mu_3)]\}$$
$$+ C.$$

where the four terms on the right-hand side respectively represent the pre-labor force, the labor force, the post-labor force, and the constant components. The α parameters represent the steepness of the descending portions of the curve, while the λ parameters represent the steepness of the ascending portions of the curve. The parameters μ_2 and μ_3 correspond to the ages upon which the labor force and post-labor force components are centered, and C is a constant related to the overall level of migration.

Model migration schedules have a number of uses. First, they provide a summary of the age pattern of migration at a given point in time. By examining how the parameters change over time, it is quite easy to see, for example, whether the retirement peak has shifted to younger or older years, or whether it has become more or less pronounced. Model schedules are also useful where the population analyst has limited information about the age pattern of migration. In many cases, educated guesses about how a region's migration varies with age can be made by using the information from model migration schedules.

The idea of using a small number of parameters to describe the general form of a demographic relationship has also been applied to the variation of fertility and mortality with age. Model life tables (see, e.g., Coale and Demeny, 1966) are often used to project mortality in regions where little reliable data are available. Similarly, Coale and Trussell (1974) describe variations in the childbearing behavior of populations by using model fertility schedules.

Migration Expectancies and Migraproduction Rates

Given the fact that age has such a strong influence upon migration rates, it is of interest to ask questions that reveal the nature of the relationship. For example, how many times can an individual now age 20 expect to move in his or her remaining lifetime? How much more moving can we expect from the baby boom cohort? Answers to such questions may be found using a table of migration expectancies. Such a table is constructed and interpreted in a manner similar to that of life tables (which we discussed in Chapter 3).

Wilber (1963) was the first to describe how to derive a table of migration expectancies. Migration expectancy at age x is defined as the number of years with moves that can be expected for a person of exact age x, assuming that the person is subjected to prevailing age-specific rates of mobility and mortality. It is calculated from

$$E_x = \left(\sum_{a=x} M_a L_a\right) / l_x \qquad (4.3)$$

where E_x is the migration expectancy at age x, M_a is the mobility rate for people in the age group beginning at exact age a, L_a is the size of the stationary population in that age group (taken from a life table), and l_x is the life table population at exact age x. The calculation is similar to that made for life expectancy in the life table; the only difference is that the value of L_a is multiplied by M_a so that only years with moves are counted. The notion of migration expectancy has been used extensively by Long (1973, 1988) to describe the age pattern of mobility in the United States.

Willekens and Rogers (1978) define the *gross migraproduction rate* (GMR) to be the number of moves that a person could expect to make during his or her lifetime, assuming current age-specific rates were to apply:

$$\text{GMR} = 5 \sum_{a=0} M_a$$

Here 5-year age groups are assumed, and M_a is the mobility rate of the age group beginning at exact age a. Note that the gross migraproduction rate is analogous to the gross reproduction rate defined for fertility rate measurement, because it is derived by summing age-specific rates. Though it is straightforward to calculate, like the gross reproduction rate, it does not account for mortality. That is, the gross migraproduction rate may be thought of as the number of moves a person at birth would make, assuming that he or she survived to the oldest age group. Willekens and Rogers go on to define a *net migraproduction rate* that is analogous to the net reproduction rate since it accounts for mortality—it is virtually identical to the migration expectancy originally defined by Wilber.

Example: Calculation of Migration Expectancies and Migraproduction Rates

Table 4.5 contains age-specific mobility rates (i.e., the fraction of the age-specific population that changes residence) in the United States for the period March 1990 to March 1991. The table also contains the l and L values from the 1988 life table of the United States.[2]

---

[2]Ideally the dates of the life table and the mobility data should match. However, there is a long lag in the tabulation of vital statistics data, and as of this writing, life tables for 1990 were still not available. Mobility data covering the period 1988–89 were not published. An alternative would have been to provide an illustration for the year 1985–86. Then, to be even more precise, a mixture of the 1985 and 1986 life tables would ideally have to be used, because mobility during the 1985–86 period includes parts of both calendar years. Since mortality exerts a relatively small influence on migration expectancies, and since changes in mortality levels from 1988 to 1990 and 1991 are likely to have been quite small, little accuracy has been lost in the illustration. Such inconsistencies are often unavoidable, and the analyst will soon come to recognize those situations where little accuracy is lost as well as those situations where the noncorrespondence of different pieces of required information may lead to intolerably large errors.

TABLE 4.5 Mobility and Life Table Data Used for Calculation of Migraproduction Rates

Age (x)	Mobility Rate	l_x	L_x
0	*	100,000	99,147[a]
1	.222	99,001	395,540[b]
5	.171	98,803	493,688
10	.136	98,683	493,155
15	.169	98,549	491,767
20	.338	98,118	489,206
25	.316	97,553	486,274
30	.210	96,597	483,035
35	.155	96,240	479,021
40	.129	95,316	473,785
45	.108	94,112	466,443
50	.091	92,335	455,194
55	.076	89,569	437,859
60	.066	85,331	411,976
65	.055	79,123	375,656
70	.042	70,779	327,120
75	.045	59,683	265,113
80	.049	46,029	190,715
85	.049	30,171	179,948

Sources: Calculated by authors from data in U.S. Bureau of the Census (1992); U.S. Department of Health and Human Services (1990b).

Notes: (*) Mobility rates are published only for the population age 1 and over.
[a] $_1L_0$
[b] $_4L_1$

The gross migraproduction rate is obtained by simply summing the age-specific mobility rates, and then multiplying by 5 (the width of the age group). In the case of the first group, we shall assume that infants less than a year old have the same mobility rate as children age 1 to 4. Thus

GMR = 5 (.222 + .171 + .136 + .169 + ... + .049) = 12.14

Wilber's migration expectancy at birth is obtained using Equation 4.3:

E_0 = [(.222)(99,147 + 395,540) + (.171)(493,688) + ...] / 100,000
    = 11.17

An "average" individual (i.e., an individual that follows the age-specific mobility and mortality rates in Table 4.5) can be expected to move roughly 12 times in his or her lifetime.

Note that the migration expectancy at birth is less than the gross migraproduction rate. This is because each of the age-specific mobility rates is now multiplied by a number slightly less than 5, reflecting the fact that not everyone in the cohort will survive 5 years to enter the next age group. Note also that the difference between E and GMR is quite small. This is readily understood by recognizing that the peak mobility years precede the years of relatively high mortality rates.

By starting the summation at later ages, the migration expectancy for individuals at a given age may be computed, again using Equation 4.3. Table 4.6 displays the results for the 1990–91 mobility data.

Migrant Selectivity

The previous subsection demonstrated the importance of age in understanding both destination choice and the level of mobility. Migration is also strongly related to a number of other characteristics of individuals. Higher levels of education are generally associated with higher rates of mobility. Renters are more mobile than homeowners. Married couples with children move less frequently than single individuals. The *selectivity* of migration refers to the fact that the characteristics of the migrants are quite different than the characteristics of the nonmover population at either the origin or the destination.

These differentials are important for at least two reasons. First, they are clearly important in a direct sense if we are to understand the nature of human behavior and decision making. People with different backgrounds will have different likelihoods of

TABLE 4.6 Migration Expectancies by Age

Age	Migration Expectancy
0	12.02
5	10.92
10	10.07
15	9.39
20	8.54
25	6.86
30	5.28
35	4.24
40	3.46
45	2.82
50	2.29
55	1.84
60	1.47
65	1.15
70	0.89
75	0.70
80	0.50
85	0.29

moving. They will also differ in the nature of the moves they make. (For instance, high-income individuals are more likely to make moves that are of longer distance than those with low incomes.) Such an understanding of the migration decisions of individuals can also be instrumental in public policy analysis (the design and implementation of mobility assistance programs, the creation of regional development programs, and the like). In addition, large-scale changes in the volume and direction of migration flows resulting from changes in population composition may be better anticipated.

Migration differentials are also important from the perspective of those interested in regional change. The selectivity of migration implies that the very nature of regions will change as a result of in- and out-migration. Central cities that experience no net migration may nevertheless change dramatically if the in-migrants have low incomes and the out-migrants have high incomes. Likewise, some rural regions age rapidly due to the combined out-migration of young adults and the return in-migration of the elderly.

The notion of migration selectivity often (but certainly not always) refers to some measure of the education, income, training, and productivity of migrants. The distinction is often made between migration that is positively selective and that which is negatively selective. When migration is positively selective, the migrants are more educated than the nonmigrants with which they are being compared (at either the origin or the destination). Migration is usually positively selective with respect to the origin. Thus migration often drains the origin region of its most productive people (hence the term *brain drain*). Blau and Duncan (1967, pp. 271–72) go on to argue that migration in the United States is also positively selective with respect to the destination and, furthermore, that the magnitude of selectivity is increasing over time.

Negatively selective migration refers to the situation where the migrants are less productive or less educated than nonmigrants. Everett Lee's "A theory of migration" (1966), for example, puts forth the hypothesis that involuntary moves that are associated with "push" factors within the home or the origin region are usually negatively selective.

Beaten Paths and the Role of Information

One of the many consequences of migration is change in the pattern of interregional information flow. Members of a nuclear family that moves from New York to Virginia retain their ties with extended family and friends back in New York after their move. Though some of the family's relationships with friends in New York will undoubtedly wither over time, significant ties with relatives and other friends are maintained. Thus the amount of information exchanged between the two places is likely to increase as a result of the move.

The increased information flow may in turn have consequences for future migration flows. Some of the family and/or friends back in New York may be convinced to move to Virginia if they receive positive information about jobs or amenities. Over the past few decades, this feedback effect has become recognized as extremely important in understanding the nature of migration patterns.

Nelson (1959) was one of the first to emphasize the important influence of information on migration flow patterns. Greenwood (1970) later developed migration models that sought to explain the migration flow between two regions as a function of previous migration that has taken place between those locations. Greenwood's justification for using previous migration flow as an explanatory variable was based upon the observation that information flows help to sustain past patterns. There is consequently a good deal of inertia that characterizes the evolution of migration flow patterns.

MacKinnon and Rogerson (1980) and Rogerson and MacKinnon (1982) developed a simulation model to emphasize the role of information. Their hypothetical multiregion system was characterized by individuals moving in response to information about job vacancies in other regions. A goal was to capture the importance of time lags in the acquisition of information, and misperceptions of where vacancies are in the job market. They found that efforts to reduce regional inequalities in unemployment and vacancy rates through selective advertising could have the unintended side effect of increasing the amplitude of temporal cycles of unemployment and vacancy rates within regions.

All of these studies point to the importance of information in understanding migration. Since information is not easy to measure, nor is information on information readily available, analysts often do not include it in their studies. Still, its importance is now universally recognized, and it should not be overlooked.

Return Migration

It is important to realize that beaten paths do not operate in a one-way fashion. Often migrants reach their destination and are disappointed in one way or another—perhaps job prospects are not as good as anticipated, or greater time than expected may be required to assimilate by meeting new friends. Return migration to the former region of residence often results. Of course, return migration may occur at any stage of the life cycle; retirement, for example, provides a natural opportunity to move back to the region of birth. As we shall see, return migration accounts for a substantial portion of the total volume of migration.

Eldridge (1965) provided a useful scheme for classifying migrants that facilitates the study of return migration. She categorized migration flows into primary, secondary, and return migrants. Primary migrants are those who are leaving their region of birth for the first time. Secondary migrants are those who are neither leaving nor arriving in their region of birth. Return migrants, as the name implies, are those who are returning to their region of birth. The U.S. Bureau of the Census collects data by state of birth; it is difficult to analyze primary, secondary, and return migration for geographic scales other than the state level in the United States.

Long (1988) notes that among individuals born in the United States and changing their state of residence between 1975 and 1980, 42 percent were primary migrants, 38.8 were secondary migrants, and the remaining 19.2 were return migrants. Thus almost one-fifth of all interstate migrants are returning to their state of birth. As Long points out, if migrants were to choose a destination state at random, only 2 percent (1/50) of interstate migrants would choose their state of birth.

Long goes on to examine the composition of particular migration streams. The New York–to–Florida stream, for example, is unusual because, among those born in the United States, 87.4 percent are primary migrants, while only 3.5 percent are migrants returning from New York to Florida.

At least some return migration is due to the migration of the elderly back to regions where they spent their childhood. To the degree that this is an important component of return migration, we can speculate that as the baby boom ages, the importance of return migration may increase during coming decades. Because many of those born during the baby boom era moved from the Frostbelt to the Sunbelt, the number of individuals at risk of migrating back to their state of birth in the Frostbelt should be increasing. Thus we might expect more significant return migration flows from the Sunbelt to the Frostbelt as the baby boom ages—though the preference of elderly retirees for warmer weather could offset the general return migration effect.

As indicated in the beginning of this section, much return migration may be attributed to individuals who are not satisfied with the destination they have recently migrated to. The notion that a significant amount of migration may be due to individuals who have only relatively short durations of residence in their current home is the subject of the next subsection.

Duration of Residence Effects and Cumulative Inertia

A long-standing empirical observation of human mobility is that within populations, there is a negative relation between duration of residence and the probability of moving (Rider and Badger, 1943; Goldstein, 1954; Taueber, 1961). Thus when we examine populations we find that subpopulations with relatively short durations of residence have higher probabilities of moving than those with longer durations of residence.

There are at least two possible reasons for this observed relationship. Perhaps the better-known mechanism is the axiom of *cumulative inertia* as defined by McGinnis (1968):

> The probability of remaining in any state of nature increases as a strict monotonic function of duration of prior residence in that state. (p. 716)

As applied to migration, cumulative inertia implies that for all individuals, the probability of moving decreases with increasing length of residence. Individuals are taken to be homogeneous along this dimension of mobility; that is, all individuals make their mobility decisions using the same curve that describes how the probability of moving declines with increasing duration of residence.

The reasons for the hypothesized decline in the individual's propensity to move have to do with the increasing ties to friends and neighbors that individuals develop over a period of time. An increasing amount of "social capital" becomes invested in a place over time, increasing the psychic costs of breaking those ties through moving.

An alternative explanation for the observed aggregate relation between duration of residence and mobility has to do with the heterogeneity of the population. Blumen, Kogan, and McCarthy (1955) observed that populations are heterogeneous in their mobility behavior; a small proportion of the population does most of the moving. Though there is in reality a continuous range of observed behavior, in its simplest manifestation, population heterogeneity in this context arises from subpopulations of highly mobile "movers" and relatively immobile "stayers." With no cumulative inertia whatsoever, the observed aggregate relation between duration of residence and probability of moving may still be observed because the movers are likely to have short durations of residence and they are likely, by definition, to soon move again. Likewise, the stayers are likely to have long durations of residence, and are less likely to move. In this scenario both movers and stayers have probabilities that, though different from one another, do not change over time (and hence there is no cumulative inertia). The aggregate relation in this instance arises entirely from the heterogeneity of the population.

To illustrate, suppose that four-tenths of the movers move in any given year, while only one-tenth of the stayers move. Further suppose that we start with an equal number of people in each subgroup. Finally, we shall assume that everyone starts with duration of residence equal to zero. During the first year, we thus observe that $[(.4 \times .5) + (.1 \times .5)]$ or 25 percent of the total population moves. Therefore, at the end of that first year, 25 percent of the population has duration of residence equal to zero, while the remaining 75 percent has duration of residence equal to one year. Of those with duration equal to zero, 80 percent of them are movers [.4 / (.4 + .1)] and the remaining 20 percent are stayers. Of those with duration equal to one year, only 40 percent are movers [.6 / (.6 + .9)], while the remaining 60 percent are stayers. During the second year, the proportion of those with duration zero that move is [(.8 × .4) + (.2 × .1)] = .34, while the proportion of those with duration one that moves is [(.4 × .4) + (.6 × .1)] = .22. Although it may appear as though cumulative inertia is operating, it is the heterogeneity of the population that yields the decline in movement with increasing duration of residence.

Clark and Huff (1977) and Pickles, Davies, and Crouchley (1982) find the evidence for heterogeneity effects to be quite strong, and the evidence for cumulative inertia to be quite weak. It seems likely that the assessment of these effects will continue to be a fertile area of research for years to come.

4.6 METHODS FOR ANALYZING GEOGRAPHIC PATTERNS OF MIGRATION

We conclude this chapter by presenting three different methods for studying the systemic structure of interregional migration streams: principal components analysis, Sonis's extremal tendencies technique, and spatial shift-share analysis. While all three of these may be used to study how migration patterns change over time, spatial shift-share analysis is the most explicitly dynamic, working with the *changes* in migration flows from one time period to the next rather than with the flows for a single time period. A further method for examining temporal change in the structure of region-to-region flows is the notion of a *causative matrix*. Because development of causative matrix techniques requires understanding the basic concepts of Markov chain analysis (which we do not present until Chapter 6), and because it has potential use for migration forecasting (whereas these other techniques should be thought of as techniques for descriptive analysis of existing patterns), we do not discuss causative matrices until Chapter 7.

Principal Components Analysis

A major challenge for the analysis of migration is to summarize compactly all the detailed information in a complete migration table. A matrix of all the gross flows between the n zones of a system of interregional migration has a total of $n(n - 1)$ elements. For example, for U.S. interstate migration there are 2,550 possible flows (if the District of Columbia is included with the 50 states). If it is only the net exchanges in the system that are being analyzed (as, for instance, in Morrill, 1988) then the total count of entries is cut in half, but there still may be a mind-numbingly large number of numbers for the analyst to work with.

Useful statistical techniques for extracting simplified pictures of the general spatial patterns of flow inherent in a full origin–destination matrix of flows include cluster analysis, factor analysis, and principal components analysis. Slater (e.g., 1976, 1981) has explored numerous cluster analysis procedures to examine the fundamental structure of migration and other types of flow tables. In this section we give a brief overview of how principal components analysis (which is also similar in form to factor analysis) can be used to group the original spatial units for which flow data are available into larger migration-based regions—or, in other words, to define a set of migration subsystems within the overall system of zone-to-zone flows.

Principal component (and factor) analyses of migration (and other flow) systems has by now a fairly lengthy pedigree within the geography literature. (For instance, see Winchester, 1977; Clayton, 1977, 1982; Davies and Musson, 1978; Plane and Isserman, 1983; and Ellis, Barff, and Renard, 1993; for specifics on how to carry out such an analysis.) Here our goal is simply to present a basic outline of how the procedure works, and to stress its uses. We give more detail about the various steps of performing a principal components or factor analysis in Chapter 10 when we describe factorial ecology studies of urban-area demographics.

The first step of a principal components migration regionalization is to treat either the rows or the columns of the migration matrix as "variables" and to compute the correlations between every pair. For example, if our matrix is a 51 × 51 origin–destination table of gross flows between U.S. states (with zeros on the diagonals), the 51 × 51 matrix of correlation coefficients (r) between each row of the table and every other row will measure the extent to which *out-migrants* from each pair of states choose in similar proportions the 49 other available destinations.[3] Similarly, if this first-step correlation matrix is computed from the columns of the table, the correlations will measure the similarity of the sources of *in-migrants*. In the subsequent discussion we shall assume that it is the out-migration-based regions that are being examined. To the extent, however, that in-migration and out-migration flows are correlated (recall our earlier discussion of this issue!), the two sets of regionalizations may be similar.

In a second step, principal components are extracted using a statistical software package such as SPSS (Statistical Package for the Social Sciences). From these, groupings of states with similar migration patterns are easily obtainable. States that load highly on the same component are those having proportionally similar destinations for their out-migrants.

The analyst then makes a decision about how many total components to retain for further processing, and whether to perform a rotation scheme to simplify the structure of loadings (as was done in, e.g., the Plane and Isserman, 1983, study).

Selected results of a principal components regionalization of the destination choices of out-migrants are illustrated in Figure 4.8. Taken from the Ellis, Barff, and Renard (1993) study, the two maps show the different regional subsystems that existed for the migration patterns of professional and managerial workers in the manufacturing sector and for operators and laborers during the period 1975–80. The authors note that the more fragmented set of subsystems obtained for the lower-status group is in accord with expectations regarding the shorter average move lengths of the lower-skilled, lower-paying occupational group.

In addition to being a useful method for comparing the structure of migration subsystems for different groups of migrants, principal components analysis is also quite useful in examining the structural stability of the regional patterns of flow. The major finding by Plane and Isserman (1983) was that there was apparently remarkable stability in the spatial structure of *gross* interstate work force migration for the periods 1960–65, 1965–70, and 1970–75—despite the dramatic turnaround in patterns of net migration over that period. The same 10 or 11 basic migration subsystems were identified in each of their six separate principal component analyses (three of the destination patterns of out-migration, three of the source patterns of in-migration). Furthermore,

---

[3]In computing the correlations, a "pairwise" deletion rule is employed since intra-unit flows are ignored in this type of analysis. For two states a and b the correlation is computed without including two pairs of corresponding elements in the rows of the migration matrix: m_{aa} and m_{ba} plus m_{ab} and m_{bb}.

these migration regions seemed to correspond relatively well with traditional cultural geographic regions of the United States, highlighting the continuing importance of regional cultural differences in influencing Americans' residential place preferences.

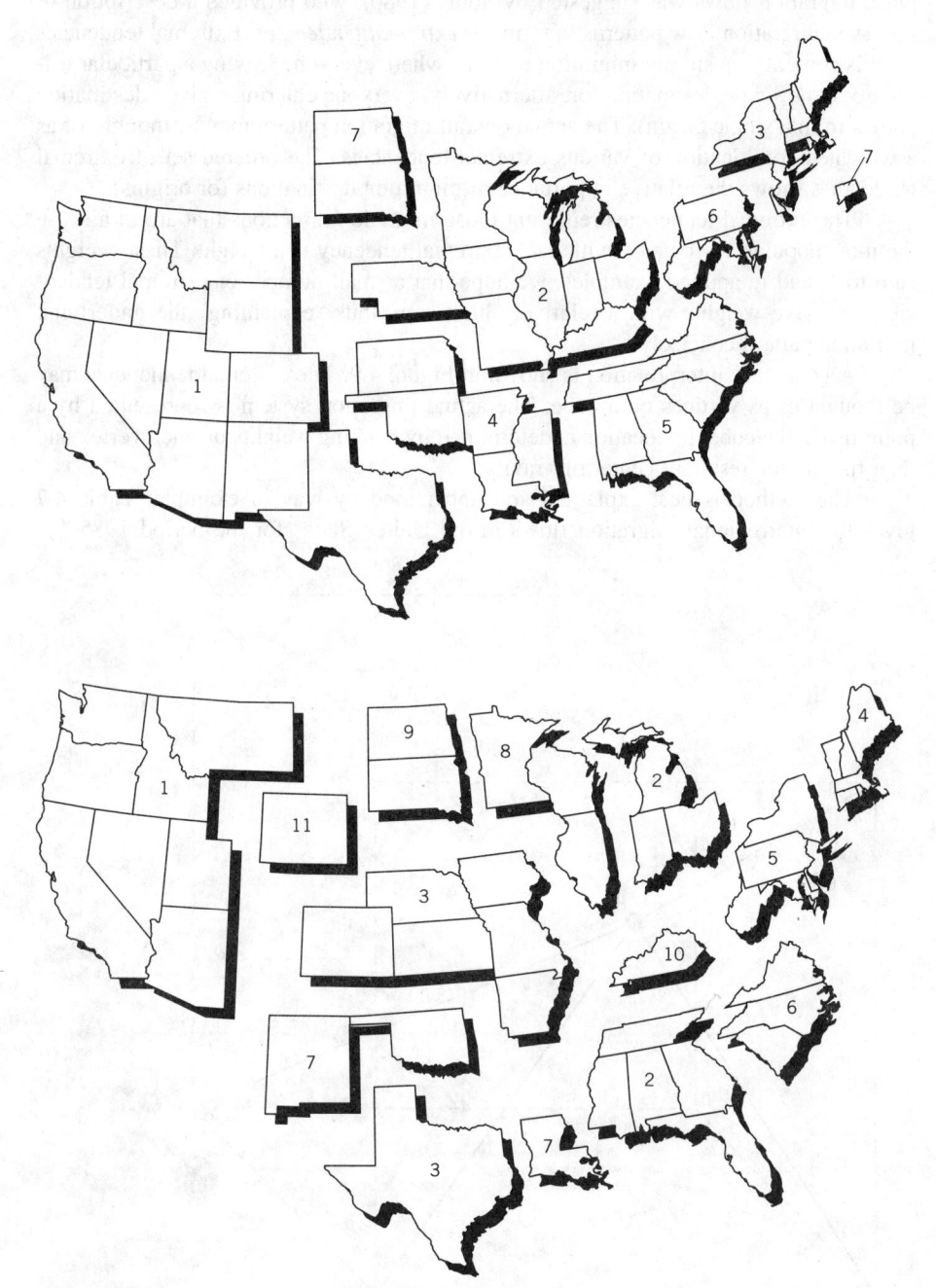

FIGURE 4.8 Principal Component Out-Migration Regions of U.S. States for Different Occupational Groups

Source: Ellis, Barff, and Renard (1993, pp. 176, 180). Reprinted by permission.

a. Professionals and Managerial Workers in the Manufacturing Sector

b. Operators and Laborers in the Manufacturing Sector

The Extremal Tendencies of Sonis

An alternative way to simplify and make sense of the complicated patterns of place-to-place migration flows was suggested by Sonis (1980), who provided a description of observed migration flow patterns in terms of *extremal tendencies*. Extremal tendencies (in this context) are simply migration patterns where everyone leaving a particular origin goes to a single destination (or, alternatively, everyone entering a given destination comes from a single origin). The actual overall migration pattern may be thought of as a weighted combination of various extremal tendencies. The ordered set of extremal tendencies shows the relative importance of migration destinations (or origins).

The extremal tendencies represent those migration directions that are in a sense the most popular. Associated with each extremal tendency is a weight. These weights sum to 1, and in applied examples, we hope that a small number of extremal tendencies will have weights with a relatively high sum, thus "explaining" the underlying migration pattern concisely.

A geometric interpretation is shown in Figure 4.9. The extremal tendencies may be thought of as vertices on a cube. The actual migration system is represented by a point inside the cube. Its location is determined by hanging weights on each vertex and then finding the resultant center of gravity.

The method is best explained and understood by way of example. Table 4.7 gives the interregional migration flows in the United States for the period 1985–90.

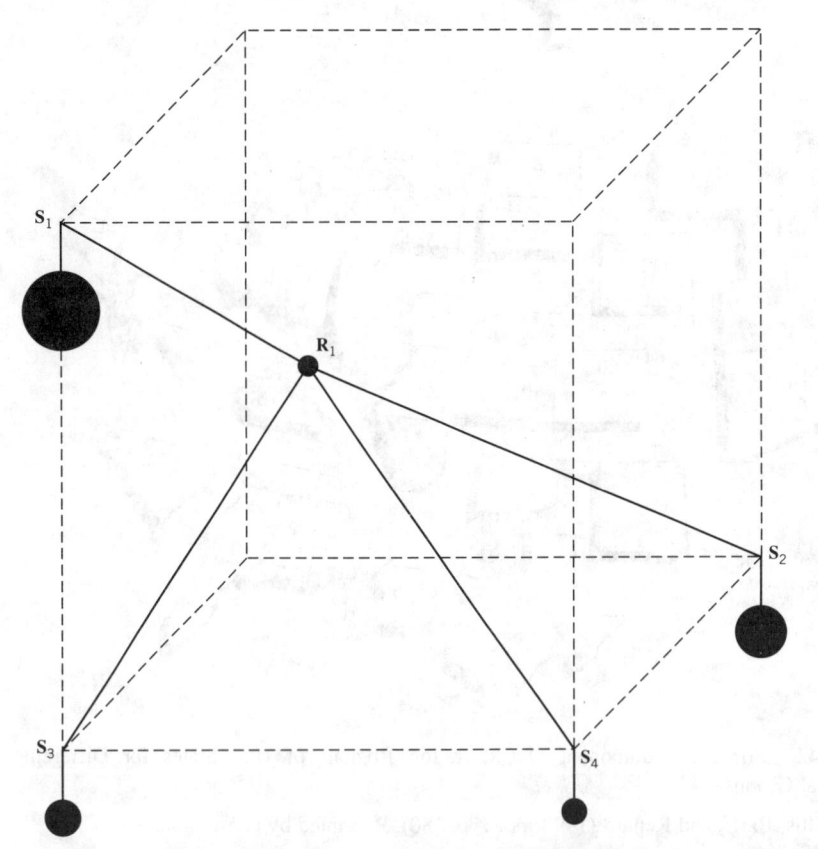

FIGURE 4.9 A Geometric Interpretation of Extremal Tendencies

Source: Sonis (1980, p. 87). Reprinted by permission.

TABLE 4.7 U.S. Interregional Migration Flows, 1985–90, and **P** Matrix Used for Sample Calculation of Extremal Tendencies

Flow Matrix, **M**

	To: Frostbelt	South	West
From:			
Frostbelt	—	3,564,643	1,579,652
South	2,096,522	—	1,247,154
West	1,109,811	1,178,772	—

P *Matrix*

	To: Frostbelt	South	West
From:			
Frostbelt	—	.6929	.3071
South	.6270	—	.3730
West	.4849	.5151	—

Intraregional flows are ignored, and the Bureau of the Census's Northeast and Midwest regions have been aggregated into a single Frostbelt region. The migration flow matrix is first transformed into a matrix **P** by dividing each element by its row sum. These quantities, shown in the lower panel of Table 4.7, represent the proportion of migrants leaving an origin region who move to a particular destination.

The first extremal tendency (S_1) is described by a matrix with 1's in the positions corresponding to the maximal element in each row:

$$\mathbf{S}_1 = \begin{bmatrix} 0 & 1 & 0 \\ 1 & 0 & 0 \\ 0 & 1 & 0 \end{bmatrix}$$

Thus the primary extremal tendency is for migrants to move from the Frostbelt to the South, for migrants leaving the South to go to the Frostbelt, and migrants leaving the West to go to the South. The weight, w_1, associated with this tendency is taken to be the minimal value among those elements of **P** that are the maximum in their rows. Thus w_1 = min {.6929, .6270, .5151} = .5151. It corresponds to what Sonis views as the "bottleneck" or "migration interdiction" in the system; it is that flow that prevents the extremal tendency from being more fully realized. We may therefore represent **P** as

$$\mathbf{P} = .5151\mathbf{S}_1 + (1 - .5151)\mathbf{R}_2 \tag{4.4}$$

where $\mathbf{R}_2$ is the remainder yet to be explained. Solving this for $\mathbf{R}_2$ yields

$$\mathbf{R}_2 = (\mathbf{P} - .5151\mathbf{S}_1) / .4849 = \begin{bmatrix} .0000 & .3667 & .6333 \\ .2308 & .0000 & .7692 \\ 1.0000 & .0000 & .0000 \end{bmatrix}$$

where the zero element in row 3, column 2 indicates the bottleneck in the system that prevents the first extremal tendency from being more fully realized. Analysis now continues through similar examination of the remainder. We thus find S_2 by using the maximal element from each row of R_2:

$$S_2 = \begin{bmatrix} 0 & 0 & 1 \\ 0 & 0 & 1 \\ 1 & 0 & 0 \end{bmatrix}$$

This extremal tendency corresponds to migration from the Frostbelt and the South to the West as well as migration from the West to the Frostbelt. It is the second most important tendency in the migration system. The weight associated with this extremal tendency is again the minimum of the row maxima; in this case the weight is .6333. Thus

$$R_2 = .6333 S_2 + (1 - .6333) R_3 \qquad (4.5)$$

Substituting this in (4.4) yields

$$P = .5151 S_1 + .3071 S_2 + .1778 R_3 \qquad (4.6)$$

To continue the derivation of the extremal tendencies, we find R_3 from (4.5):

$$R_3 = \begin{bmatrix} .0000 & 1.0000 & .0000 \\ .6294 & .0000 & .3706 \\ 1.0000 & .0000 & .0000 \end{bmatrix}$$

The extremal tendency in the R_3 matrix is

$$S_3 = \begin{bmatrix} 0 & 1 & 0 \\ 1 & 0 & 0 \\ 1 & 0 & 0 \end{bmatrix}$$

The minimum among the maximum elements of each row in R_3 is .6294; thus

$$R_3 = .6294 S_3 + (1 - .6294) R_4 \qquad (4.7)$$

Solving (4.7) for R_4 yields

$$R_4 = S_4 = \begin{bmatrix} 0 & 1 & 0 \\ 0 & 0 & 1 \\ 0 & 1 & 0 \end{bmatrix}$$

Substituting (4.7) in (4.6) yields the final decomposition of the observed migration matrix:

$$P = .5151 S_1 + .3071 S_2 + .1119 S_3 + .0659 S_4$$

The extremal tendency method provides a decomposition of migration patterns into a set of patterns that each have a much simpler structure. By comparing extremal tendency decompositions for different time periods, the geographical analyst of population can gain insight into how the migration system is fundamentally shifting over time—how some regions may be gaining in significance in the grand scheme of things, while others are losing importance.

Spatial Shift-Share Analysis

The final technique we present in this chapter for analyzing the spatial patterns of migration is, like the extremal tendency technique, a method for the decomposition of migration patterns. Unlike either the principal component regionalization technique or extremal tendencies, however, the goal of spatial shift-share analysis is to provide better understanding of the *changes* in the volumes of flow between the regions of a migration system.

The spatial shift-share technique for migration analysis is a straightforward extension of the widely used shift-share method used by regional economists and economic geographers to understand changes over time in the sectoral structure of employment. The migration variation was first set forth in Plane (1987) and has been used subsequently to examine the changing age structure of migration in the United States as the baby boom generation has aged (Plane, 1992), as well as migration pattern changes in Japan (Ishikawa, 1992) and Britain (Green, 1994, forthcoming).

To illustrate the method we shall use the two 4 × 4 matrices of U.S. interregional migration for 1975–80 and 1985–90 obtained from the past two decennial censuses (Table 4.8). The diagonal elements here include both nonmovers and persons moving within each of the four regions. Spatial shift-share analysis posits that the differences in each of the migration flows from one time period to the next (ΔM_{ij}, as shown in the lowest panel of Table 4.8) can be seen as the sum of three separate quantities: a *population base component* (B_{ij}), a *mobility component* (U_{ij}), and a *geographic distribution component* (G_{ij}):

$$\Delta M_{ij} = B_{ij} + U_{ij} + G_{ij} \qquad (4.8)$$

These may be further subdivided into various subcomponents if the data are available by age. (See Plane, 1992.) However, to facilitate basic understanding of the method we shall work here with this aggregate form of the method.

The first of the three components, the population base component, reflects the differences in the volumes of interregional flow that could be expected based on the changing size of the population "at risk" of becoming migrants. It is computed by the formula

$$B_{ij} = (\Delta P_i / P_i) M_{ij} \qquad (4.9)$$

This simply says that we apply the population growth rate for the origin region (the expression in parentheses) to the first period's volume of flow to obtain the expected change in flow. An alternative way to understand the component is to rewrite it as

$$B_{ij} = (M_{ij} / P_i) \Delta P_i$$

which says to apply the first time period's migration rate to the number of additional (or fewer) people at risk of migrating during the second time period.

To compute population base components from the data in Table 4.8, we must first derive the population in the origin region we are interested in, say, the Northeast ($i = 1$) in 1975 and in 1985 that survives to be counted in *any* of the four regions on the 1980 and 1990 censuses. Recalling the rules of demographic accounting discussed

TABLE 4.8 U.S. Interregional Migration Flow Matrices, 1975–80 and 1985–90, and Changes in Flows, 1975–80 to 1985–90 (ΔM_{ij} Terms Used for Calculation of Spatial Shift-Share Components)

1975–80 flows (in thousands)

	From: Northeast	Midwest	South	West
To:				
Northeast	46,052	465	1,817	777
Midwest	360	51,850	1,878	1,267
South	654	1,029	64,021	1,069
West	261	630	1,044	35,356

1985–90 flows (in thousands)

	From: Northeast	Midwest	South	West
To:				
Northeast	44,517	360	1,815	545
Midwest	373	52,393	1,769	1,029
South	845	1,246	73,062	1,253
West	386	718	1,185	43,841

Changes in flows from 1975–80 to 1985–90, $\{\Delta M_{ij}\}$

	From: Northeast	Midwest	South	West
To:				
Northeast	—	−105	−2	−232
Midwest	+13	—	−109	−238
South	+191	+217	—	+184
West	+125	+88	+141	—

Source: Calculated by the authors from 1980 and 1990 U. S. decennial census data.

in Chapter 3, the reader should be convinced that such population totals are found by adding to the diagonal elements the number of interregional out-migrants (i.e., by summing across the row of the relevant table).[4] The reader should verify that the appropriate 1975 and 1985 surviving populations (in thousands) for the Northeast are 49,111 and 47,237.

Since the at-risk population declined from 1975 to 1985 in the Northeast, the population base components will be negative. Using Equation 4.9, the population base component for the change in migration to the South ($\Delta M_{13} = -2$) is found as

$$B_{13} = [(47{,}237 - 49{,}111) / 49{,}111]\, 1817 = -69$$

---

[4]Equally, we could sum down the columns of the table to obtain the census-year regional populations of persons! Note, however, that these populations would be less than the actual total census-enumerated populations of the regions because they would include only those people who were also living in one of the four regions in 1975.

Notice that since the change in Northeast to South migration was only a decrease of 2,000, and the population base component would imply a decrease of 69,000, either the mobility of Northeasterners must have increased (i.e., positive U_{ij}) or the South must have become an even more favored destination for Northeast out-migrants (positive G_{ij}) or both.

The mobility component shows how many additional (or fewer) migrants beyond those attributable to the population base effect would have been found in the stream from region i to region j if that stream increased (or decreased) proportionately to the rate of change of migration to all destinations. The formula is

$$U_{ij} = [(\Delta M_{i*} / M_{i*}) - (\Delta P_i / P_i)] M_{ij}$$

where M_{i*} is the total number of out-migrants from i to all destinations. Computing this quantity for the change in Northeast to South migration,

$$U_{13} = \{[(2{,}720 - 3{,}059) / 3{,}059] - [(47{,}237 - 49{,}111) / 49{,}111]\} 1{,}817$$
$$= (-201) - (-69) = -132.$$

Thus the mobility component is also negative, meaning that had migration to the South from the Northeast decreased by as much as it did to all regions, the volume of Northeast-to-South flow would have been even lower than it actually was in 1985–90.

With the information we have now calculated about the population base and mobility components, it is easily seen that the geographic distribution component for Northeast-to-South migration must have been positive and of substantial magnitude. The third component may be found from Equation 4.8 as the residual:

$$G_{ij} = \Delta M_{ij} - B_{ij} - U_{ij}$$

so

$$G_{13} = (-2) - (-69) - (-132) = +203$$

Vis-à-vis other regions, the South was a relatively much more common destination choice for Northeastern out-migrants in 1985–90 than it had been in 1975–80. Intuitively this should make sense from a comparison of the substantially reduced number of Northeast-to-Midwest and Northeast-to-West migrants shown in Table 4.8.

Although it is easiest to compute the geographic distribution component in this fashion as a residual, some additional intuitive understanding of its meaning may be gained from studying the expression for its general form:

$$G_{ij} = [(\Delta M_{ij} / M_{ij}) - (\Delta M_{i*} / M_{i*})] M_{ij}$$

The geographic distribution component gives the portion of the change in migration from i to j that is attributable to increases (decreases) in the destination-specific flow that go beyond the part of the change that could be accounted for by increases (decreases) in mobility from region i.

We leave as an exercise for the reader the calculation of the components of interregional migration change for various other region-to-region–specific flows in the 1975–80 and 1985–90 migration matrices. All $12 \times 3 = 36$ shift-share components are shown in the answers to the Chapter 4 set of exercises. See how well your intuition works in predicting the sign and general magnitude of each of these from simply examining the numbers in Table 4.8. For instance, it should make sense to you that, because of the large 1975–85 increase in population in the West region, its population base components will be large and positive, and that the mobility components for the other two Northeast outflows will, like that to the South, be negative.

4.7 SUMMARY

The migration component of population change is often the largest contributor to shifts in the population of local areas. As was just illustrated, in fact, by the results of our example of how to compute spatial shift-share components of the changes in migration flow from the Northeast to the South, it tends to be temporally quite volatile. In-migration and out-migration often swing considerably faster and more dramatically than either of the two components of natural increase: births or deaths.

In this chapter we have attempted to provide some foundations for the measurement of migration and some fundamental background information about its geographical properties. In addition we have shown several methods to help us better understand the structure inherent in large numbers of elements in a migration flow matrix.

In the next two chapters we build on the foundations provided in this and the previous two chapters in examining practical methods for estimating and projecting regional populations. Following our general exposure of population projection methodology, we return to migration once again, looking at approaches that have been used to model and forecast it. Whereas relatively accurate estimates and projections may be made with existing techniques for the natural increase component of population change, the temporal evolution of migration systems remains an area in need of considerably more basic research. It is in forecasting the future course of migration into and out of regions that practicing geographic analysts of population quite often face their greatest challenges. Hence our special, in-depth attention to migration in this book.

EXERCISES

1. Using the gross migration flows reported in Table 4.8, calculate a matrix of transition probabilities for U.S. interregional migration in the period 1975–80. Compare and comment on the differences between the elements of this matrix and those shown in Equation (4.2) for the 1985–90 period.

2. Calculate 1975–80 in-, out-, and net migration rates for the four U.S. regions. Use row and column sums from Table 4.8 to estimate the midperiod (October 1, 1977) populations required for these rates. Why are these populations different from the actual midperiod populations as well as from the true populations "at risk" of migrating? Compare and comment on the differences between these rates and those for 1985–90 in Table 4.2.

3. Ledent (1993) gives some data on in- and out-migration to and from Indonesian provinces during 1975–80. He breaks these totals down by the place of birth of migrants. For the provinces of East and West Java the figures (in thousands) are:

	East Java	West Java
In-migration	188	505
Primary	106	345
Secondary	32	82
Return	50	79
Out-migration	569	465
Primary	519	399
Secondary	23	37
Return	27	30

(a) Calculate the percentage distributions of in- and out-migration by type for each province and compare these to Long's figures on U.S. interstate migration reported in the text.

(b) Calculate the demographic effectiveness (E_j) of total, primary, secondary, and return migration for East and West Java.

4. (a) Calculate the E_j demographic effectiveness percentages for U.S. interregional migration in 1975–80 and compare these to the 1985–90 percentages reported in the text.

(b) Calculate the e_{ij} stream-effectiveness values for U.S. interregional migration in 1975–80 and 1985–90 and note the largest changes in these that give rise to the differences in the E_j values for the two time periods.

5. For net migration change (ΔNM) between 1975–80 and 1985–90 for each of the U.S. regions, calculate the in- and out-migration contributions: IMC and OMC.

6. Find the row (out-migration) and column (in-migration) entropies for U.S. interregional migration in the 1985–90 period.

7. (a) Using the 1975–80 interregional migration flows in Table 4.8, combine the Northeast and Midwest regions into a single "Frostbelt" region and derive the (3 × 3) Markov transition probability matrix, **P**, of destination choice probabilities used to find extremal tendencies.

(b) Find the extremal tendencies and the weights associated with the matrix derived in (a).

(c) Compare the weights found in (b) with those reported in the text for interregional flows in the period 1985–90.

8. Following the same steps shown in the text, derive the three shift-share components of Δm_{ij} for interregional migration from the Midwest to the West regions. The required data are in Table 4.8.

> **The greatest of all gifts is the power to estimate things at their true worth.**
>
> —La Rochefoucauld, *Reflexions: ou sentences et maximes morales*
>   *(Reflections: or aphorisms and ethical maxims)*

CHAPTER 5
POPULATION ESTIMATION

5.1 Introduction

5.2 Simple Interpolation and Extrapolation

5.3 The Regression or Ratio-Correlation Method

5.4 Component Methods

5.5 Housing-Unit Methods

5.6 Evaluating Estimates

5.7 Summary

5.1 INTRODUCTION

Among the most commonly encountered problems in the geographical analysis of population are those involving the need to *estimate* the population—or perhaps one or more subpopulations—of a particular geographic area for one or several points in time when no census data exist. Census data may not be available because the resources do not exist to conduct a special census or because the desired time point reference is in the past. Long (1993) noted that in fiscal year 1989 population estimates were the basis for allocating over $10 billion in federal funds to states and localities.

Estimation problems take many forms. Thus—not surprisingly—a variety of techniques have been developed to solve such problems. Before describing the mechanics and practical aspects of the applications of a number of the most commonly used techniques, we shall examine the various types of estimation problems that the geographical analyst is likely to encounter. The appropriate technique to use in any given situation depends on the characteristics of the problem faced as well as on the specific data resources available for the study area and time period.

The Parlance of Estimates

The term *estimate* has had a more restricted meaning in demographic analysis than in common usage. In particular, a distinction is made between an estimate and a *projection*. An estimate refers to a point in time in the immediate present or in the past. A projection, on the other hand, refers to figures computed for a future point in time. A further distinction has frequently been made between a population projection and a population *forecast*; the former represents a mechanical extrapolation of trends given a set of easily specified assumptions whereas the latter implies that the projected time series is the analyst's best guess of the actual future course of population growth.

In some cases the mechanics of estimates and projections may be quite similar, and the distinction is largely the semantic one of the time reference. In other cases, however, the difference in time reference implies very different sets of potential data to work from, and thus radically different methodologies may be suggested for a projection than for an estimate.

Symptomatic Variables

Most estimation methods involve the use of some sort of *symptomatic variable(s)*. As we shall use the term, a symptomatic variable is one that changes over time according to a predictable and logical relationship with population. Estimation techniques are generally predicated on variables that become available on a timely and routine basis, often for other government purposes. A fundamental difference between a population projection technique and an estimation technique is this: estimates use actual data on variables that are measured in order to estimate a variable (population) that is not directly measured. If the same kinds of variables related to population change are also to be used in projections, however, then these variables must themselves be projected before the analyst can derive from them predictions about future population.

Among the more frequently used symptomatic variables in population estimation are new residential building permits issued within a political jurisdiction, the number of active residential utility connections in a service area, the number of auto license registrations issued to persons with addresses in a particular residential area, and the number of tax returns filed with addresses in such an area. In the case of, for example, certain *vital events* methods, one or more components of population change (recorded resident births and/or deaths) are even used as indicators of the other, less easily measured components (gross in- and out-migration). So too, if an assumption regarding the stability of the demographic mix of an area is possible, one of the more easily measured segments of the overall population (e.g., the number of children enrolled in schools or elderly persons enrolled in government health-care programs) may serve as a variable symptomatic of changes in the broader population.

Intercensal and Postcensal Estimates

In countries such as the United States and Canada, where large-scale, high-quality censuses have been regularly undertaken, estimates usually are either *intercensal* or *postcensal*. An intercensal estimate refers to filling in a population figure for a point in time between two different census years (the censuses being either decennial, quinquennial, or special ones), whereas a postcensal estimate is for a time point after the date of the last available census.

Figure 5.1 shows the conceptual differences between intercensal estimates, postcensal estimates, and projections as well as the availability of symptomatic variables for carrying out each type of application. Without wishing to create confusion, we note

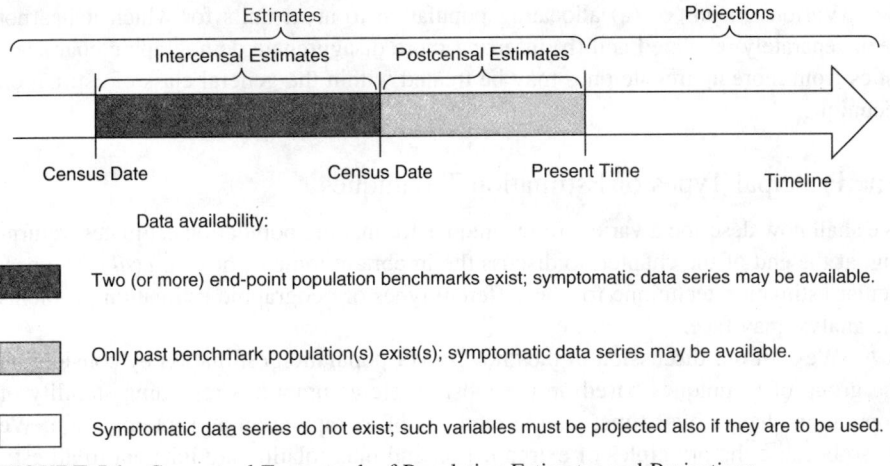

FIGURE 5.1 Conceptual Framework of Population Estimates and Projections

that postcensal estimates are made during the time we find ourselves *between* censuses, while intercensal estimates are commonly made shortly *following* a census!

Many commonly encountered planning and business problems involve trying to obtain the best guess of the current numbers of persons residing in a planning or market area and belong in the realm of postcensal estimation. As we shall see, the game in such applications is generally to update a decennial or special census benchmark population (assumed to be largely accurate) based on what has occurred in the interim period since the census was taken. On the other hand, intercensal estimation problems are of practical interest to far more people than just historical researchers because current and future planning and business decisions may be better informed if the time trend of past demographic change is closely examined. In the words of Abraham Lincoln (which have been prominently displayed as a motto, of sorts, in the headquarters building of the U.S. Bureau of the Census in Suitland, Maryland):

> If we could first know who we are and whither we are tending we could better judge what to do and how to do it.

"Synthetic" Estimates

A further type of estimation problem is that of synthesizing a population figure for a geographic area and/or subgroup of the overall population of an area for which no hard census-type information exists, but for which good-quality, contemporaneous data do exist at a higher level of geographic or demographic aggregation. For example, to estimate the geographic distribution of indoor and landscape water use demands for a private or municipal water company, a planner may need to derive data from the most recent decennial census on the number of households of various sizes living in the different housing types (single-family, duplex, multifamily apartment units, etc.) within the company's service area. By going to block-level census data, estimates of total population and total numbers of housing units may be found that allow close assignment to the actual service area boundaries; however, the crucial information for indoor demand—household size—and that for landscape watering—housing type—are probably obtainable only for grosser geographic units (such as census tracts or block groups).

Various means of (a) allocating population to areal units for which it has not been separately tabulated and (b) inferring more disaggregate demographic characteristics from more aggregate ones may be treated within the general class of estimation techniques.

The Principal Types of Estimation Techniques

We shall now describe a variety of techniques for making population estimates, returning, at the end of the chapter, to discuss the important topic of how to *evaluate* a particular estimation technique for the different types of geographic estimation problems the analyst may face.

We start our discussion of the "how-to" of population estimation by considering the group of techniques based on the most heroic assumptions regarding stability of population change over time: simple methods of extrapolation and interpolation. We do so because the principles of extrapolation and interpolation are inherent to all estimation techniques.

We then turn to the three methods most widely employed by the U.S. Bureau of the Census in its county estimates program: the regression (or ratio-correlation) method, the component method II, and the administrative records method. Each illustrates the use of symptomatic variables to proxy population change, though they do so in different ways. The regression technique is quite distinct from the other two, which may be conveniently viewed as variations on a single general approach.

Finally, we discuss housing-unit methods, which probably account for the greatest share of all postcensal estimates prepared by planners and business people in the field.

5.2 SIMPLE INTERPOLATION AND EXTRAPOLATION

Albert Einstein issued the dictum to "make everything as simple as possible, but not simpler." Among the simplest of methods for estimating population are those based on treating population in the aggregate and assuming that the amount or rate of population change observed over one time period will be a good indicator of change for another period.

These techniques are possibly the most parsimonious with respect to data input requirements and are simple to explain and compute. By dealing with an aggregate quantity (total population change), rather than with pieces of that quantity, there may be some inherent accuracy advantages. However, the one key assumption—that population change is a temporally smooth process—is quite problematic and not well supported by the evidence for many subnational geographic units. In particular, the in- and out-migration components of local-area population change often respond dramatically to cyclical as well as longer-term economic events, both within the study area itself and in all the potential source areas for in-migrants and destination areas for out-migrants.

In general, postcensal *extrapolation* of aggregate population should not be used except for short intervals after a census or in cases where data constraints rule out better methods. Intercensal *interpolation* may be used with somewhat more confidence; the time path of population change, after all, will thus be anchored at both ends so that the error for any point in between should not be too large.

Linear Interpolation

Simple *linear* interpolation is used frequently for a common practical problem in U.S. population analysis. This use of interpolation is a legitimate use since it involves short estimation periods. All the recent U.S. decennial censuses have been conducted for an April 1 enumeration date, when more people are likely to be at their usual place of residence than during other seasons of the year.[1] Estimates, by contrast, are almost always prepared for a hypothetical July 1 (midyear) population. The goal is not, in fact, to judge how many people are actually present on that date (since many persons may then be away from their usual abodes), but rather to represent an average over the entire year of the area's resident population. To see how linear interpolation is useful to adjust the dates for which data are available, consider the following example.

Example: The "4/5th–1/5th Adjustment"

To prepare a historical time series for graphing annual growth rates suppose a researcher needs to have a July 1, 1980 estimate for Pima County, Arizona (which is coterminous with the Tucson Metropolitan Statistical Area). The county had a revised April 1, 1980, census figure of 531,443. The Bureau of the Census's first postcensal population estimate prepared after that date was 551,400, for July 1, 1981. As illustrated in Figure 5.2, assuming that the course of population growth between the census date and the estimate date was a straight line, we would compute this simply as

$$P_{7/1/80} = P_{4/1/80} + (1/5) \times (P_{7/1/81} - P_{4/1/80})$$
$$= 531,443 + 3,991 = 535,400 \text{ (to the nearest hundred)}[2]$$

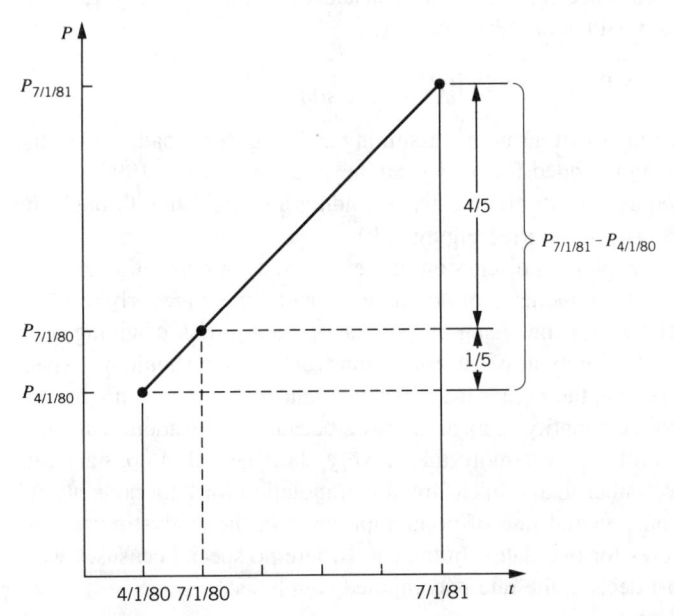

FIGURE 5.2 Graphical Representation of the 4/5th–1/5th Linear Interpolation Technique

---

[1]Special censuses, however, may be conducted at any time during the year as local resources permit. This was the case, for instance, with the 1985 round of mid-decade special censuses paid for largely by local governments, but conducted under the technical supervision of the U.S. Bureau of the Census.

[2]Note that although we report our estimate of county population only to the nearest 100 persons, because this is the convention that the Bureau of the Census adopted for one of the two inputs, intermediate calculations should be carried out to several places beyond the decimal point to avoid rounding errors.

Notice that the same answer can be obtained by taking 4/5ths of the census population figure and adding to it 1/5 of the first postcensal estimate: 425,154 + 110,280 = 535,400 (to the nearest hundred). The implicit assumption here is that population grew during the first quarter of a year after the census by the same amount as in each of the next four quarters, there being a full five quarters between the census date and that of the first available estimate. Because this calculation is so frequently performed by population analysts, it has been given a special name: the 4/5th–1/5th adjustment procedure.

Note, however, that the same logic (but not the same 4/5th and 1/5th weights!) can be used to interpolate between any two population figures regardless of the time span between them and for where in the interval the interpolated estimate is needed. When working with special census figures, the appropriate weighting factors may be based on a consideration of the number of months or even days elapsing between census and estimate dates.

Alternative Extrapolative Techniques

Analogous to the interpolation problem we've just considered, *linear extrapolation* might also be used for postcensal estimates, though another, more refined form of extrapolation might be more advisable if the time interval beyond the last known population count is very great.

As an example, consider the current situation for any geographic unit for which no official estimate has been made. The 1990 census figure is now known, but several years have already elapsed since that count. For instance, to estimate July 1, 1993, population a crude guess would simply be

$$P_{7/1/93} = P_{4/1/90} + 3.25\ [(P_{4/1/90} - P_{4/1/80}) / 10]$$

The average annual increment of growth—assuming a linear time path—over the decade of the 1980s is simply added for 3.25 years beyond the April 1, 1990, census date. The procedure is equivalent to graphically extending a straight line through the populations at the two census dates. (See Figure 5.3.)

As we indicated in Chapter 3's discussion of the dynamics of population growth, however, the birth and death components of population change more properly result in a nonlinear dynamic reflective of the changing population base at risk of giving birth or dying. So, too, the absolute amount of out-migration from an area should be expected, *ceteris paribus*, to rise as the area's base expands, and in-migration might also logically increase in absolute quantity as an area grows because of the additional stock of jobs turning over. Therefore, over more than a very short period, a constant rate extrapolation is advisable rather than a linear trend extrapolation for total population.

To find a continuous, annual rate of population growth, the analyst must have available population figures for two dates. In the case where no special censuses were conducted during the past decade, the rate is computed simply as

$$r = [\ln(P_t / P_{t-10})] / 10 \qquad (5.1)$$

where "ln" denotes the natural logarithm of the ratio of the two census populations. This formula arises directly from the continuous rate of growth model as presented in Chapter 3. Readers needing a brushup on logarithms should consult Appendix C.

To use the continuous growth rate model for practical estimation problems, the analyst first calculates the relevant r from Equation 5.1. The procedure is similar regardless of whether a post- or intercensal estimate is required.

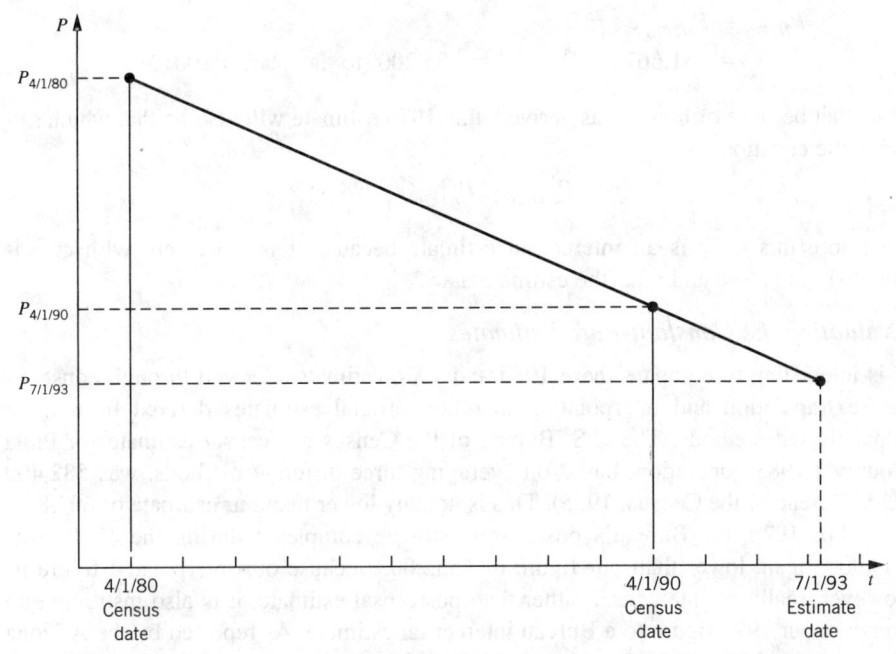

FIGURE 5.3 Graphical Representation of a Linearly Extrapolated Postcensal Estimate

Example: A Postcensal Rate-Based Extrapolated Estimate

Consider the historical problem faced after the 1980 census of deriving a July 1, 1983, postcensal estimate for Pima County, Arizona. We would use the continuous growth rate model from Chapter 3 in the following form:

$$P_{7/1/83} = P_{4/1/80} \, e^{3.25 \, r}$$

Here $P_{4/1/80}$ is the 1980 census count and the growth rate, r, would be calculated for the April 1, 1970 to April 1, 1980, base period using Equation 5.1:

$$r = [\ln(531{,}443 \, / \, 351{,}667)] \, / \, 10 = .0413 \text{ (approximately)}$$

Thus, plugging in this value and the actual 1980 census figure,

$$P_{7/1/83} = 531{,}443 \, e^{\, 3.25 \, (.0413)} = 607{,}800 \text{ (to the nearest 100)}$$

Note that this form of extrapolative, postcensal estimate is equivalent to making a short-term population projection. (Chapter 6 discusses the use of constant population growth rate assumptions for projection purposes.)

Example: An Intercensal Constant-Rate Interpolated Estimate

To use constant-rate *interpolation* to find an estimated population between two census dates, use the two decadal counts to derive r and then use the standard continuous-growth-rate model formula. For example, to derive a Pima County, July 1, 1976, estimate, we use the same value of r previously calculated from the 1970 and 1980 censuses thusly:

$$P_{7/1/76} = P_{4/1/70}\, e^{\,6.25\, r}$$
$$= 351{,}667\, e^{\,6.25\,(.0413)} = 455{,}200 \text{ (to the nearest 100)}$$

Note that because of how r was derived, this 1976 estimate will also be that which satisfies the equation

$$P_{4/1/80} = P_{7/1/76}\, e^{\,3.75\, r}$$

Therefore this truly is an intercensal estimate because it is consistent with census counts both before and after the estimate date.

Evaluating the Constant-Rate Estimates

It is interesting to compare these 1983 and 1976 estimates derived through constant-rate extrapolation and interpolation to other, official estimates derived from more sophisticated methods. The U.S. Bureau of the Census postcensal estimate for Pima County's 1983 population, based on averaging three different methods, was 582,400 (U.S. Bureau of the Census, 1985). This is notably lower than our estimate of 607,800.

For 1976, the Bureau's postcensal estimate completed during the 1970s was 453,900, again lower than our figure of 455,200. Because our interpolated figure is, however, really an intercensal rather than postcensal estimate, it is also instructive to compare our 1976 figure to a Bureau intercensal estimate. As reported by the Arizona Department of Economic Security (1985), the federal intercensal estimate, when proportionally adjusted to take account of a revised state-level control total, would be 471,200, which is higher than our interpolated figure.

During the 1970s, the Bureau of the Census's postcensal estimates for Arizona failed to very accurately gauge the extremely rapid growth experienced during the decade. On the other hand, the value of using contemporaneous symptomatic variables for population estimation is also highlighted by these comparisons. The fact that our extrapolated figure for 1983 is higher than the postcensal estimate is probably because we are using a 1970s growth rate that was unrealistic for the early 1980s, when, in fact, southern Arizona was suffering significant economic woes as a result, in part, of abnormally low world copper prices—which caused many mines in the region to close or significantly cut back production. Simple extrapolative techniques cannot pick up the demographic impacts of specific economic events that may be very clearly detectable in the symptomatic variables used in other estimation techniques.[3]

In the next two sections we examine the techniques used most widely by the U.S. Bureau of the Census for making state- and county-level estimates, highlighting how symptomatic data series are used in these techniques to judge the actual course of population change over time.

5.3 THE REGRESSION OR RATIO-CORRELATION METHOD

Although data on population may not be available for a particular area during noncensus years, various other types of data that have a clear connection to population are collected on a routine and more frequent basis. The number of auto registrations recorded, the number of hunting and fishing licenses issued, and counts of school enrollment are examples of symptomatic variables that are updated monthly or annually.

---

[3]For a discussion and comparative test of a wide variety of extrapolative techniques, see Isserman (1977).

The principal concept underlying the ratio-correlation method of population estimation is to relate changes in the symptomatic variables to changes in population. If school enrollments go up, there is often good reason to believe that the population of the area has increased. Any decline in auto registration is likely to have occurred because out-migration has reduced the population residing in the area.

Multiple regression analysis is used to relate population change to changes in the symptomatic variables. The reader unfamiliar with multiple regression can consult any good statistics book, such as Griffith and Amrhein's (1991) treatment of the use of statistical techniques for geographic analysis. To gain a conceptual understanding of the ratio-correlation method, however, it is not necessary to understand all the statistical apsects of regression modeling.

The basic idea is that the *changing shares* of population in a subregion are assumed to be a function of that subregion's *changing shares* of the other symptomatic variables:

$$Y = a + b_1 X_1 + b_2 X_2 + \ldots + b_n X_n \qquad (5.2)$$

In fact, the population variable (Y) and the symptomatic variables ($X_1, X_2, \ldots, X_n$) are expressed in somewhat unusual form. In this equation Y is the ratio of a subregion's population share in one year (say, t) to that same ratio in some previous base year (0):

$$Y_i = \frac{P_i^t / P_*^t}{P_i^0 / P_*^0} \qquad (5.3)$$

where i designates the subregion and P_* is the total population of the larger region that contains it. The n symptomatic variables are defined in an analogous manner. If X_1 represents the use of auto registrations (A) as a symptomatic variable,

$$X_{1i} = \frac{A_i^t / A_*^t}{A_i^0 / A_*^0} \qquad (5.4)$$

The variable X_1 thus represents subregion i's changing share of total auto registrations for the entire region of which it is a part. Thus the method actually employs ratios of ratios!

Regression analysis is used to estimate the coefficients ($a, b_1, b_2, \ldots, b_n$). This requires that data on Y and all the X's be available for all the subregions (for instance, for all the counties of a state). Because these variables themselves are expressed as ratios, population and symptomatic variable data are required for two points in time. Commonly data from the two previous censuses are used to estimate the regression equation. This, however, is less than ideal because information from the distant past is being used to estimate a relationship to be applied to the present. Thus population estimates made in 1999 rely on regression coefficients estimated with data from 1980 and 1990. The validity of this method rests on whether the *structural relationships* between the symptomatic variables and population are longstanding ones.

If an estimated regression equation is already available, an updated set of population estimates for the subregions may be computed as soon as all the symptomatic data series become available for estimate year t. The $X_1, X_2, \ldots, X_n$ ratios of ratios are calculated (using Equation 5.4) and plugged into Equation 5.2 to obtain each subregion's value of Y. Then, rearranging Equation 5.3, the estimated population, P_i^t, of any subregion may be found:

$$P_i^t = (P_*^t / P_*^0) (P_i^0) Y_i$$

Examine the first three terms on the right-hand side: the ratio of estimate to base year population multiplied by the subregion's base year population would give us an estimate based on the subregion growing at an identical rate as the overall study area. Note that without the Y term, this is the same as saying that the study area would retain its base year share of the study area's total population:

$$P_i^t = (P_i^0 / P_*^0)(P_*^t)$$

The actual estimate, however, is adjusted upward or downward by Y_i, which reflects the composite shifts in the study area's shares of the symptomatic variables.

A Real World Ratio-Correlation Equation

Let's look at a real world example of a ratio-correlation equation. The following regression equation was estimated with data for the counties of the state of Kentucky for the period 1980–85 by the U.S. Bureau of the Census (U.S. Bureau of the Census, 1988):

$$Y = -.190 + .301X_1 + .190X_2 + .108X_3 + .094X_4 + .142X_5 + .349X_6$$

where:

X_1 = the ratio of ratios based on federal income tax exemptions claimed by persons filing with addresses in the county.

X_2 = the ratio of ratios of numbers of federal Medicare program enrollees.

X_3 = the ratio of ratios of resident births.

X_4 = the ratio of ratios of resident deaths.

X_5 = the ratio of ratios of auto registrations.

X_6 = the ratio of ratios of school enrollment in grades 1 through 8.

Note that the coefficients sum to a number close to 1 (actually 0.994). This is a characteristic of all ratio-correlation regression equations, not just a peculiarity of this particular example. To see why this should be so, consider the following. If there were no change in the symptomatic variables for a particular county, the X variables will all equal 1 (because the share in one year equals the share in the other year, implying that numerator and denominator are equal). The equation would then imply that this county will have a population share equal to 99.4 percent of its population share in the base year. We should expect the ratio-correlation method to produce subregion population estimates that display little or no change in the share of total population when there is no change in the share of the symptomatic variables. Regression coefficients that sum to 1 are clearly both desirable and intuitive. If the sum differs much from 1, significant changes in population share might be predicted even for those instances where there is no change in the share of symptomatic variables.

Problems with Ratio-Correlation Estimates

The ratio-correlation method suffers from several problems in addition to the fact that dated information is often used in regression equation estimation. First, there is no reason to expect that all of the subregions' estimated population shares will sum to 1. There is always the need then to perform some ad hoc adjustment to the resulting population share estimates. This can be done in a fairly simple, straightforward manner. For example, each estimated share could be multiplied by the inverse of the total estimated share. If the estimated population shares total 1.1, each subregion's share is multiplied by 1.0 / 1.1 to produce new shares that total to 1. Though straightforward, this additional step takes the estimates one step further away from being interpretable via the link to symptomatic variables.

Perhaps an even more severe problem results from the fact that in almost all applications the length of the period over which the regression estimates are made differs from the length of the period to which the regression estimates are applied. As just noted, coefficients are typically estimated using data from 2 census years that are 10 years apart. Virtually all postcensal applications use the most recent census as the base year (0), with the current year of interest (t) necessarily less than 10 years away. There is no a priori reason for believing that the coefficients that hold over a 10-year period are the same as those that will hold over a period of less than 10 years. Thus, for example, a rise in a subregion's share of births may be more meaningful when it occurs over a 10-year period than when it occurs over a 1- or 2-year period. Despite these problems, there is evidence that the ratio-correlation method performs well when compared with other methods. (See, e.g., Shryock and Siegel, 1976.)

5.4 COMPONENT METHODS

Both the extrapolative and ratio-correlation methods estimate total population. An alternative approach is to use the demographic accounting equation relationship discussed in Chapter 3 between population change and its components: births, deaths, and net migration.

Component Method II

A widely used method for estimating a region's population during a noncensus year is the U.S. Bureau of the Census's "component method II." The technique is dubbed a component method because births and deaths are taken directly from vital statistics data, whereas the net migration component of population change is indirectly estimated. The underlying idea of component method II is to relate changes in the population of school-age children to changes in the region's total population due to migration. A comparison is first made between actual enrollment in the estimate year and that enrollment expected by applying survival ratios to the children who were in the region at the time of the previous census. The difference is attributed to the net migration of school-age children. A factor that relates mobility for the total population to mobility for school-age children is then applied to obtain the net migration for the region's population.

Sample Calculations

Tables 5.1 and 5.2 give examples of the basic data and steps required to produce an estimate using component method II. The problem we tackle is to compute a December 31, 1983, estimate of the population of the City of Evanston, Illinois, from the data in Table 5.1. Working through this example should give you a feeling for the level of care that must be taken at each step of the way in producing a good population estimate.

The calculations in Table 5.2 begin with the most recent census population count. Natural increase is usually straightforward to estimate because state health

TABLE 5.1 Data Needed for Sample Component Method II Calculation: Evanston, Illinois

1. Age	Population	2. Date	Enrollment (K–8)
<1 year	814	10/1/79	7,061
1–2 years	1,413	10/1/80	6,840
3–4	1,276	10/1/81	6,503
5	672	10/1/82	6,272
6	663	10/1/83	6,125
7–9	2,327	10/1/84	6,076
10–13	3,543	10/1/85	6,073
14	885		
15	960	*3. Year*	*Deaths*
16	1,019	1980	809
17	1,019	1981	759
18	1,779	1982	765
19	2,393	1983	713
20	2,249	1984	696
21	2,070		
22–24	5,112	*4. Year*	*Births*
25–29	7,856	1980	927
30–34	6,259	1981	930
35–44	8,028	1982	951
45–54	6,738	1983	1,077
55–59	3,232	1984	996
60–61	1,226		
62–64	1,763		
65–74	5,146		
75–84	3,746		
85 years and over	1,518		

Note: Population-by-age data for 4/1/80 from U. S. decennial census; enrollments for public schools only.

departments collect data on births and deaths at least annually, and on a place of residence basis. Unless these data are broken down by month or quarter, an assumption regarding the distribution of vital events during the year is often needed. It will usually suffice to assume that these events are distributed uniformly throughout the year. For example, the number of births between April 1, 1980, and December 31, 1980, may be taken as 3/4 of the total number of births during 1980 (927 × .75 = 695). Lines 1a and 1b of Table 5.2 are computed using this assumption.

In steps 2 and 3 the estimates of school-age population at the time of the estimate and the expected size of the school-age population under the assumption of no net migration are determined. In step 2, the school-age population on the estimate date is determined by assuming that the ratio of school-age population to enrollment that held in the census year is applicable in the year for which the estimate is being prepared. Steps 2a through 2c calculate this ratio for the census year. Students in grades K through 8 on April 1, 1980, are assumed to be between 5 and 14 years old on January 1, 1980. Therefore they are between 5.25 and 14.25 years old on April 1, as indicated by the entry in 2a. Again we have used an assumption that births are evenly spread throughout the year; the line 2a figure includes three-quarters of the 5-year-olds, and one-quarter of the 14-year-olds enumerated in the 4/1/80 census.[4]

The ratio in 2c should be close to 1; most school-age children are enrolled in school! However, it will differ from 1 for a number of reasons. First, enrollment data are typically gathered for public schools only; the extent of private school enrollment will influence the amount by which this ratio differs from 1. Second, school district boundaries may not exactly coincide with the boundaries of the subregion for which the population is being estimated. The estimated population of school age at the time of the estimate (2e) is then found by multiplying this ratio by the enrollment figure (2d) at the time of the estimate.

In step 3, the population cohort that was between 1.25 years old and 10.25 years old at the time of the census is aged forward to the time of the estimate; at that point in time the cohort contains individuals in the age interval 5 through 14. Survival ratios

TABLE 5.2 A Sample Component Method II Computation: Evanston, Illinois

1. Total resident population (April 1, 1980)	73,706
a. Births (4/1/80 to 12/31/83)	3,653
b. Deaths (4/1/80 to 12/31/83)	2,844
2. Estimated school-age population	
a. Population ages 5.25 to 14.25 on 4/1/80	7,258
b. Total enrollment, grades K through 8 on 4/1/80	6,950
c. Ratio of population to enrollment on 4/1/80 (2a / 2b)	1.0443
d. Enrollment, grades K through 8 on 12/31/83	6,113
e. Estimated population ages 5 through 14 on 12/31/83 (2c × 2d)	6,384
3. Expected population with no migration	
a. Population ages 1.25 through 10.25 on 4/1/80	6,396
b. Survival ratio	0.9978
c. Expected population ages 5 through 14 on 12/31/83 (3a × 3b)	6,382
4. Net migration of school-age children (2e − 3c)	+2
5. Net migration rate for school-age children (4 / 3a)	+.00031
6. Mobility rate for all ages / mobility rate of school-age children	1,191
7. Net migration rate for all ages (5 × 6)	+.00037
8. Population on 4/1/80 + 1/2 births − 1/2 deaths	74,111
9. Net migration of persons of all ages (7 × 8)	+27
10. Population on 12/31/83 = 1 + 1a − 1b + 9	74,138

---

[4]This assumes that January 1, 1980, is the cutoff point for determining what grade students are in. It is also common to use grades 1 through 8 rather than K through 8 to avoid potential complications introduced by irregular kindergarten enrollments.

are obtained from a life table. (See the discussion of these in Chapter 3.) Data for the life table may come from either the region in question or some larger geographic unit.

Steps 4 and 5 are straightforward; the estimated net migration of school-age children is found as the difference between the results in steps 2 and 3, and this is converted into a rate by dividing by the initial size of the cohort. In step 6, the ratio between the net mobility rate for the total population and the rate for school-age children is given. This ratio is often obtained by using national data for the most recent time period available. (See, e.g., *Current Population Reports*, Series P-20.) The local net mobility rate for the total population (item 7) can then be determined by multiplying this ratio by the local net mobility rate for school-age children. Item 8 is an estimate of the survived midperiod population; it includes one-half of all births and deaths that have occurred between the census and the time of the estimate. Total net migration is determined in (9) by multiplying the net migration rate by this midperiod population. In the final step, the population accounting equation is used to determine the population estimate.

Key Assumptions

The two key assumptions in the component method II are (1) that the ratio of school enrollment to school-age population does not change over time and (2) that the national ratio of total to school-age mobility applies to the region being examined. This latter assumption is perhaps most difficult to accept. One study, for example, showed that three-fourths of all states had ratios that fell outside of the range 1.00 to 1.24 when the national ratio was 1.12. (See Shryock and Siegel, 1976.)

The Administrative Records Method

An alternative component method of population estimation has been developed by the U.S. Bureau of the Census. It was employed throughout the 1980s as one of the Bureau's preferred (and, in some cases, only) methods for producing annual, substate estimates.

As in the component method II, natural change is directly measured from data on resident births and deaths. Net migration of population under age 65 (not in group quarters), however, is inferred not from changes in school enrollment, but rather from matches of federal income tax returns from year to year. In addition to tax forms, the method uses changes in the number of persons covered by the federal Medicare program to estimate net migration of persons 65 and over[5]—thus the technique's name: the administrative records method. The final step in the administrative records method is to add in changes in group-quarters population, immigration from abroad, and net movement between the military and civilian population for the area being estimated.

Because the method is conceptually so similar to the component method II, we shall not give a detailed example of its calculation. We shall, however, briefly describe how the migration data are obtained from matching individuals' income tax returns. These data have become increasingly used to monitor state-to-state and county-to-county migration patterns on an annual basis (see also Engels and Healy, 1981; Isserman, Plane, and McMillen, 1982; Rogerson and Plane, 1985), and are a good resource

---

[5]In fact, in recent years, the component method II has also used changes in Medicare enrollment to estimate the elderly population. Because it is persons under 65 years of age who are most likely to move along with elementary school age children, bifurcating the net migration component in this fashion is logically defensible.

for many other population analysis applications. Other nations, such as Canada, have also developed "synthetic" migration data in the same general fashion.

The Bureau of the Census's IRS Migration Data

Each year following the tax-return-filing season, the Internal Revenue Service sends to the Bureau of the Census a small amount of summary information coded off individuals' (1040 series) income tax returns. Included is the Social Security number of the filer, the listed address, and information about exemptions claimed for dependents and for being 65 years or older. Census personnel then carry out a massive computer sorting and matching process to pair up the Social Security numbers of filers in consecutive years. For those returns for which a match is successfully obtained, the listed addresses are then compared. If the addresses differ, a household containing migrants may be indicated. Geocoding information is employed to determine the political jurisdictions within which the addresses are located. The number of movers allocated from matches that show filing addresses in different jurisdictions is determined from the number of tax exemptions claimed for spouses and dependent children, with the assumption that such persons move along with the filer. The exemptions for persons 65 and over are used to eliminate such persons in producing the net migration rates for persons 64 years and under required by the administrative records method.

The migration data synthesized in this fashion do not perfectly reflect actual population movements. Coverage of the total population, though high, is still less than complete due to persons with income too low to be required to file a tax return and, of course, because of tax cheaters who do not file. This less-than-complete coverage problem may not be as serious as it might seem, however, because in the administrative records method a *rate* rather than an absolute number of migrants is estimated from the matched returns. The question is really to what extent the uncovered groups migrate differently from the aggregate, covered population.

A second kind of problem is that the address listed on the return is not always the filer's actual place of residence. For instance, a filer may be temporarily away from home during the period of the year when filing takes place or when refund checks are mailed out.[6]

A third issue is the extent to which persons claimed as exemptions also live and move with the filer of the return. For example, college students whose usual place of residence should be their university dormitory or apartment might still be eligible to be claimed by their parents as dependents.

Despite these difficulties, the Bureau of the Census has worked hard at improving the administrative records matching procedures. Census personnel now believe the method to be outstanding and use it widely in their own estimates programs. Local government and business analysts may also find the migration estimates useful in their work, though these data have not always been publicly released on as timely a basis as some other symptomatic data series typically used for contemporaneous estimates.[7]

We turn now to the category of housing-unit methods, which are widely employed for making estimates when the reference point in time is at, or quite close to, the present.

---

[6]Another, more unusual example of this was noted at a State of Arizona Population Technical Advisory Committee Meeting, when the representative of the Navajo Indian Nation asserted that a sizable number of tax returns for persons living on the Arizona portion of the reservation were being filed with addresses listed for car dealers and other merchants across the state border in New Mexico; these merchants are then able to have the refund checks signed over to them in lieu of cash payment.

[7]The figures may be obtained from state data centers and the Statistics of Income Division of the Internal Revenue Service.

5.5 HOUSING-UNIT METHODS

Among the most widely used methods for estimating local-area populations for business, planning, and other purposes are those based on tallying up total numbers of occupied housing units and multiplying by an average number of persons per unit. Conceptually this is just the opposite of the old joke about how a farmer estimates the number of cattle in a field: by counting the legs and dividing by four!

Housing unit methods are appealing for a number of reasons, not the least of which is a very practical one: It is easier to find relatively large, generally geographically stationary housing units than the highly mobile persons who inhabit them. Furthermore, most local governments routinely keep records of new housing units in the form of building permits. Such data are publicly accessible at the local level and become available almost immediately. Alternative sources of information for monitoring the total stock of housing (as opposed to just net additions to it) include (1) records on the number of active, residential utility service connections, (2) direct field enumeration (if the study area is small enough to make this a cost-effective option), and (3) air photo interpretation.

An important distinction to make in deriving a housing unit estimate of total population is that between *group-quarters* population and *household* population (which we mentioned in passing in describing the administrative records method). The group-quarters component includes persons living in institutions such as inmates in jails and penitentiaries, patients in nursing homes, students in college dormitories, and military personnel in barracks. In addition, there is often an additional, noninstitutional group-quarters population (for instance, people living in halfway houses, communes, and various types of shelters). Keep in mind that the Bureau of the Census's definition of group-quarters population includes only persons living in group housing for which there are 10 or more individuals. Thus the figures obtainable from the Bureau and the state-level affiliated agency may well reflect this threshold-size criterion. Group-quarters population is usually fairly easily estimable by direct checking based on local knowledge of the universe of such facilities within the study area. Therefore housing-unit methods are aimed at estimating the second, *household* component of total population, which is generally the larger of the two. Group-quarters population is added in as a final step in the estimation procedure.

The basic formula for computing a housing-unit estimate is

$$P = (H \times \text{OCC} \times \text{PPH}) + \text{GQP}$$

where P is estimated total resident population, H is the total number of (nongroup) housing units in the study area, OCC is an occupancy rate factor (equal to 1 minus the vacancy rate factor), PPH is the average number of persons per occupied housing unit, and GQP is the independently obtained total of persons in group quarters.

We shall now make some observations about each of the three constituent elements of the household population component. Because a fairly accurate estimate of GQP can usually be obtained, and because it is generally considerably smaller than household population, most of the pitfalls in using the housing-unit method concern the measurement of H, OCC, and PPH.

Monitoring Temporal Change in the Number of Housing Units

In a postcensal application of a housing-unit method it is wise to benchmark the measure of total housing units against the most recently obtained census inventory count of total housing units. For example, if using building permits, the analyst would likely compute the housing stock at estimate date, t, as

$$H_t = H_0 + (\text{PERMITS}_{0, t-\text{lag}} \times \text{BFACT}) - \text{DEMOL}_{0,t}$$

To the census base-year stock of housing units found in the area, H_0, we add the total number of new residential units listed on building permits taken out between the census date, 0, and a time, t – lag, shortly before the estimate date. It is advisable to take into account a construction time lag representative of the average time between the approval of the building permits and when housing units ultimately are completed and become ready for occupancy. In addition, common practice is to assume that only a certain proportion of units for which building permits are issued, BFACT, in fact, get built.

Use of such assumptions about average construction time lags and buildout factors can sometimes be avoided entirely if ongoing resources exist to monitor the actual progress of new construction in the community. Metropolitan areas may have long-term, commercially or university-based *land use study projects* that involve the "driving of permits" to establish both the current state of construction activity and the current inventory of housing product becoming available in various market segments (e.g., single-family, duplex, townhouse/condominium, apartment complexes). For smaller communities that have had relatively little new residential construction since the past census, it may be possible for the analyst to do the field checking. In certain jurisdictions, *occupancy certificates* are required in addition to building permits. In this case, it is advisable to use these certificates rather than building permits for preparing estimates; because occupancy certificates are issued only when units are ready for human habitation, they obviate the need to estimate the overall proportion of permitted units that become available for residency.

It is also necessary to subtract the total number of residential units demolished since the past census; local conventions for permitting demolitions should be investigated before proceeding with the estimation process.[8]

Measuring Housing Units through Field Counts or Air Photos

If instead of using building permits (or occupancy certificates), the analyst intends to enumerate the total housing stock through either direct field counts or air photo interpretation, the Bureau of the Census's conventions regarding what constitutes a housing unit should be carefully examined prior to specifying the procedures to be followed. In air photo interpretation, thorny problems may be encountered involving how to distinguish the number of separate living quarters within individual buildings, as well as how to distinguish buildings used for human habitation from those in use for other purposes.[9] In the latter case, for example, it sometimes becomes a judgment call as to whether an outbuilding constitutes a separate, full-time residential unit.

---

[8] It may also be advisable to pay particular attention to how mobile home units are or are not picked up through building permit records. For instance, in some jurisdictions permits are required for new mobile home *pads*, which may not actually contain a housing unit. A further practical caution relating to the use of building permits is to pay particular care to how many actual separate units are listed on any given permit; sometimes separate structures are shown rather than the total number of households for which each is intended.

[9] In a trial application of color air photo interpretation in Arizona, the number of evaporative (or "swamp") cooling units on the roofs of multifamily housing units was suggested as part of the interpreters' decision criteria. Making use of specific local circumstances can often improve estimation methodology!

Measuring Housing Units with Utility Data

Probably the major alternative method to using building permits to derive H_t involves obtaining access to utility company records. Government officials and planners working for public agencies may generally obtain permission to access the billing record data base of local utilities. Sometimes, too, procedures will exist or may be established so that such data may also be obtained by private-sector demographic analysts. Most typically data on the number of active residential electricity meters is the preferred utility-based variable to try to obtain. Water or sewer connections may sometimes be more easily obtained, but the coverage rates are typically not as good—an issue to which we shall return shortly. In using either electricity or water company records, care should be taken to explore the nature of, and separate out, *group meters*. Group metering occurs when two or more households are served through a single meter; in some apartment buildings, for instance, the overall bill is paid by the landlord who factors in some estimate of a unit's average monthly usage in setting each tenant's rent.

Again, as in the case of using building permits, an important principle is to *benchmark* the utility-based measure of the total housing stock against actual census counts. Because it is the total stock that is being proxied, rather than net additions to stock as in the case of building permits, a good procedure to use here might be to compute H_t through the ratio adjustment:

$$H_t = (U_t / U_0) H_0.$$

Here U_t is the number of metered units at the estimate date (after adjusting for any group metering) and U_0 is the number at the census date.

It is critical to benchmark utility connections against a known inventory count of housing units because (1) the utility data may not cover all housing units in the study area and (2) the number of utility connections may include some to units intended for occupance by seasonal population rather than full-time residents. Although electrical service, water, and sewers are now centrally provided to most households in the United States and Canada, all housing units are still not served. The reason why we stated earlier that electrical connections rather than water or sewer connections are the preferable utility-based variable to use is that few homes have their own electrical generators, whereas in many communities a sizable proportion of housing units have their own wells or septic systems. By using the ratio procedure to benchmark to the housing-unit inventory count of the most recent census, we assume, in part, that the ratio of actual enumerable total units to those with utility connections has remained constant. If the analyst has good reason to believe otherwise, an additional adjustment factor could be added.

An advantage of using utility records rather than building permits for estimation purposes is that it may be possible to use information on monthly usage of electricity (or water since sewer flows are rarely separately metered!) to distinguish *active* from *inactive* accounts. We can thereby produce a somewhat sophisticated indicator of vacancy rates from the same data source from which total units are being monitored. This, too, may be one way to take into account the problem important in some communities of separating seasonally occupied units from those of the resident population.

We turn now to a discussion of OCC, the important second component of the housing-unit method equation.

Measuring Occupancy Rates

In many local housing markets, vacancy rates are significant in size and may change substantially over time. Vacancy rates include a *structural* component reflecting the normal temporary period when a unit is unoccupied as its ownership or rental is turned over. They also depend on both cyclical fluctuations and longer-term trends in the local economy. In communities not already fully developed, a "boom-bust" psychology has typified the residential real estate industry in recent decades. A critical issue in applications of housing-unit methods is thus to choose a reasonably correct downward adjustment to apply to the total stock of *available* units in order to more accurately represent those really *occupied* at the time of the estimate. Although there are no hard and fast rules about how this is to be done, the ultimate accuracy of the overall population estimate depends significantly on the truthfulness of the multiplier value, OCC, that is chosen.

As previously mentioned, the ratio of active to inactive residential utility accounts may provide one symptomatic indicator of true vacancy rates. Also, as earlier described, the area for which the estimate is to be derived may be included within the study area of an ongoing land use survey project. Such projects are often undertaken largely to inform the investment decisions of residential developers with up-to-date measures of the current tightness or slackness of the local housing market.

Alternatively, the analyst might be wise to devote some resources to generate data on vacancy rates through sampling techniques (e.g., by placing phone calls to the managers of apartment complexes or by driving streets and observing apparent vacancies). Multiple-listing service data on homes for sale, newspaper advertisements, and postal delivery information might also be examined. Expert judgment is sometimes a valid source of up-to-the-minute estimates of vacancy rates, although in evaluating an estimate thereby produced, an interesting question to ask is whether the expert(s) consulted had a reason to shy on either the high or low side. In the absence of other, more contemporaneous information, the occupancy rate at the census date may be all that is available for use.

An extremely important point to keep in mind when deriving or evaluating occupancy rate factors is that for the specific purpose of population estimation, for a unit to be considered as occupied it must be occupied by persons *at their usual place of residence*. Quite frequently the vacancy rates correctly taken from the census may seem high when compared to the kinds of vacancy rates tossed about in real estate circles. This is because we usually wish to exclude from our estimated population persons who are only temporarily or seasonally present in the study area. Care should be taken, for example, in deciding how to use rates based on active utility accounts. In the case of many communities in, for instance, states such as Florida and Arizona with significant winter ("snowbird") population, or in the case of summer resort communities elsewhere, the monthly pattern of a unit's usage contains clues as to whether it is being maintained for the use of resident or seasonal population. Again, the need to benchmark to census data is stressed.

The Persons-per-Housing-Unit Factor

All the careful attention devoted to accurately estimating housing stock and occupancy rates can be for nought if a sloppy job is done in choosing the final of the three constituent elements of our household population estimate. Even a small percent error in the PPH factor can throw off the overall estimate of population by an unacceptable absolute amount. The problem of choosing this factor can be conceptualized as involving two considerations: (1) to what extent has *household structure* changed in general and in the study area specifically? and (2) to what extent has the study area's *housing* mix shifted?

Recently, average family size has been declining in the United States. Geographic dimensions of this and other aspects of household structure are detailed in

Chapter 9. The general decline, however, does not mean that every community has been experiencing the same sorts of shifts. Demographic differences across the regions of the United States have exhibited rather surprising persistence, as detailed by Morrill (1993).

At the local scale, the second consideration can be an extremely useful one upon which to base an adjustment to a census date PPH in order to reflect the reality at the estimate date. The various types of housing units such as single-family, duplex, triplex, and apartment units are characteristically occupied by rather different average numbers of persons. If the study area experiences significant changes between the census and estimate dates in its housing mix, the PPH factor should be adjusted accordingly. In the course of compiling information on the net change in H (the overall housing stock), it may be possible to maintain separate tabulations for each type, and then to multiply by *housing-type-specific* persons-per-housing-unit factors. These may be more stable than the overall PPH, which can be conceptualized as consisting of a weighted average of type-specific components.

A final practical observation is that a sampling procedure can be employed to establish a fairly accurate estimate of current household size. In addition to sample size and sampling error, attention should be paid to such issues as the representativeness of the sample depending on how it is to be reached (e.g., does a phone survey miss a segment of the overall population with distinguishably different household size?) and whether respondents are answering truthfully (e.g., does the study area contain a large number of illegal aliens whom the respondents might not include in their reported tally of residents of the housing unit?). Of course, if the estimate is for some time period other than the approximate present, a sampling approach would not be feasible.

5.6 EVALUATING ESTIMATES

Thus far this chapter has been a fairly practical, detailed accounting of how actually to make population estimates. In the final section we wish to step back from the welter of details that have typified the earlier sections and talk about some bigger-picture issues connected with judging the utility of population estimates produced by one or more of these methods.

An essential element of "faith" is inherent in any population-estimating technique. And no estimate is ever totally "correct." The problem, after all, is to derive a figure for which true empirical verification (actually going out and counting) is impossible. Thus the accuracy of the art of estimation depends on how plausible the underlying assumptions are for any particular method and for any particular application of that method.

The fundamental assumptions made for population estimation generally boil down to presumptions of (a) some level of temporal stability in population change (or one or more of its components) or (b) some level of temporal and structural stability between population change (or one or more of its components) and other symptomatic variables.

Two Methods for "Testing" Estimation Procedures

Whereas the assumptions underlying any estimation method may never be fully verified (or falsified) through empirical testing, some scientific analysis of their plausibility is possible. For example, a method under consideration for use in a county's postcensal population estimates during the 1990s may be checked as to its ability to accurately hit the county's 1990 census count when a 1980 census base population is used together with the 1980s values of the same symptomatic variables under consideration for use in the 1990s. Such a procedure is called "in sample" testing. Just because the method would have worked well during the past decade does not necessarily mean, however, that it will perform adequately in the present one. The relationship between the symptomatic variables and population may not remain identical across the two decades, or, as is often the case, the methods and quality of data collection may change over time, rendering such test results misleading.

A similar form of testing may be done after the fact. When the year 2000 census results become available, the typically more accurate methods for intercensal estimation may be applied to derive a historical series for the 1990s. The analyst then can study the ability of the postcensal estimates that were derived during the 1990s to track the newly derived, higher-quality intercensal estimates.

Our main point, though, is that any figure derived for a noncensus year in the 1990s can never be totally checked; at best we can compare it against a better estimate made when more data become available than was the case when it first needed to be computed.

Some Rules of Thumb to Improve Estimates

Having convinced you, we hope, that no estimation technique is ever fully testable, we want to stress several commonly accepted practices that should improve the accuracy of an estimate. Even more important than knowing how to make estimates is the ability, as the quotation at the beginning of this chapter suggests, to know how to evaluate their worth. The following rules of thumb are not magic formulas for success. Rather, good common sense and a fundamental understanding of the geographical aspects of population processes are required of the analyst. To give you the latter is our major goal throughout this book. We know of no easy way to acquire additional doses of the former.

> To know how to apply the general to the particular is an additional "natural gift," the want of which is "ordinarily called stupidity," according to Kant, who adds, "And for such a failing there is no remedy."
>
> —Hannah Arendt (article in *The New Yorker*, November 28, 1977, p. 114)

Aggregate Unit Estimates Are More Accurate

More and better types of data generally become available for making estimates for larger geographic units (e.g., states and counties) than for smaller ones (e.g., minor civil divisions or census tracts). Sampling theory tells us, too, that imperfectly collected data should be more accurate for geographically aggregate units than for disaggregate ones; the chances for data collection errors to compensate for one another become better as aggregation proceeds to higher-level geographic units. As the examples in Table 5.3 demonstrate, a reporting or tabulation error for a single subunit that may critically affect the accuracy of an estimate for that subunit will probably, in the grand scheme of things, have just a small effect on the accuracy of the overall region's estimate.

On the other hand, keep in mind the principle of the least common denominator. Sometimes a smaller geographic unit may have a unique data source that the savvy

TABLE 5.3 Impact of a Data-Reporting Error for a Subregion on the Accuracy of an Estimate for the Overall Region

Percent Error for Subregion	Subregion's (Actual) Percentage of Overall Region's Population	Percent Error Induced in Region's Estimate
10%	1%	0.1%
	5	0.5
	10	1.0
20	1	0.2
	5	1.0
	10	2.0
50	1	0.5
	5	2.5
	10	5.0

analyst can exploit for making that area's estimates. For instance, a county may require a certificate of occupancy before new residential housing units may be moved into. If other counties only require building permits at the start of construction, state-level estimates derived from a housing-unit method must rely on the somewhat less satisfactory latter type of government records. The U.S. Bureau of the Census's postcensal county-level estimates rely on different methods for different states, tailoring the choice of methodology to the strengths of the data resources available for each.

Data Access, Cost, and Time Pressure Are Important Considerations

A second practical observation to bear in mind, important when defining the set of feasible methods, is that certain types of data are available only to certain government officials. A good case in point is the increasing use by the Estimates Branch of the Population Division of the U.S. Bureau of the Census of data derived from matching federal income tax returns. Certain summary tabulations derived in the course of this process are now made available to the public (through the Statistics of Income program of the Internal Revenue Service). However, the sensitive nature of the data source—involving interesting philosophical distinctions regarding privacy rights versus the common good—means that only sworn employees of the federal government may make full use of these data, on a timely basis, for estimates.

Connected to the issue of access is that of the cost and time involved in the acquisition of data. Quite frequently the users of estimates are as much, if not more, concerned with having the figures available quickly than they are in what might be only small marginal increases in accuracy brought about by waiting for additional symptomatic variables to become available.

Unfortunately, a component of cost is sometimes simply the level of technical expertise needed to carry out one of the more mechanically involved estimation procedures. While it has been remarked that estimation methods are far from rocket science, those of you who are making the effort to read and understand this chapter will be far more technically competent to carry out and, even more importantly, *evaluate* population estimations than are the vast majority of persons who actually get called upon to do them.

The matter of the time available to carry out an estimate is a significant one. The common American desire for up-to-the minute information means that for many estimation problems the choice of methods is reduced to going with what is available. Thus housing-unit methods are perhaps those most commonly employed for local-area estimates because the data sources upon which they are based are at the local level and become available almost immediately; other methods are predicated on data that must be retrieved from more centralized reporting and tabulating systems.

Conceptual Clarity and Defensibility Should Be Considered

After the spectrum of available options is narrowed because of practical exigencies, how then to proceed in choosing a method? As we suggested earlier, the evaluation process ultimately comes down to a matter of expert judgment regarding the plausibility of the assumptions upon which the method—and its application to the problem at hand—are based. As a general rule of thumb, we would contest that the closer a method comes to actually proxying the mechanisms of population change, the more conceptually pleasing, and thus defensible, it is likely to be. The basic demographic accounting equation we discussed in Chapter 3 (and which forms the basis for component methods) should be kept indelibly etched in mind when debating the merits of one method versus those of another.

Related to the notion of conceptual correctness and clarity is the ease with which a method can be explained to the public- or private-sector officials who are likely to be those who commission the estimating job. The more that a method resembles a complicated black box, with lots of bells and whistles, the easier it may be for a critic to cast aspersions upon its results. Thus the ratio-correlation method developed by the U.S. Bureau of the Census may be treated more skeptically by persons in the field than, say, a housing-unit method, even though the specific application of the housing-unit method may rest on heroically assumed multipliers for occupancy rates and household size.

Beyond these general observations and principles that can be borne in mind when evaluating alternative methods, there are two mechanistic devices that wise population analysts use whenever possible to improve the accuracy and defensibility of their work: (1) they use controls totals and (2) they combine the results independently obtained from two or more methods. Both rules of thumb have their basis in sound numerical modeling experience.

Use Control Totals

As we discussed at the outset of this discussion of how to evaluate estimates, percentage accuracy levels are generally greater for estimates of aggregate units than disaggregate units. The corollary is that total population usually may be more accurately estimated than constituent subpopulations.

It is generally accepted (and the results of recent decennial censuses have largely confirmed) that national-scale estimates of the U.S. population are quite good. The level of faith erodes (and the post hoc evidence becomes more troubling) as we proceed down the geographic scale to the state and thence to county and local levels. The Bureau of the Census, as an important part of its procedures, routinely adjusts state-level estimates to add up to the independently derived national number. The wise analyst preparing local-area estimates does likewise whenever possible.

The most commonly employed adjustment procedure to ensure that subunit estimates sum to the control total for the aggregate unit is called *proportional* or *pro rata adjustment*. Consider the case of estimating the population, P_i, of all the $i = 1, 2, \ldots, n$ counties of a state whose independently estimated total population is P_*. As the final step of the county estimation process, we simply multiply any given method's set of estimates, $\{P_i\}$, by the proportionality constant:

$$\rho \equiv \frac{P_*}{\sum_{i=1}^{n} P_i}$$

For example, the July 1, 1986, Bureau of the Census postcensal estimate of Delaware's population was 633,000. Suppose we are experimenting with a new method that results in initial 1986 estimates of Delaware's three counties' populations of $P_1 = 104,346$, $P_2 = 425,488$, and $P_3 = 110,166$. (Here $i = 1$ indicates Kent County; $i = 2$, New Castle County; and $i = 3$, Sussex.) The correct pro rata adjustment proportionality constant is thus $\rho = 633 / (104.3 + 425.5 + 110.2) = .9890625$ (approximately) and the adjusted (and rounded)[10] final estimates are 103,200, 420,800, and 109,000, respectively. Some estimation methods inherently build in *top-down* modeling principles before any final proportional adjustment procedure is carried out.

Combine Results of Different Estimation Techniques

Combining the results of two or more estimating procedures reflects a belief that each method has certain strengths and weaknesses and, therefore, the possibilities of being far off from the actual population will be diminished by averaging or otherwise splitting the differences among results. The principle is analogous to the practice in the somewhat subjective worlds of scoring diving or figure skating competitions: "reality" is presumed to be best proxied by throwing out extreme cases and averaging the scores in the middle.

In population estimation, however, sometimes we really do have strong reason to believe that one method is clearly superior to all other potentially usable ones. If this is the case, then we should not pollute that method's results by combining them with other, less reliable ones. More typically, however, our beliefs about the reliability of alternative methods are not so clearcut; we may feel that a certain procedure might be the most accurate, but that a plausible case can be made also for certain alternative techniques. This then raises the spectre of unequal weighting schemes. We end this chapter with a brief case study reflecting on the use of differential weights.

A Case Study in the Pitfalls of Unequal Weighting

One of this book's authors served on the State of Arizona's Population Technical Advisory Committee (POPTAC), an oversight body with representation from the state's various regional Councils of Government, from each cabinet-level state agency, and from each of the state's three universities. Annual postcensal population estimates for each county and incorporated city and town in the state must be made by December, with the previous July as the reference date. These official State of Arizona estimates are used for such state government purposes as allocating lottery revenues to local governments and setting expenditure limitations for counties and municipalities.

For the 1988 round of county-level estimates prepared by the state's Population Statistics Unit (housed in the Department of Economic Security) it was decided that a

---

[10]Note that the Bureau of the Census rounds state estimates to the nearest thousand and county estimates to the closest hundred persons. In working with proportional adjustment factors, we should start with unrounded initial estimates, take the adustment factor out to as many decimal points as possible, and do whatever rounding is desired as the final step of the overall estimation procedure.

combination of the same three methods would be used for each county. It was further agreed, however, that these could be weighted differently for each county given the perceived county-specific quality of the inputs to, and results from, each of the three methods. The maximum weight that could be assigned to any method was fixed at 60 percent and the minimum at 20 percent.

Not too surprisingly, in retrospect, because of the local (Council of Government) representation on the panel, the upshot of a very lengthy POPTAC review meeting was to place 60 percent weights on any given county's highest figure and lesser weights on each of the lower two. Although we could cynically decry such an outcome because of its evident political, rather than scientific, motivation, an opposing viewpoint with possible legitimacy is that when it is worse (in terms, e.g., of fairness of service provision) to err on the low side, then perhaps we should put the greatest stock in the highest number generated by a widely accepted, conscientiously applied method. In the example cited here, the use of population estimates to impose state-mandated caps on government expenditures at the county and local levels may well fall into this category of being as equitable as possible in the face of scientific uncertainty. On the other hand, the use of the estimates for allocating a fixed, statewide pool of revenue from the lottery makes POPTAC's estimation problem, in part, a zero sum game: Any additional population going to one county through ad hoc juggling of the weights results ultimately in diverting some amount of lottery funds from each other county.

5.7 SUMMARY

In this chapter we have built upon the substantive base of knowledge about population distribution, population growth processes, and the components of population change that formed the material for Chapters 2, 3 and 4. In tackling one of the most common forms of applied population analysis—the preparation of population estimates—we have stressed that diverse approaches exist, with the most suitable technique depending on the setting and characteristics of the specific task.

In Chapter 6 we turn to the related set of problems involved with projecting or forecasting the *future* course of population growth or decline. Whereas many of the same considerations are present for projection applications as for estimation, the unavailability of contemporaneous data on symptomatic variables implies additional uncertainties and poses even greater challenges for the geographical analyst of population.

EXERCISES

1. *Linear interpolation.* Suppose the 1990 decennial census population count for Columbine County was 49,173. The July 1, 1991, estimate for the county is 51,400. Interpolate a July 1, 1990, population.

2. *Constant continuous rate extrapolation.* Suppose the July 1, 1991, estimate for Columbine County reported in Exercise 1 was calculated using the exponential growth formula with an annual continuous rate of growth, r, derived from the county's 1980 and 1990 decennial census counts. What value of r was used in the extrapolation?

3. Determine Columbine County's 1980 decennial census count using the answer from Exercise 2 and the data given in Exercise 1.

152 Chapter 5 • Population Estimation

4. *The ratio-correlation method.* The accompanying table shows annual data on hunting licenses issued, recorded deaths of residents, and automobile and pickup truck registrations for both Saltlick Township and Deerkill County from 1980 to 1992. The population figures represent midyear estimates based on 1980 and 1990 decennial census counts and the official state-prepared county-level annual estimates.

	Deerkill County				Saltlick Township			
Year	Population	Licenses	Deaths	Registrations	Population	Licenses	Deaths	Registrations
1980	25,680	9,482	404	17,200	5,766	486	89	3,116
1981	24,986	8,787	402	16,569		473	92	3,208
1982	24,981	8,210	417	16,234		412	85	3,002
1983	25,544	8,912	423	16,608		460	70	3,001
1984	25,707	9,500	421	16,940		458	81	3,108
1985	25,824	9,511	434	17,001		458	90	3,213
1986	27,005	9,722	470	17,223		454	74	3,204
1987	27,043	9,087	468	17,308		460	81	3,318
1988	30,158	11,113	529	19,476		468	80	3,370
1989	31,333	12,009	517	19,482		469	79	3,391
1990	31,714	11,790	538	20,617	4,805	445	64	3,112
1991	32,340	12,442	547	20,808		436	58	2,979
1992	32,101	11,642	526	21,005		422	55	2,944

The staff of the Deerkill County Planning Board has estimated the following regression equation using data from all the townships of the county for 1980 and 1990:

$$Y = -.082 + .421X_1 + .177X_2 + .415X_3$$

where X_1 is the number of hunting licenses issued to residents, X_2 is the number of resident deaths, and X_3 is the number of auto and pickup truck registrations. Choosing the appropriate numbers from the table to compute the ratios required for the ratio-correlation method, estimate Saltlick Township's 1992 midyear population.

5. What additional step would you use to refine the estimate derived in Exercise 4 if you had all the data available to the Deerkill Planning Board staff?

6. How well does the regression equation estimate Saltlick Township's census-based 1990 population figure listed in the table?

7. *Component method II.* Using the data in Table 5.1, calculate an estimate of the population of Evanston, Illinois, as of December 31, 1984. Use the same mobility ratio as on Table 5.2, line 6, but assume a survival ratio of .9972 in place of that shown on line 3b.

8. *Housing-unit method.* The accompanying tables show monthly data assembled by consultants working on the Metropolitan Alphaville Land Use Study.

Exercises   153

Total residential units on building permits issued by all jurisdictions in the Alphaville MSA and total units demolished

	1989		1990		1991		1992	
Month	Number of Building Permits	Number of Units Demolished	Number of Building Permits	Number of Units Demolished	Number of Building Permits	Number of Units Demolished	Number of Building Permits	Number of Units Demolished
January	N.A.		46	0	23	9	137	0
February	N.A.		58	0	16	0	159	0
March	N.A.		60	1	24	0		
April	N.A.		72	3	30	2		
May	N.A.		84	8	33	0		
June	N.A.		57	0	17	1		
July	34	0	40	0	11	1		
August	50	0	32	3	13	11		
September	16	1	48	0	20	0		
October	20	14	112	0	29	28		
November	31	0	39	128	42	0		
December	40	6	45	2	58	0		

Alphaville MSA 1990 decennial census data

Total population:	662,114
Total housing units:	277,532
Vacant housing units:	11,430
Group quarters population:	1,512

The Land Use Study requires quarterly population and housing-unit estimates for the Alphaville MSA. Compute estimates of population and housing-unit inventory for January 1, 1990, through January 1, 1992. Assume a 6-month construction time lag, a 95 percent buildout factor, and no change in the group-quarters population and the census-derived vacancy rate.

9. *Pro rata adjustment*. The Deerkill County Planning Board staff has completed the results of 3 different types of estimates for the July 1993 populations of the county's five townships. Averaging these results, the townships' populations would be

Cooperville Township	10,197
Falling Leaf Township	8,953
Red Jacket Township	2,010
Saltlick Township	4,405
Sucker Brook Township	8,244

An astute staff member has just discovered, however, that these figures do not add up to the independently derived 1993 County estimate of 32,006. Calculate a pro rata adjustment factor, ρ, and adjust the township populations to sum to the county total.

> **This is like déjà vu all over again.**
> —attributed to Yogi Berra

> **I've seen the future, and it's a lot like the present, only longer.**
> —a former major league baseball pitcher

CHAPTER 6
POPULATION PROJECTIONS

6.1 Introduction
6.2 Elementary Extrapolative Methods
6.3 The Single-Region Cohort Component Model
6.4 Cohort Component Models with Migration
6.5 School Enrollment Projections
6.6 Projecting the Demographic Structure of Organizations
6.7 Summary

6.1 INTRODUCTION

The previous chapter addressed methods and issues associated with estimating the size of past and present populations. In this chapter, we turn to the problem of projecting the size, location, and demographic structure of future populations.

The accurate determination of future population levels and characteristics is essential for many planning activities in both the public and private sectors. For example, businesses wish to forecast sales potential and market shares, local governments desire plans for the adequate provision of public services, and industries require accurate assessments of future labor needs. Population projections are a fundamental input to all of these planning activities.

As indicated in the past chapter, estimates of current population are often used in funding allocation formulas. Long (1993) noted that in funding year 1989, federal programs allocated over $10 billion to states and substate regions on this basis. Projections of population are also often used as input to funding allocation formulas. An argument can be made that population projections are more relevant than estimates for both funding decisions and other planning problems. An area that is expected to grow will need the revenue, services, and infrastructure to handle increased demands. It is certainly better to anticipate needs than to attempt to deal with them as they arise. In the early 1970s, for example, the Environmental Protection Agency allocated money for the construction of sewage treatment plants partially on the basis of population projections. Treatment plants were thus more likely to be built in those areas that would need them in the future.

However, the counterargument that "planning is a self-fulfilling prophecy" can also be made. That is, if money and facilities are directed toward those regions that are expected to grow, the very act of allocating makes it more likely that these regions will grow. Likewise, if money and facilities are not allocated to those regions projected to decline, it will almost certainly be more difficult for those regions so designated to exhibit much growth.

Some planning agencies explicitly recognize this in preparing their population projections. Representatives from local communities within a county or metropolitan region are brought together, and are in effect asked what they would like their population projection to be! The local projections must of course sum to the county or metropolitan total. Some communities may wish to choose projections that are conservatively low, with the hope that recent growth and in-migration—and any attendant problems—may abate. Communities desiring growth would choose high population targets, thinking that the increased allocation of money and facilities may induce development. Here population projections play an entirely different role in the planning process—they are used more as a tool to guide development and foster growth control than as true projections based upon specific assumptions about the demographic and economic forces that lead to population change.

In this chapter, we examine a number of alternative methods for projecting population. These methods range from simple extrapolative techniques to more sophisticated approaches that account for the age and location of the population. In the latter part of the chapter, we also demonstrate how these methods have a variety of other applications to planning. Specific examples are given of their use in the preparation of school enrollment projections and in labor force planning.

6.2 ELEMENTARY EXTRAPOLATIVE METHODS

Suppose that we must estimate the future population of a region. If our only piece of data is its current population, the number of ways to proceed is severely constrained by lack of available data! About the only alternative in this instance would be to assume that the population remains constant over time. Though perhaps crude, the assumption does lead to a statement regarding future population and therefore may be treated as a projection.

More often, of course, some historical data are available, and we may be able to discern some temporal trend in regional population levels. In this instance, it is straightforward to plot the recent population values as a function of time. A particularly simple approach to projection is to extrapolate the line (or the geometric or exponential curve) that best fits the current data into the future. Figure 6.1 depicts a simple example of linear extrapolation.

We would not generally expect these elementary methods of extrapolation to provide accurate forecasts in light of the small amount of information used. They should only be used in those instances where no information, other than past total population levels, is available. In some cases, it may be necessary to generate a projection quickly; there may not be sufficient time to collect other, more detailed data even

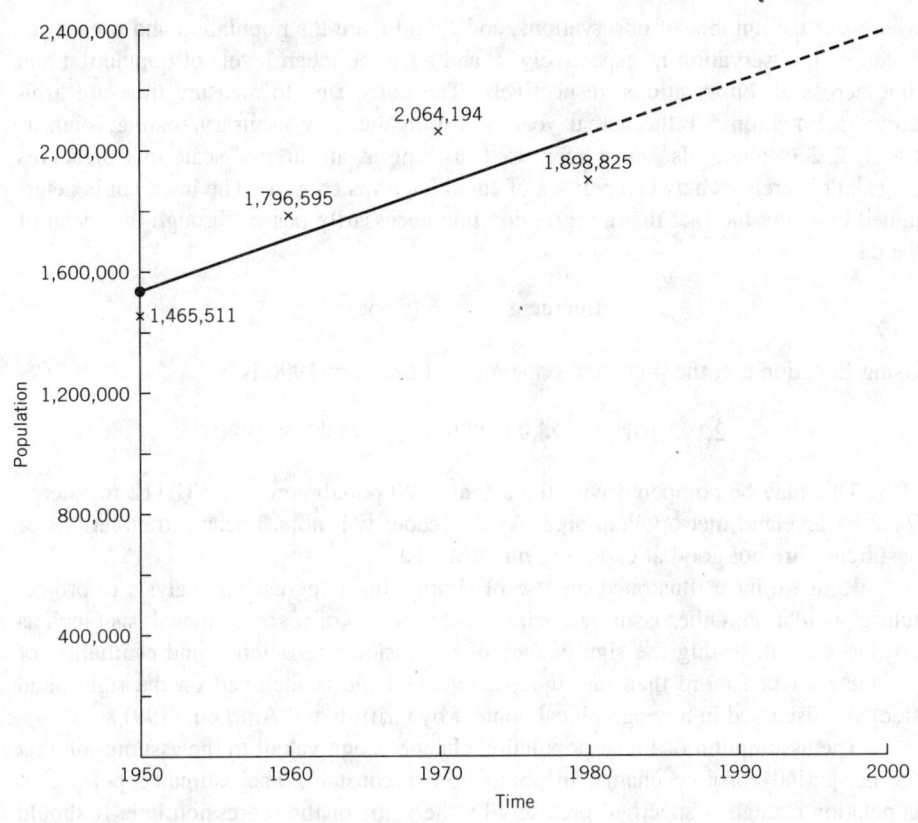

FIGURE 6.1 Linear Extrapolation of Population for the Cleveland Metropolitan Area

though they exist. In these instances, curve-fitting methods suffice to give a "quick-and-dirty" projection. Finally, simple curve-fitting procedures may sometimes be justified on the basis of providing a comparative perspective—they illustrate what would happen if current trends in total population were to hold in the future.

Linear Extrapolation

The data plotted on the graph in Figure 6.1 show population data for the Cleveland Metropolitan Area from the 1950–80 decennial censuses. A best-fitting line through these four observations is given by the equation

$$P_t = -28{,}995{,}889 + 15{,}675.41 \times t \tag{6.1}$$

where 15,675.41 is the slope, $-28{,}995{,}899$ is the intercept, t is time given in years (e.g., 1,990 is used to represent the year 1990), and P_t is the population in year t. These slope and intercept parameters are found by carrying out a simple linear regression analysis. The slope is found from

$$\frac{\sum_{i=1}^{n}(P_i - \overline{P})(t_i - \overline{t})}{\sum_{i=1}^{n}(t_i - \overline{t})^2} \tag{6.2}$$

where n is the number of observations, and P_i and t_i are the population and time associated with observation i, respectively. $\bar{P}$ and $\bar{t}$ are the mean levels of population and time across all observations, respectively. The units used to measure time are arbitrary; in Equation 6.1 the actual year was used, but any arbitrary scaling, such as $t = 1, 2, 3, 4$ could also have been used (as long as an interval scale that preserves constant differences between periods of equal length is chosen). The intercept is determined by using the fact that a regression line necessarily passes through the mean of the data:

$$\text{Intercept} = \bar{P} - (\text{slope})\bar{t}$$

Using Equation 6.1, the predicted population at time $t = 1990$ is

$$2{,}198{,}167 = -28{,}995{,}899 + 15{,}675.41 \times 1990$$

This may be compared with the actual 1990 population of 1,831,112 registered for the Cleveland metropolitan area. As the reader will note, linear extrapolations of past trends are not good at capturing turnarounds!

Here we have illustrated the use of simple linear regression analysis to project future population. Other issues associated with the use of regression analysis (such as goodness of fit, testing the significance of regression coefficients, and estimation of parameters when more than one independent variable is included on the right-hand side) are discussed in a geographical context by Griffith and Amrhein (1991).

The assumption of linear population change is equivalent to the assumption that the per-period absolute change in population is constant. The estimated per-period population change is specified precisely by the slope of the regression line. It should be noted that the regression line is fitted to observed data, and then the line is extrapolated into the future. The last observed point is not the take-off point for the extrapolation. If the regression line does not provide a good estimate of the last observed population value, short-run projections will not be consistent with the most recent population figure. One rather ad hoc and arbitrary way to fix this would be to use the slope estimated from the regression equation to project forward in time from the last observed population value. In our example, this would yield a better projection of $1{,}898{,}825 + 15{,}675.41 = 1{,}914{,}500$ (after rounding to the nearest integer).

Geometric and Exponential Extrapolation

It is often more realistic to assume that the population will continue to grow according to its recently observed *rate of growth,* at least in the short run. As noted in Chapter 3, this would lead either to geometric growth when the rate of growth, r, is specified for a discrete time period (Equation 3.5) or to exponential growth when the rate is assumed to be compounded continuously over time (Equation 3.6).

Like the linear extrapolation just described, the geometric growth curve may also be fit by regression analysis. Taking the natural logarithms of both sides of Equation 3.5,

$$\ln P_t = \ln P_0 + t \ln(1 + r) \tag{6.3}$$

where ln denotes the natural logarithm. This is the equation of a straight line, with slope equal to $\ln(1 + r)$ and intercept equal to $\ln P_0$. The slope of the graph of $\ln(P_t)$ versus time may be estimated as:

$$\frac{\sum_{i=1}^{n}(\ln P_i - \overline{\ln P_i})(t_i - \bar{t})}{\sum_{i=1}^{n}(t_i - \bar{t})^2} \tag{6.4}$$

The intercept is

$$\overline{\ln P_i} - \bar{t}\,(\text{slope}) \tag{6.5}$$

For the data in Figure 6.1, we find a slope of .000353296 and an intercept of 13.779. These results may then be retransformed back into the terms of Equation 3.5. Using the fact that the slope equals $\ln(1 + r)$, we find that r equals $e^{.000353296} - 1 =$ .0003534. This means that our best estimate of the annual geometric growth rate over the 1950–80 period for the Cleveland area is 0.03534 percent. Using the fact that the intercept equals $\ln P_0$, we find that $P_0 = e^{13.779} = 964,009.5$. The projected population for 1990 using this approach is

$$1,945,910 = 964,009.5\,(1 + .0003534)^{1990}$$

In the case of exponential growth expressed by Equation 3.6, we have, upon logarithmic transformation,

$$\ln P_t = \ln P_0 + rt$$

We now proceed to fit a straight line to a plot of $\ln P_t$ on the vertical axis, and t on the horizontal axis. The slope of the line is r, and the intercept equals $\ln P_0$. We therefore find that $r = .00035330$ since the calculation of the slope is identical to what we just performed (because we are still plotting $\ln P_i$ versus t). The projected population growth for 1990 under the assumption of exponential growth is

$$1,947,224 = 964,009.5 e^{(.00035330)(1990)}$$

This projection is similar to our projection that assumed geometric growth; in fact the projections should be identical, and the difference between them can be attributed to roundoff error. Note that in both the exponential and geometric cases, these projections are based upon the best-fitting line and do not use the last observed population figure. Alternative projections could be based upon the last observed population values and the estimated slope. For the geometric case, the projected population in 1990 would be $1,898,825(1.0003534) = 1,899,496$, and for the exponential case, the projected population would be $1,898,825 e^{(.00035330)(10)} = 1,905,545$.

The only difference between the cases of exponential growth and geometric growth is the value of r, the per-period growth rate. In the geometric case, the slope is $\ln(1 + r)$; here in the exponential case the slope is r. Differences between the fitted values of r are quite small, and the implications for population growth differ only in the very long run. Note that, for reasons we discussed in Chapter 3, the exponential or continuous rate of annual growth is lower than the geometric or simple annual rate because of the additional "compounding" posited in this model.

If an elementary method of population projection is to be chosen, which should it be? Certainly fixed-rate (geometric and exponential growth) projections are more prevalent in practice than are projections based upon linear extrapolation. Doubling times are somewhat easier to calculate with the exponential model, but the geometric model can make use of a growth rate based upon a period length (e.g., 1 or 10 years) that is consistent with the way that data are collected. However, these are not particularly compelling reasons to choose one over the other. Since geometric and exponential models produce projections that are nearly identical, the choice between geometric and exponential models is primarily a matter of personal choice and convenience.

There is, in fact, little evidence that one particular method results in projections that are better than any other. In any particular scenario, the linear projection may generate projections that are more accurate than either the geometric or exponential projections. Evidence from other studies (e.g., Isserman, 1977; Pant and Starbuck, 1990; Smith and Sincich, 1992) indicates that more complex and sophisticated methods do not perform any better than elementary methods at the task of projection. But more sophisticated models are often preferable since they have the considerable advantages of providing additional information (such as age composition), and they may also be used for analyzing demographic and economic processes (e.g., effects of changes in migration destination choices on future population growth).

6.3 THE SINGLE-REGION COHORT COMPONENT MODEL

In addition to population totals, other demographic information often is available. The most common forms of such additional detail concern the age and sex structure of the population and recent rates of fertility and mortality. Information is also sometimes available on migration, in the form of either net migration, in- and out-migration, or origin-to-destination migration flows. Isserman (1993) notes that in the cohort component model, "[t]he guiding theoretical insight is that demographic events, such as death or migration, happen to people with probabilities that vary systematically with age and sex (and race or ethnic group)." We defer treatment of migration in population projections until the next section; here we focus upon the single-region projection where net migration is assumed to be zero.

As the name implies, there are two principal concepts underlying the cohort component approach. First, the population is subdivided into groups of individuals or cohorts. Typically, the population is disaggregated by age and sex. Most often, 5-year age groups are used. It is common to carry out projections to age 85 and over, implying 18 separate 5-year age groups. Some software and some data tabulations use 75 and over as the last age group, but it is usually desirable to have more detail regarding the age structure of the elderly population, particularly in light of the aging of the baby boom. Cohorts are often further subdivided by race or ethnicity.

With the cohort component approach to projection, there is also a focus on the components of change, since the population change in each cohort is accounted for through births, deaths, and migration. In the case where net migration is assumed equal to zero, survival rates are used to survive cohorts forward in time. In addition, age-specific rates of childbearing are used together with age-disaggregated figures for the size of the female population to project the youngest age group. In Figure 6.2, two successive population pyramids depict this cohort component approach. The population pyramid depicted in the figure shows how cohorts progress from one age group to the next over the course of a projection period. Also conveyed by the figure is the way in which the childbearing age groups in one period influence, through age-specific fertility rates, the size of the youngest cohort in the next period.

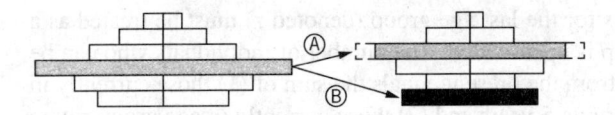

Ⓐ All cohorts, such as ▭, are aged by moving them up one level. Some fraction of each cohort is killed off (▭) by applying age-specific survival rates.

Ⓑ A new cohort, ■, is created by applying age-specific fertility rates to cohorts in the childbearing ages, such as ▭.

FIGURE 6.2 The Cohort Component Concept

Clearly, the task of projection will be easiest when the width of the age group equals the length of the projection period. For example, suppose that the length of the projection period and the age group width both equal their usual value of 5. The number of 10 to 14 year-olds that can be expected in a population 5 years from now (neglecting the effects of migration) is simply the present number of 5 to 9 year-olds, multiplied by the appropriate 5-year survival probability. More generally,

$$p_{x+5}(t+5) = s_x p_x(t) \tag{6.6}$$

where $p_x(t)$ denotes the population in age group $(x, x+4)$ at time t, and where s_x denotes the 5-year survival probability for individuals in age group $(x, x+4)$. Referring back to our discussion of notation in Chapter 3, a more precise notation for these quantities would be $_h p_x(t)$ and $_h s_x$. Because we have the added notation associated with time periods here, in what follows we shall assume $h = 5$ and drop the left-handed h subscript from the notation.

Note also that the survival rates are not a function of time; they are assumed to remain constant for the duration of the projection period.

The only data required to produce a cohort component projection are (1) population data disaggregated by age and sex, (2) data on deaths disaggregated by age and sex, and (3) data on births by age of mother and sex. These data are readily available at the local level, but a few caveats are in order. The source of population data at the local level is typically the decennial census, and hence the base year population data will sometimes be out of date. State health departments typically collect vital statistics down to the census tract level of geography, though there may be significant publication lags.

To compute the survival probabilities, the L terms from a life table are used. These may be calculated from data on annual age-specific death rates for the relevant age groups. Alternatively, life tables from some higher level of geography are sometimes adopted. For example, state-level life tables are sometimes used when none exist at the county level.

The age-specific probability that a member of the population survives for 5 years (recall the discussion in Section 3.5) is

$$s_x = L_{x+5} / L_x$$

Since the population is assumed to be homogeneous (i.e., all individuals are assumed to have the same survival probabilities), s_x may also be interpreted as the proportion of individuals age x to $x + 5$ who survive 5 years to be counted in the next age group.

162   Chapter 6 • Population Projections

- The survival probabilities for the last age group (denoted z) must be treated as a special case since the age group is open-ended. The number of individuals who can be expected in this group 5 years from the present equals the sum of (1) those currently in the previous age group who survive 5 years and (2) those currently in age group z who survive 5 years. (They will still be in the same age group, since it is open-ended.) Suppose that the oldest age group consists of those 80 and over. Then

$$p_{80+}(t+5) = s_{75} p_{75}(t) + s_{80+} p_{80+}(t) \tag{6.7}$$

where the 5-year survival probability for those now age 75 to 79 is specified by using the ratio of the stationary life table population age 80 to 84 to the stationary life table population age 75 to 79:

$$s_{75} = L_{80} / L_{75} \tag{6.8}$$

The 5-year survival probability for those now age 80 and over is specified by using the ratio of the stationary life table population age 85 and over to the stationary life table population age 80 to 84 and over:

$$s_{80+} = T_{85} / T_{80} \tag{6.9}$$

If, however, life table data are only available to age 80 and above, then Equations 6.7 through 6.9 are not applicable, and we may use the following as an alternative:

$$p_{80+}(t+5) = s_{75+} p_{75+}(t) \tag{6.10}$$

where

$$s_{75+} = T_{80} / T_{75}$$

There appears to be much confusion over the proper treatment of the last age group. Many popular texts assume that projection of the last age group is no different from other age groups:

$$p_{80+}(t+5) = s_{75-79} p_{75-79}(t) \tag{6.11}$$

where

$$s_{75-79} = L_{80+} / L_{75-79} \tag{6.12}$$

This approach may lead to serious inaccuracies. First, the survival "probability" given by Equation 6.12 may yield an uninterpretable value greater than 1. Second, Equations 6.11 and 6.12 often yield short-run projections of the oldest age group that are much too high. The international population projections by Keyfitz and Flieger (1991), for example, use this approach; the projections of the oldest age group (85+) are often more than 15 percent too high. (See Table 6.1.)

Special attention also needs to be given to the projection of the youngest age group. (See Figure 6.2.) The number of people who will be alive in the 0 to 4 age group 5 years from now are those infants and toddlers who were born during the projection period who survive to the end of the period. The projected population in the 0 to 4 age group is derived by summing the products of age-specific birth rates and age-specific female populations over all childbearing age groups:

Table 6.1 Percent Errors in 5-Year Projections of Elderly Female Population

Country	Projected Female Population 85 years and over (000s)	Percent Error
Australia	111	10.9%
Austria	82	11.4
Belgium (1989)	114	15.8
Canada	219	18.9
Denmark	60	12.5
England/Wales	594	10.6
Federal Republic of Germany	751	16.2
Finland	43	16.8
France	691	13.1
German Democratic Republic	135	4.5
Greece	71	8.4
Hungary	68	6.9
Italy (1988)	296	16.8
Netherlands	150	18.3
Norway	52	14.3
Puerto Rico	21	8.1
Spain (1988)	296	16.8
Sweden	112	15.1
Switzerland (1991)	95	18.5
Scotland	55	11.1
Soviet Union (1992)	2,947	35.0
United States	2,486	13.4

Note: Projection year is 1990 unless otherwise indicated.

Source: Keyfitz and Flieger (1991) and authors' calculations

$$p_0(t+5) = \sum_{x=\alpha}^{\beta} b_x p_x(t) \qquad (6.13)$$

where b_x is the age-specific birth rate (usually assumed to be constant over time) for females in the 5-year age group beginning with age x, and $p_0(t+5)$ is the notation for the population age 0 to 4 at time $t+5$. The first and last childbearing age groups are denoted by α and β, respectively. Equation 6.13 reflects the fact that the projected population of the first age group is the sum of the contributions to births from females in all childbearing age groups.

How is b_x derived from available data on fertility? We need to account for the fact that toward the beginning of the period, most of the cohort age $(x, x+4)$ will be bearing children at an annual rate given by the age-specific fertility rate, F_x. However, toward the end of the period, most of the female population that has survived will be bearing children at the rate given by F_{x+5}. In practice, we take an average of these two rates

$$\frac{1}{2}\left(F_x + \frac{L_{x+5}}{L_x} F_{x+5}\right) \tag{6.14}$$

where the ratio of L values accounts for the survival of women from one childbearing age to the next.

Since the F's are *annual* rates, we need to multiply Equation 6.14 by 5, to reflect the fact that women in age group $(x, x+4)$ are bearing children at this average rate throughout the projection period. Finally, the survival of the infants born over the 5-year period until the end of the period must be accounted for. From Chapter 3, the appropriate survival term is $L_0 / 5l_0$. Therefore, we have

$$b_x = \frac{L_0}{2l_0} \sum_x \left[F_x + \frac{L_{x+5}}{L_x} F_{x+5}\right]$$

where the 5s have canceled. To summarize, the projected population age 0 to 4 is

$$p_0(t+5) = \frac{L_0}{2l_0} \sum_x \left[F_x + F_{x+5}\left(\frac{L_{x+5}}{L_x}\right)\right] p_x(t) = \sum_x b_x p_x(t) \tag{6.15}$$

Equations 6.6 and 6.15 may now be taken together with either Equation 6.7 or 6.10 as a 5-year, age-disaggregated projection of the population, which ignores the effects of migration. The projected population may then be used together with the same age-specific fertility and mortality rates to generate a projection of the population for the following 5-year period. This iterative procedure may be carried out as long as desired to generate longer-run projections of population.

It is important to recognize that projections are usually produced separately for each sex since males and females have very different sets of survival rates. In carrying out the projection of the size of the youngest age cohort, the result found from Equation 6.15 needs to be multiplied by the proportion of all births that are of the sex being projected.

A particularly desirable characteristic of the cohort component model is the relatively small amount of data required for input. The only data requirements are (a) an age-disaggregated initial population, (b) age-specific fertility rates, and (c) age-specific mortality rates.

Matrix Form of the Cohort Component Model

It is often desirable to express the cohort component model in matrix terms. (See Appendix D for an overview of matrices.) The populations of each of the n age groups at time t are arranged in order from youngest to oldest in a column vector with n elements which is denoted $\mathbf{p}(t)$. The growth matrix $\mathbf{G}$, has n rows and n columns, and contains the birth rates (b_x) and the survival probabilities (s_x).[1] The structure of the $\mathbf{G}$ matrix is given in Figure 6.3. The first row of the matrix contains the birth rates, with

---
[1]This matrix is also sometimes known as the *Leslie matrix* after its originator (Leslie, 1945).

the b_x terms occupying the columns corresponding to the childbearing age groups. Note that the first several columns and the last several columns contain zeros, reflecting the fact that childbearing occurs only over a limited number of years during the middle of a female's life. Multiplying the first row of **G** by the column vector **p** implies summing the products $b_x p_x$, which results in the projected population age 0 to 4 at time $t + 5$.

The survival probabilities are arranged in a diagonal fashion, immediately below the main diagonal of the matrix. For $i > 1$, the element in row i and column $i - 1$ contains the probability of survival over the 5-year period, from age group $i - 1$ to age group i (where age groups are numbered in order, with 1 representing the youngest age group, and n the oldest). For example, the element in row 3, column 2 is the probability of survival from age group 2 (5 to 9 years old) to age group 3 (10 to 14 years old). The element in the last row and last column represents the five-year survival probability for those beginning the projection period in the last age group. All other entries in each row are set equal to zero (since people do not age into younger age groups, nor is there a way to skip an age group).

The advantage of the matrix form is not only the compactness of its expression, but also the ability to readily generate projections with longer time horizons. Using the rules of matrix multiplication,

$$\mathbf{p}(t + 5) = \mathbf{G}\mathbf{p}(t) \tag{6.16}$$

where $\mathbf{p}(t + 5)$ is the age-disaggregated population vector at time $t + 5$ and **G** is the matrix of five-year birth rates and survival ratios. Multiplying any row of **G** by the column vector **p** using the rules of matrix multiplication will result in a projection of a particular age group for time $t + 5$. In fact, the matrix form in Equation 6.16 may be viewed as a compact way of expressing the set of equations in Equations 6.6 and 6.15.

Also,

$$\mathbf{p}(t + 10) = \mathbf{G}\mathbf{p}(t + 5) = \mathbf{G}\mathbf{G}\,\mathbf{p}(t) = \mathbf{G}^2\mathbf{p}(t)$$

and, in general,

$$\mathbf{p}(t + 5u) = \mathbf{G}^u \mathbf{p}(t)$$

Until now, we have focused upon the use of the cohort component model as a short-run projection device (though we have yet to treat the migration component of population change). Planning agencies often use the model to make projections for

$$\begin{bmatrix} 0 & b_5 & b_{10} & b_{15}\ldots b_{45} & 0\ldots & 0 \\ s_0=\frac{L_5}{L_0} & 0 & 0 & 0\ldots\ldots & & 0 \\ 0 & s_5=\frac{L_{10}}{L_5} & 0 & 0\ldots\ldots & & 0 \\ 0 & 0 & s_{10}=\frac{L_{15}}{L_{10}} & 0\ldots\ldots & & 0 \\ 0 & 0 & 0 & s_{15}=\frac{L_{20}}{L_{15}} & & \vdots \\ \vdots & \vdots & & & \ddots & \vdots \\ 0 & 0\ldots\ldots & & & s_{75}=\frac{L_{80}}{L_{75}} & s_{80+}=\frac{T_{85}}{T_{80}} \end{bmatrix}$$

FIGURE 6.3 The Leslie Matrix (**G**) Used in Single-Region Cohort Component Projections

periods of around 20 years. Longer time horizons are sometimes used, but one must then call into question the validity of the assumptions of fixed regimes of fertility and mortality. (See, e.g., Keyfitz, 1981.)

Still, the cohort component model exhibits interesting behavior in the long run, and it provides an indication of where the population is headed, should present rates continue. If the fertility and mortality rates are assumed to hold over a long period of time, the population will eventually converge to a stable age structure (i.e., an age structure that does not change over time). Though this age structure does not serve well as a long-term population projection, it may be viewed as a target toward which the current population system is headed. Of course, as the underlying fertility and mortality rates change, the location of the target changes.

An additional characteristic of the long-run equilibrium age structure is that it is entirely dependent upon the elements of the growth matrix **G** (that is, the fertility and mortality rates) and is entirely independent of the initial age composition. In addition, the population will eventually grow (or decline) at a constant per-period rate. Thus in the long run, a population characterized by constant age-specific fertility and mortality rates will grow or decline geometrically.

Example

Suppose that a planner is provided with the data given in the first four columns of Table 6.2. The planner is asked to produce a 5-year projection of the population, ignoring the effects of migration (or equivalently, assuming that net migration for each age group over the 5-year period is zero). Use the following steps to construct the projection. (We limit our illustration to the female population only.)

1. Survival probabilities are computed for each age group by forming the ratio of the value of L in the next age group to the value of L in the age group of interest. That is, $s_x = {}_5L_{x+5} / {}_5L_x$. The 5-year survival probability for individuals age 70 and over is found from $S_{70+} = T_{75} / T_{70} = L_{75+} / (L_{75} + {}_5L_{70}) = 643{,}765 / 1{,}015{,}092 = .63419$. The result of this step is given in column 5.

2. The b_x terms are computed using Equation 6.15. Column 6 displays the results of this step. For example, $b_{20} = (495{,}349 / 200{,}000)[.1082 + .1092(.99846)] = .53803$ (where l_0 is 100,000). The terms are calculated similarly for other age groups. It is worth noting that the population currently age 5 to 9 produces children during the 5-year period since they age into the 10 to 14 year-old age group:

$$b_5 = (495{,}349/200{,}000)[0 + (.0013)(.99943)] = .00322$$

For the last childbearing age group,

$$b_{45} = (495{,}349 / 200{,}000)[.0002 + (0)(.98630)] = .0005$$

3. The projected female population age 0 to 4 is calculated using Equation 6.15 and multiplying by the proportion of total births that are female:

$$[(8{,}600)(.00322) + (8{,}000)(.14182) + \ldots + (6{,}300)(.00050)] / 2 = 9{,}067$$

where we have assumed here that half of all births are female.

4. The projected female population in each of the remaining age groups is derived by multiplying the current female population by the appropriate survival probability. For

TABLE 6.2 Illustrative Cohort Survival Projection

Age (1)	Female Population (2)	$_5L_x$ (3)	Age-Specific Fertility Rate (4)	s_x (5)	b_x (6)	Projected Population (7)
<5	8,900	495,349	—	—	.99904	9,067
5–9	8,600	494,875	—	.99943	.00322	8,891
10–14	8,000	494,593	.0013	.99928	.14182	8,595
15–19	9,000	494,239	.0560	.99888	.40638	7,994
20–24	9,900	493,686	.1082	.99846	.53803	8,990
25–29	10,900	492,927	.1092	.99796	.44101	9,885
30–34	10,600	491,923	.0069	.99709	.23016	10,878
35–39	9,400	490,493	.0024	.99537	.06955	10,569
40–44	7,900	488,223	.0041	.99191	.01065	9,356
45–49	6,300	484,272	.0002	.98630	.00050	7,836
50–54	5,600	477,637	—	—	.97752	6,214
55–59	5,800	466,900	—	—	.96283	5,474
60–64	5,800	449,546	—	—	.93481	5,584
65–69	5,200	420,241	—	—	.88360	5,422
70–74	4,600	371,327	—	—	.63419	4,595
75+	7,800	643,765	—	—	—	7,864
Total	124,300					127,214

example, the 5-year projection for age group 25 to 29 is $(9,900)(.99846) = 9,885$. Column 7 portrays the results, with all figures rounded to the nearest integer. The projected population in the last, open-ended age group is made using Equation 6.10 (with 75+ instead of 80+ representing the last age group):

$$p_{80+}(t + 5) = p_{75+}(t)(T_{75}/T_{70}) = (4,600 + 7,800)(.63419) = 7,869$$

We repeat here our earlier suggestion to retain enough significant digits throughout the calculation. Rounding off fertility and survival rates to two or three digits may lead to substantial roundoff errors, and these errors will cumulate over succeeding projection periods.

Figure 6.4 shows how the fertility and survival terms are arranged in the growth matrix, **G**. A projection for year 10 may be derived by premultiplying the 5-year projection by **G**. A projection for year 15 may be derived by premultiplying the 10-year projection by **G**, and so on. The reader may (or may not!) wish to verify that if this process is repeated for a sufficiently long period of time, the age distribution of the population approaches the proportions in Table 6.3. The birth rate, however, is not sufficiently high to sustain long-run population growth, and after a short-term rise, the population eventually dwindles in size to zero. This may be verified by noting that the total fertility rate, which equals five times the sum of the age-specific fertility rates, is 1.86.

$$
\begin{bmatrix}
0 & .00322 & .14182 & .40638 & .53803 & .44101 & .23016 & .06955 & .01065 & .00050 & 0 & 0 & 0 & 0 & 0 & 0 \\
.99904 & 0 & 0 & 0 & 0 & 0 & 0 & 0 & 0 & 0 & 0 & 0 & 0 & 0 & 0 & 0 \\
0 & .99943 & 0 & 0 & 0 & 0 & 0 & 0 & 0 & 0 & 0 & 0 & 0 & 0 & 0 & 0 \\
0 & 0 & .99928 & 0 & 0 & 0 & 0 & 0 & 0 & 0 & 0 & 0 & 0 & 0 & 0 & 0 \\
0 & 0 & 0 & .99888 & 0 & 0 & 0 & 0 & 0 & 0 & 0 & 0 & 0 & 0 & 0 & 0 \\
0 & 0 & 0 & 0 & .99846 & 0 & 0 & 0 & 0 & 0 & 0 & 0 & 0 & 0 & 0 & 0 \\
0 & 0 & 0 & 0 & 0 & .99796 & 0 & 0 & 0 & 0 & 0 & 0 & 0 & 0 & 0 & 0 \\
0 & 0 & 0 & 0 & 0 & 0 & .99709 & 0 & 0 & 0 & 0 & 0 & 0 & 0 & 0 & 0 \\
0 & 0 & 0 & 0 & 0 & 0 & 0 & .99537 & 0 & 0 & 0 & 0 & 0 & 0 & 0 & 0 \\
0 & 0 & 0 & 0 & 0 & 0 & 0 & 0 & .99191 & 0 & 0 & 0 & 0 & 0 & 0 & 0 \\
0 & 0 & 0 & 0 & 0 & 0 & 0 & 0 & 0 & .98630 & 0 & 0 & 0 & 0 & 0 & 0 \\
0 & 0 & 0 & 0 & 0 & 0 & 0 & 0 & 0 & 0 & .97752 & 0 & 0 & 0 & 0 & 0 \\
0 & 0 & 0 & 0 & 0 & 0 & 0 & 0 & 0 & 0 & 0 & .96283 & 0 & 0 & 0 & 0 \\
0 & 0 & 0 & 0 & 0 & 0 & 0 & 0 & 0 & 0 & 0 & 0 & .93481 & 0 & 0 & 0 \\
0 & 0 & 0 & 0 & 0 & 0 & 0 & 0 & 0 & 0 & 0 & 0 & 0 & .88360 & 0 & 0 \\
0 & 0 & 0 & 0 & 0 & 0 & 0 & 0 & 0 & 0 & 0 & 0 & 0 & 0 & .63419 & .63419
\end{bmatrix}
$$

FIGURE 6.4 The Leslie Matrix Derived from the Hypothetical Data in Table 6.2

TABLE 6.3 Long-Run Age Distribution of the Hypothetical Population

Age	Long-run Proportions	Age	Long-run Proportions
<5	.0562	40–44	.0632
5–9	.0571	45–49	.0637
10–14	.0580	50–54	.0639
15–19	.0589	55–59	.0635
20–24	.0598	60–64	.0622
25–29	.0607	65–69	.0591
30–34	.0616	70–74	.0531
35–39	.0625	75+	.0963

6.4 COHORT COMPONENT MODELS WITH MIGRATION

A feature that is clearly missing from the previous discussion of the cohort survival model is migration. Since migration is often the most important component of population change at the local level, it must be accounted for.

Cohort Survival with Constant Net Migration Rates

One of the most straightforward ways to include migration in a population projection model is to add age-specific net migration rates to the survival rates in the cohort survival model. Figure 6.5 shows how the growth matrix of the cohort survival model is modified. The matrix is identical to that in Figure 6.3, with the exception that net migration rate terms have been added to the subdiagonal elements. The rate in the second row and first column, for example, represents the net migration rate for individuals who are in the first age group at the beginning of the period and in the second age group at the end of the period. Estimates of the net migration rate are derived by dividing the net migration during a period for individuals who are in a specified age group by the population of that age group at the beginning of the period:

$$n_x = N_x / p_x$$

where n_x is the net migration rate for age group x, N_x is the net migration for individuals in age group x at the beginning of the period, and p_x is the population of age group x at the beginning of the period.

This model as it is usually presented by, for example, Rogers (1971) and Wilson (1974) contains no provision for the net migration of individuals born during the period. This, however, could be achieved by adding a net migration rate term to the birth rate terms in the first row of Figure 6.5.

As Rogers (1990) and Isserman (1993) note, the constant net migration rate approach is logically inconsistent since the net migration rate is not based upon the proper at-risk population. Dividing net migration to a region by the region's population is equivalent to dividing both in- and out-migration by the region's population. But the population at risk of migrating in consists of all people *outside* of the region, and not people within it. The effect of this shortcoming is that growing regions continue to grow despite the fact that the population supply in other regions is declining

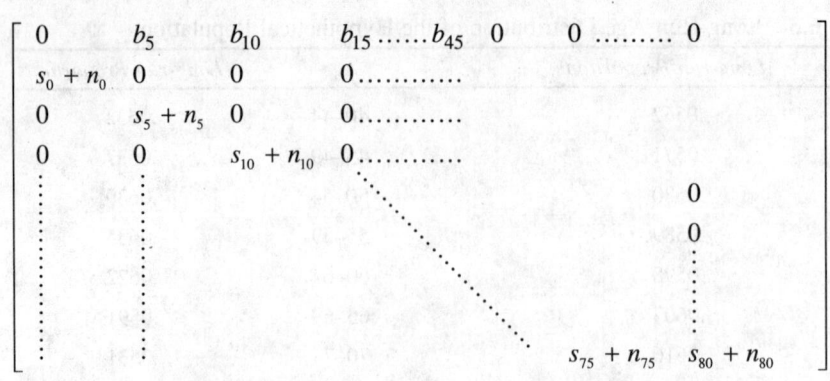

$$\begin{bmatrix} 0 & b_5 & b_{10} & b_{15} \cdots b_{45} & 0 & 0 \cdots \cdots 0 \\ s_0+n_0 & 0 & 0 & 0 \cdots \cdots \cdots & & \\ 0 & s_5+n_5 & 0 & 0 \cdots \cdots \cdots & & \\ 0 & 0 & s_{10}+n_{10} & 0 \cdots \cdots \cdots & & \\ \vdots & \vdots & & \ddots & & 0 \\ & & & & & 0 \\ \vdots & \vdots & & & & \vdots \\ & & & & s_{75}+n_{75} & s_{80}+n_{80} \end{bmatrix}$$

FIGURE 6.5 The Leslie Matrix Modified to Include Net Migration Rates

(due to the net migration to the growing region). Likewise, declining regions continue to decline, despite the increased supply of people (and hence potential migrants) in growing regions. We discuss this further in our treatment of migration in Chapter 7.

A related drawback of the net migration rate model is that it fails to account for interactions between regions. In the next subsection we describe the Markov model, which focuses on the destination choice probabilities of individuals. In that approach, data on the migration flows between regions are required, and hence interactions between regions are captured.

The Markov Model for Population Redistribution

In this section we temporarily ignore the age structure of the population to focus on the internal redistribution of individuals between regions. As in Section 6.3, where migration and population redistribution were ignored to focus on age distribution, this analysis should be considered a partial one. These partial analyses provide important insights into changing age and geographical distributions. In the following subsection we present an interregional cohort component model that combines migration and aging.

Consider a system divided into n regions. As described in Chapter 4, we denote the probability that an individual migrates from region i to region j by p_{ij}. (Recall that p_{ii} represents the probability that an individual in region i at the beginning of the period is also in region i at the end of the period, and that this group includes both nonmigrants and intraregional movers.) Let $P_i(t)$ represent the population of region i at time t. Then, ignoring births and deaths to focus entirely on population redistribution, the population in region i at time $t + 1$ is given as the population staying in i, plus the sum of the populations moving from other regions j:

$$P_i(t+1) = \sum_{j=1}^{n} P_j(t) p_{ji}$$

where the sum over j includes i.

Alternatively, using matrix notation, if **p** is a row vector containing the n regional populations at time t, and **P** is an $n \times n$ matrix containing the set of transition probabilities p_{ij}, then

$$\mathbf{p}(t+1) = \mathbf{p}(t)\mathbf{P}$$

6.4 Cohort Component Models with Migration

Again, one advantage of this notation is its conciseness. Also, the generation of regional population projections is straightforward. Thus the population two periods ahead may be found by postmultiplying the base period population by the square of the transition probability matrix:

$$\mathbf{p}(t + 2) = \mathbf{p}(t + 1)\, \mathbf{P} = \mathbf{p}(t)\, \mathbf{P}^2$$

More generally, the population u periods ahead is found by postmultiplying the base period population vector by the transition matrix raised to the power u:

$$\mathbf{p}(t + u) = \mathbf{p}(t)\mathbf{P}^u$$

Note that this Markov approach assumes that (1) the probabilities of movement between regions do not change over time, (2) the population is homogeneous in the sense that each individual is governed by the same set of probabilities, (3) the probabilities apply to a fixed period of time, and (4) the Markov property holds. The Markov property states that the next destination chosen depends only on the present region and not on the location of prior residences.

Of course, many of these assumptions do not seem realistic. We know that probabilities of interregional movement change over time. Likewise, different groups of people within a region surely have very different likelihoods of moving to various destinations. We also are aware that the destinations chosen by migrants are often strongly tied to their past regions of residence. (Recall our discussion in Chapter 4.) Return migration of individuals to former regions of residence is a well-documented phenomenon so we do not expect the fourth assumption to be satisfied.

What then is the purpose of carrying out an exercise where the assumptions do not precisely hold? First, the approach can still give reasonable short-run forecasts (at least compared with other projections) even when the assumptions are *not* satisfied. Either the deviations from the assumptions are not large, or the effect of the deviations on the projections is not large enough to produce projections that are wildly out of line with projections produced by other methods. Second, the Markov model can provide us with a baseline projection against which we may compare other forecasts (Rogers and Woodward, 1991).

In cases where observed migration flow data are available for more than one time period, it is common practice to adopt the most recent flows in computing migration probabilities. In other cases, however, we may know that interregional migration probabilities are changing, and it is important to model and project such changes. Methods designed to account for changes in migration probabilities are detailed in Chapter 7.

Example

The data in Table 6.4 depict 1987–88 migration flows. For purposes of illustration, we take the initial 1987 population of each region to equal the row sums of Table 6.4. (Note that the actual initial populations of each region would include others, such as those who died during the period and those who emigrated from the country.) The diagonal elements include both individuals who did not move and individuals who moved within the region. The transition probability matrix (**P**) in Table 6.5 is derived by dividing each element by its row sum.

The projected population in 1988 is given by the column sum, since we are neglecting the effects of natural increase and international migration. Note that the

TABLE 6.4 Interregional Migration Flows in the United States, 1987–88

Origin	Northeast	Midwest	South	West	Total
Northeast	50,110,517	108,543	484,351	149,590	50,853,001
Midwest	118,800	58,564,428	513,225	285,537	59,481,990
South	318,247	448,968	81,836,449	415,819	83,019,483
West	133,593	235,434	375,717	48,718,974	49,462,718
Total	50,681,157	59,357,373	83,209,742	49,569,920	242,817,192

Source: Internal Revenue Service.
Note: Because of incomplete reporting, estimates of migration flows from the IRS data are too low. The flows here have been adjusted by the authors to be consistent with statewide population totals in the origin year.

1988 population may also be reproduced by postmultiplying the row vector of 1987 populations by the transition probability matrix in Table 6.5:

$$\mathbf{p}(1988) = \mathbf{p}(1987)\ \mathbf{P} = [50{,}681{,}157\ \ 59{,}357{,}373\ \ 83{,}209{,}742\ \ 49{,}569{,}920]$$

Projection of the 1989 population proceeds by postmultiplying the row vector of 1988 populations by the 1987–88 probabilities, thereby assuming that they remain fixed for the period 1988–89:

$$\mathbf{p}(1989) = \mathbf{p}(1988)\ \mathbf{P} = [50{,}512{,}589\ \ 59{,}235{,}846\ \ 83{,}395{,}385\ \ 49{,}674{,}372]$$

Note that this same result could also be found by

$$\mathbf{p}(1989) = \mathbf{p}(1987)\ \mathbf{P}^2$$

The elements of $\mathbf{P}^2$ have a clear interpretation, as do the elements of higher powers of $\mathbf{P}$. Specifically, the element in row i and column j of the matrix $\mathbf{P}^s$ is the probability that an individual starting in region i is residing in region j after s periods. Alternatively, it may be viewed as the proportion of people in i during the base period who reside in j after s periods.

Carrying this out one more step, the projected regional distribution of the 1990 population is

$$\mathbf{p}(1990) = \mathbf{p}(1989)\mathbf{P} = \mathbf{p}(1987)\mathbf{P}^3 = [50{,}347{,}234\ \ 59{,}117{,}335\ \ 83{,}576{,}521\ \ 49{,}777{,}102]$$

If this iterative procedure is carried out for enough into the future, the regional populations approach the following steady-state or equilibrium distribution:

$$[41{,}614{,}919\ \ 54{,}756{,}238\ \ 90{,}689{,}682\ \ 55{,}757{,}353]$$

Since the populations converge in the long run, so too must the powers of $\mathbf{P}$. To see this,

$$\mathbf{p}(t + u) = \mathbf{p}(t)\ \mathbf{P}^u$$
$$\mathbf{p}(t + u + 1) = \mathbf{p}(t)\ \mathbf{P}^{u+1}$$

Because the left-hand sides are approximately equal in the long run, the right-hand sides must also be. Furthermore, the rows of $\mathbf{P}^u$ are identical, and they contain the equilibrium proportions of population in each region. In our example, for large values of u,

$$\mathbf{P}^u \approx \begin{bmatrix} .1714 & .2255 & .3735 & .2296 \\ .1714 & .2255 & .3735 & .2296 \\ .1714 & .2255 & .3735 & .2296 \\ .1714 & .2255 & .3735 & .2296 \end{bmatrix}$$

Thus the percentage distributions of population across regions 1 through 4 in the long run are 17.14 percent, 22.55 percent, 37.35 percent, and 22.96 percent, respectively.

More on the Long-Run Distribution

We have characterized the long-run distribution of the population across regions as one that is in equilibrium since the long-run proportions of the population in each region approach constant values. The equilibrium is a dynamic one in the sense that people still move between regions, but the net migration for each region is zero.

We have seen that one way to find the equilibrium distribution is to simply carry out the iterative projection procedure for a large number of periods. We now explore another way to find the long-run equilibrium distribution. The purpose of this exploration is twofold: to describe a more efficient method and to convey a deeper understanding of the meaning of equilibrium.

Suppose that we are examining the population distribution within a metropolitan area, and that we are interested in migration flows between the central city and the suburbs. Our data consist of a 1×2 row vector of central city and suburban populations, and a 2×2 transition probability matrix describing the probabilities of transitions between the two regions. Since we are ignoring the effects of birth, death, and migration between the metropolitan area and the outside world, the row sums of $\mathbf{P}$ equal 1, and the total population of the metropolitan area does not change.

We begin by dividing the central city and suburban populations by the total metropolitan population to express their populations as fractions. These population shares are expressed in the row vector $[y_{cc} \; y_s]$. At equilibrium these population shares do not change over time. Therefore in the long run

$$[y_{cc} \; y_s] = [y_{cc} \; y_s] \begin{bmatrix} p_{cc.cc} & p_{cc.s} \\ p_{s.cc} & p_{s.s} \end{bmatrix}$$

TABLE 6.5 Interstate Transition Probabilities, 1987–88

	\multicolumn{4}{c}{*Destination*}
Origin	*Northeast*	*Midwest*	*South*	*West*
Northeast	.98540	.00213	.00952	.00294
Midwest	.00200	.98457	.00863	.00480
South	.00383	.00541	.98575	.00501
West	.00270	.00476	.00760	.98494

Writing this matrix equation out yields two equations:

$$y_{cc} = y_{cc}p_{cc,cc} + y_s p_{s,cc}$$
$$y_s = y_{cc}p_{cc,s} + y_s p_{s,s}$$

We also know that the population shares must total; that is, $y_{cc} + y_s = 1$. Since the transition probabilities are known, there are three equations and two unknowns (y_{cc} and y_s). Because the first two equations are redundant (for example, we do not really need the second equation, since by substituting $p_{cc,s} = 1 - p_{cc,cc}$ and $p_{s,s} = 1 - p_{s,cc}$ in the second equation, we obtain the first equation), we can use either of the first two equations plus the identity $y_{cc} + y_s = 1$ to find the two unknowns in terms of the transition probabilities. By substituting $y_s = 1 - y_{cc}$ into the first equation, we find that

$$y_{cc} = p_{s,cc} / (1 + p_{s,cc} - p_{cc,cc})$$

The proportion y_s may be found either from $y_s = 1 - y_{cc}$, or from

$$y_s = p_{cc,s} / (1 + p_{cc,s} - p_{ss})$$

For example, suppose that the transition probability matrix is

To:	Central city	Suburbs
From:		
Central city	.8	.2
Suburbs	.1	.9

The long-run proportion of the population in the central city is $.1 / (1 + .1 - .8) = 1/3$, and the long-run proportion in the suburbs is $1 - 1/3 = 2/3$.

For examples with three regions, a similar procedure is followed. The 3×3 transition matrix and the 1×3 vector of population shares results in three equations, one of which is redundant (since each row sums to 1, if we know any two elements in the row, we also know the third). Adding the identity that expresses the fact that the population shares must add to 1 yields three equations and the three unknown population shares. For more than three regions, a computer program would be used to solve the system of equations.

The Interregional Cohort Component Model

Rogers (1971, 1975, 1985) developed a multiregional cohort survival model that combines the features of the models discussed in the previous two sections. The model uses both the cohort survival concept to age individuals from one age group to the next, and Markov transition probabilities to model the flow of individuals between regions.

Figure 6.6 (from Isserman, 1993) nicely summarizes how the single region cohort component model (Figure 6.2) may be generalized to more than one region. In the figure, other regions have been combined into a single category called "rest of the United States." This so-called biregional projection example portrays the flows of individuals between age categories and regions. It is straightforward to extend this to

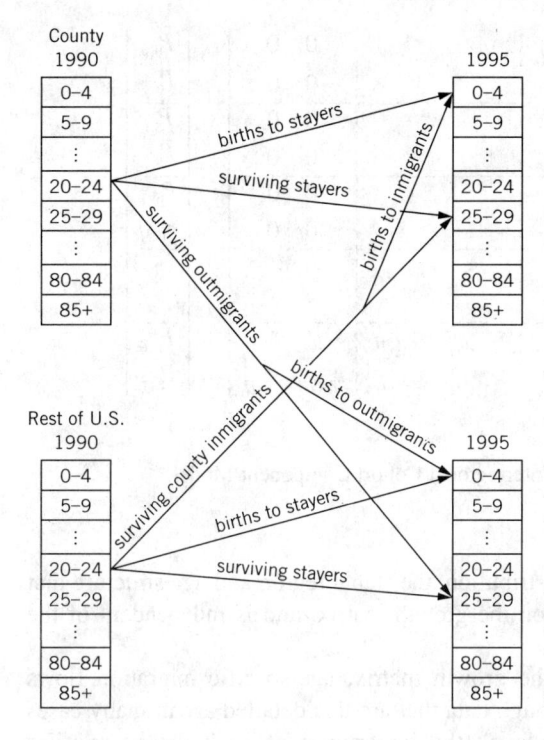

FIGURE 6.6 The Interregional Cohort Component Concept

Source: Isserman (1993)
Reprinted by permission.

include flows between more than two regions (though the diagram would of course be more complex and would lose much of its pedagogic value).

Figure 6.7 summarizes the matrix form of the multiregional model. The population vector is subdivided first into z age groups; for each age group there is a set of populations disaggregated by region (for purposes of illustration, Figure 6.7 shows two regions—i and j). In the population vector, we have used the first subscript to denote the age group, and the second to denote the region.

The structure of the growth matrix also results from a simultaneous consideration of the age and location of individuals. Note that the structure of the submatrices is identical to that of the cohort survival model, with terms representing birth rates in the top row of submatrices, and terms representing survival along the subdiagonal set of submatrices. Here, however, migration is also accounted for. Within each 2×2 submatrix, the diagonal terms represent nonmigrants, and the nondiagonal terms represent migrants. The first subscript on the s terms represents the age group at the beginning of the period (where we have denoted the beginning age of the age group, and have omitted the $h = 5$ subscript), and the last two subscripts represent the regions of destination and origin, respectively. The structure of the matrix as it is portrayed assumes that the width of the age group equals the length of the projection period.

The procedure for carrying out short-run projections is precisely the same as it was in the previous two sections—the growth matrix is assumed to remain constant to derive projected population vectors.

Projections derived via the multiregional model also ultimately attain an equilibrium state. In the long run, the proportion of the total population in any particular age–region category becomes constant. Thus the age structure and the spatial distribution of the population become stable. As is the case for the cohort survival model and

$$\begin{bmatrix} 0 & 0 & b_{5,ii} & b_{5,ji} & b_{10,ii} & b_{10,ji} & & & 0 & 0 \\ 0 & 0 & b_{5,ij} & b_{5,jj} & b_{10,ij} & b_{10,jj} & \cdots & & 0 & 0 \\ s_{5,ii} & s_{5,ji} & 0 & 0 & 0 & 0 & & & 0 & 0 \\ s_{5,ij} & s_{5,jj} & 0 & 0 & 0 & 0 & \cdots & & 0 & 0 \\ 0 & 0 & s_{10,ii} & s_{10,ji} & 0 & 0 & & & 0 & 0 \\ 0 & 0 & s_{10,ij} & s_{10,jj} & 0 & 0 & & & 0 & 0 \\ & & & & & \ddots & & & & \\ 0 & 0 & 0 & 0 & & \cdots & s_{z-1,ii} & s_{z-1,ji} & s_{z,ii} & s_{z,ji} \\ 0 & 0 & 0 & 0 & & & s_{z-1,ij} & s_{z-1,jj} & s_{z,ij} & s_{z,jj} \end{bmatrix} \begin{bmatrix} P_{0,i} \\ P_{0,j} \\ P_{5,i} \\ P_{5,j} \\ P_{10,i} \\ P_{10,j} \\ \vdots \\ P_{z,i} \\ P_{z,j} \end{bmatrix}$$

G   P

FIGURE 6.7 The Matrix Form of the Interregional Cohort Component Model

the Markov model of population redistribution, the stable region and age structure that is achieved is entirely dependent upon the growth matrix, and is independent of the initial population vector.

To estimate the elements of the growth matrix, age-specific migration flows between locations are required. Although data that are this detailed are in many cases unavailable, especially at the local level, the importance of such data cannot be overemphasized. We must remember that regions should not be viewed as isolated entities. Hence projection models that assume constant rates of net migration, while they will continue to be used, contain no mechanism for accounting for the interactions between regions. And, as Smith (1986) and Isserman (1993) demonstrate, the biases inherent in constant net migration rate projections (e.g., the tendency for growing regions to have projected growth that is too rapid) can be significant.

Accounting-Based Population Projections

Until this point in the chapter we have avoided a full treatment of the issues associated with population accounting that we examined in Chapter 3.

In the cohort survival model, life table quantities were used to derive the sub-diagonal survival probabilities and to find the proportion of newborns who survived until the end of the period. Those quantities, the ratio of successive "L" values from the life table, represent the probability that an individual starting in one age group at the beginning of the period is alive and in the next age group at the beginning of the following period. This is precisely the quantity that should be used in the cohort survival model. However, recall that the probabilities of dying (the q values) were derived from death rates that were in turn found by dividing deaths by midperiod population. To find the probability of dying, the midperiod population was converted to an initial period population by adding 1/2 of all deaths to the midperiod population. A more precise method would call for other terms to be added to and subtracted from the midperiod population to account for both migration into and out of the region, and the joint events of migrating and dying. Interested readers should consult Rees and Wilson (1977) for additional details.

In the Markov model, a transition probability between two regions is defined as the number of surviving migrants to a destination region, divided by the initial population of the origin region. Since migration data typically refer only to survivors, and exclude the migrations of nonsurvivors, this allows for the computation of correct transition probabilities.

In Chapter 7 we discuss alternative ways to represent interregional migration linkages.

Other Applications of the Cohort Component Concept

The use of the cohort component concept is not limited to the generation of population projections. The concept is frequently applied to a wide variety of other planning applications. Indeed, this point echoes the more general theme that many of the techniques and methods described in this book may be applied not only to one specific situation, but rather to a multifarious assortment of problems. It is therefore essential to understand the concepts underlying the methods, as well as the methods themselves, to be fully prepared to use the methods in other circumstances.

In the remainder of this chapter, we review two examples illustrating alternative uses of the cohort component concept. The first is an application to school enrollment projection. The second example demonstrates how the concepts may be used in understanding the evolution of the demographic structure of organizations.

6.5 SCHOOL ENROLLMENT PROJECTIONS

Virtually all school districts prepare grade-specific enrollment projections as part of their planning process. A historical record of grade-specific enrollment in the district forms the basis of most projections (Table 6.6). Two alternative projection methods are commonly employed: a method based on *grade progression ratios* and the *housing-unit method*.

TABLE 6.6 Historical School District Enrollments for Amherst, New York.

Grade	1985	1986	1987	1988	1989	1990	1991
K	197	164	217	173	158	205	202
1	207	218	189	256	203	186	232
2	210	191	226	194	254	201	197
3	190	210	203	228	192	242	208
4	180	193	202	208	225	186	244
5	196	171	189	205	202	215	180
6	196	197	164	185	208	193	218
7	206	201	196	174	186	210	202
8	209	213	196	199	179	185	210
9	268	228	202	201	204	192	188
10	274	254	220	206	205	211	183
11	309	266	251	202	199	205	205
12	295	317	269	261	210	204	207

The Grade Progression Ratio Method

A grade progression ratio is the ratio of the number of children in a specific grade during a given year to the number of children in the previous grade during the previous year:

$$r_{g,t} = e_{g,t} / e_{g-1,t-1}$$

where $r_{g,t}$ is the grade progression ratio for grade g in year t, and $e_{g,t}$ represents enrollment in grade g in year t.

The grade progression ratio thus closely resembles the construction of cohort survival probabilities. With no migration and mortality, and no transfers from one school to another among nonmigrants (and also assuming that all are promoted!), the number of children in fourth grade this year should equal the number of children in third grade last year (meaning the grade progression ratio equals 1). Unlike survival probabilities, grade progression ratios may be greater than 1, and indeed will be in districts with sufficient net in-migration to offset any mortality.

In practice, a grade progression ratio is calculated for each grade by taking an average of the grade progression ratios that have been observed in recent years:

$$r_g = \sum_{i=0}^{T} \frac{r_{g,t-i}}{T}$$

where $T + 1$ is the number of years of historical data used. For projection purposes, grade progression ratios are usually assumed to remain constant throughout the projection period. This is equivalent to the assumption that the demographic forces that act to determine the grade progression ratios—namely, mortality and migration—will not change over the projection period (or they will change in such a way that their effects on the grade progression ratio precisely offset one another). An alternative would be to give more weight to recent data:

$$r_g = \sum_{i=0}^{T} a_{t-i} r_{g,t-i} \bigg/ \sum_{i=0}^{T} a_{t-i}$$

where the a_{t-i} are the weights (with more recent data being assigned higher weights). The process of assigning weights is inevitably somewhat arbitrary, but experimenting with alternative sets of weights on data from previous periods may be helpful.

Grade progression ratios are sufficient to produce school enrollment projections for all but kindergarten. Projections of kindergarten enrollment must necessarily be based upon other considerations. Many school districts frequently canvas neighborhoods to prepare censuses of school-age children. However, these may be notoriously inaccurate. Table 6.7, for example, provides data from Amherst, New York, that imply a large difference in population size at each single year of age between those age less than 5, and those age 5 and over. It is possible, though highly improbable, that the differences are real; but more likely, the relatively small number of children under 5 is simply due to undercounting in the special census. The magnitude of the undercount associated with the census of soon-to-be–school-age children is simply too large, at least in this example, to make the data useful.

Further evidence that the special census undercount is substantial comes from data on births (Table 6.8) and data from the 1990 Census (Table 6.9). Both tables con-

TABLE 6.7 Amherst Central School District 1988 School Census (as of August 30, 1988)

Age	Number of Children	Age	Number of Children
<1	70	9	271
1	163	10	252
2	189	11	258
3	247	12	254
4	201	13	247
5	257	14	242
6	316	15	266
7	269	16	262
8	294	17	309

firm that the number of preschoolers in the district should not be declining over time. In fact, these tables indicate that the number of preschoolers may be growing. For example, the relatively high number of births during the period 1985–88 implies that the 1988 special census should have counted a relatively high number of preschool children. Likewise, the 1990 census data show a relatively large number of children age 3 to 4, and this too should have been captured by the special census of 1988.

Given the inadequacy of special censuses of preschool children, the data in Tables 6.8 and 6.9 may instead be used to project kindergarten enrollment. The number of births in a given year should be related to the number of children in kindergarten 5 years later. One possible assumption would be to assume that the ratio of kindergartners in a year to births 5 years earlier (calculated on the basis of historical data) is constant over time. The ratio would then be multiplied by data on births in a particular year to project kindergarten enrollment 5 years later.

Census data on the number of preschool children may also be useful in preparing enrollment projections, at least during the early years of a decade. Projections of kindergarten enrollment for the second half of a decade depend upon births during the decade, and not upon the number of preschool children alive at the beginning of the decade. The age structure of the population data in Table 6.9 implies that kindergarten enrollments may rise during the period 1991–94, before declining in 1995.

We could also assume that the ratio of 1990 kindergarten enrollment to the 1990 census count of 5-year-olds could be used to multiply the number of 4-year-olds to derive a projection of the kindergarten enrollment in 1991. Likewise, the ratio would be multiplied by the number of 3-year-olds to derive the kindergarten enrollment in 1992, and so on. Alternatively, the ratio used could be derived by including other information (e.g., by comparing the number of 6-year-olds counted in the census with first-grade enrollment in 1990).

The reader should get the impression that these methods are by no means sophisticated. Precisely how to project kindergarten enrollment depends upon many factors, such as the stability of the region's overall population change (i.e., whether growth is dynamic or stagnant) and stability of the school system (e.g., whether there are likely to be changes in the public/private school mix). The analyst needs to assess which of the somewhat ad hoc approaches stands the best chance of being most accurate.

TABLE 6.8 Births, 1980–89

Year	Births
1980	327
1981	345
1982	384
1983	327
1984	326
1985	361
1986	366
1987	389
1988	373
1989	350

Source: Birth data are from the New York State Department of Health.

Note: Data are collected for census tracts that correspond roughly to the Amherst School District boundary.

Illustration

The process of enrollment projection will now be illustrated using the data in Tables 6.6 through 6.9. Projected enrollment in grade 1 in 1992 is derived by multiplying the 1991 kindergarten enrollment (202) by the appropriate grade progression ratio. The grade progression ratio is calculated from a historical average of the ratio of first graders one year to the number of kindergartners in the previous year:

$$\left(\frac{218}{197} + \frac{189}{164} + \frac{256}{217} + \frac{203}{173} + \frac{186}{158} + \frac{232}{205}\right)\bigg/6 = 1.154$$

In this district the ratio is much greater than 1 because the public schools have only a half-day kindergarten, and many parents choose to send their children to private school for kindergarten, before entering the public school system in first grade.

The 1992 projection for first grade is, therefore, $202 \times 1.154 = 233$. This compares favorably with the actual 1992 first grade enrollment of 235 (see Exercise 8). Projections for the other grades, with the exception of kindergarten, are carried out similarly. To project the size of the kindergarten class using the birth data, we first determine the average historical ratio of kindergarten enrollment to births 5 years later:

$$\left(\frac{197}{327} + \frac{164}{345} + \frac{217}{384} + \frac{173}{327} + \frac{158}{326} + \frac{205}{361} + \frac{202}{366}\right)\bigg/7 = .5395$$

This ratio is substantially less than 1 because (a) not all children in the district go to public school, especially in kindergarten, (b) the census tract area, for which the population data are tabulated, is slightly larger than the school district area, and (c) there are possible effects of net out-migration. With this approach, kindergarten enrollment for 1992 is projected to be $.5395 \times 389 = 210$.

Using the population data, the ratio of 1990 kindergarten enrollment to the 1990 census population age 5 is $205 / 292 = .7021$. The projected size of the 1992 kindergarten class based upon an assumption of no net migration is .7021 × (642 / 2) = 225. The actual enrollment in kindergarten in 1992 was 236. The first of the two methods under-projects enrollment by 26—roughly the size of a classroom. The vagaries of school enrollment trends are well illustrated by this example and by Exercise 8b, where the reader is asked to compare actual and projected kindergarten enrollment for 1993.

The Housing-Unit Method

The grade progression ratio and the use of birth and population data as just illustrated account to some degree for recent migration trends within a school district. However, changes in migration levels will lead to changes in both grade progression ratios and the relationship between kindergarten enrollment and data on births and population. If net migration is increasing over time, actual grade progression ratios will be greater than their historical averages, and the actual size of kindergarten classes will be greater than the expected size derived by using data on either births or population.

When rapid population change occurs, it is common to relate changes in the number of housing units to changes in enrollment. The general strategy remains the same—the analyst first examines the historical relationship and then assumes that the historical relationship will not change in the short run, so that it may be used for projection.

Our example again comes from the Amherst, New York, school district. Table 6.10 displays the needed historical data on enrollment, housing units, and the enrollment-to-housing ratio. The table also depicts yearly changes in the ratio. To produce a projection the local planning agency first extrapolated the enrollment-to-housing ratio. This was done by assuming a geometric decline in the temporal change of the ratio. Specifically, change in the ratio during the first year was predicted to equal half of the change in the previous year (−.0016 is roughly 1/2 of −.0031). During each succeeding year, the change is halved for purposes of projection. The approach may be par-

TABLE 6.9 1990 Population Data

Age	Population
<1	241
1–2	680
3–4	642
5	292
6	247
7–9	805
10–11	494
12–13	485
14	234
15	235
16	248
17	270

Source: U.S. Bureau of the Census.

Note: Decennial census data are for census tracts that correspond roughly to the Amherst Central School District boundary.

TABLE 6.10 Housing-Unit Method for School Enrollment Projection

Year	Enrollment	Housing Units	Enrollment/Housing Ratio	Change in Ratio
Historical Data:				
1980	3,750	10,376	0.3614	—
1981	3,564	10,378	0.3434	−0.0180
1982	3,302	10,381	0.3181	−0.0253
1983	3,164	10,391	0.3045	−0.0136
1984	2,969	10,396	0.2856	−0.0189
1985	2,937	10,401	0.2824	−0.0032
1986	2,823	10,417	0.2710	−0.0114
1987	2,724	10,420	0.2614	−0.0096
1988	2,692	10,422	0.2583	−0.0031
Projection:				
1989	2,677	10,428	0.2567	−0.0016
1990	2,670	10,434	0.2559	−0.0008
1991	2,667	10,440	0.2555	−0.0004
1992	2,667	10,446	0.2553	−0.0002
1993	2,667	10,452	0.2552	−0.0001

Source: Data provided by Erie and Niagara Counties Regional Planning Board.

tially justified by claiming that because our degree of uncertainty about the future is sufficiently high, we should ensure that the projection eventually assumes no change in the enrollment-to-housing ratio. This process of extrapolating the present situation to an eventual situation of no change, though clearly ad hoc, is frequently used in demographic applications. The U.S. Bureau of the Census employed a similar approach in assuming that geographic and racial differences in fertility and mortality rates will decline over time.

Once the enrollment-to-housing ratio has been extrapolated, the next step is to project the number of housing units. This projection may be based upon trends in building permits, upon the amount of open space in an area, and upon local comprehensive plans. The final step is to simply multiply the projected enrollment-to-housing ratio by the projected number of housing units to derive projected enrollment.

There is little open space in the Amherst district, and consequently there has been little change in the number of housing units. Because no change in the number of housing units was projected, no change in enrollment was projected!

This illustration provides a good example of some pitfalls in the housing-unit method. Housing turnover in the district (with young adult couples replacing elderly individuals) combined with an increase in the birth rate has led to a recent significant rise in the number of preschool children. (Recall Tables 6.8 and 6.9.) To assume no change in enrollment simply because there is no expected change in housing stock seems naive. In this instance, reliance on birth data and census population data seems more appropriate. The housing-unit method is best suited for districts experiencing a lot of new construction, with relatively small (or at least predictable) change in the nature of household composition (which would affect the enrollment-to-housing ratio).

6.6 PROJECTING THE DEMOGRAPHIC STRUCTURE OF ORGANIZATIONS

The enrollment projections described in the previous subsection are invaluable in planning for the demands that will be placed upon classroom space, for the number of teachers and amount of supplies and resources that will be needed and so on. School officials should also be aware of the demographic structure of their faculty. Enrollment projections are perhaps made more routinely than are projections of the age structure of the faculty. Yet we can easily argue that the latter are just as important, because the age structure of the faculty has significant implications for such important areas as financial planning and anticipated hiring levels. Indeed, the same comment applies to virtually all organizations.

Quite a large literature has developed on the subject of how to model the demographic structure of organizations. Bartholomew and Forbes (1979) and Grinold and Marshall (1977) provide relatively advanced treatments; these two works are notable because they focus upon the use of the transition probability concept.

To illustrate the types of questions that may be addressed by using the Markov model approach to organizational structure, consider a university faculty divided into three ranks: assistant, associate, and full professor. The university currently has 200 assistant professors, 200 associate professors, and 1,000 full professors. To keep our example simple, we assume that the long-run size of the faculty is to be kept fixed at 1,400. The promotion of professors from one rank to the next is analogous to the survival of individuals from one age category to the next in the cohort component model. Likewise, new hires are analogous to births, while retirements, deaths, and quits are analogous to deaths in the cohort component model since they represent exits from the system. Suppose that a study of personnel records from the past few years reveals the following annual transition probability matrix, **P**:

To:	Assistant	Associate	Full	Exit
From:				
Assistant	.50	.15	0	.35
Associate	0	.60	.10	.30
Full	0	0	.85	.15

Note that in this case **P** is written as a 3 × 3 matrix, with row sums less than 1, due to the possibility that some professors will leave the system. The exit probabilities from each rank are derived by subtracting each row sum from 1. Thus in any given year, 35 percent of assistant professors, 30 percent of associates, and 15 percent of full professors leave their posts. Straightforward application of the by now familiar projection process reveals that in the absence of hiring, the rank distribution of the university's faculty over the next 2 years will be

$$\mathbf{p}(0) = [200\ 200\ 1{,}000]$$
$$\mathbf{p}(1) = \mathbf{p}(0)\mathbf{P} = [100\ 150\ 870]$$
$$\mathbf{p}(2) = \mathbf{p}(1)\mathbf{P} = \mathbf{p}(0)\mathbf{P}^2 = [50\ 105\ 755]$$

We may now examine the effects of alternative hiring policies. Suppose that the university faces a difficult and uncertain financial future, and decides that 90 percent of all hires must be at the assistant level, with the remaining 10 percent spread evenly across the remaining two ranks. What then is the projected distribution of the faculty across ranks for the next few years? Each year the number of retirees (and hence the number of new hires needed) is $\mathbf{p}(t)\mathbf{r}(t)$, where $\mathbf{r}(t)$ is a column vector formed from the annual rank-specific probabilities of retiring. In our example,

$$\mathbf{r}(t) = \begin{bmatrix} .35 \\ .30 \\ .15 \end{bmatrix}$$

Under the hiring policy described, we may write the projected distribution one year ahead as the sum of the distribution across ranks of the faculty there throughout the period, plus the distribution across ranks of new hires:

$$\mathbf{p}(t+1) = \mathbf{p}(t)\,\mathbf{P} + \mathbf{p}(t)\,\mathbf{r}(t)\,\mathbf{h}(t)$$

where $\mathbf{h}(t)$ is a row vector describing the proportions of professors hired at each rank. Continuing with our example, we have

$$[352\ 178\ 870] = [100\ 150\ 870] + [200\ 200\ 1{,}000]\begin{bmatrix} .35 \\ .30 \\ .15 \end{bmatrix}[.9\ .1\ 0]$$

Note that faculty size remains fixed at 1,400. With this framework, it is not difficult to compare short-run and long-run effects of alternative hiring policies.

6.7 SUMMARY

In this chapter we have described several approaches to projection. Many principles that apply to the projection of populations may be fruitfully applied to the projection of other quantities of interest, such as school enrollments and the number of employees at specific ranks within organizations.

The reader should be aware that the multiregional methods we have discussed so far essentially focus on the destination choices of individuals at specific origins. This perspective—that there are a set of probabilities governing how individuals at each origin choose their destinations—is by no means the only one that can be adopted. In the next chapter we examine models that focus on the role played by both origins and destinations in determining migration flows, and on the interactions between regions.

EXERCISES

1. **a.** Fit linear, geometric, and exponential trends to the following population data for the metropolitan area of Toledo, Ohio:

Year	Population
1950	395,551
1960	456,931
1970	692,751
1980	791,599

b. Extrapolate the trend in each case to predict Toledo's population for the years 1990 and 2000. Use the estimated slope and the actual 1980 population to do the extrapolation.

2. Use the data and answers for the data accompanying Exercises 8 through 15 of Chapter 3 to develop a cohort survival projection of New York State's female population for 1985 and 1990.

3. The following migration flows characterized the United States during the period 1985–90:

To:	Northeast	Midwest	South	West
From:				
Northeast	44,517,000	360,000	1,815,000	545,000
Midwest	373,000	52,393,000	1,769,000	1,029,000
South	845,000	1,246,000	73,062,000	1,253,000
West	386,000	718,000	1,185,000	43,841,000

Using the column sums as the 1990 populations, find the 1995 and 2000 populations by region, assuming that the 1985–90 probability matrix remains fixed. Ignore the effects of natural increase and international migration.

4. *Manpower planning.* A university currently employs 2,200 faculty members. Of these, 600 are untenured and 1,600 are tenured. University officials have decided that this is a desirable faculty size, but they are concerned about the short- and long-run distributions of faculty across the two ranks.

A study finds that 65 percent of all untenured faculty will be in the same rank during the following year, and 20 percent will gain tenure. The remaining 15 percent of the untenured faculty leave the university in any given year. The study also finds that 80 percent of all tenured faculty remain tenured and at the university, while the other 20 percent leave the university system. These probabilities have remained stable over the years, and it is assumed that they will remain constant in the future.

The university has decided to replace all untenured professors who leave with new untenured professors. It is now attempting to decide what fraction of the new hires made to replace departing tenured faculty should be made at the untenured level.

Draw a graph depicting the number of untenured faculty over each of the next 4 years for each of four separate policies. An example of a policy would be to have 50 percent of all new hires made to replace departing tenured faculty at the untenured level. Depict all four trajectories on the same piece of graph paper.

5. Using the data for Exercise 4, graph the long-run proportion that will be tenured against the fraction of new hires made to replace tenured faculty that are made at the untenured level. Then interpolate to answer the following:

a. What should the university's policy be if it wishes to have 60 percent tenured in the long run?

b. What should the university's policy be if it wishes to have 75 percent tenured in the long run?

6. Polk Middle School has the following recent enrollment history:

186   Chapter 6 • Population Projections

	1988	1989	1990	1991	1992	1993
Grade						
6	60	62	70	63	62	59
7	52	63	69	77	62	67
8	68	58	65	76	77	69

a. Compute the historical averages of the grade progression ratios from grade 6 to grade 7 and grade 7 to grade 8.

b. An analysis of elementary school enrollments yields the following projected 6th-grade school enrollment:

1994:   58
1995:   57
1996:   59

Using the solution to (a), project enrollment for Polk Middle School, by grade, for 1994–96.

7. Tyler High School is in a rapidly growing suburban area. School officials are debating whether to open a new school. They decide that their first step will be to produce a 5-year projection of total enrollment. Use the housing-unit method and the following data to derive enrollment/housing ratios for each year. Then use the average ratio and the projected number of housing units to derive projected enrollment.

Year	Enrollment	Housing Units
1988	2,000	10,000
1989	2,056	10,200
1990	2,098	10,400
1991	2,110	10,600
1992	2,160	10,900
1993	2,200	11,000

Year	Housing Units
1994	11,300
1995	11,600
1996	11,900
1997	12,000
1998	12,100

8. **a.** Use average grade-progression ratios from Table 6.6 to project the enrollment of grades 2–12 for Amherst Central Schools in 1992.

b. Use the data from Tables 6.8 and 6.9 to prepare two alternative projections of kindergarten enrollment in 1993.

c. Compare your solutions to (a) and (b) with the actual enrollments given below. What are some of the possible reasons for discrepancies?

	1992	1993
K	236	205
1	235	244
2	223	248
3	196	225
4	205	204
5	234	195
6	178	239
7	220	187
8	207	209
9	215	220
10	200	231
11	186	191
12	209	186

> **Those who know enough to forecast migration know better than to try.**
>
> —*Peter Morrison (1973)*

CHAPTER 7

MODELING AND FORECASTING MIGRATION

7.1 Introduction
7.2 What Migration Quantity to Model?
7.3 The Gravity Model
7.4 The Intervening Opportunities Model
7.5 Temporally Varying Transition Probabilities
7.6 Economic Gravity Models
7.7 Toward Economic–Demographic Models
7.8 Summary

7.1 INTRODUCTION

In Chapters 5 and 6 on population estimation and projection we emphasized that at the local and regional scales of demographic analysis the *migration* component of population change typically proves extremely difficult to model. For many of the planning and business applications that users of this book are most interested in, however, it is vital to understand how migration into and out of an area is likely to evolve over time.

This chapter covers several different approaches to modeling and forecasting interregional migration. We begin by discussing the inappropriateness of modeling net migration, and the advantages of focusing on the systemic structure of interregional or interarea flows. The discussion that follows centers on the interregional transition probability framework introduced in Chapter 6 and on two classic conceptualizations of spatial interaction theory: the *gravity model* and the *intervening opportunities model.*

We then summarize the voluminous literature that has arisen on "econometric gravity models," in which aggregate economic variables are introduced as *determinants* of migration. Analogous to the Chapter 5 discussion of population estimation methodology, the use here of "symptomatic" variables is predicated on the ability of the analyst to (1) obtain accurate measurements of variables that trend well with migration and (2) specify robust structural relationships between such variables and migration. In other words, to be useful for *forecasting* purposes, the economic variables used as determinants of migration must themselves be able to be modeled with a fair measure of success.

Following the discussion of extended, economic gravity models, we set forth approaches to deriving combined economic–demographic migration models that unite the elegant transition probability framework with behavioral migration theory. The chapter concludes with a brief description of attempts to build integrated economic–demographic modeling systems. The primary impact of such models has been in helping researchers learn more about the complex dynamics of interregional migration systems; no approach, to date, has emerged as a fully satisfactory, general purpose forecasting technique.

Migration modeling and forecasting remains part art and part science. Migration experts, as noted by Morrison in the quote that begins this chapter, have come to appreciate the complexity of the dynamics of interregional migration systems. Human migration arises as a response to an enormous number of ever-changing conditions and perceptions. Modern regional economies are quite complex and many of the traditional barriers to movement have been greatly reduced. It is perhaps not surprising, therefore, that forecasting the geographic pattern of interregional migration proves to be at least as complex a task as forecasting the (largely deterministic) pattern of weather systems. Just as there are many real reasons why we may want to know what the weather will be in the next few days, there will always be plenty of reasons why planners, government officials, and businesspeople will want to have the most informed guesses at their fingertips about the future course of migration and, thus, population change.

7.2 WHAT MIGRATION QUANTITY TO MODEL?

Back in Section 4.3 we examined different ways to aggregate all the individual moves that people make into quantities called migration. Migration modelers have attempted to specify functional forms that abstractly replicate everything from the individual decision process of the potential migrant up to macro measures of the total volume of movement within an entire interregional system of flows. Before deciding on the appropriate technique to use in modeling migration, the geographical analyst of population must decide how the *dependent* variable, migration, is to be measured.

Sometimes the choice will be highly constrained by the lack of detailed migration tabulations. But other times the analyst must choose: Should net migration (N_k), gross in- and out-migration (IM_k, OM_k), or region-to-region or local-area-to-local-area streams of movement (m_{ij}) be the focus of analysis? Sometimes disaggregated flows must be produced to satisfy the demands of specific problems. For example, take the case of a residential development firm in the Sunbelt that would like to have an idea of how many prospective home buyers will be elderly retirees from northern states versus how many will be young labor force persons coming to the local area from nearby states. The type of housing that the developer decides to provide can be better targeted with reasonable forecasts on the geographic origins of future in-migrants.

For other problems, however, the rationale for using a more disaggregate migration measure may be related more to the *defensibility* of the forecasts that result. On theoretical grounds it is most often advisable to model gross in- and gross out-migration in preference to net migration, or, even better whenever possible, to model the entire web of place-to-place streams of movement within a migration system. To see why this is so, consider first the inherent volatility of net migration and then the "fallacy" of working with net migration *rates*.

The Volatility of Net Migration

Figure 7.1 shows the annual time series of net migration for the state of Texas from 1981 to 1988. As a result of the oil glut and consequent recession, the upward trend of net in-migration during the first two years was sharply reversed, with dwindling numbers of net in-migrants during the middle years of the period, until 1986–87 and 1988–89 when the bottom dropped out of the state's economy and sharp net out-migration occurred.

Figure 7.2 illustrates the fact that the enormous swings in net migration for Texas over this period came about as a result of the relatively lesser percent changes in the volumes of gross in- and out-migration to the state. Notice that slackening in-migration was a greater contributor to the reversal of the direction of net migration than was increased out-migration.

Further understanding of these 1980s trends can be garnered from examining geographic patterns of state-to-state interchanges. Figure 7.3 starkly illustrates (through use of the demographic effectiveness measure introduced in Chapter 4) that the sources of the large gross in-movement during the early 1980s and the destinations of the net out-movement at the end of the decade were startlingly different.

Any attempt to model net migration to Texas solely on the basis of changing conditions in Texas fails to adequately portray the fact that individual migration decisions reflect the consideration of conditions in the current region of residence *relative* to conditions (or, at least the *perception* of conditions) in potential migration destination regions. Just as it would be foolish to base weather forecasts only on what is happening in the atmosphere over the local area, so too the amount of migration to a local area can only be understood in relation to what is happening elsewhere, as well as in the region itself.

The folly of adopting a *uniregional* rather than a *multiregional* perspective on migration is illustrated by a migration forecasting technique that has, unfortunately, been all too frequently employed in actual practice: the net migration rate method.

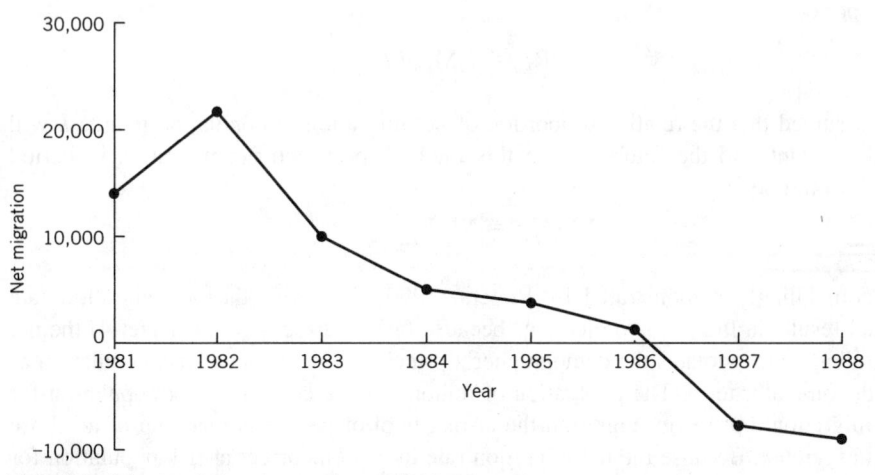

FIGURE 7.1 Net Migration to Texas, 1981–88

Source: Graphed by authors from U.S. Bureau of the Census/Internal Revenue Service migration data.

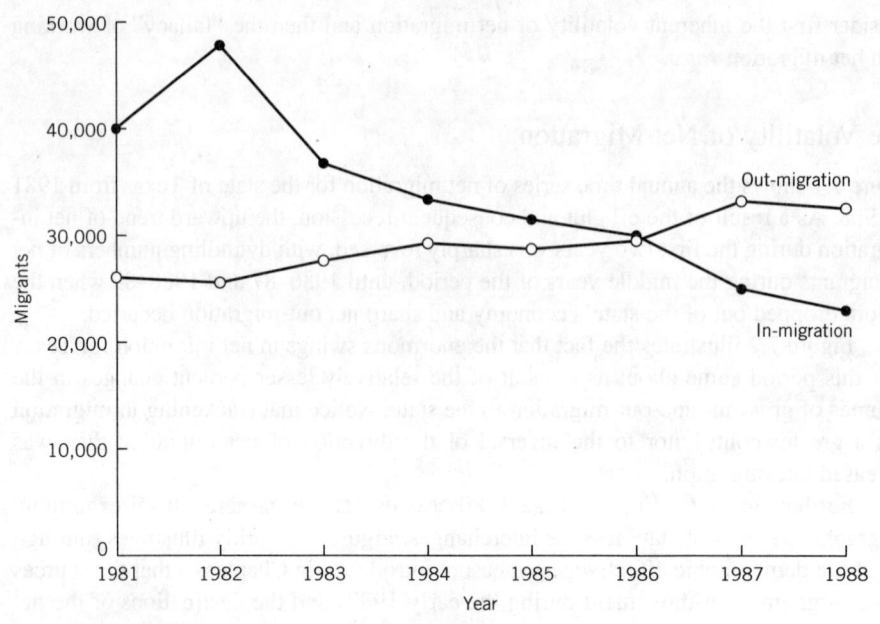

FIGURE 7.2 Gross Migration to Texas, 1981–88

Source: Graphed by authors from U.S. Bureau of the Census/Internal Revenue Service migration data.

The Fallacy of Using Net Migration Rates for Forecasting

Seemingly one of the easiest ways to model and forecast migration might be to make assumptions about how *net migration* to an area will change over time. As one of the components of population change, it is tempting to try to treat net migration in a similar fashion to births and deaths. Quite frequently births and deaths are modeled by assuming temporal constancy in the rates at which they occur within a population. Population analysts have frequently, but misguidedly, taken a net migration rate for a base period:

$$NMR_{k,t} = NM_{k,t} / P_{k,t-1}$$

and assumed that the relative proportion of net migration to population in area k will hold constant into the future. Using this method, projected net migration in period $t + 1$ would be

$$NM_{k,t+1} = NMR_{k,t} \times P_{k,t}$$

As compellingly demonstrated by Rogers (1990), however, the net migration rate model results in fallacious projections because (as we discussed in Chapter 4) the net migration "rate" violates the demographer's principle of using an at-risk population as the denominator term. The population of region k is the correct at-risk population for out-migration, but for in-migration the at-risk pool of people is made up of all those *not* in region k. Because the net migration rate uses an incorrect at-risk population for in-migration, employing it to make population projections means that fundamental accounting identities will be violated for the migration system as a whole. It turns out,

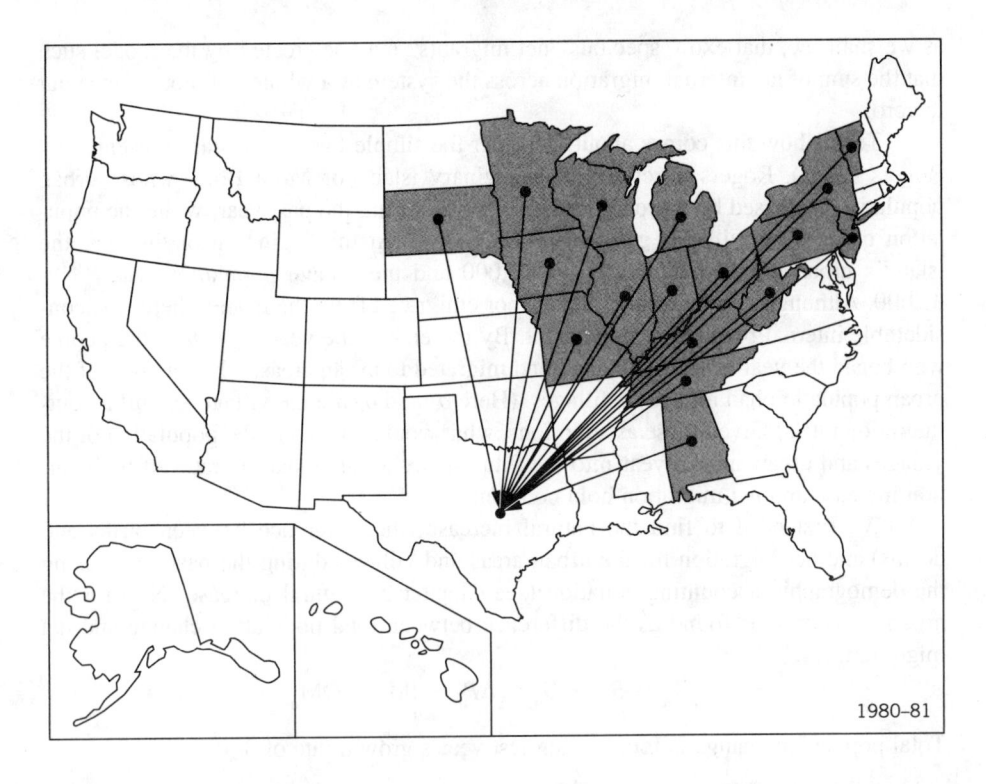

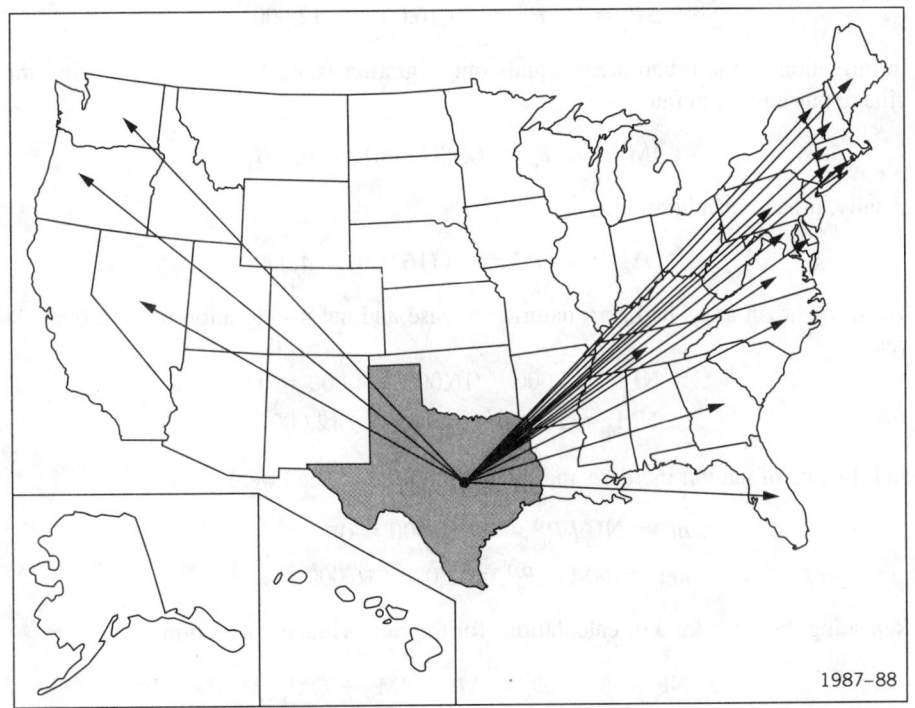

FIGURE 7.3 Changing Pattern of the Most Demographically Effective Migration Streams for Texas, 1980–81 and 1987–88

Source: Mapped by authors from U.S. Bureau of the Census/Internal Revenue Service migration data.

as we shall see, that extra, specious "net migrants" may be created by the model such that the sum of net internal migration across the system as a whole will not balance out to zero.

To see how this comes about, consider the simple two-region case presented by Rogers (1990). Rogers describes the imaginary island of Mora Bora, whose urban population increased by a growth rate of $r_u^0 = ¾$ during the past year, while the population of the rural villages grew by only $r_v^0 = ⅛$. At the beginning of the year the island's urban population was $P_u^0 = 16,000$ and the village population was $P_v^0 = 32,000$. Although no one immigrated to nor emigrated from the island, there was considerable internal population movement. By the end of the year, $o_v = ½$ of the people who began the year as rural villagers had migrated to urban areas, while $o_u = ¼$ of the urban population had moved to villages. (Here o_v and o_u are the village and urban out-migration rates.) Given these assumptions, what would happen to the population of the villages and urban areas a year into the future if we assume that the rates of both natural increase and *net* migration hold constant?

We first need to find the natural increase (the difference between births and deaths) and net migration for the urban areas and villages during the past year. Using the demographic accounting equation (see Chapter 3), natural increase (NI_u) for the urban areas may be found as the difference between total population change and net migration:

$$NI_u = B_u - D_u = \Delta P_u - IM_u + OM_u$$

Total population change is found using last year's growth rate of 3/4:

$$\Delta P_u = r_u^0 \, P_u^0 = (¾)(16,000) = 12,000$$

In-migration to the urban areas equals out-migration from the villages, so using the village out-migration rate,

$$IM_u = o_v \, P_v^0 = (½)(32,000) = 16,000$$

Finally, urban out-migration is

$$OM_u = o_u \, P_u^0 = (¼)(16,000) = 4,000$$

Thus, the urban areas had zero natural increase and net in-migration of +12,000 last year:

$$NI_u = 12,000 - 16,000 + 4,000 = 0$$
$$NM_u = 16,000 - 4,000 = +12,000$$

and the rate of natural increase and the net migration "rate" were

$$ni_u = NI_u / P_u^0 = 0 / 16,000 = 0$$
$$nm_u = NM_u / P_u^0 = 12,000 / 16,000 = ¾$$

Repeating the same kind of calculations for the rural village population,

$$NI_v = B_v - D_v = \Delta P_v - IM_v + OM_v$$
$$\Delta P_v = r_v^0 \, P_v^0 = (⅛)(32,000) = 4,000$$
$$IM_v = o_u \, P_u^0 = (¼)(16,000) = 4,000$$
$$OM_v = o_v \, P_v^0 = (½)(32,000) = 16,000$$

Thus the village population's natural increase and net migration were

$$NI_v = 4,000 - 4,000 + 16,000 = 16,000$$
$$NM_v = 4,000 - 16,000 = -12,000$$

with associated rates:

$$ni_v = NI_v / P_v^0 = 16,000 / 32,000 = \tfrac{1}{2}$$
$$nm_v = NM_v / P_v^0 = -12,000 / 32,000 = -\tfrac{3}{8}.$$

To do a projection based on constant net migration rates we begin from the current populations:

$$P_u^1 = P_u^0 + \Delta P_u = 16,000 + 12,000 = 28,000$$
$$P_v^1 = P_v^0 + \Delta P_v = 32,000 + 4,000 = 36,000$$

and then apply the respective natural increase and net migration rates:

$$\Delta P_u = ni_u\, P_u^1 + nm_u\, P_u^1$$
$$\Delta P_u = (0)(28,000) + (\tfrac{3}{4})(28,000) = +21,000$$
$$P_u^2 = P_u^1 + \Delta P_u = 28,000 + 21,000 = 49,000$$
$$\Delta P_v = ni_v\, P_v^1 + nm_v\, P_v^1$$
$$\Delta P_v = (\tfrac{1}{2})(36,000) - (\tfrac{3}{8})(36,000) = 18,000 - 13,500 = +4,500$$
$$P_v^2 = P_v^1 + \Delta P_v = 36,000 + 4,500 = 40,500$$

Recall what our purpose was for repeating all these calculations from Rogers (1990)! We are interested in showing that there is something wrong with the Mora Boran population figures derived from net migration rate projections. To see what this problem is, examine the projected population of the *entire* island. The calculations show the island's total population growing from $P_*^0 = 48,000$ to $P_*^1 = 64,000$ and then to $P_*^2 = 89,500$. Natural increase, however, is only +18,000 from time $t = 1$ to $t = 2$. The correctly projected island population for time $t = 2$ should be $P_*^2 = 82,000$. So where do the extra 7,500 people come from? They are spurious "net migrants" that were created by our model because the net migration multipliers that we have used (nm_u and nm_v) are not true demographic rates. Note that the net migration loss projected for the villages of 13,500 is not equal to the net migration gain for the urban areas of 21,000. The fundamental accounting constraint that net internal migration within the system must sum to zero is violated when we use net migration rates. By the way, if the Mora Boran example is computed out further into the future, the error of closure becomes quite large; for the 24th projection period there are more than 8 billion spurious net migrants (Plane, 1993a).

Unfortunately the net migration rate method is used frequently by population analysts who are concerned with only a single area and not with making consistent projections for an entire population system. Many U.S. states project their populations in this fashion. For example, the State of Arizona's official population projections prepared by its Department of Economic Security and released on September 20, 1991, were generated with a demographic cohort-survival model that employed net migration rates. Although Arizona has been a fast-growing state, these projections are artificially inflated because of the structural problem with using net migration rates that we have just examined. A significant portion of the projected population increase from

3.8 million in 1991 to 9.4 million in 2040 is due to spurious net migrants that would not be included if a model were used that more correctly based in-migration to Arizona on the changing future pools of population in *origin* regions (i.e., the other 49 states, Mexico, and foreign countries) rather than in Arizona itself.

The Markov model that we examined in Chapter 6, by contrast, provides a demographically correct representation of the changing influence of the pools of population "at risk" of becoming migrants in an interregional population system. When temporally *fixed* transition probabilities are used in that model, however, the predictions about migration system dynamics are also suspect. Such predictions are fundamentally incompatible with most structural geographic representations of migration processes (Plane, 1993a). Whereas the population size of a region is certainly a major influence on how many migrants are likely to originate in that region, the pattern of destination choice is also shaped strongly by the number of opportunities available in the alternative destination regions of the migration system, and "opportunities" have often been proxied quite successfully with destination region population. By including the population of origin regions, but not the populations (or opportunity "mass") of the destinations, the standard fixed-transition-probability model gives a lopsided view of the dynamics of a migration system.

We now examine two classic models of spatial interaction: the *gravity model* and the *intervening opportunities model*. Both have been developed to represent the dual roles of origin "at-risk" populations and destination opportunities. Both seek to replicate the entire web of place-to-place migration flows $\{m_{ij}\}$ rather than total gross in- and out-migration for a region (IM_k and OM_k) or simply its net migration (NM_k). Whenever possible, the analyst is well advised to model the streams of movement rather than to use shortcut techniques to model the more aggregate quantities. Even when the scope of the problem calls for only modeling population change in a single area, good advice is always to at least keep in mind a *multiregional perspective*. By this we mean to think about how the study area functions in relation to the rest of the migration system and to model in- and out-migration keeping in mind the conditions in, and the study area's connections to, all other areas.

7.3 THE GRAVITY MODEL

For more than a century *gravity* models have been employed in migration analysis.[1] Ravenstein, in his seminal paper "The Laws of Migration," noted that the analyst

> must take into account the number of natives of each county which furnishes the migrants, as also the population of the towns or districts that absorb them. (1885, p. 198)

He also noted that most migrants move only a short distance, whereas long-distance moves are relatively infrequent. These two concepts—(1) the proportionality of the flows of migrants in various migration streams to the population sizes of origin and destination regions and (2) the inverse proportionality of the size of the streams to the distances separating destinations from origins—are the constituent elements of the gravity model.

---

[1] Hua and Porell (1979) give a full historical overview of the use of gravity models in regional research.

The original rationale for gravity modeling of spatial interaction phenomena was simply an argument by analogy to the physicist's gravity model, although more recently a number of alternative interpretations have been offered, including the entropy-maximizing derivation (Wilson, 1970) and the elegant "cost-efficiency principle" (Smith 1978). The univeral law of gravitation was set forth by Isaac Newton when he was a 23-year-old student of natural philosophy sent home from Cambridge University during an outbreak of the plague. In equation form,

$$F_g = G \frac{m_1 m_2}{d_{12}^2} \tag{7.1}$$

The gravitational force, F_g, between two objects is directly proportional to their masses, m_1 and m_2, and inversely proportional to the square of the distance d_{12} between them, where G is the universal gravitational *constant* (meaning that it has the same value for all pairs of objects).

When applied to migration, the "force" of attraction between two regions is measured by the volume of migration interchange between them. The population of each of the two regions is most commonly used as the proxy for "mass." The distance term is raised to a power that may be somewhat different from 2 (and it is, as we shall see, sometimes included in a different functional form than the negative power function just shown). Finally, the value of the constant of proportionality is specific to the migration system being analyzed.

Consider, for example, the problem of modeling migration during the period 1985–90 between each pair of the six New England states (Connecticut, Maine, Massachusetts, New Hampshire, Rhode Island, and Vermont). A simple gravity model would estimate the flows to be

$$M_{ij} = k \frac{P_i P_j}{d_{ij}^b}; \quad i = 1, 2, \ldots, 6; \quad j = 1, 2, \ldots, 6; \quad i \neq j \tag{7.2}$$

Here M_{ij} is the model-*estimated* flow of migrants from origin state i to destination state j (whereas m_{ij} is the *actual* flow), d_{ij} is the distance between the two states (i.e., the distance between the states' population centroids—see our discussion in Chapter 2), k is a constant of proportionality, and b is a *distance deterrence* or *attenuation parameter*.

Distance Deterrence

The b parameter expresses the extent to which the volume of migration falls off with increasing distance. There is no reason to assume that b for a migration gravity model should equal 2 as in the physics model. As shown (in stylized form) in Figure 7.4, the greater the value of b, the more "bent" will be the relationship between M_{ij} and d_{ij}; the closer b is to zero, the flatter and more horizontal will be the relationship between distances and expected flows. When distance deterrence is high, the model distributes most migrants to nearby destinations; when distance is less of a factor (vis-à-vis other considerations potential migrants face), the relative size of destinations becomes the dominant term in distributing migrants among destinations.

An alternative form of the model is often employed, in which the volume of migration is not inversely proportional to distance raised to some power, b, but rather

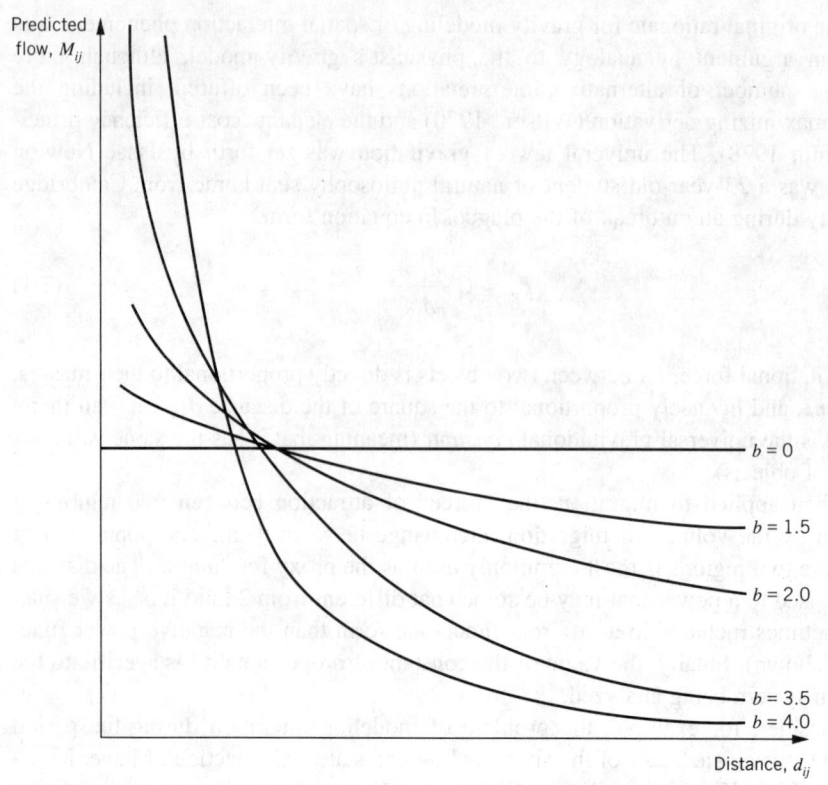

FIGURE 7.4 Stylized Relationship between Predicted Flows and Distance for Various Values of the Distance Deterrence Parameter, *b*, in the Power Function Gravity Model

to the base of the natural logarithms, *e* (approximately 2.7141), raised to the product of distance and a parameter, β. The *negative exponential* gravity model is written

$$M_{ij} = \kappa\, P_i P_j\, e^{-\beta d_{ij}} \tag{7.3}$$

For an explanation of why the negative exponential form of distance attenuation may be preferred to the negative power function, see Wilson (1970) and Smith (1978). In brief, the matter rests on whether people in fact respond directly to distance or to the logarithm of distance in evaluating alternative destination choices.

Regardless of whether the power function or exponential form is used, there is an interesting inverse relationship between the value of the distance deterrence parameter and the mean distance that the migrants in the system will be assigned to move. Figure 7.5 shows a curve tracing out this relationship for different values of β in a negative exponential model.

Gravity models are sometimes used, in fact, to study the general effect that distance has on impeding migration behavior. If the mean length of moves has been creeping up over time, we expect a gravity model fit to more recent data to have a lower (closer to zero) *b* or β parameter than one fit for an earlier time period. As travel becomes easier and information diffuses more rapidly, the localized characteristic of migration noted by Ravenstein (1885) diminishes, although as we saw in our Chapter

4 discussion of the principal component technique for migration matrix analysis, there are still distinct regional subsystems of movement in the United States.

Constrained and Unconstrained Gravity Models

We now have some intuitive understanding of the meaning of the b (or β) parameters of the gravity model. What about the proportionality constant, k or κ? In discussing this term of the model we note an important general distinction made in spatial interaction modeling between *unconstrained, singly constrained,* and *doubly constrained formulations.* In a singly constrained model, a separate value of the parameter is chosen for each region included in the system. In the doubly constrained formulation, two proportionality parameters are found for each region.

Before discussing these more sophisticated forms of the gravity model and the role of the constant terms in them, let us begin with the most direct migration analogy to the classical gravity formulation of physics shown in Equation 7.1, the "unconstrained" model. It is called unconstrained because—just as there is no guarantee that its prediction of any given migration flow may not exactly replicate reality—there is also no necessary reason why the total number of migrants that it ends up predicting for the system as a whole will be identical to the actual number.

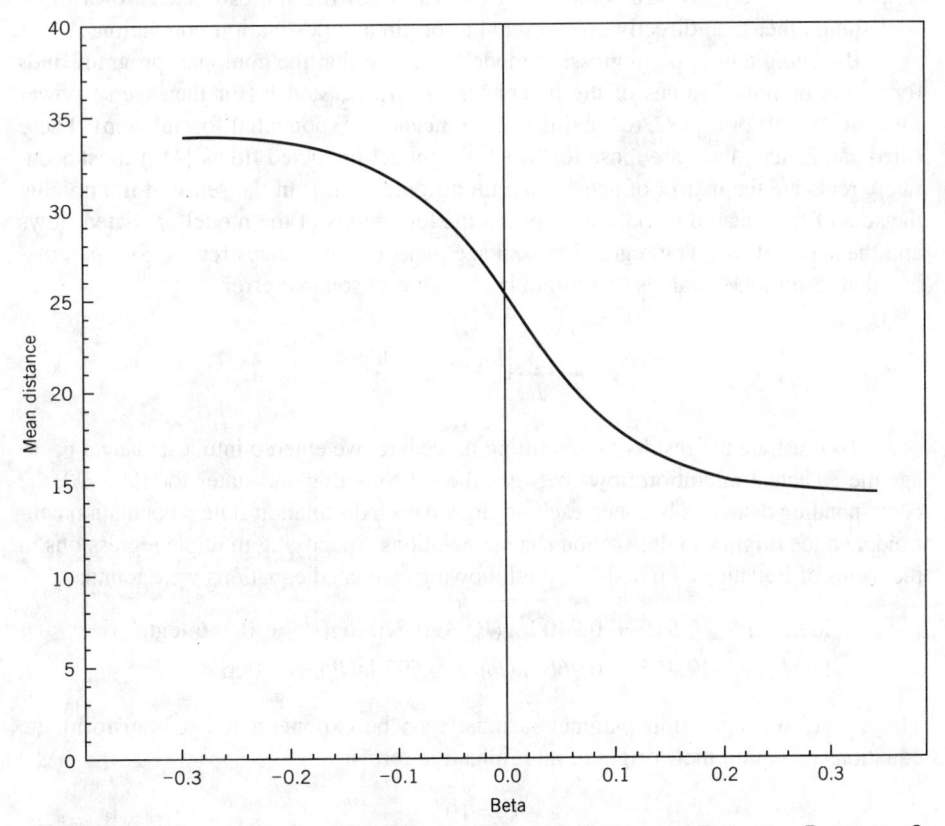

FIGURE 7.5 Relationship between Mean Distance and the Distance Deterrence Parameter, β, in a Negative Exponential Gravity Model

Source: Adapted from Plane (1981). Reprinted by permission.

Fitting an Unconstrained Model

To operationalize the unconstrained models in Equations 7.2 and 7.3, the most common approach is to use log-linear regression analysis.[2] We first transform the variables in the equations by taking logarithms of both sides:

$$\log(M_{ij}) = \log(k) + \log(P_i) + \log(P_j) - b \log(d_{ij}) \qquad (7.4)$$

$$\ln(M_{ij}) = \ln(\kappa) + \ln(P_i) + \ln(P_j) - \beta \, d_{ij} \qquad (7.5)$$

In the latter case natural logarithms (logs to base e) are used, whereas in the former equation common logarithms (e.g., those to base 10) suffice. Note that the impact of this step is to turn our formerly multiplicative models into additive ones. (If you need to, see the review of logarithms in Appendix C.)

Next, using any standard computerized statistical package that supports multiple regression applications, we "fit" the models. The equations typically estimated by the regression progam are slightly modified versions of (7.4) and (7.5):

$$\log(M_{ij}) = a_0 + a_1 \log(P_i) + a_2 \log(P_j) - b \log(d_{ij}) \qquad (7.6)$$

$$\ln(M_{ij}) = \alpha_0 + \alpha_1 \ln(P_i) + \alpha_2 \ln(P_j) - \beta \, d_{ij} \qquad (7.7)$$

Here $a_0 \equiv \log(k)$ and $\alpha_0 \equiv \ln(\kappa)$. Also, two new parameters have been added on the population variables that make the models a bit more flexible. The a_1 and a_2 or α_1 and α_2 parameters may diverge somewhat from 1.0, allowing the estimated flows to be something other than directly proportional to origin and destination populations.

By *fitting* a multiple regression model we mean that the computer program finds for us the optimum values of the parameters a_0, a_1, a_2, and b (for the inverse power function model) or α_0, α_1, α_2, and β (for the negative exponential formulation). These fitted parameter values are those for which the model-predicted flows $\{M_{ij}\}$ most accurately replicate the matrix of actual migration figures $\{m_{ij}\}$, in the sense of minimizing the sum of the squared deviations between the logarithms of the model-predicted flows and the actual flows. The regression package picks out the values for the four parameters that result in M_{ij} values that minimize the sum of squared errors:

$$\text{SSE} = \sum_{i=1}^{r} \sum_{j=1}^{r} \left[\log(m_{ij}) - \log(M_{ij}) \right]^2$$

To illustrate the results of this fitting procedure, we entered into a statistical package the 30 actual migration flows between the six New England states for 1985–90, the corresponding distances between each origin and each destination state's population centroid, and the origin and destination state populations. Specifying multiple regressions in the forms of Equations 7.6 and 7.7, the following estimated equations were found:

$$\log(M_{ij}) = -3.919 + 0.940 \log(P_i) + 0.570 \log(P_j) - 0.746 \log(d_{ij})$$

$$\ln(M_{ij}) = -12.465 + 0.966 \ln(P_i) + 0.595 \ln(P_j) - 0.006 \, d_{ij}$$

The a_0 and α_0 regression parameters must then be exponentiated to transform the equations back into their original, multiplicative forms:

$$k = 10^{a_0}$$
$$\kappa = e^{\alpha_0}$$

---

[2] But see Haworth and Vincent (1979) for a statistical criticism and refinement of this technique.

Our fitted gravity models are thus

$$M_{ij} = (0.000120503)\, P_i^{0.940}\, P_j^{0.570} / d_{ij}^{(0.746)}$$
$$M_{ij} = (0.00000385939)\, P_i^{0.966}\, P_j^{0.595}\, e^{-(0.006)d_{ij}}$$

Note that we obtained rather low values for the b and β distance deterrence parameters, suggesting that among these small, nearby states the relative distances to potential destinations is not all that important in determining the size of the migration flows. Note, too, that while the origin population exponents are both close to 1.0, the destination ones are significantly below. This suggests that whereas outmigrants are generated in fairly rough proportions to the populations of the origin states, the "drawing power" of destinations are not exactly proportional to current population. In fact, two of the most popular migration destinations over this period in New England were Maine and New Hampshire, each having only a little more than a million persons as of the 1990 census. A property of the basic gravity model set forth in Equations 7.2 or 7.3 is that it predicts the same flow in each direction—in other words, zero net exchange between any origin–destination pair (i, j). However, with the addition of differential parameters for origin and destination population such as we have done here (a_1 and a_2 or α_1 and α_2), this is no longer the case.

Table 7.1 shows the model-estimated flows as well as the actual flows. The model-generated flows shown in the table fit the census migration data reasonably well. The R^2 values (which indicate the proportion of variance in the original, logarithmically transformed flows "explained" by the estimated flows) are .764 and .770, respectively.

Of course, a gravity model only builds in two of the determinants of the volumes of flow in actual migration streams, namely population and distance. One of the best reasons for fitting a simple gravity model like we have just done is, in fact, to examine *residuals:* (the differences between actual flows and the flows the model predicts). The migration residual (or "error") for origin i and destination j is

$$E_{ij} = m_{ij} - M_{ij}$$

A positive residual indicates that the actual flow was greater than the estimated one (an underprediction); a negative value means the actual flow was smaller (an overprediction).

Table 7.1 also lists these residuals for the 30 migration flows. After examining these residuals, can you think of some other variables that could be included in a refined form of the model to more closely proxy other determinants of migration? The net exchanges suggest that in New England there is movement toward lower-density areas. Part of this trend is picked up in the exponents on destination population, but note the big positive residual for the flow from Massachusetts to New Hampshire.

Similarly, if we estimated regression equations for the entire 51 × 51 matrix of state-to-state flows it might be noted that many of the most underpredicted flows are those from Frostbelt to Sunbelt states, and the most overpredicted are flows in the opposite direction. Many researchers have recently included climatic and other variables expressing the role of *amenities* in influencing migration. (See, e.g., Porell, 1982.) Later in this chapter we examine the voluminous literature that has arisen on economic gravity models that feature additional variables in a multiplicative expression that expands on the simple intuitive formulations of Equations 7.2 and 7.3.

The gravity model formulation answers a criticism that we leveled at the net migration rate and Markov fixed-transition-probability models. A role is posited for both origin and destination population. Note, however, that the model is a cross-

TABLE 7.1 Unconstrained Gravity Model Estimates of State-to-State Migration in New England, 1985–90

				Power Function Model		Negative Exponential Model	
Origin	Destination	d_{ij}	m_{ij}	M_{ij}	(Residual)	M_{ij}	(Residual)
CT	ME	262	9,621	7,464	(2,157)	5,922	(3,699)
CT	MA	94	40,283	39,689	(594)	45,172	(−4,889)
CT	NH	137	9,012	11,429	(−2,417)	12,503	(−3,491)
CT	RI	76	11,042	16,759	(−5,717)	17,470	(−6,428)
CT	VT	177	7,018	6,412	(606)	6,446	(572)
ME	CT	262	4,563	5,184	(−621)	4,111	(452)
ME	MA	174	13,800	9,931	(3,869)	10,404	(3,396)
ME	NH	127	11,258	4,792	(6,466)	5,151	(6,107)
ME	RI	206	1,757	3,154	(−1,397)	2,913	(−1,156)
ME	VT	169	2,276	2,630	(−354)	2,622	(−346)
MA	CT	94	31,303	49,643	(−18,340)	56,520	(−25,217)
MA	ME	174	27,743	17,885	(9,858)	18,753	(8,990)
MA	NH	63	83,484	36,031	(47,453)	36,167	(47,317)
MA	RI	36	29,790	51,676	(−21,886)	40,567	(−10,777)
MA	VT	141	13,632	13,413	(219)	14,587	(−955)
NH	CT	137	4,613	7,645	(−3,032)	8,358	(−3,745)
NH	ME	127	15,393	4,615	(10,778)	4,961	(10,432)
NH	MA	63	25,890	19,269	(6,621)	19,322	(6,568)
NH	RI	98	1,978	4,992	(−3,014)	5,306	(−3,328)
NH	VT	89	7,814	3,858	(3,956)	3,985	(3,829)
RI	CT	76	7,413	10,801	(−3,388)	11,252	(−3,839)
RI	ME	206	2,677	2,927	(−250)	2,703	(−26)
RI	MA	36	21,437	26,629	(−5,192)	20,882	(555)
RI	NH	98	3,242	4,810	(−1,568)	5,112	(−1,870)
RI	VT	172	1,264	2,147	(−883)	2,116	(−852)
VT	CT	177	2,822	3,336	(−514)	3,350	(−528)
VT	ME	160	2,655	2,052	(603)	2,081	(574)
VT	MA	141	7,998	5,580	(2,418)	6,059	(1,939)
VT	NH	89	8,016	3,001	(5,015)	3,099	(4,917)
VT	RI	172	695	1,733	(−1,038)	1,707	(−1,012)

Populations (1990 Census)

Connecticut (CT)	3,287,116
Maine (ME)	1,227,928
Massachusetts (MA)	6,016,425
New Hampshire (NH)	1,109,252
Rhode Island (RI)	1,003,464
Vermont (VT)	562,758

sectional one rather than a time-series formulation; that is, it applies to one time period rather than to *changes* in the volumes of migration over time. As we shall see later in the chapter, it is possible to unite the two approaches. But in its unconstrained form, the gravity model lacks the careful attention to accounting principles inherent in the demographer's approach to population analysis.

Although used far less frequently to date in migration analysis than in, for example, transportation and shopping modeling, the singly constrained and doubly constrained gravity model formulations embed accounting constraints that should be taken into account in modeling a complete system of flows.

Singly Constrained Gravity Models

The careful reader may have noted that when we first specified a gravity model for migration in Equation 7.2, we omitted the diagonal elements of a migration matrix (i.e., we did not try to estimate the actual values m_{ii}). We did not include *intrastate* migration when we computed the regression estimates of the parameters for New England migration streams, nor did we attempt in our model to treat *nonmigrants* (those persons not leaving their current county of residence). In reality these are both important numerically. For example, of all intercounty migrants tallied in the Current Population Survey data for moves made in 1986–87 (U.S. Bureau of the Census, 1989), 84.5 percent moved within the same state. Furthermore, 93.5 percent of the U.S. population were nonmigrants. (Of the 18.1 percent of the population that moved from one housing unit to another, 63.9 percent moved within the same county.) To form a gravity model that seeks to explain the magnitudes of state-to-state transition probabilities, we must include these important diagonal elements and ensure that we meet the fundamental accounting constraint that all such probabilities for a given origin must sum to 1.

It is relatively easy to write down the formula for an *origin-constrained* gravity model for a migration system with r total regions:

$$M_{ij} = a_i P_i P_j f(d_{ij}) \quad \text{for } i = 1, \ldots, r; \quad j = 1, \ldots, r \tag{7.8}$$

Here $f(d_{ij})$ is a generalized expression for the distance attenuation component of the model; in practice it may take the negative power function, negative exponential, or even some other form. The major difference between this expression and the earlier, unconstrained model of Equations 7.2 and 7.3 is that values are now computed when $i = j$ and there is a different proportionality constant, a_i, for each origin region. In fact, these constants are now called *balancing factors* because their values are chosen to assure that the sum of the numbers of migrants to each of the destinations plus the number of nonmigrants and intraregional migrants (M_{ii}) must equal the population that starts out in each region i:

$$\sum_{j=1}^{r} M_{ij} = P_i, \quad i = 1, \ldots, r$$

The value of a_i is obtained simply by summing both of the terms in Equation 7.8 having a j subscript and then taking the reciprocal of this quantity:

$$a_i = \frac{1}{\sum_{j=1}^{r} \left[P_j f(d_{ij}) \right]}, \quad i = 1, \ldots, r \tag{7.9}$$

Note, in fact, that this singly constrained gravity model provides estimates of transition probabilities that could be used in a Markov model. Substituting the value for the balancing factor in Equation 7.9 into 7.8 and dividing both sides by the population of the origin region, P_i, gives us

$$p_{ij} = \frac{P_j f(d_{ij})}{\sum_{j=1}^{r}\left[P_j f(d_{ij})\right]}, \quad \text{for } i = 1,\ldots,r; j = 1,\ldots,r \tag{7.10}$$

By adding the balancing factor terms to the gravity model, we ensure that the same number of people are distributed as existed to begin with.

In practice it is difficult to estimate a singly constrained model of the type shown by (7.8) or (7.10) because of the difficulty of selecting an appropriate set of values to use for the "intrazonal" distances, d_{ii}. A variety of rules of thumb may be used to estimate a distance if the analyst wishes to model within-region movement, but not also the overall propensity of people to move. For example, one approach is to take "half the nearest neighbor distance" (i.e., to use as the diagonal element of the distance matrix 0.5 times the smallest distance found elsewhere in each of the rows). Given the irregular shapes of areal units, this might roughly approximate the average distance between any two points within each region. Alternatively the average distance from the population centroid of each region and its boundary is sometimes used. For a fuller discussion of this issue, see Rogerson (1990b).

If the gravity model is to be used to generate a complete set of transition probabilities, then it is good practice to undertake a separate migration "generation" phase of analysis to model the propensity to migrate, and then to use the gravity formula to *distribute* migrants to destinations, including destinations within the current region of residence. This two-phased procedure is the approach typically taken in urban transportation planning applications of the gravity model, as we discuss in Chapter 8 when we address forecasting transportation flows on the basis of population and demographic characteristics. The reason we must include intraregional flows in the distribution phase of the analysis is that the relative proportion of these to interregional flows will vary greatly if the areal units used for the analysis differ widely in size. Because the transition probabilities commingle intraregional movement with non-movement, and the diagonal elements of migration matrices are quite large vis-à-vis the off-diagonal elements, the overall accuracy of the modeling effort depends to a great extent on how well the p_{ii} elements are modeled.

Other Forms of the Gravity Model

The unconstrained and origin-constrained gravity models are those most often seen in migration research. Mitchneck (1990), however, used a destination-constrained (or "attraction-constrained") model for analyzing movement patterns in the Soviet Union. Plane (1981) experimented with a *net-constrained* version. Tobler (e.g., 1983; see also Dorigo and Tobler, 1983) suggested using an *additive* version rather than the standard multiplicative form.

Plane (1984) experimented with an *inverse* gravity model, in which a "migration space" is mapped by asking what set of distances, when fed into a *doubly constrained* gravity model, would exactly replicate the actual pattern of movement. A doubly constrained model is one in which the total estimated numbers of out-migrants from each origin region and in-migrants to each destination region must add up to the actual numbers. As shown in Figure 7.6, in the case of some U.S. states, migration space represents a large warping of actual physical separation. This cartographic technique is, in essence, a way of visually presenting the residuals from a gravity model. As we said earlier, the residuals include the composite influences of variables other than actual physical separation on the volumes of the state-to-state migration flows.

Doubly constrained models, commonly used in urban transportation research, are relatively infrequently seen in migration analysis. Stillwell (1978) explored intercounty migration flows in Britain using a variety of distance deterrence specifications in a doubly constrained model. Plane (1982) applied such a model to U.S. interstate migration for 1978–79 and found that the doubly constrained model had an average relative error of 45.8 percent in replicating the 51 × 51 matrix of flows. Obviously factors other than distance are significant in determining the spatial distribution of migrants. Before examining extended, economic gravity models, we first present an alternative to using distance as a primary determinant of migrant choice.

(a) Location of population centroids in "physical" space

(b) Their location based on in-migration flows to Arizona...

(c) ...based on in-migration to Washington

(d) ...and on out-migration from New York.

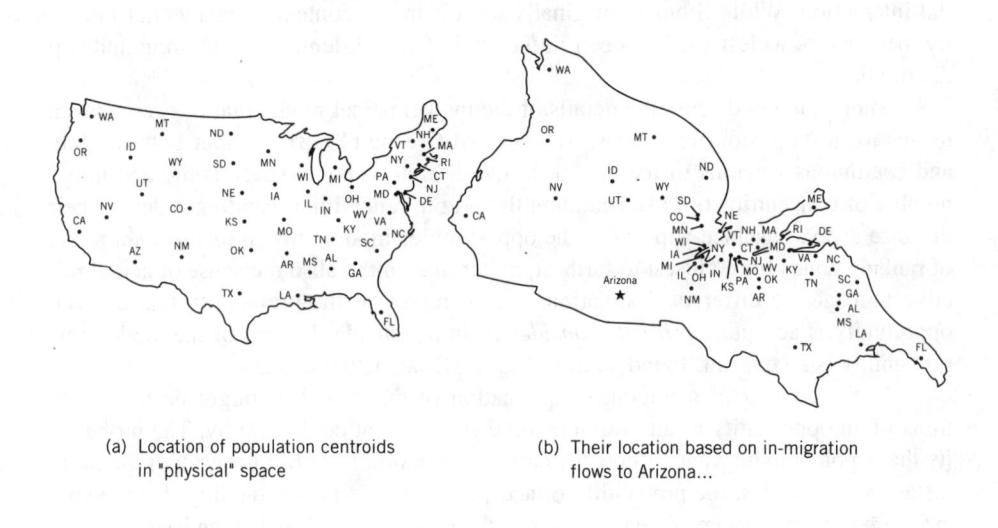

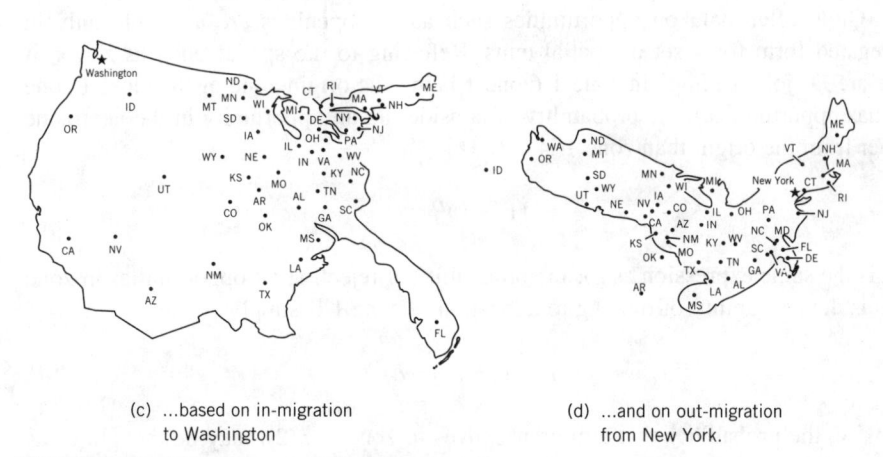

FIGURE 7.6 U.S. Interstate "Migration Space"

Source: Plane (1982). Reprinted by permission.

7.4 THE INTERVENING OPPORTUNITIES MODEL

After noting the important role played by distance in geographic interaction, Stouffer (1940) noted that there is "no necessary relationship between mobility and distance." Instead, Stouffer argued, the amount of interaction between two locations is dependent upon the type and number of opportunities that may be considered as intervening between the two locations:

> The number of persons going a given distance is directly proportional to the number of opportunities at that distance and inversely proportional to the number of intervening opportunities. (p. 846)

Stouffer tested his ideas and found considerable support for them using residential mobility data for the metropolitan area of Cleveland.

The theory of intervening opportunities has been applied to many forms of spatial interaction. While Stouffer originally used it in the context of residential mobility, perhaps its widest use has been in the area of travel demand forecasting and trip distribution.

Many authors discuss the details of the mathematical model that has served as a means for testing Stouffer's ideas. Schmitt and Greene (1978) consider both discrete and continuous versions of the model. In the discrete version, space is divided into a number of concentric zones surrounding the origin, ranked in ascending order of their distance away. Distance appears in the opportunities model only in this *ordinal* sense of ranking zones from closest to farthest, rather than in the absolute sense of actual relative mileages to different destinations. Let L represent the probability that a given opportunity is accepted, *if it gets considered*. In the standard forms of the model, this probability is a constant, found so as to best replicate actual data.

In the context of a migration application of the model, it might be helpful to think of an opportunity as an available job that a migrant could occupy. The probability that a potential migrant accepts the closest job is simply L. But for each of the more distant opportunities, the probability of acceptance equals the probability that the specific opportunity is accepted *conditional on* all closer opportunities having been rejected. Thus for the second-closest opportunity the overall probability that it will be accepted is $L(1 - L)$, for the third-closest, it is $L(1 - L)(1 - L)$, and so forth.

Quite often data on opportunities such as job openings are available only in aggregated form for a set of spatial units. Referring to the spatial units as zones, if there are D_1 job openings in zone 1 (zone 1 being the origin zone or the closest zone that has opportunities), the probability of considering an opportunity in the next zone farther from the origin than zone 1 is

$$(1 - L)^{D_1}$$

This is the same expression as for the probability of rejecting all opportunities in zone 1. Thus the probability of moving to a location *in* zone 1 is simply

$$p(1) = 1 - (1 - L)^{D_1}$$

Likewise, the probability that a migrant moves to zone 2, $p(2)$, may be thought of as the probability that the migrant goes beyond zone 1 (i.e., that opportunities in zone 2 come to be considered) minus the probability that he or she goes beyond zone 2 (i.e., that all opportunities in both zones 2 and 1 are rejected):

$$p(2) = (1 - L)^{D_1} - (1 - L)^{D_1 + D_2}$$

This leads to a general expression for the probability that a migrant leaving a given origin fills an opportunity in any specific destination zone, j:

$$p(j) = (1 - L)^{\sum_{i=1}^{j-1} D_i} - (1 - L)^{\sum_{i=1}^{j} D_i}$$

Methods for estimating the L parameter depend in part on the form of the data. If each of the opportunities is ordered in terms of distance away from the origin, and if observations are collected on the specific destinations chosen by individuals, L may be estimated as

$$L = \frac{N}{\sum_{i=1}^{N} n_i} \qquad (7.11)$$

where N is the total number of individuals for which there are data and n_i is the number rank of the opportunity actually chosen by individual i. For example, suppose there are $N = 3$ individuals, who choose the three closest opportunities. The first individual chooses the third-farthest opportunity, the second accepts the closest, and the third individual selects the middle one. Thus $n_1 = 3$, $n_2 = 1$, $n_3 = 2$, and

$$L = \frac{3}{\sum n_i} = \frac{3}{(3 + 1 + 2)} = 1/2$$

So the probability that any potential migrant chooses the opportunities are $p(1) = .5$, $p(2) = .25$, and $p(3) = .125$. The astute reader will note that these probabilities do not add to 1.0. With $L = .5$ the model will predict positive probabilities of moving to opportunities beyond the one ranked as third-closest. The formula assumes, in fact, that there are an infinite number of opportunities to choose from. In practice, as long as there are a large number of opportunities, the probabilities will sum quite close to 1.0. An adjustment procedure could be used to ensure that the probabilities do sum to 1 over all the opportunities (θ) available to potential migrants. To do so, each of the probabilites $p(1)$, $p(2)$, $p(3)$, ... , $p(\theta)$ would be multiplied by an adjustment factor equal to

$$\rho = \frac{1}{\sum_{j=1}^{\theta} p(j)}$$

In most real cases, where θ is sizable, the adjustment factor ρ is quite small.

As Schmitt and Greene point out, the parameter L is not only interpretable as the probability of accepting a given opportunity, but its inverse may be interpreted as (approximately) the average number of opportunities passed over. This interpretation can be understood by examining the formula for estimating L (Equation 7.11). The number of opportunities "skipped over" by an individual equals $n_i - 1$, one less than the rank of the opportunity accepted. So the average number of opportunities skipped over is

$$S = \frac{\sum_{i=1}^{N}(n_i - 1)}{N} = \frac{1}{L} - 1 \tag{7.12}$$

Since L is usually much less than 1, there will be negligible difference between the values of L and $1/S$ as computed with Equations 7.11 and 7.12.

More often than having base period observations for individual migrants, the data are organized into a set of subregions or zones and information is not obtainable at the individual level. Ruiter (1967) discusses one method for estimating the parameter L in this case. The method involves graphing on the vertical axis the natural logarithm of the observed proportion of trips ending either prior to, or within, zone j, and the total number of opportunities within zone j, or closer to the origin, on the horizontal axis. Data from each zone will generate one point on the graph, and the slope of the best-fitting straight line provides an estimate of L. In practice, rough estimates of the slope are often made by eye; regression analysis can be used to obtain a more precise estimate.

An Example: The Opportunity and Constrained Gravity Models Compared

To illustrate the numerical calculation of the migration flows predicted by the opportunity model, and to compare its predictions to those obtained from an origin-constrained gravity model, consider the hypothetical scenario in Figure 7.7. On the imaginary peninsula of Black Rock are four planning zones: Promonty, Norrfeld, Sudfield, and Mainspur. Rogerson and Plane Shoe Company officials have just announced the closure of their Promonty factory. In all, 400 shoemakers will be laid off. Mitigating the impact of these job losses somewhat, the Black Rock Economic Development Commission forecasts that 200 jobs for workers with these skills will come open in the next year throughout the region, distributed among the four regions as shown in the figure.

To apply the opportunities model, assume that $L = .002$ is the probability that a former Rogerson and Plane employee will land any one of these jobs if the opportunity is "considered." Using the opportunity model formula, the probabilities that any one of the laid-off workers will migrate to the four possible Black Rock destination zones are

$p(1) = 1 - (1 - .002)^{40} = 1 - .9230 = .0770$

$p(2) = (1 - .002)^{40} - (1 - .002)^{40 + 50} = .9230 - .8351 = .0879$

$p(3) = (1 - .002)^{40 + 50} - (1 - .002)^{40 + 50 + 50} = .8351 - .7556 = .0795$

$p(4) = (1 - .002)^{40 + 50 + 50} - (1 - .002)^{40 + 50 + 50 + 60} = .7556 - .6701 = .0855$

These individual-level probabilities are then multiplied by the 400 total workers laid off to figure out the estimated migration flows:

$$M_{11} = O_1\, p(1) = 400\,(.0770) = 30.8$$

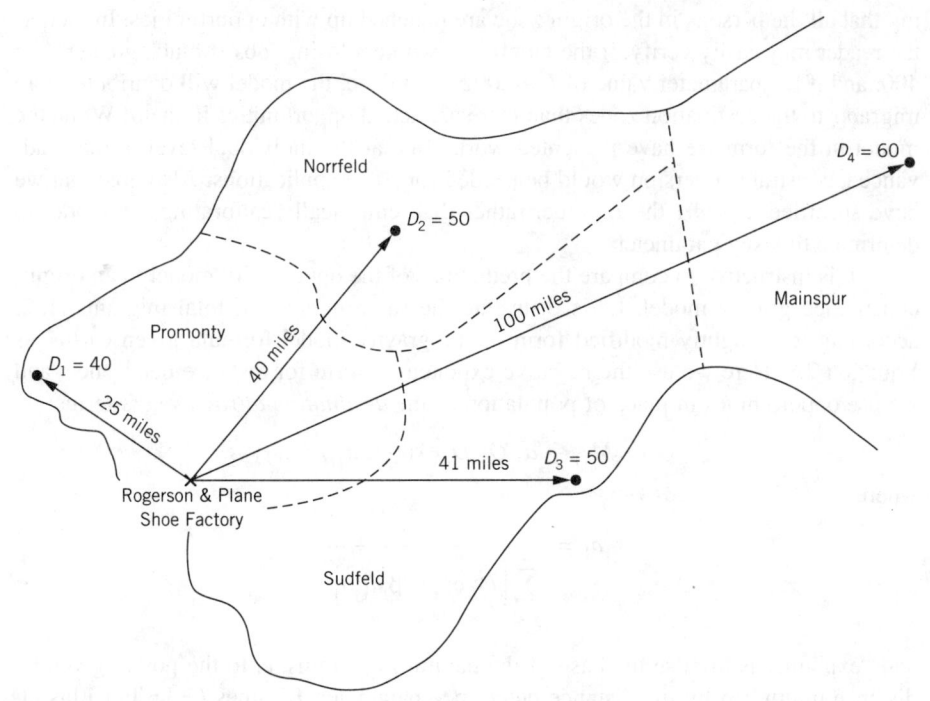

FIGURE 7.7 Location of Job Opportunities on the Black Rock Peninsula for Former Employees of the Rogerson and Plane Shoe Factory

or 31 when rounded to the closest whole person! Similarly,

$$M_{12} = 400(.0879) = 35.16$$
$$M_{13} = 400(.0795) = 31.80$$
$$M_{14} = 400(.0855) = 34.20$$

So our estimated migration flows are 31, 35, 32, and 34.

Note that the model does predict a form of distance decay. There are fewer migrants to the more distant zone of Sudfeld than to the nearer zone of Norrfeld (both having the same number of job openings). And the number of migrants to Mainspur is less than it would be if this zone—which has the most opportunities—were one of the nearer ones to the origin. As in the gravity model, there is a trade-off between distance and the attractiveness (number of opportunities) of the destination zones. Because in this example the more distant zones have more opportunities than the closer ones, the final model predictions of migrants to each are roughly similar in magnitude. But another way to look at the results would be to examine the *fraction* of available jobs in the zones expected to be filled by the laid-off workers from Promonty. Standardizing in this fashion, we find that the percentages decline from 77 percent for the new jobs opening up in Promonty, to 50 percent of the jobs in Mainspur.

One other thing to note about the opportunity model's results is that of the 400 workers to be laid off, only 132 have been placed in jobs on the peninsula. The 268 others may move to even more distant locations than shown here, move to one of the other zones without filling a job opening, or remain in Promonty unemployed. (Such is the nature of layoffs!) The model as used here is not constrained in the sense of assuring that all the persons in the origin zone are matched up with opportunities. In fact, as the reader may easily verify, if the number of workers losing jobs is much greater than 400, and if the parameter value of $L = .002$ is retained, the model will distribute more migrants to the destination zones than there are actual opportunities in them! While the model in the form we have presented works fine at the individual level, a more advanced, constrained version would be needed for some applications. Also note that we have specified *a priori* the L value, rather than empirically calibrating the model to determine this key parameter.

It is instructive to compare the predictions of the opportunity model to an origin-constrained gravity model. Let's distribute the same number of total migrants, 132, according to a slightly modified form of the gravity model formula given earlier as Equation 7.8. Here we use the negative exponential form for distance deterrence, and we use opportunities in place of population as the *destination attractiveness* terms:

$$M_{1j} = a_1 \, O_1 \, D_j \exp(-\beta \, d_{1j})$$

where

$$a_1 = \frac{1}{\sum_{j=1}^{4}\left[D_j \exp(-\beta \, d_{1j})\right]}$$

and "exp" means to raise the base of the natural logarithms, e, to the power given by distance multiplied by the distance deterrence parameter, β, times (-1). For illustrative purposes we shall use a parameter value of $\beta = .02$, although in a real application, empirical calibration of the model would be required.

Computing first the value of the balancing factor,

$$a_1 = 1 \,/\, \{(40) \exp[(-.02)(25)] + (50) \exp[(-.02)(40)] + (50) \exp[(-.02)(41)]$$
$$+ (60) \exp[(-.02)(100)]\}$$
$$= 1 \,/\, 76.87 = .0130$$

It is then a simple matter to derive the estimated migration flows:

$$M_{11} = (.0130)\,(123)\,(40)\,(.6065) = 38.79$$
$$M_{12} = (.0130)\,(123)\,(50)\,(.4493) = 35.92$$
$$M_{13} = (.0130)\,(123)\,(50)\,(.4404) = 35.21$$
$$M_{14} = (.0130)\,(123)\,(60)\,(.1353) = 12.98$$

Thus the origin-constrained gravity model, with the negative exponential version of distance deterrence and a $\beta = .02$ parameter value, predicts a distribution of 39, 36, 35, and 13 persons to the four destination zones. Recall that the opportunity model, with $L = .002$, resulted in a distribution of 31, 35, 32, and 34 laid-off workers to these same zones. The differences illustrate the different ways in which distance enters the two models.

Compare, first, the predictions of M_{12} and M_{13}. Because zone 3 is only one mile farther away from the origin zone than zone 2 (and they have the same number of job openings), the gravity model predicts almost exactly the same number of migrants to both zones. In the case of the opportunities model, however, with distance as an ordinal concept, *all* the job openings in zone 2 are evaluated before any of those in zone 3, so there is a somewhat greater difference between the distribution of trips to these two zones.

The biggest difference between the two forecasts, however, is for moves to zone 4. The gravity model distributes relatively few trips to this zone because it is so much farther away, in terms of absolute distance, than zone 3. The opportunities model simply begins to evaluate job openings in zone 4 after all those in zone 3 have been considered and rejected. Its predictions would be identical if zone 4 were 42 miles rather than 100 miles from the Rogerson and Plane factory!

The results of a large number of empirical applications of the gravity and intervening opportunities models have convinced us that neither is an exact replication of how people evaluate distance in their spatial decision making. In many cases, real behavior may well represent a compromise between viewing distance in ordinal and cardinal ways, with the relative weighting of the two dependent on the specific context.

7.5 TEMPORALLY VARYING TRANSITION PROBABILITIES

We began this chapter's discussion of migration modeling with the classical gravity model. It includes, in addition to distance, the population of origin and destination zones as independent variables used to estimate the size of migration streams. In the previous chapter on population projections, the Markov model of migration was presented. In the Markov model, origin population enters in the determination of future levels of forecasted migration flow because it measures the changing population "at risk" of possibly becoming migrants, but destination population has no role.

As pointed out in "Requiem for the Fixed-Transition-Probability Migrant" (Plane, 1993a), the fixed-transition-probability Markov model and the gravity model represent fundamentally different views of long-term dynamic change in a migration system. The Markov model states that migrants continue to leave origin regions and go to the different destination regions in fixed proportions for an indefinite period into the future. As we have seen in Chapter 6 and indeed throughout various preceding sections, it is common practice to assume that the parameters describing demographic rates and probabilities are constant over time. This greatly facilitates the process of projection, and is in fact a justifiable approach when either there is little reason to expect change or there is a substantial degree of uncertainty regarding the direction of change. Migration, however, is a fundamentally different phenomenon than either birth or death in that the outcome of the process involves *destination* choice. As argued in Plane (1993a), fixed transition probabilities do not represent a correct behavioral representation of a migration system. As the gravity model suggests, the changing distribution of population should, itself, be a determinant of future patterns of destination choice. Before examining models that extend the fixed-transition-probability framework to build in a role for destination population, we examine the issue of temporal stability of transition probabilities through the notion of *causative matrices*.

The Causative Matrix Model

When the analyst has available a number of matrices representing past migration patterns, how can that information be used to project future patterns? Several alternatives are available. It is common to assume that the most recently observed rates of migrating between origins and destinations will remain constant. Collins (1972), however, suggests pooling the data together to estimate an average pattern; this average pattern could then be used for projection purposes.

When change is rapid, the assumption of fixed rates is no longer satisfactory, and an alternative must be found. One choice is to assume that migration is a function of independent, explanatory variables. We shall explore this approach in Section 7.6 when we take up the matter of estimating economic gravity models. Once estimated, the parameters of the functional relationship are assumed constant. Change in migration patterns then arises when the independent variables change. The accuracy of such projections depends upon the adequacy of the assumption of parameter stationarity, as well as the accuracy of the projected independent variables. Gale (1972) terms this approach *functional stationarity*, and Spilerman (1972) provides an application to mobility.

Alternatively, *differential stationarity* may be defined where the relationship between transition probabilities in one period is related in some constant way to transition probabilities in the previous period. For example, Rogerson (1979) examines the case where transition probabilities change linearly over time. Another type of differential stationarity would have constant transition probabilities with constant ratios from one period to the next.

Rogerson and Plane (1984) and Plane and Rogerson (1986) use a constant matrix (**C**) to transform one probability matrix into the next:

$$\mathbf{P}(t + 1) = \mathbf{P}(t)\,\mathbf{C} \tag{7.13}$$

The matrix **C** is termed a constant causative matrix; in this section we focus upon its use and interpretation.

For those familiar with regression analysis, Equation 7.13 is a model that is equivalent to its matrix version. The causative matrix is analogous to the notion of slope, and is assumed to remain constant over time. Once estimated, the matrix **C** is used to project the transition probabilities forward in time. Or perhaps an analogy with physics may be informative. Rather than to assume velocity (represented by the transition matrix) is constant, certain situations require that the rate of change of velocity, or acceleration (represented by the causative matrix), be constant. Although the probabilities of moving between origins and destinations are not constant, the rates of change in these probabilities (as captured by **C**) are constant.

The estimation of the causative matrix is straightforward when two transition matrices are available. Both sides of (7.13) are premultiplied by the inverse[3] of $\mathbf{P}(t)$ to obtain

$$\mathbf{C} = \mathbf{P}(t)^{-1}\,\mathbf{P}(t + 1) \tag{7.14}$$

The fit in this case is necessarily perfect; it is analogous to fitting a regression line to two points. Fitting the "matrix slope" **C** to more than two matrices is more complex. The problem becomes one of nonlinear least squares; Rogerson and Plane (1984) give details regarding the estimation of **C** in this instance.[4]

---

[3]The reader not familiar with matrix algebra may read this section to pick up the flavor of the technique. For people interested in working with causative matrix methods, excellent introductory treatments of linear algebra that go beyond the rudiments presented in our Appendix D may be found in Chapter 1 of Miller (1972) and in Rogers (1971).

[4]Also note that a different causative matrix can be obtained by premultiplying $\mathbf{P}(t + 1)$ rather than postmultiplying as in Equation 7.13. The "right" causative matrix shown here, however, is more commonly found in the literature than the alternative "left" matrix. The interpretation of $\mathbf{C}^R$ is also somewhat more intuitive than that of $\mathbf{C}^L$. The latter represents a competing-origins rather than competing-destinations perspective. Plane and Rogerson (1986) cover alternative interpretations of the two.

Table 7.2 was used to demonstrate the different nature of long-term migration system change in U.S. interregional migration during the 1960s and 1970s than during the two previous decades. This comparison exposes the dramatic shift (or clean break, as some authors dubbed it) that took place from a longstanding pattern of population *concentration* in the traditional "core" region of the nation (the Northeast and Midwest) to the current *deconcentration* regime of Frostbelt-to-Sunbelt net migration. As we discussed in Chapter 1, this pattern of interregional flow began in the late 1960s and reached extraordinary heights in the 1970s—when the U.S. baby boom generation flooded regional labor markets.

The elements of the **C** matrices in Table 7.2 represent the relative competitive changes in the abilities of the various regions of the United States to "compete" with one another as attractors of migrants. Positive values indicate increased competitiveness. The largest off-diagonal elements are those representing the competition of the South with the Northeast, and the South with the Midwest. The South was the only region of the four to experience increasing relative attractiveness with respect to all other regions during both the 1935–40 to 1955–60 and 1965–70 to 1975–80 periods.

The *column sums* of the causative matrices may be interpreted as overall measures of changes in the destination attractiveness of the regions. Values greater than 1.0 indicate increased overall attractiveness vis-à-vis the other regions. Note that the South is the only region to have such values for both time periods. It should be mentioned that since the **C** matrices are based only on the internal migration movements among the four census divisions of the United States, the changing focus of foreign immigration from the East Coast to West Coast over this period is not picked up as an important contributor to the Sunbelt population growth phenomenon of recent times.[5]

The constant causative matrix is primarily useful as a tool for analyzing the nature of change that has taken place between time periods in actual transition probabilities, although it could also be used for forecasting purposes. To do so, the analyst would first find **C** from two historical matrices:

$$\mathbf{C} = \mathbf{P}(t-1)^{-1}\,\mathbf{P}(t)$$

and then use Equation 7.13 to forecast a future matrix, $\mathbf{P}(t+1)$. As we have noted earlier, this is an advanced form of trend extrapolation. The interpretation of **C** matrices as measures of interregional competition is, however, relatively complicated and this approach has therefore not yet been used in many actual forecasting applications.

We turn now to simpler methods for positing a role for destination attractiveness—while still maintaining the mathematical elegance and demographic accounting logic of the transition-probability approach. There have been a number of attempts to allow migration transition probabilities to change, as a function of changes in destina-

---

[5]One final measure that may be derived from the causative matrices deserves mention, though its computation requires an understanding of matrix algebra considerably beyond that expected for the average reader of this book. The *maximum eigenvalue* of the **C** matrix, λ^*, measures the convergence properties of the migration system. If its value is 1.0, a *convergent* period is indicated; a value greater than 1.0 suggests *divergent* temporal change (i.e., disruptive dynamics or a break with past trends). The eigenvector for the $\mathbf{C}_{40,60}$ matrix is $\lambda = (1.0000, .9684, .9637, .9561)$, indicating convergent dynamics for the period from 1935–40 to 1955–60. For the $\mathbf{C}_{60,80}$, however, the eigenvector is $\lambda = (1.0017, 1.0000, .9955, .9775)$; because λ^* is greater than 1.0 the 1970s are indicated as a period of disruption in the previously established trends. This causative matrix evidence accords well with the previous discussion about the net migration changes that took place during the 1970s.

TABLE 7.2 Transition Probability Matrices, **P**, and Causative Matrices, **C**, for U.S. Interregional Migration, 1935–40 to 1975–80

P$_{40}$

	NE	MW	S	W
NE	.98498	.00452	.00741	.00309
MW	.00386	.97124	.00875	.01615
S	.00462	.00981	.97593	.00964
W	.00328	.00944	.00860	.97868

P$_{60}$

	NE	MW	S	W
NE	.95728	.00898	.02241	.01133
MW	.00692	.94396	.02398	.02515
S	.01203	.02044	.94852	.01901
W	.00719	.01703	.02315	.95264

P$_{80}$

	NE	MW	S	W
NE	.93487	.00991	.03868	.01655
MW	.00650	.93669	.03392	.02289
S	.00980	.01542	.95878	.01601
W	.00700	.01690	.02799	.94811

C$_{40,60}$

	NE	MW	S	W
NE	.97179	.00455	.01532	.00833
MW	.00313	.97166	.01563	.00959
S	.00766	.01108	.97154	.00973
W	.00399	.00792	.01491	.97319
Column sums:	.98657	.99521	1.01740	1.00084

C$_{60,80}$

	NE	MW	S	W
NE	.97664	.00116	.01661	.00560
MW	−.00022	.99242	.01003	−.00223
S	−.00205	−.00515	1.01030	−.00309
W	.00003	.00012	.00453	.99532
Column sums:	.97440	.98855	1.04146	.99560

Source: Plane and Rogerson, 1986, p. 99. Reprinted by permission.

Note: NE = Northeast; MW=Midwest; S=South; W=West. Regions are those defined by the U.S. Bureau of the Census. Subscripts of the **P** and **C** matrices denote the census years.

tion attractiveness. To conclude this section we present several related models that embed destination population as a measure of regional attractiveness. In Section 7.6 we examine models that have used more explicit measures of economic opportunity.

Feeney's Model

Feeney (1973) used destination population as a surrogate for economic opportunity, reasoning that more people require more services (and hence enhance employment growth prospects). This opportunity-size effect is further enhanced by agglomeration economies, which make larger regions more attractive to employers to locate still more jobs in them. Feeney suggested that the Markovian transition probabilities be adjusted to account for changes in the distribution of destination populations:

$$p_{ij}^t = p_{ij}^0 \frac{p_j^t \Big/ \sum_{k \neq i} p_k^t}{p_j^0 \Big/ \sum_{k \neq i} p_k^0} \qquad (7.15)$$

where the superscript 0 is used to denote observed populations and transition probabilities in some base year, and the superscript t is used for the forecast year. Equation 7.15 states that the likelihood of going from region i to region j during year t is directly related to region j's changing share of the total population residing in all locations other than i. Thus if region j has an increasing percentage of the total systemwide population (excluding region i, since Feeney considers only movement to locations other than i), we would expect the likelihood of going from i to j to increase between year 0 and year t. Note that if the population share does not change, the population share terms in the numerator and the denominator of the right-hand side of (7.15) are equal, so the transition probability is projected to remain unchanged.

Under Feeney's approach, the diagonal elements of the transition probability matrix are computed as residuals; after all the other probabilities have been found, the analyst simply derives them from the formula

$$p_{ii} = 1 - \sum_{j \neq i} p_{ij} \qquad (7.16)$$

To see how the model may be used for projection, consider the hypothetical three-region system in Figure 7.8. The interregional migration flow data is first used to derive the transition probability matrix. (See Chapters 4 and 6.) The resulting matrix is

To:	Region $j = 1$	Region $j = 2$	Region $j = 3$
From:			
Region $i = 1$	.8000	.1000	.1000
Region $i = 2$	.0500	.8500	.1000
Region $i = 3$	.0300	.0600	.9100

It is clear from Figure 7.8 that region 1 is the only region receiving positive net in-migration. Using the standard multiregional projection method discussed in Chapter 6, application of this transition matrix to the starting population yields a new population of 2,200 in region 1, 5,900 in region 2, and 9,900 in region 3.

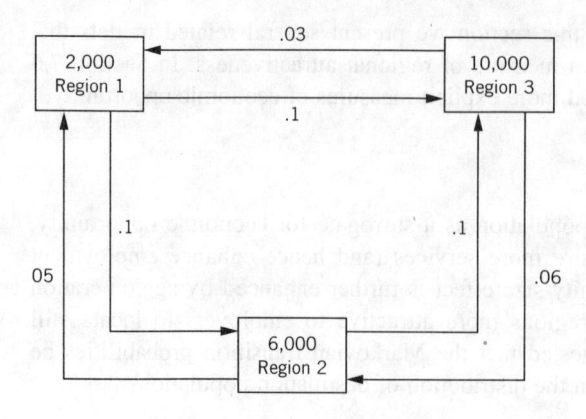

216   Chapter 7 • Modeling and Forecasting Migration

```
         .03
  ┌─────────┐ ←──── ┌──────────┐
  │  2,000  │       │  10,000  │
  │ Region 1│       │ Region 3 │
  └─────────┘ ────→ └──────────┘
               .1
       ↑              ↑
    .1 │           .1 │
  .05  │              │  .06
       │  ┌─────────┐ │
       └──│  6,000  │─┘
          │ Region 2│
          └─────────┘
```

FIGURE 7.8 Populations and Transition Probabilities for a Hypothetical Three-Region System.

Note: Numbers in boxes are populations; numbers along arrows are transition probabilities.

The resultant increase in region 1's population in turn gives rise to increased probabilities of moving to region 1. Using Equation 7.15, the new transition probability matrix is found to be

To:	Region $j = 1$	Region $j = 2$	Region $j = 3$
From:			
Region $i = 1$	.8002	.0096	.1002
Region $i = 2$	.0545	.8473	.0982
Region $i = 3$	.0326	.0583	.9091

For example, the probability of moving from region 1 to region 2 equals the original probability (.1) multiplied by region 2's changing share of the total destination population, as viewed from region 1 (which is equivalent to the population of region 2 plus the population of region 3). Thus

$$p_{12}^{t} = .1000 \frac{5,900 \ / \ (5,900 + 9,900)}{6,000 \ / \ (6,000 + 10,000)} = .0096$$

Table 7.3 displays the projected regional populations, and compares them with the Markov, constant-transition-probability projections. Note that a long-run equilibrium is reached—one that is quite different from the Markovian scenario.

Why is an equilibrium reached at all with Feeney's model? If the growing regions by definition become more attractive destinations, why doesn't the most attractive region of all contain the entire system's population in the long run? The reason is that, even though the attractive region (in this case, region 1) initially draws an increasing number of migrants, it will subsequently be drawing from a smaller population elsewhere at risk of moving. At the same time, the growth of the attractive region will also result in a higher base population at risk of moving out. Note that because the origin region, i, is excluded as a potential destination for migrants, the *out-migration rates* stay constant over time—it is only the distribution of migrants to (external) destinations that is influenced by the changing population distribution within the system.

In this regard, the Feeney model is somewhat like the constant-rate/fixed-share-of-pool method formerly used by the U.S. Bureau of the Census (1979) to make state-level population projections. In that approach, all migrants resulting from applying age-specific base period out-migration rates to the initial populations of each state were assigned first to a national migrant "clearinghouse" and then allocated back to each state according to that state's historical share of the base period's migrant pool. A

TABLE 7.3 Markov and Feeney Model Population Projections Compared for a Hypothetical Three-Region System

	Region 1		Region 2		Region 3	
Time	Markov	Feeney	Markov	Feeney	Markov	Feeney
1	2,200	2,200	5,900	5,900	9,900	9,900
2	2,352	2,405	5,829	5,795	9,819	9,800
3	2,468	2,612	5,779	5,686	9,753	9,701
4	2,556	2,820	5,744	5,576	9,700	9,604
.	.	.	.	.	.	.
.	.	.	.	.	.	.
.	.	.	.	.	.	.
13	2,815	4,373	5,677	4,707	9,508	8,920
.	.	.	.	.	.	.
.	.	.	.	.	.	.
.	.	.	.	.	.	.
Long run	2,842	5,479	5,684	3,913	9,474	8,609

politically expedient property of such a method is that over time states that are losing population in the base period will have their rates of loss diminished (while growth will also slow down in growing states). Regardless of what actually takes place during the projection period, population numbers based on these kind of dynamics are probably "safer" ones for bureaucrats to issue than those showing continuous or accelerating population losses in declining states.

With the Feeney model there may be instances in which some regions are completely "drained" of their entire population.[6] Even worse, however, there is no internal mechanism within the model to ensure that the projected transition probabilities don't exceed 1.0. Under some values of base transition probabilities, ultimately more people may be moved to some specific destination j from origin i than there were people in region i to begin with! Though the actual magnitudes of p_{ij} terms are unlikely to cause such a result in practice, a considerably more likely possibility is that the sum of the transition probabilities over all destination regions will exceed 1. As we see in Equation 7.16, the model could thus forecast negative diagonal elements and even negative populations for some regions.

Does the Feeney model generate projections that significantly improve upon Markov projections? The answer must necessarily depend upon the specific empirical application and the nature of the temporal change taking place in migration patterns during the projection period. There may be some situations where the Feeney model does better, and others where it does not. We used Internal Revenue Service data on annual migration flows between the four census regions over the period 1980–88 to

---

[6]The reader may wish to verify that the long-run population of region 1 will be zero when the initial populations in Figure 7.8 are used with the following initial transition probability matrix:

To:	Region $j = 1$	Region $j = 2$	Region $j = 3$
From:			
Region $i = 1$	.6000	.2000	.2000
Region $i = 2$	.0500	.8000	.1500
Region $i = 3$	.0300	.0600	.9100

project destination choice probabilities for the following year, using both Feeney and Markov models. In six out of the seven comparisons, the Markov destination choice probabilities were closer to the actual probabilities than were those derived via Feeney's model. At least for this geographic scale over this time period, when examining only one-year-ahead projections, the modification of transition probabilities by changing population shares did not improve upon the Markov projection.

The Destination Population Weighted (DPW) Model

This model was formulated by Plane (1982) to circumvent the potential accounting problems with Feeney's model. It may be stated as

$$p_{ij}^t = a_i p_{ij}^0 (P_j^t / P_j^0)$$

where a_i is a balancing factor, similar to those we have seen for origin-constrained gravity models, designed to ensure that the transition probabilities sum to 1 when added across all regions, and that all future-year probabilities have values between 0 and 1. This constant term may be calculated as

$$a_i = 1 \bigg/ \sum_{j=1}^{r} \left[p_{ij}^0 \left(P_j^t / P_j^0 \right) \right], \quad i = 1, \ldots, r \tag{7.17}$$

One difference between the DPW and Feeney models is the treatment of the origin population. In the Feeney model, the probability of staying within the origin region is derived as a residual; in the DPW model, this probability is treated in the same way as the flows to potential destinations. (Note that the summation in Equation 7.17 is over all destinations, including the origin.) While including the origin as a potential destination ensures the derivation of appropriate transition probabilities in the DPW model, and while it makes sense to consider the possibility of *intrazonal* movement in addition to moves outside the origin, the "improvement" comes at a cost. The DPW model possesses considerably stronger positive feedback than the Feeney model. The out-migration probability is itself a function of the origin region's changing relative share of systemwide population. Eventually all elements of the transition matrix will go either to 0 or 1, with the undesirable result of everyone in one region and no one in the other regions.

Table 7.4 gives the results of DPW model projections using the hypothetical three-region system in Figure 7.8. A comparison of these results with those in Table 7.3 reveals the stronger positive feedback inherent in the DPW model.

Improving on the Basic DPW Model

Plane (1993a) sets forth some modified versions of the DPW model in which the strong feedback inherent in the basic model is *dampened*. It can be contested that during a portion of the regional development process, greater population is an advantage because of agglomeration economies. But at some stage the diseconomies and disamenities of overcongestion outweigh these initial advantages. At later stages, growth ought to slow down on its own accord. For example, we saw that in the case of 1985–90 migration among the New England states with their mature, postindustrial economies, there was a current of net movement away from highly urbanized Connecticut and Massachusetts toward less densely settled Maine and New Hampshire. Considerable recent research has focused on the notions of "counterurbanization,"

TABLE 7.4 DPW Model Population Projections for a Hypothetical Three-Region System

Time	Region 1	Region 2	Region 3
1	2,200	5,900	9,900
2	2,454	5,771	9,775
3	2,777	5,605	9,618
4	3,190	5,391	9,419
.	.	.	.
.	.	.	.
.	.	.	.
13	14,698	289	3,014
.	.	.	.
.	.	.	.
.	.	.	.
Long run	18,000	0	0

periphery-to-core/core-to-periphery net migration turnarounds, and *polarization reversals*. (See, e.g., Champion, 1989; Berry, 1988; Vining and Strauss, 1977; Vining and Kontuly, 1978; Vining and Pallone, 1982; Richardson, 1980; Brown and Lawson, 1989.) *Density-dampened DPW models* (Plane, 1993a) can be employed to proxy such views of development dynamics.

Another important issue raised by our discussion of the Feeney model versus the DPW model is that in modeling temporal changes in the diagonal elements of transition probability matrices, attention should be paid to the distinction between *nonmovers* and *within-region* movers. Both types of persons are typically included, and the basic DPW model essentially treats both types in the same fashion—as if motivated by the same forces as those that determine the flows to the other regions of the system. Considerable behavioral migration research, however, has focused on the distinction between the *decision to move* versus the *decision about where to move*. See Plane (1993a) for more details about DPW-type models that embed elements of a *mover–stayer* framework, and for comparisons based on the inclusion of density-dampening terms.

The Feeney model, the DPW model, and its variants all attempt to weld together the powerful conceptual framework of multistate demographic accounting techniques (as introduced in Chapters 3 and 6) with the considerable body of spatial interaction literature that has arisen in geographic migration research. The role of origin and destination variables has been extensively explored in the regional science literature. A role is accorded to both origin and destination changes in attractiveness, unlike in the standard fixed-transition-probability Markov model. We turn now to examine models of migration that seek to extend the notion of "attractiveness" to encompass economic factors and a variety of other behavioral variables

7.6 ECONOMIC GRAVITY MODELS

At about the same time, Lowry (1966) and Rogers (1967) published results of extended versions of the basic gravity model positing a role for economic conditions in both origin and destination regions. Shortly thereafter Greenwood (1969) contributed an influential study on the economic variables affecting labor mobility in the United States. Since then, a large, lively literature has sprung forth on the econometric estimation of relationships between migration and its determinants. A number of review surveys are now available to guide the interested reader through this regional science literature on migration (Shaw, 1975; Greenwood, 1975, 1985), and several research prospective papers attempt to set new directions for this type of migration modeling (Rogerson, 1984; Plane and Rogerson, 1990; Greenwood et al., 1991).

The theoretical underpinnings of these models are often traced to a seminal paper by Sjaastad (1962) setting forth the framework of analyzing the economic *costs* faced by migrants versus the economic *gains* (or returns) that the migration decision can bring, treating the process in the context of long-term utility-maximizing behavior. This *human capital* perspective grew out of neoclassical theories of regional growth. As suggested by Borts and Stein (1964), wage rate differentials induce individuals to move between regions. Migration thus acts as an equilibrating mechanism bringing labor supply and demand into balance. According to this view, in the long term, migration is an important mechanism for reducing interregional wage differentials.

Simultaneous with Lowry's and Rogers' seminal economic modeling exercises, Lansing and Mueller (1967) published the voluminous results of a survey analyzing the motivations and characteristics of labor force migrants. Their findings have since been much cited and used to justify the inclusion of particular variables in migration modeling applications.

Lowry's Model

The extended gravity model framework set forth by Lowry (1966) has exerted a powerful influence on subsequent work, although, as we saw in Chapter 4, some of his empirical results (such as that about the relative role of destination versus origin economic conditions) have generated considerable controversy. The model was estimated on census data for movement between metropolitan areas from 1955 to 1960. It took the form

$$M_{ij} = a \, (u_i^{c_1} / u_j^{c_2}) \, (w_j^{c_3} / w_i^{c_4}) \, L_i^{c_5} L_j^{c_6} / d_{ij}^{b} \tag{7.18}$$

Here M_{ij} is the estimated number of migrants moving from origin metropolitan area i to destination area j, and d_{ij} is the air distance between the two areas. Taking the place of total populations as the origin and destination "mass" terms are the sizes of the respective (nonagricultural) labor forces, L_i and L_j. The conceptually new variables are u_i and u_j, the unemployed percentages of the labor force at origin and destination, and the respective hourly manufacturing wage rates, w_i and w_j.

The quantities a, b, and c_1 to c_6 are the regression coefficients that result from the best fit of the model to the actual 1955–60 migration data. To determine these, Lowry used the same log-linear transformation of Equation 7.18 that we described earlier for the basic gravity model. That is, he actually estimated the parameters α, β, and γ_1 to γ_6 for the equation

$$\begin{aligned}\log M_{ij} = \alpha &+ \gamma_1 \log u_i + \gamma_2 \log u_j + \gamma_3 \log w_j \\ &+ \gamma_4 \log w_i + \gamma_5 \log L_i + \gamma_6 \log L_j + \beta \log d_{ij}\end{aligned} \tag{7.19}$$

The economic rationale for the model accords well with our intuition. Its dual assumptions are: (1) individuals have an incentive to move from low-wage to high-wage areas, so the gross flows should be greater in those directions than the opposite ways, and (2) unemployed persons in areas with surplus labor—that is, where the unemployment rate is high—should seek out opportunities where labor shortages exist and consequently unemployment rates are low.

Note that Lowry set up the original model (7.18) so that all the parameters (a, b, c_1, c_2, c_3, c_4, c_5, c_6) would be expected to have positive signs. High origin unemployment should serve as a push factor, so u_i is shown in the numerator of Equation 7.18; we would expect to find a positive sign for the γ_1 coefficient in the logarithmically transformed version (7.19). On the other hand, there should be an inverse relationship between destination unemployment and the size of migration flows, with smaller than average movement to areas that already have many people out of work. Consequently u_j is shown as a denominator variable in the model and a negative sign would be expected for γ_2 in (7.19). Origin and destination wages enter the model in just the opposite fashion. Higher than average wages at the destination should lure more migrants, while higher wages at the origin should serve to retain existing members of the labor force and diminish out-migration streams. Thus a positive γ_3 and a negative γ_4 should result when the model is estimated.

Lowry's results (Table 7.5) indicated that adding the economic condition variables marginally improved the statistical fit compared to a standard gravity model having only L_i, L_j, and d_{ij} as independent variables. Of the four new variables, however, only destination unemployment was statistically significant. High rates of unemployment did seem to discourage in-migration. The wage variables had negligible effects, and origin unemployment, though statistically insignificant, actually had a counter-intuitive, negative sign—meaning that the higher the unemployment rate, the less likely would it be for people to migrate out from their current metropolitan areas of residence.

In the years since Lowry published his somewhat surprising findings, considerable research has been done on the issues raised, and we now have a much better

TABLE 7.5 Lowry's Regression Results for an Economic Gravity Model of Migration Between 90 U.S. Metropolitan Areas, 1955–60

Basic Gravity Model		Economic Gravity Model	
Variable	Coefficient	Variable	Coefficient
Intercept	−7.9098	Intercept (α)	−12.74999
—	—	log u_i (γ_1)	−0.13304
—	—	log u_j (γ_2)	−1.29351***
—	—	log w_j (γ_3)	0.24230
—	—	log w_i (γ_4)	−0.02657
log L_i	1.01863***	log L_i (γ_5)	1.04734***
log L_j	1.02273***	log L_j (γ_6)	1.08592***
log d_{ij}	0.25711	log d_{ij} (β)	−0.49311***

Source: Adapted from Lowry (1966), Table 1 (p. 15) and Table 2 (p. 16).

***Indicates the coefficient is significantly different from zero at the .999 level of confidence; coefficients unmarked were not significant at either the .999, .99, or .95 confidence levels.

understanding of why aggregate measures of unemployment and wages are not always related to the volumes of migration flows in the ways we would first expect. Unemployment rates are typically high both in economically depressed regions, where layoffs are common and few jobs are open, and in relatively healthy places, where considerable *job turnover* is taking place—where people are willing to quit less desirable jobs in expectation of finding better ones after a relatively short search period. To use demographic parlance, the unemployment rate is a stock measure, derived from several different *flow* variables such as new hires, quits, and layoffs. The flow measures may be more logical and "cleaner" measures of migration probabilities than stock variables. (See Figure 7.9.)

Another important point is that, even during severe recessions, most of the labor force remains employed. For employed persons the probability of movement goes down when the health of a regional economy is poor, and geographic labor mobility increases during expansionary periods. It is largely during good times when firms are able to engage in creating new regional capital, thus opening new job opportunities for persons seeking to upgrade their economic lot; during recessions people tend more to hold onto existing jobs and take fewer risks in going into the tight job market.

The preponderance of labor force migration in the United States and other developed countries may be classified as *contracted* rather than *speculative*. Contracted migrants are those who have a new job contract in hand before moving, whereas speculative migrants move first and then engage in the job search process. (Silvers, 1977, analyzes the very different optimal behavior for these two different types of movements.)

The Role of Amenities

The failure of wage variables to perform strongly in the basic economic gravity model led to numerous attempts to improve upon the specification of the model given by Equation 7.18. It was felt that by entering additional control variables, the pure effects of wages could be brought to light. In this connection, the body of literature on *hedo-*

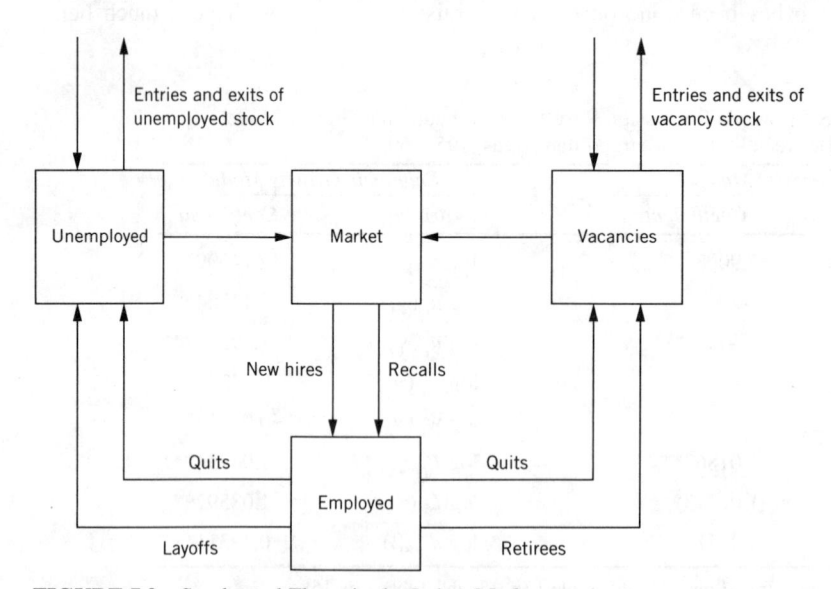

FIGURE 7.9 Stocks and Flows in the Labor Market

Source: Plane and Rogerson (1985, p. 192). Reprinted by permission.

nic prices assumes importance for the population analyst. The idea is that people derive satisfaction from various sources—some monetary and some nonmonetary. Some sources of *psychic income* are highly place-specific—for example, climatic conditions and the presence of cultural or recreational opportunities (e.g., art museums, symphony orchestras, major league ball teams, skiing resorts, hiking trails, and lakes).

Several notable works have sought to estimate the value of amenities in the migration decision using a regression modeling approach (Graves, 1980; Graves and Linneman, 1979; Porell, 1982). The premise of this school of migration research is an *equilibrium* viewpoint that takes issue with the disequilibrium, wage-adjustment perspective of the Borts and Stein (1964) school of thought. Similar to the predominant approach used in urban economics research, migration is viewed as facilitating short-term adjustments to new equilibrium states. The impetus for migration can arise, according to this perspective, as the demand for amenities changes in response to, for example, rising levels of affluence (Greenwood, 1985). In other words, as money income levels have increased, people may have become willing to forgo a portion of the maximum money income they could earn in order to receive the psychic income of, for instance, more sunny days in Tucson, Arizona, than in Buffalo, New York. Employers, therefore, whose bottom line is the money profit motive, may find it rational to pay the costs of relocating jobs to the Sunbelt. There they can attract a skilled work force for less than they have to pay to keep labor "home" in the metropolitan areas of the traditional American Manufacturing Belt. In this traditional "core" region of the Northeastern United States, past labor agreements have pushed wages to higher levels than in the formerly lagging regions: the South and West.

Table 7.6 shows some results obtained by Porell (1982) from assuming that migration patterns may be used to infer the values migrants accord to various amenities and disamenities. The packages of amenities and disamenities in various areas are valued so as to compensate for differences in regional wages.

Adding further complexity to this type of analysis, housing and land markets are also strongly influenced by amenities and migration. Housing costs are of critical concern to a potential migrant in evaluating the feasibility of moving. (See Graves, 1983.)

TABLE 7.6 Porell's Computed Trade-Offs of Weekly Wages versus Selected Other Determinants of Intermetropolitan Migration

Variable	*Compensating Change in Weekly Wages*
+1 inch of snowfall per year	+26.3¢
+1 inch of rainfall per year	+28.9¢
+1 day of sunshine per year	−14.2¢
+1 public swimming pool per 100,000 population	−16.0¢
+1 park and recreational acre per 1000 population	−25.1¢
+1 public tennis court per 100,000 population	−1.3¢
+1 microgram / cubic meter concentration of SO_2	+12.7¢
+1 day annual inversion frequency	+6.4¢

Source: Adapted from Porell (1982, p. 154).

Note: These variables were selected from the 23 total variables included in Porell's regression model estimates.

The Need for a Micro Perspective in Behavioral Migration Research

> My life is made of patterns that can scarcely be controlled.
>
> —Paul Simon, "Patterns"

Considerable progress has occurred since the early attempts to estimate the effects of aggregate economic conditions on the volumes of flow in a system of migration streams. The migration decision process is currently being explored with individual-level data in which *life cycle* variables can be controlled for, as well as family relationships and other types of *place ties*. And recent macro-scale work also reemphasizes the importance of age composition of a population for regulating migration patterns (e.g., Plane and Rogerson, 1991; Plane, 1992).

During recent decades, as many as two-thirds of the net interstate migration exchanges have pointed in the direction from higher to lower average wages. It does not necessarily follow, however, that the migrants themselves are assuming lower-paying jobs in their new states of residence. As we saw in Chapter 4, U.S. migration tends to be *positively selective* with respect to education, age, and occupation.

Perhaps the next logical step in behavioral migration research would be to apply the new techniques of *experimental economics* to try to understand the decision process (Greenwood et al., 1991). The only attempt to date to bring this exciting new branch of economic analysis to migration studies appears to be a project done using prospective retirement-related migrants (Louviere et al., 1989). The marketing research technique known as *conjoint analysis* was used to examine how the interactions between various place attributes are assessed. By having respondents sort through a set of descriptions of possible retirement destinations and then rank the desirability of the destinations, the researchers were able to figure out which characteristics of the places were those that actually were most important.

An extremely important distinction between the migration process as it occurs at the individual level and aggregate models concerns the *spatial choice set* (Greenwood et al., 1991, pp. 248–58). Spatial interaction models typically assume that *all* destinations have a positive probability of being selected (cf. our discussion of the opportunities model), but Lansing and Mueller (1967) reported that most migrants in their study had *considered* no other—or at most one other—destination than the one actually chosen. Most conceptualizations of the migration process emphasize the role of free choice across a broad set of spatial alternatives, but reality emphasizes the importance of *constraints* in shaping our spatial choice sets. Remember that the decisions of public- and private-sector officials in locating job opportunities set the backdrop against which migrant decision making is played out. At the same time, however, perhaps too little emphasis is given to certain kinds of migration in itself fueling demand for new jobs. Consider, for instance, the impacts of retirement migration on the service sectors of many Sunbelt communities.

Research on the types of individual decision factors and constraints just discussed lies outside the purview of this book; further developments in behavioral migration research, however, will lead to better practical methods for modeling and forecasting aggregate migration patterns.

An Important Accounting Constraint: Milne's "Seemingly Unrelated Regression Approach"

The Lowry model and many other regression models of migration have been deficient in failing to control for another type of constraint—the accounting identity that net migration in a system of interregional flows must sum to zero. In unconstrained gravity models with more complicated specifications than the simple ones we examined earlier, there is no assurance that this fundamental constraint must hold.

Milne (1981; see also Foot and Milne, 1989) has been influential in pointing out the importance of embedding this system-level constraint in econometric models of migration. He suggested using a method due to Zellner (1962) when fitting gravity-type migration models. Called *seemingly unrelated regression*, the technique imposes a number of cross-variable restrictions and results in somewhat different flow estimates than would regular, unconstrained multiple regression.

The transition probability family of models—and constrained spatial interaction models, in general—do not suffer from this undesirable feature. As Plane (1982) argues, if migration models are derived from an *information-theoretic* perspective (see, e.g., Snickars and Weibull, 1977), it is unlikely that fundamental constraints will be forgotten. We turn now to consider extensions of the destination population weighted projection models that we earlier examined. These models embed measures of relative regional attractiveness within the demographic accounting framework set forth in Chapters 3 and 6. And they build on the experience gained in economic gravity modeling.

7.7 TOWARD ECONOMIC–DEMOGRAPHIC MODELS

In an attempt to improve upon the foregoing models, Plane and Rogerson (1985; see also Isserman et al., 1985) made two changes to the DPW approach. First, they suggested that a link between migration flows and changes in economic conditions at the destination be included by allowing the "destination weights" to be a variable (or variables) other than population. Second, they added a parameter, γ, that is related to the elasticity of the migration response to changes in economic conditions at the destination.

The generalized destination weighted (GDW) model may be stated as

$$p_{ij}^t = a_i p_{ij}^b (X_j^t / X_j^b)^\gamma$$

where X_j represents some measure of the economic attractiveness of destination j, and γ is the migration elasticity parameter to be estimated. The balancing factor, a_i, for this model is

$$a_i = \frac{1}{\sum_{k=1}^{r} p_{ik}^b \left(X_k^t / X_k^b \right)^\gamma}$$

Implementation of the GDW model leads to several questions. What economic variable should be chosen to represent destination conditions? Presumably we should be guided by theory in this choice, and, for interregional migration models, this suggests the use of variables such as wages, unemployment, change in employment, or labor force size. Plane and Rogerson (1985), in their application to U.S. interstate migration, used a measure that consisted of a ratio of economic opportunity to competition:

$$X_j^t = (\Delta E_j^{t-1} + s^{t-1} E_j^{t-1}) / (U_j^{t-1} + s^{t-1} E_j^{t-1})$$

where s^{t-1} is the rate of job separation during period $t - 1$; E_j^{t-1} and U_j^{t-1} are the average levels of employment and unemployment, respectively, during period $t - 1$; and ΔE_j^{t-1} is the change in employment in region j during period $t - 1$.

The numerator represents a measure of job opportunity; it includes the increase (or decrease) in total employment opportunity as well as the number of job separations (since most vacant positions will be filled). The denominator represents a measure of job competition; it includes not only the number of unemployed, but also those who have recently separated from their jobs (since most of them will either be moving, or wishing to move, into other employment).

The choice of an economic attractiveness measure will usually be guided by theory, constrained by data availability, and modified by exploration of what seems to work and what does not. Thus Plane and Rogerson originally found that destination wages did not give reasonable results in their study. The sign of the parameter was "wrong" (i.e., people seemed to go in the direction of the lower-wage state), and parameters varied markedly when the model was fit for the destination choices of particular origin states. Experimentation led to the employment opportunity/competition attractiveness measure.

This modeling experience also leads to the suggestion that the ratio of economic attractiveness from one time period to the next should be a ratio that does not have a wide range of variability (i.e., that it not vary too far from 1). Otherwise, the base year transition probabilities will be modified too much, and the parameter will have to take on values over a wide range to bring the transition probabilities back toward reasonable values.

Figure 7.10 shows the results of using the GDW model as opposed to the fixed-transition-probability Markov model to "forecast" interstate migration for 18-to-44–year-old males in 1978–79. The error distribution with respect to net migration for each of the 50 states (plus the District of Columbia) is illustrated. The dots in quadrant 1 correspond to states for which both the Markov and GDW models overpredict; quadrant 4 contains dots for those states underpredicted by both models. The dotted lines separate the cases for which the GDW model performed better than the Markov model (sections B and D) from those for which fixed transition probabilities performed better (sections A and C).

This test was encouraging. For 31 states the GDW model gave superior results. To obtain the results, however, the *actual* values of economic variables were used to compute the employment opportunity to competition ratios, $\{X_j\}$. If this—or any economic–demographic method—is to be used for making true population projections, the variables used as economic determinants themselves need to be forecast. Forecast errors in estimating each X_j term would presumably (but not necessarily) lead to greater errors in projected migration flows.

Comprehensive Economic–Demographic Modeling Frameworks

The GDW model was developed at the U.S. Bureau of the Census as the migration component for a much larger population forecast modeling exercise. The ECESIS

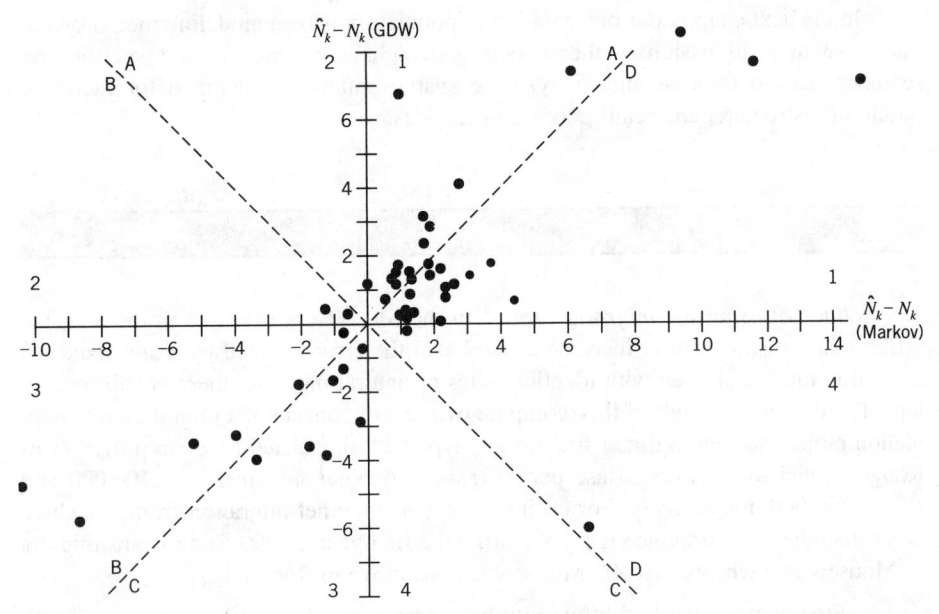

FIGURE 7.10 Comparison of Net Migration Forecast Errors for the Markov and GDW Models.
Source: Plane and Rogerson (1985). Reprinted by permission.

model (Beaumont, 1984, 1989) sought to simultaneously determine economic and demographic variables for each U.S. state. Individual state econometric and cohort–component models were linked (à la Lawrence Klein's Project Link world econometric model) via equations for migration and trade relationships. The entire modeling system had 7,400 endogenous (model-forecasted) and 884 exogenous variables plus matrices of 2,550 state-to-state migration flows for each of 10 age–sex groups. While ECESIS proved to be too unwieldy for use for census projection purposes, it has been successfully employed for simulation studies. (See Beaumont, 1989.)

ECESIS is just one of a number of integrated economic models that have been developed. A book edited by Isserman (1986) contains a number of interesting summaries of alternative approaches plus an assessment of the state of the art of economic–demographic forecasting of regional population change—a field still in its infancy with many important contributions still to come.

7.8 SUMMARY

Migration is typically the most difficult of the components of population change to model and forecast. In this chapter we dealt with several conceptual and practical problems that confront the geographical analyst of population who wishes to model or forecast migration. In addition we presented the two major spatial interaction methods used in this type of work: the gravity model and the opportunities model. We highlighted the importance of combining the expertise of the demographer, the regional economist, and the geographer in developing theoretically sound explanatory models of the migration process.

In the next chapter our presentation of population-driven modeling methods continues—with a discussion of the models geographers, planners, and operations researchers use to forecast and analyze the spatial patterns of demand for facilities, public infrastructure, and retail goods and services.

EXERCISES

1. *Fallacy of using net migration rates.* In the Mora Bora example in the text, the differential rates of natural increase exacerbated the error induced by using fixed net migration rates. But even with identical rates of natural increase, there is still a problem. To convince yourself of this, compute two sets of constant net migration rate population projections out to time $t = 2$ for the hypothetical, isolated northern provinces of Rongovia and Moosewood. Base period ($t = 0$) populations are $P_R^0 = 100,000$ and $P_M^0 = 200,000$, respectively. For each province assume net migration from anywhere other than the other province is zero. During the first year 20,000 Rongovians migrate to Moosewood, whereas 10,000 Moosewoodians move to Rongovia.

   a. First assume that both provinces have zero rates of natural increase ($ni_R = 0$; $ni_M = 0$). Compare the sum of both provinces' projected populations in year $t = 2$ to 300,000 (the unchanged combined population if there is no natural increase).

   b. Now assume that $ni_R = +0.200$ and $ni_M = -0.100$ (a rate of natural increase of 200 per thousand for Rongovia and a rate of natural decrease of 100 per thousand for Moosewood). Once again, compare the combined projected populations for year $t = 2$, this time to the total resulting from a constant transition probability model. Comment on the differences between this answer and what you found in part (a).

2. *Gravity model estimation.* An unconstrained power function gravity model was estimated for 1985–90 migration flows into Arizona from each of the other 49 states using log-linear regression ($R^2 = .692$). Note that because in this case only one destination is being considered, the destination population term is not used. (The population of Arizona is the same for each of the 50 flows being considered, so this variable is redundant with the intercept or constant term, k). The estimated regression equation is

$$\log M_{ij} = 5.4 + (.668) \log P_i + (-1.238) \log d_{ij}$$

   a. Calculate the value of the constant, k, from the estimate of the intercept ($a_0 = 5.4$).

   b. Find the model's estimates of the number of migrants moving to Arizona from California, Georgia, Illinois, Minnesota, Nevada, and New York. These states' respective 1990 census populations (in thousands) were 29,760; 6,478; 11,431; 4,375; 1,202; and 17,990. The distances (in great circle miles) to Arizona's population centroid from their respective centroids are 468; 1,604; 1,386; 1,263; 414; and 2,088.

   c. The actual numbers of migrants moving to Arizona from these states during 1985–90 were

California	136,465	Georgia	5,354
Illinois	39,171	Minnesota	17,350
Nevada	10,934	New York	23,753

Calculate the residuals (E_{ij}) for these flows and comment on their values.

3. *Origin-constrained gravity model*

   a. Calculate the matrix of transition probabilities and the matrix of internal migration flows that would result from an origin-constrained, negative exponential gravity model for a hypothetical three-region nation. Assume regional populations of 100,000; 200,000; and 300,000 for the three regions. Use a distance deterrence parameter of $\beta = .004$ and the following matrix of interregional distances (in kilometers):

   $$\mathbf{D} = \begin{bmatrix} 20 & 500 & 800 \\ 500 & 40 & 600 \\ 800 & 600 & 50 \end{bmatrix}$$

   Include the intrazonal flows (encompassing both non- and intraregional migrants).

   b. Repeat the calculations and derive a new transition probability matrix using a distance deterrence parameter of $\beta = .008$; explain the differences between these probabilities and those obtained in part (a).

4. *Intervening opportunity model.* The intraurban residential relocation decisions of 10 households are observed. The ranks of the houses selected from among all those listed on the market, when ordered in terms of the distances away from each household's original location, are 32, 45, 21, 3, 56, 67, 34, 2, 75, and 12.

   a. Estimate the parameter L used in the model.

   b. Assume now that we have data only for the number of listings by neighborhood. In neighborhood 1 there are 37 houses for sale. In neighborhood 2 (the next-closest neighborhood to 1) there are 31 opportunities. Using the L value just found, use the intervening opportunities model to find the probability that a household currently living in neighborhood 1 will choose a house outside that neighborhood. What is the probability that this household will relocate to neighborhood 2?

5. Suppose in the Rogerson and Plane Shoe Company layoff example given in the text that the value of L is .001 rather than .002. Recompute the probabilities and expected numbers of workers relocating to each of the four planning zones.

6. *Causative matrices.* If the causative matrix $\mathbf{C}_{60,80}$ in Table 7.2 applies over the period 1980–2000, calculate an updated transition probability matrix for year 2000.

7. *DPW model.* Calculate the 3 × 3 matrix of DPW model-generated transition probabilities used to obtain the time 2 populations shown in Table 7.4. (Use the base period probabilities and time period 0 populations in Figure 7.8 and the time period 1 populations in Table 7.4.)

> If you build it, he will come.
>
> —W. P. Kinsella, Shoeless Joe

CHAPTER 8
THE ROLE OF POPULATION IN INFRASTRUCTURE PLANNING

8.1 Introduction
8.2 Urban Travel Forecasting
8.3 Recreation Facilities Planning
8.4 Site Location
8.5 Impediments to the Production of Improved Forecasts
8.6 Optimal Demand Assignment Methods
8.7 The Use of Forecasts in Decision Making
8.8 Summary

8.1 INTRODUCTION

It is often desirable to have a forecast of the future demand for public or private facilities. For example, the long-range expansion plans of a supermarket chain depend critically upon the demand that can be expected. Similarly, in anticipating where additions to a highway network will be needed, a forecast of trip flows is necessary. Such forecasts are crucial in deciding which facilities need extra capacity and in suggesting potential locations for additional facilities.[1]

Most demand forecasting problems may be conveniently divided into two steps: *demand generation* and *demand distribution*. A forecast of the demand generated by the population is first prepared by considering the preferences, composition, and location of that population. This is followed by a forecast of how demand will be distributed across potential facilities. This latter step is carried out by considering the characteristics of the facilities and the location of the population with respect to the location of the facilities.

---
[1]Of course, in some instances where a region is in decline, interest will be in where to diminish capacity, and in which facilities to close.

The remainder of the chapter is devoted to explaining these concepts and methods. Sections 8.2 and 8.3 elaborate on the methodology through applications to transportation and recreation planning, respectively. Section 8.4 discusses the use of demand forecasts in site location. Section 8.5 summarizes some impediments to the production of more accurate forecasts. In section 8.6 we illustrate a method for the optimal allocation of demand to public facilities through a school districting problem. Finally, Section 8.7 more generally covers the use of demand forecasts in planning.

The methods discussed in this chapter are general in the sense that they may be applied to a broad variety of problems. Though the focus is on applications to the planning of facilities for recreation, transportation, and education, the approach taken in this chapter may easily be modified to match supply and demand for other types of infrastructure (e.g., housing, sewage treatment facilities, health care facilities, and utilities). The methods are often equally applicable to problems associated with the markets for consumer goods.

We now turn to two applications of demand forecasting: transportation planning and recreation planning. Through these examples, we introduce the essentials of demand forecasting and location analysis, and we point out the additional features of such problems that often must be considered in practice.

8.2 URBAN TRAVEL FORECASTING

A widely used procedure for forecasting traffic flows on the streets of an urban transportation network employs a considerable amount of data on current and future urban population and the spatially disaggregated demographic characteristics of such populations. Thus transportation planners are among those persons who need and commonly use information about the geographic distribution of population.

One of the more highly developed forms of demographic demand analysis is used to forecast future usage and additional roadway capacity that will be needed in different parts of a metropolitan area. This section details two of the six steps of the traditional urban transportation planning process. The trip generation and the travel demand forecasting stages of that process use methods broadly similar to those in other areas of population-based demand analysis. We also examine, in lesser detail, the other four steps of the process. These are either steps that do not explicitly involve the services of a population analyst, or, in the case of the population forecasting stage, steps that we have already covered in considerable detail. (See Chapter 6.) Before describing these steps, however, we first note one distinction between urban travel demand analysis and more general forms of demand analysis in economics.

The Role of Free Choice and Constraints in Urban Travel Demand Analysis

Geographers have been careful over the years to note that within a broad class of aggregate spatial choice models, we lump together not only many different individual reasons for choosing from among alternative geographical destinations, but also the net impacts of many constraints on individual behavior. Pure economic theory pertaining to demand analysis often assumes that aggregate demand results simply from adding up the free choices made by a number of individually unconstrained consumers choosing from among alternative economic goods. Within urban travel demand analysis the aggregate spatial patterns of traffic flow that we observe and model likely reflect not simply free choice, but also differential possibilities that people have, given their economic and social situations, to choose from among alternative destinations in different parts of the urban area. For this reason, the demographic and socioeconomic characteristics of population are critical for understanding future travel demand.

There are, in particular, feedback mechanisms between our transportation planning and the ultimate constraints and, thus, choices that people will make. Take, for instance, a case similar to that offered by Sheppard (1986, pp. 92–93). Suppose there is a poor inner-city residential area where car ownership is not very widespread. Few employed persons in this area are thus likely to be able to commute to jobs in a large suburban factory not reachable by other than private automobile. On the other hand, those jobs are likely to be taken largely by blue collar persons living in more auto-oriented suburban communities. Thus, to the transportation modeler a need will be evident based on existing travel patterns for intrasuburban transportation but not for inner-city–to–suburban alternative-mode transportation. To the extent that our planning thus reflects the current status of constraints on behavior, it thus also serves to perpetuate, in many cases, existing inequities. Olsson notes this tendency in his philosophical treatise on the nature of human geography, *Eggs in Bird*, when he speaks about the use by planners of one of our primary forms of distance deterrence functions, the negative exponential.

> Unveiling the negative exponential is to grasp how a model can present
> itself as a means not only for understanding the world but for changing it
> as well. (1980, p. 31e)

The interested reader should see Sheppard (1980) for a fuller discussion.

Steps in the Urban Travel Forecasting Process

The traditional process is often conceptualized to consist of six distinct phases (Figure 8.1). Within each phase, particular models or techniques have been widely employed. Current and recent research, however, has focused on methods for combining and simultaneously deriving the outputs of two or more of these steps. (See, for example, Boyce, 1984.) Thus in real world applications the modeling process may not follow the rigid, linear progression in which we shall present the steps, but may rather endogenize some of the *feedbacks* that characterize actual urban travel patterns.

Step 1: Population and Economic Forecasts

The first step of the process is a nongeographic one in the sense that it typically involves generating projections of population and economic growth for the entire metropolitan study area. For urban travel forecasting, there is typically a need to forecast not only total population, but often population within different income categories, different family-size categories, and so forth, which are thought to be the crucial determinants of trip-making propensities. In addition, at this first step of the travel demand forecasting process, the modeler will also desire geographically aggregate projections of land use change—for example, how many new residential units of different types (single-family, duplex, apartment complexes), how many total square feet of new retail shopping facilities—as well as aggregate predictions of new economic opportunity in the form of jobs in different sectors of the economy.

This stage of the process is extremely important to the ultimate success or failure of the total transportation modeling exercise. If the analyst starts out assuming that

> 1. Population and economic forecasts
>   Project population, economic activities, and land use totals for the study area.
>
> 2. Land use forecasts
>   Allocate projected population, economic activities, and land use traffic to zones.
>
> 3. Trip generation
>   Determine how many future trips of each type will likely leave each traffic zone.
>
> 4. Trip distribution
>   Allocate trips generated in each origin zone to possible destination zones.
>
> 5. Modal split
>   Separate automobile trips from public transit trips from alternative-mode trips and estimate the proportion of travel between each origin–destination pair of zones.
>
> 6. Traffic assignment
>   Assign trips to specific route segments that link up origin and destination zones.

FIGURE 8.1 The Six-Step Urban Travel Forecasting Process

future growth will only be half of what actually is to occur, then the transportation planning is highly likely to be inadequate regardless of how well the remaining steps of the procedure are carried out. As a result, there may be an inherent tendency to "be safe rather than sorry" and thus to base the remaining steps of the process on higher, rather than lower, projections of future aggregate growth. In some cases, this can be carried too far—there may also be a tendency for elected officials and interest groups to make unrealistically optimistic assumptions about population and land use change, based upon wishful thinking. Very commonly, alternative projection series are used to study the implications for the region's transportation needs of, perhaps, low, middle, or high amounts of future growth.

Step 2: Land Use Forecasts

Once the analyst has generated control figures for the entire regional study area for such things as total population growth, new jobs, and new square footage of retail and office space, the geographic part of the process begins. The second step is to allocate such growth to specific parts of the metropolitan area. Usual practice is to first break the overall study area down into a large number of (usually very small) traffic zones. Traditional, big-city transportation studies may involve thousands of individual traffic zones. Following traditional practice, in the forthcoming discussion we reference any particular traffic zone with the nominal variable i, for an origin zone, and with j for a destination zone.

Zones may be designated somewhat arbitrarily using simple grid zones of, say, one mile on a side, or they may be chosen to reflect the traffic-generating functions of different parts of town. Thus a residential subdivision with fairly similar houses may be designated as one zone, a major employment center such as a factory as a separate zone, a university as another, and a shopping center as a fourth. Sometimes, too, to facilitate the use of census data the boundaries of census units such as census tracts or block groups may also be used as those defining the limits of traffic zones. (In fact, city and regional transportation planners are often involved on the local committees that designate new tracts and block groups prior to each decennial census.)

After the traffic zones themselves have been identified, the major part of Step 2 is to designate the actual future distribution of population and economic activity within the study area. As in Step 1, this often involves a fair bit of seat-of-the-pants analysis with the judgment of local experts often playing a leading role. But planners engaged in generating future land use plans often engage in extensive data collection efforts to examine (1) past locational trends (because the dynamics of urban location patterns reflect, in part, an element of geographic diffusion in the form of "growth momentum"), (2) availability of developable land, and (3) existing zoning and land use plans.

The third category of background information is suggestive, in fact, of one of the unique aspects of the land use forecasting stage of urban travel demand forecasting, for it is at this stage that decision makers can influence the travel patterns that result from subsequent steps of the modeling process. It is probably critical at this second step that the analyst insists on policy makers providing inputs into the generation of a number of distinctly different growth scenarios. Because most of the ensuing steps of the process are highly automated (computerized), we can look at the potential consequences of alternative patterns that future growth could be encouraged to assume. In fact, this is one of the main advantages of a modeling approach. We call such analyses *simulation studies*. The goal is really not to forecast the most probable future, but rather to promote a better understanding of the complex relationships between the various elements of an urban system that will ultimately shape its future growth. The hope is that, with such additional understanding, policy makers will be more able to influence a desirable future course for growth to assume. Once money has been invested in the labor-intensive early phases of model development and data collection, it is relatively cheap to run off a variety of "what if" scenarios. Indeed, development costs of modeling exercises are most correctly charged off against the far bigger costs of making inefficient and socially suboptimal planning decisions if such a decision-making tool were not to exist.

Step 3: Trip Generation

Once we know (or, more correctly, have a variety of estimates about) where new activity is to go in the metropolitan planning region, the next phase is to determine how much traffic will potentially be "produced" within each traffic zone.

It is desirable at the trip generation phase to model different trip types separately. The two most common types of trips in metropolitan areas are journey-to-work and shopping trips. In fact, much of the interest in modeling urban travel focuses on the journey to work because the two daily peak periods of commuting define the maximum need for urban transportation facilities. For commuting trips, however, the trip generation stage is not the key step of the modeling process because it is sensible to simply take employment in various zones as the measure of the number of work-to-home trips being created (instead of modeling the number of journey-to-work trips). The large amount of symmetry present in flow patterns can then be used to infer the number of (typically morning) journeys to work. The harder step is to figure out the patterns of journey-to-home flows to all the various residential zones, a determination typically left for modeling in the trip distribution stage of analysis. It is expensive to do the sort of survey research required to accurately map the complete pattern of where each employer's employees live. Typically big employers and a random sample of small employers could be surveyed to determine average characteristics of the journey to home/journey to work, but the actual matching of origins with destinations is most cost-effectively modeled. The trip generation inputs to such modeling are thus at the employment end, and represent largely the results of the previous step of the modeling process rather than a large amount of additional analysis.

We shall focus our discussion of trip generation, therefore, on shopping trips originating in residential zones. Here it is the home-based origin that is conceptualized as the trip generation "anchor," with the pattern of store-destination choices left to the subsequent distribution step of the process. Most generation analyses are based on either multiple regression or category analysis models. The analyst examines the number of trips made during a typical day by households to see how, on an average basis, they correspond to observed characteristics of the households.

In regression models, each residential zone's trip generation ratio (representing average daily trips per household) is computed based on a mathematical formula estimated for the entire study area. The trip generation ratio for a specific zone represents the specific values of the included variables in the model, weighted by each variable's areawide average contributions to shopping-trip–making propensities. Regression methods are preferable when the basic data on the economic and demographic characteristics of zones take on continuous values. An example would be residential density, for which a negative regression coefficient would be predicted because of the expected inverse relationship between density and the average distances consumers must travel to shop.

A category analysis, on the other hand, makes use of the "demographics" of neighborhoods in much the same fashion as we detailed in Chapter 2. We cross-tabulate the characteristics of households and, through surveying, determine the average propensities across the entire study area of each category of household to make shopping trips. The number of trips generated in any specific origin zone, i, is then simply the weighted average of these category-specific generation ratios multiplied by the number of households of each specific type found in the zone. Many of the variables that are thought to most strongly influence trip-making behavior are, in fact, categorical rather than continuous; at the individual household level they take on discrete, integer values only. Examples include the number of persons living in the household and the number of automobiles available to household members. Other significant variables for trip-making behavior that are conceptually continuous, such as income, may in reality only be collected and reported for broad ranges.

Tables 8.1 and 8.2 demonstrate, for a highly simplified example, the basics of the category analysis approach. Table 8.1 reports data for a base year and for the entire study area, whereas Table 8.2 contains projected data for a future time period for a single selected origin zone.

The top of Table 8.1 cross-tabulates the Hypoville Metropolitan Area households according to income level and auto ownership. Such data might be obtained from census or other survey data. The second part of the table shows the estimated numbers of shopping trips on a typical weekday made by Hypoville households. Typically estimates like these are derived from special surveys carried out so as to be representative of the overall population. (If necessary, differential postsurvey weighting may be carried out so that the sample responses represent the overall characteristics of the study region.) The table then gives the computed trip generation ratios for each category of household in Hypoville. Note that higher-income households generally made more trips during the base period than did lower-income ones, and that the average number of daily shopping trips goes up with increasing numbers of vehicles available, but the ratios are separately computed for each combination of income and auto ownership class. If no cars are available, even high-income households make many fewer trips than the average for all households in Hypoville.

TABLE 8.1 The Hypoville Example of the Calculation of Base Year Trip Generation Ratios Using Category Analysis

a. Number of Households in Hypoville for Base Year (from census data)

| | Automobile Ownership | | | |
Income	None	1	2 or more	Total
Less than $10,000	1,000	800	200	2,000
$10,000–$20,000	1,000	5,000	2,000	8,000
More than $20,000	200	1,000	800	2,000
Total	2,200	6,800	3,000	12,000

b. Number of Daily Shopping Trips in Hypoville for Base Year (estimated from a survey)

| | Automobile Ownership | | | |
Income	None	1	2 or more	Total
Less than $10,000	800	1,000	400	2,200
$10,000–$20,000	800	7,000	4,200	12,000
More than $20,000	200	2,000	1,800	4,000
Total	1,800	10,000	6,400	18,200

c. Average Daily Trips per Household—Base Year Trip Generation Ratios

| | Automobile Ownership | | | |
Income	None	1	2 or more	Total
Less than $10,000	0.80	1.25	2.00	1.10
$10,000–$20,000	0.80	1.40	2.10	1.50
More than $20,000	1.00	2.00	2.25	2.00
Total	0.82	1.47	2.13	1.52

These areawide trip generation ratios are then used for forecasting future shopping trip generation in any particular residential traffic zone, such as the Snob Hill neighborhood in Table 8.2. As indicated in the top part of the table, a forecast is first needed of the number of households expected to be found in the neighborhood in the forecast year for each income/auto-ownership category. Such forecasts may be considered part of the previous, land use forecasting phase of analysis. The numbers in each of the nine cells of cross-tabulated categories of income and auto ownership are then multiplied by the nine trip generation ratios previously computed from the base year survey phase of the analysis to derive the predicted forecast year trips shown in the bottom part of Table 8.2.

It should be obvious from this discussion that the base period and future geographical distributions of demographic and socioeconomic characteristics are what drive the trip generation phase of urban transportation modeling. The cross-classified "attributes" of individuals and the households that they live in—demographics—are crucial information for use in transportation planning problems. Before turning to the trip distribution stage of analysis, in which demographics again play a key role and in which geographical models of spatial interaction are used that extend the concepts of accessibility analysis discussed in Chapter 2, we should note two criticisms of the simple, standard type of generation analysis that we have shown.

TABLE 8.2 The Snob Hill Neighborhood Example of the Use of Constant Ratios and Category Analysis for Forecasting Trip Generation

a. Projected Number of Households in Snob Hill Neighborhood (estimated in Step 2 of the urban travel demand modeling process)

	Automobile Ownership			
Income	None	1	2 or more	Total
Less than $10,000	0	0	0	0
$10,000–$20,000	0	25	50	75
More than $20,000	40	75	120	235
Total	40	100	170	310

b. Forecasted Number of Daily Shopping Trips from Snob Hill (Using Table 8.1 Trip generation ratios applied to number of households shown above.)

	Automobile Ownership			
Income	None	1	2 or more	Total
Less than $10,000	0	0	0	0
$10,000–$20,000	0	35	105	140
More than $20,000	40	150	270	460
Total	40	185	375	600

First, note that we have used generation ratios for a base period and assumed that these will stay constant into the future. In fact, we know that during recent decades, shopping trips by car have increased significantly. The method, however, is easily modified to take into account projected changes in propensities to shop. A simple, equal proportional adjustment could be made to all base year ratios, or more elaborate assumptions regarding differential changes for different categories of households could be embedded in the ratios used for the forecast period.

The second problem is rather more serious. This concerns the fact that trip-making behavior for the entire study area is used to derive the base year ratios. All households in a particular economic/demographic category are thus assumed to have the same propensities to make shopping trips in the forecast year, regardless of where the homes are predicted to be located. It is known that in reality, however, the distribution of trip lengths to potential shopping opportunities also plays a role in trip-making frequency. If every trip involves a long drive, then consumers tend to make more of their total purchases during each trip, and thus to make fewer trips, overall, than do persons in households with the same socioeconomic/demographic characteristics for whom stores are closer to home. Indeed, this is a reason why population density of residential areas is frequently included as one of the variables in trip generation studies; lower density implies, *ceteris paribus*, geographically more extensive market areas for stores and thus longer average trips for consumers. For this reason, too, recent research has explored combined methods for trip generation and trip distribution.

Step 4: Trip Distribution

Once the modeler has predicted how many trips are going to leave each traffic zone in the system, the next step is to determine where those trips are going, that is, to distribute the trips to destination traffic zones, as shown in Figure 8.2. We say that the sizes of these flows, which we designate with the notation T_{ij}, are constrained to add up to the previously determined number of trips, O_i, originating in origin zone i. We write such an origin constraint

$$O_i = \sum_j T_{ij} \tag{8.1}$$

For example, for the four destinations shown in the figure, Equation 8.1 implies:

$$O_1 = T_{11} + T_{12} + T_{13} + T_{14}$$

Note that some trips have zone 1 as both their origin and destination. Such trips are referred to as intrazonal flows.

The most widely employed model for trip distribution is the gravity model, which Chapter 7 covered in the context of migration modeling.

We now use these same concepts to demonstrate how the distribution of urban traffic flows may be predicted. Once again we focus on shopping trips (though gravity models are also widely used to model the journey to work and other trip types). Suppose we have the hypothetical situation pictured in Figure 8.3. The Snob Hill Neighborhood, origin zone $i = 1$, is surrounded by three shopping centers of different sizes (GLA indicates "gross leasable floor area") and thus different levels of functional complexity (which, in a marketing study, might be designated as centers at different hierarchical levels). A consumer making a trip from zone i could buy more types of items at the regional-level center, but it is farther away. We assume that the role of destination "mass" is played by the total amount of GLA in each potential destination zone ($j = 2, 3, 4$, in this example) and denote it as F_j. For the origin mass term in our gravity model of shopping we simply use the number of trips generated in an origin zone, O_i. The singly constrained gravity model combining these terms is

$$T_{ij} = a_i \frac{O_i F_j}{d_{ij}^\beta}$$

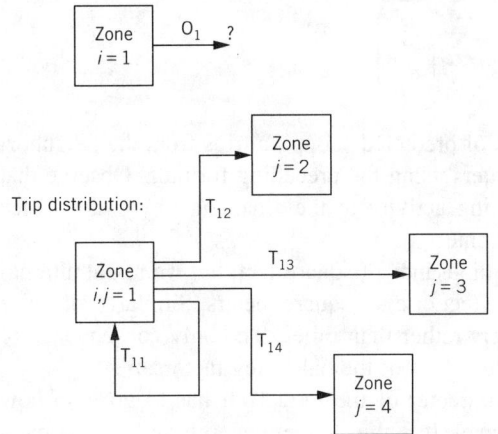

FIGURE 8.2 The Relationship between Trip Generation and Trip Distribution

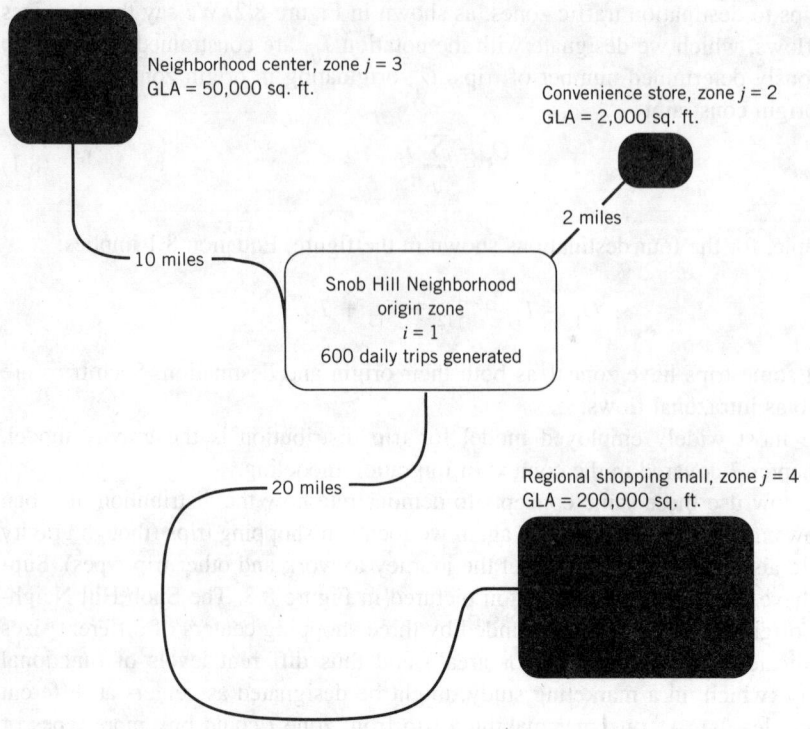

FIGURE 8.3 Hypothetical Arrangement of Shopping (Destination) Zones around the Snob Hill Neighborhood (Origin) Zone

Note that sometimes the exponent on distance is not taken to be 2 as in the Newtonian gravity model. Hence it is shown in the preceding equation as a general parameter, β. Sometimes, also, a functional form other than the negative power function shown here is used. (The negative exponential is perhaps the most widely used alternative.) But for demonstration purposes, let us retain this functional form and use $\beta = 2$.

Note that there is a separate balancing factor constant, a_i, for each origin zone i in which shopping trips originate. The formula for computing a_i is

$$a_i = \frac{1}{\sum_j \left(\dfrac{F_j}{d_{ij}^{\beta}}\right)}$$

Table 8.3 shows the computation of predicted shopping trips from the neighborhood to each of the three shopping centers using the preceding formula. Observe that we predict the same level of trip-making activity to the small, nearby convenience store as to the large, faraway regional center.

Note that Table 8.3 lists GLA in thousands of square feet, but we might alternatively have done so using simply square feet or even square meters. Similarly, distance might have been measured in kilometers rather than miles. The only consequence of such a rescaling would be to change the values of the balancing factors, a_i.

What role is played by the β parameter of the model? It has to do with how strongly distance retards or impedes travel. It is thus sometimes referred to as a measure of the "friction of distance" or a "distance deterrence" parameter. The higher the value β, the lower will be the fraction of trips predicted to go to the more distant destinations. Recall the inverse relationship between the parameter and the average distance moved discussed in Chapter 7 (Figure 7.4). Generally for shopping trips the analyst would want to employ a higher value of β than for work commuting trips. (And in detailed shopping studies, different values should be chosen for each type of shopping trip. Trips to the supermarket tend to be shorter ones, on average, than trips to shop for a new car.)

Given our focus in this book on demographics, we should note that average trip lengths for the journey to work vary with race, sex, and income level (among other characteristics). Thus a detailed trip distribution modeling strategy should also employ different distance deterrence parameters for each group of trip makers for which the generation step forecasted total originating trips; we would, for example, expect to use a higher value of ß for lower-income than for higher-income households.

Until now, we have neglected to indicate how to estimate β. One common method is to choose a value of β that maximizes the fit between the model and a set of observed flows for some base period. For example, to return to the Hypoville example, suppose that a survey of 100 households finds that 40 go to the neighborhood center, 30 go to the convenience store, and 30 go to the regional shopping mall. What β can we choose for the origin-constrained gravity model that will yield predicted flows that are as close as possible to the observed flows? A simple, "brute force" method is to simply try many different values of β and choose that one that minimizes the sum of squared deviations between observed and predicted flows. For Hypoville, we find that β = 1.969 minimizes the sum of squared errors. (See Figure 8.4.) Using this value of β in the origin-constrained gravity model yields predictions of 33.6 trips to the neighborhood center, 32.2 trips to the convenience store, and 34.2 trips to the regional shopping mall. Once β has been estimated, it may be used in projecting future flows by assuming that its value remains constant over time.

In the approach we have just examined, projected trips between origin and destination zones were determined by making use of (a) projected trip generation by origin zones, (b) a measure of the anticipated attractiveness of destination zones, and (c) a measure of distance deterrence, β, that is assumed to remain constant over time.

Another technique for projecting future flow patterns makes use of (a) projected trip generation by origin zones, (b) the projected number of trips ending in each zone,

TABLE 8.3 Sample Calculation of Origin Balancing Factor for Gravity Model Forecast of Shopping Trip Distribution in Hypoville

Origin Zone i	Dest. Zone j	O_i	F_j	d_{ij}^2	$O_i F_j$	$\dfrac{F_j}{d_{ij}^2}$	$\dfrac{O_i F_j}{d_{ij}^2}$	$T_{ij} = \dfrac{a_i O_i F_j}{d_{ij}^2}$
1	2	600	2	4	1,200	0.5	300	200
1	3	600	50	100	30,000	0.5	300	200
1	4	600	200	400	120,000	0.5	300	200
						$1/a_j = 1.5$		600

Note: $a_i = \dfrac{1}{\sum_j \left(\dfrac{F_j}{d_{ij}^2}\right)} = 1/1.5 = .6667$ (approximately).

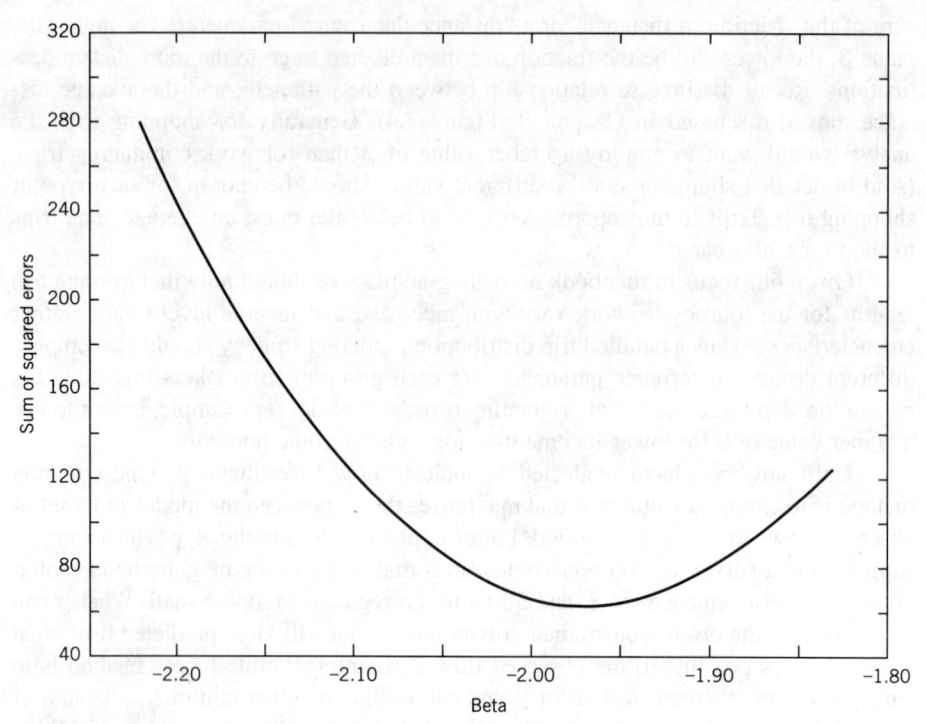

FIGURE 8.4 Sum of Squared Errors as a Function of Beta

and (c) a base period pattern of trips between origins and destinations. This method is often used for journey-to-work trips, where the numbers of trips ending in zones may be readily projected on the basis of projected employment.

To illustrate, suppose that current data reveal the following journey-to-work commuting flows between three subregions within a metropolitan region:

	To:	Central City	Suburbs	Rural	Total
From:					
Central city		1,000	500	200	1,700
Suburbs		1,500	1,000	300	2,800
Rural		800	600	400	1,800
Total		3,300	2,100	900	6,300

The projected employed population and employment is:

	Population	Employment
Central city	1,800	3,000
Suburbs	3,100	2,700
Rural	1,700	900

What is the projected commuting pattern between origin and destination zones? Fratar (1954) describes a procedure that yields a flow matrix that is consistent with the projected population at origin and the projected employment at destination, and that is as close as possible to the flow pattern encapsulated in the base period flow matrix.[2]

The procedure, now known as the Cross–Fratar method, is an iterative one. It is best illustrated through a description of the stages to be followed:

1. Multiply the elements in the original flow matrix by the ratios of the projected row sums to the base period row sums. For example, elements in the first row of the original table are multiplied by 1,800 / 1,700, elements of the second row by 3,100 / 2,800, and the elements of the last row by 1,700 / 1,800. This results in:

To:	Central City	Suburbs	Rural	Total
From:				
Central city	1,059	529	212	1,800
Suburbs	1,661	1,107	332	3,100
Rural	755	567	378	1,700
Total	3,475	2,203	922	6,600

Notice that although the row totals match the projections, the column totals do not. This matrix is now called the *current matrix*.

2. Next, multiply the elements of the current matrix by the ratios of the projected column sums to the column sums in the current matrix. Thus we multiply elements in the first column of the current matrix by 3,000 / 3,475, elements of the second column by 2,700 / 2,203, and elements of the last column by 900 / 922. The result is:

To:	Central City	Suburbs	Rural	Total
From:				
Central city	914	649	207	1,770
Suburbs	1,434	1,357	324	3,115
Rural	652	694	369	1,715
Total	3,000	2,700	900	6,600

Now the column sums match their projected values, but the row sums do not. This becomes the current matrix.

3. The elements of the current matrix are multiplied by the ratios of the projected row totals to the row totals in the current matrix. This is precisely the same as the first step, except that the current matrix is used instead of the base period matrix. Again, the result is denoted as the current matrix. In our example, the reader may verify that the result is:

---

[2]The procedure is now commonly called the Cross–Fratar method, but it, or minor variants of it, have reappeared many times under various names.

To:	Central City	Suburbs	Rural	Total
From:				
Central city	930	660	210	1,800
Suburbs	1,427	1,350	323	3,100
Rural	646	688	366	1,700
Total	3,003	2,698	899	6,600

The row totals match the projected values, and the column totals almost do.

Stages 2 and 3 are next applied successively, until there is little change in the result. In our example, if we carry out one further iteration (in this case by applying stage 2 to the current matrix), we obtain:

To:	Central City	Suburbs	Rural	Total
From:				
Central city	929	660	211	1,800
Suburbs	1,425	1,351	323	3,099
Rural	646	689	366	1,701
Total	3,000	2,700	900	6,600

which is "close enough for government work," as the saying goes.

Step 5: Modal Split

We shall not go into much detail about this step of the urban travel forecasting process, but we do wish to set out its purpose and tell how demographic considerations are critical to the modeling. As suggested by the name, modal split models are used to divide up trips from each origin zone i to each destination zone j by mode of travel taken. The ratios of public transportation trips to automobile trips will likely vary considerably from one study area to another, and, within a single study area, from one traffic zone to another. Models are usually based on such variables as household income levels and the percentage of car ownership within each residential zone, as well as on the forecasted quality of transit services anticipated to be available for residents of the zone.

Step 6: Traffic Assignment

The final step of the process is to assign the origin-to-destination flows of trip makers traveling by each transport mode to specific route segments in the urban street and public transit networks. Thus the end result of the six-step modeling process will be detailed predictions of the volumes of forecasted travel demand on each of the highways, streets, subway lines, and so forth in the study-area transport system.

Methods for this step are typically based on shortest-path computer models with the entire street and highway network coded up into a large matrix representative of the links between intersections. Critical at this stage of the forecasting process are assumptions about how congestion will alter the route choices made by individuals. Because of this consideration, modern techniques actually do not always relegate traffic assignment to the final step of the process but rather build in feedback mechanisms with the earlier steps of the process. Combined trip distribution and traffic assignment methods recognize the interdependency not only between congestion and route choice, but also the equally fundamental one between congestion and destination selection. (One is reminded of Yogi Berra's observation, "That place is so crowded, nobody goes there any more.") Conceptually, similar feedbacks exist to the trip generation and modal split stages as well; individuals may be dissuaded from making certain trips, may make them outside of peak hours, or may take alternative transport modes because of anticipated congestion on specific links in the street and highway network. Although demographic and socioeconomic characteristics are generally not used in traffic assignment models, because of the real world feedbacks, there is a direct conceptual connection between the forecasted demographics of the various residential zones and the amount of traffic demand ultimately predicted for any given link in the future transportation network.

In the next section, we examine an actual application of these ideas to recreation planning in New York State.

8.3 RECREATION FACILITIES PLANNING

The federal government of the United States requires states to prepare and update statewide comprehensive outdoor recreation plans to be eligible for funds from the U.S. Department of the Interior. These plans are developed independently by states, and no standard format governs their preparation. Though there are often similarities in the techniques adopted, there is inevitably some variation in the methods used to forecast future demand for outdoor recreation activities. The discussion here follows the procedure adopted by New York State in preparing its 1978 comprehensive plan.[3]

Figure 8.5 provides an overview of the state's recreation planning process. Data acquisition constitutes the first step in the process. Three basic types of data are required to develop forecasts and to establish plans, goals, and objectives. First, supply inventories are necessary to estimate current capacities for each outdoor recreation activity. Second, surveys are particularly useful for estimating current demand and for establishing the relationship between particular subgroups of the population and their preferences for various recreation activities. Finally, basic data on land use and the size and characteristics of the population are also needed to forecast demand. Once these data have been assembled, they are used to match recreation supply with anticipated demand. This matching process identifies regions that are deficient in their ability to provide recreation services, and naturally suggests priority levels for the various regions and outdoor recreation activities.

Of course, factors other than relative need are important in prioritizing recreation projects. Environmental and economic impacts are examples of such factors. Figure 8.6 displays the ranking system used by the state to prioritize recreation projects. We are particularly concerned in this section with item A2a, the index of relative intensity of need. This index combines a measure of demand relative to supply (the use/capacity ratio) with a measure of cost-effectiveness. Recreation activities that are

---

[3]At the time this book was written, statewide plans were also available for 1983 and 1988. The description of the planning methodology, as well as the amount of information presented in tables and maps, was, however, more detailed in 1978. Though changes have occurred in the planning process, the magnitude of these changes has been relatively small, and the methods to be described capture the important concepts that underlie the planning process.

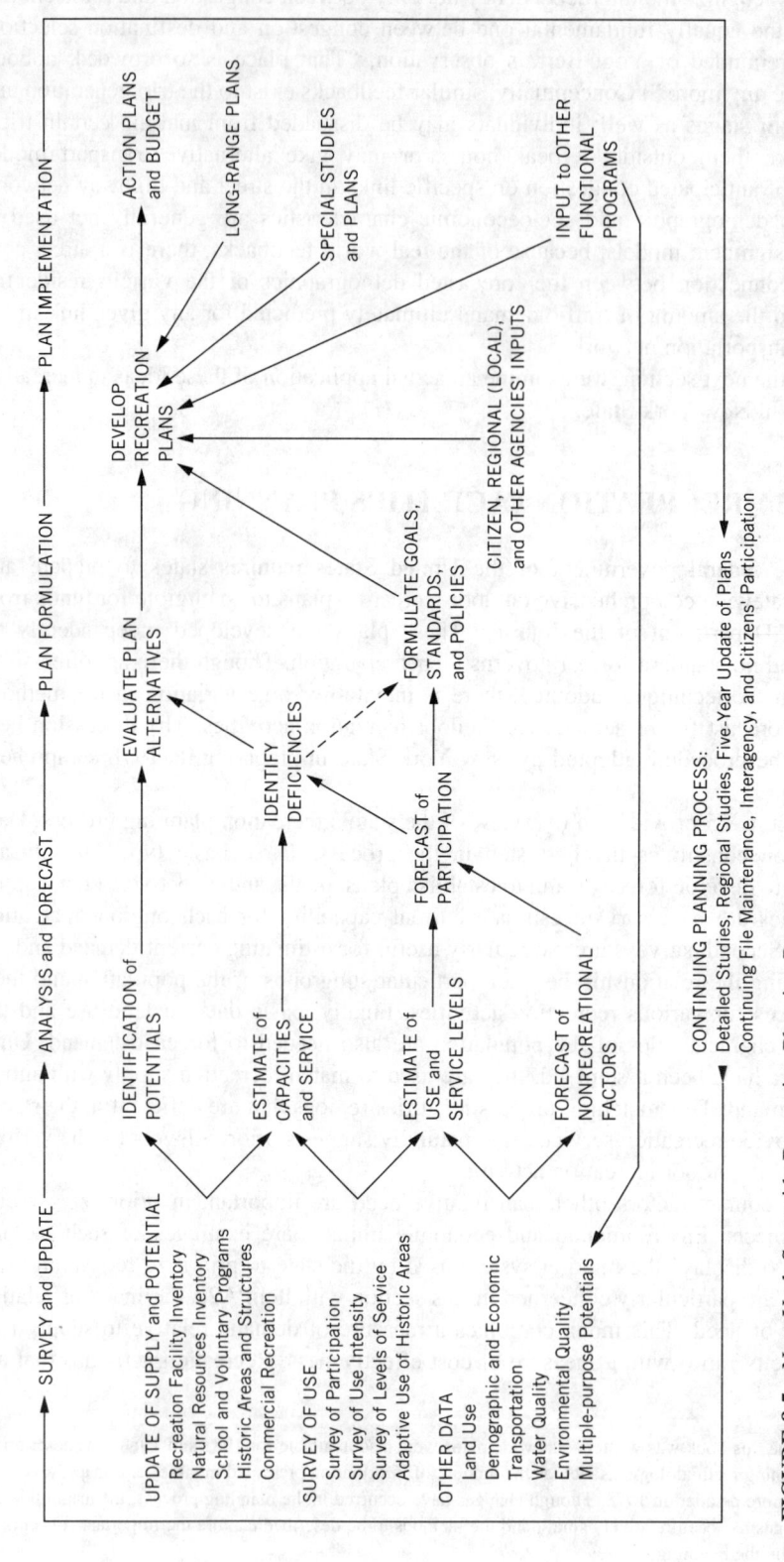

246 Chapter 8 • *The Role of Population in Infrastructure Planning*

FIGURE 8.5 New York Statewide Recreation Planning Process

Source: New York State Office of Parks and Recreation (1978)

PROJECT NO. _____

Park Name _____
Sponsor _____   Town-City _____
                        Village _____
Location: County _____
Land Acquisition Cost _____
Acres Acquired _____
Construction Cost _____
Total Project Cost _____
Phase (Specify) _____
Population of Sponsoring Area _____
Total Cost of Projects by the Sponsor funded with LWCF and/or State Grants:
   within the last three years: _____
   since beginning of LWCF grants: _____

A. PHYSICAL AND RECREATIONAL FACTORS
  1. STATEWIDE AND LOCAL OPEN SPACE AND RECREATION PLANS
     +1 Contributes to the implementation of a local open space or recreation plan and/or specifically contributes to the implementation of SCRP, other state, national, or regional plans
      0 Of only marginal contribution   _____
  2a. RECREATION PROJECTS: INDEX OF RELATIVE INTENSITY OF NEED
     capacity ratio   relative activity
     2000 forecast use/   for the major project
     capacity ratio   ×   activity   × effectiveness factor   =   _____
     (from SCRP)
  2b. OPEN SPACE/NATURAL RESOURCE PROJECTS: DEGREE OF URBANIZATION/NEED FOR OPEN SPACE OF THE PROJECT AREA (from directory of population densities—1970)   _____
  3. LENGTH OF OPERATIONAL SEASON
     +1 Extends system's operational season
      0 Does not extend system's operational season
     -1 Only reinforces peak use periods   _____
  4. ADDS TO CAPACITY AND/OR UNIQUE SERVICE PROVIDED
     +1 Substantially increases capacity of system and/or provides unique service
      0 No substantial added service   _____

  5. ACCESSIBILITY
     +1 Improves access to or within an open space or recreation facility and/or provides overnight accommodations that increases the accessibility of remote open space areas to distant populations
      0 No significant contribution   _____

B. SOCIAL AND ECONOMIC FACTORS
  1. VARIETY OF POPULATION TO BE REACHED
     +1 A wide spectrum (age, income, race) of population reached
      0 Benefits confined to a limited group   _____
  2. CONSISTENT WITH EXISTING PUBLIC AND PRIVATE FACILITIES
      0 No conflict
     -2 Will compete with existing facilities or will tend to inhibit expansion of private supply to meet recreation demand _____
  3. LOCAL ECONOMIC IMPACT
     +1 Will be of substantial importance to local economy
      0 Only of marginal value to local economy   _____
  4. TIME SPECIFICITY OF PROJECT
     +1 A critical acquisition or rehab project that must be done now
      0 Project can be postponed   _____

C. ENVIRONMENTAL CONTRIBUTION
  1. RESOURCE PROTECTION
     +2 Protects or improves a *unique* regional ecological, historical, and/or open space resource
     +1 Protects or improves a significant ecological, historical, and/or open space resource
      0 No significant contribution
     -1 May cause some environmental problems
     -2 Serious conflict with ecological, historical, and/or environmental resource   _____
  2. ENVIRONMENTAL EDUCATION
     +1 Provides a program and/or facility for an educational or learning experience
      0 No program   _____

                                                    TOTAL _____

COMMENTS:

FIGURE 8.6 New York SCRP Priority System

Source: New York State Office of Parks and Recreation (1978).

both cost-effective and in high demand relative to the supply will score highly on this index. Since the use/capacity ratio can be as high as 2 or 3, and since the relative-activity–effectiveness factor (a crude ordinal measure of the cost-effectiveness of providing capacity for particular recreation activities) can be as high as 5, scores in item A2a can be quite high relative to the possible scores on other factors.[4] Thus the demand/supply analysis is particularly influential in determining which recreation projects are funded. We now turn to a closer examination of this demand/supply analysis.

Recreation Supply

The statewide outdoor recreation facilities inventory is a comprehensive data base that now contains information on approximately 13,000 facilities. The inventory contains a high percentage of the available supply of organized recreation opportunities in the state. Very different roles are played by different operators of recreation facilities. Though the federal government does not maintain many facilities within the state, they are by far the biggest facilities when measured by average size. The number of facilities located in and operated by cities and towns is much greater, but, of course, their average size is much smaller. The state's role lies between these two extremes: Their facilities are smaller than those operated by the federal government, and the number of facilities is much smaller than the number run by cities and towns. However, the total acreage of state facilities is far greater than that of any other recreation operator, indicating the importance of the state's role as a provider of recreation opportunity.

There are also significant variations in the geographic distribution of recreation supply. Seventy percent of all recreation acreage lies in rural counties. Figure 8.7 depicts the state recreation planning regions, and Table 8.4 shows the spatial variations in supply. As we might expect, the biggest facilities are found in the more rural regions of the state and the smaller facilities are found in the more urban regions.

Clearly there are other recreation sites that are excluded from the inventory; examples include backyard picnic tables and unorganized hiking trails. But this analysis is aimed at the matching of recreation supply and demand at organized facilities (particularly state facilities), so these omissions are not particularly serious for planning purposes.

Recreation Demand

Recreation demand was estimated and projected for 14 recreation activity groups for the years 1975 and 2000. Though many factors influence demand for recreation, population size is one of the most significant. The population of the state was at the time projected to rise 6.6 percent during the 25-year planning period. This increase alone is responsible for some of the increased pressure expected to be placed on recreation facilities by 2000. The issue grows more complex when we realize that the changing composition of the population implies additional effects on demand. For example, an aging population will reduce per capita demand for field games such as baseball, but increased per capita demand may be expected for activities such as golf. We now discuss factors associated with the generation of recreation demand.

---

[4]The relative-activity–effectiveness factor is an activity-specific measure of the cost-effectiveness of providing additional supply that is related to use per unit cost.

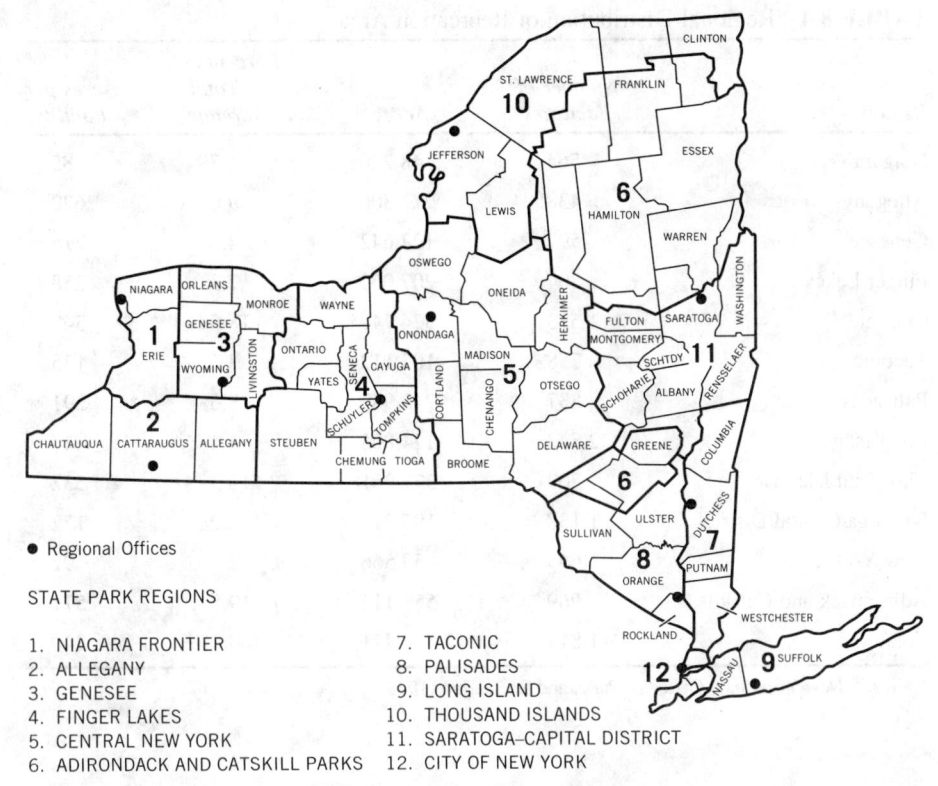

FIGURE 8.7 New York State Park Regions and Regional Offices

Source: New York State Office of Parks and Recreation (1978)

Demand Generation

Three factors were chosen as independent variables exerting a significant influence on recreation preferences: (1) age, (2) income, and (3) population density. The last of these deserves further explanation. The price paid for recreation includes not only costs associated with the expenses incurred at the site, but also the travel time associated with the trip to and from the site. Since price is difficult to measure directly, population density is used as a surrogate or proxy for the price variable. Presumably participation in activities that normally take place in urban areas (e.g., court games—primarily basketball and handball) should increase with population density because of the relatively low cost. People in rural regions will be unwilling to travel the long average distances to sources of supply and will therefore be unwilling to pay the high price associated with these activities. For similar reasons, activities usually taking place in more rural regions (such as camping) should exhibit participation rates that decline with increasing population density since areas of high density are associated with longer travel distances and higher cost.

Information on annual participation frequencies for each of the 14 recreation activities was obtained from a survey of some 7,500 state residents. These frequencies were then related to population density and the respondent's age and income using linear regression with dummy variables. (See, e.g., Draper and Smith, 1981.) Dummy variable regression is appropriate whenever the explanatory variables are categorical. Dummy variables take on a value of 1 if the individual is in a particular category and

TABLE 8.4 Regional Distribution of Recreation Area

Region	Number of Facilities	Acres	Percentage of Total Acreage	Acres per Facility
Niagara	564	48,220	1.7%	85
Allegany	438	272,300	9.6	622
Genesee	597	127,642	4.5	214
Finger Lakes	804	207,062	7.3	258
Central	1,513	524,747	18.5	347
Taconic	758	102,112	3.6	135
Palisades	837	243,937	8.6	291
Long Island	2,842	124,805	4.4	44
Thousand Islands	709	377,251	13.3	532
Saratoga/Capital District	1,138	195,717	6.9	172
New York City	645	59,566	2.1	92
Adirondack and Catskill	969	553,112	19.5	571
Total	11,814	2,836,471	100.0%	240

Source: New York State Office of Parks and Recreation (1978).

they take on a value of 0 if the individual is not in the category. For each set of variables, one category must be omitted. This is a technical requirement of regression analyses that make use of dummy variables.

Coefficients were estimated for the following equation:

$$f = a + b_1 I1 + b_2 I2 + b_3 I3 + b_4 I4 + b_5 A1 + b_6 A2 + b_7 A3 + b_8 A4 + b_9 D1 + b_{10} D2 + b_{11} D3 \qquad (8.2)$$

where a is the intercept or base participation rate, and I1 through I4 are dummy variables representing low, low-middle, middle, and high-middle income categories, respectively. A1 through A4 are dummy variables representing youth, young adult, adult, and middle-aged groups, and D1 through D3 are dummy variables representing rural, exurban, and suburban locations, respectively. It should be noted that the highest income, age, and population density categories have been omitted here, though the choice of the omitted category is entirely arbitrary. The dependent variable f is the number of days during the year in which the individual participated in the activity.

Results of the analysis are shown in Table 8.5. Zero is the implied coefficient value for omitted categories. The table may be used in two ways. First, it may be used to examine how participation varies with each of the independent variables. Second, it may be used to estimate the frequency of participation for individuals in particular groups residing in particular locations.

For most activities, participation rates decline with age, as would be expected. These declines are reflected by the decreases in coefficient values that occur as age increases. Exceptions exist for picnicking, golf, fishing, skiing, and snowmobiling. Consistent with a priori expectations, most activities witness higher participation rates

TABLE 8.5 Recreation Participation Coefficients

Activity	Base Participation Rate[a]	Income Low	Income Low Middle	Income Middle	Income High Middle	Age Youth	Age Young Adult	Age Adult	Age Middle Aged	Population Density Rural	Population Density Exurban	Population Density Sub-urban
Swimming	2.45	−5.55	−4.37	.73	3.22	22.83	9.83	8.94	2.91	8.51	5.98	4.78
Biking	.06	−.94	−.11	−.92	.07	21.63	6.21	3.82	1.17	1.64	2.95	1.31
Court games	1.34	−.55	.63	.43	1.31	16.41	5.76	1.65	.12	−3.65	−2.11	−1.54
Camping	.27	−.13	−.01	.34	.39	1.93	.44	−.01	−.19	1.25	.41	.80
Tennis	.74	−2.30	−2.42	−2.45	−.46	7.76	4.28	3.31	.61	1.48	1.78	2.14
Picnicking	.66	−.15	.31	.48	.67	1.87	2.23	1.84	.67	2.30	.97	1.10
Golf	.93	−1.44	−1.38	−.71	−.58	−.30	−.30	.28	.70	1.02	1.34	.42
Fishing	.21	1.35	.17	.57	.83	3.36	1.26	2.01	.72	1.77	1.41	1.17
Hiking	1.31	.22	.20	−.04	.77	2.49	.65	.47	.18	.53	.54	.30
Boating	.24	−1.06	−.77	−.09	.45	3.58	1.42	1.28	.50	2.36	1.68	1.86
Field games	−.17	−.39	−.19	1.53	.86	8.22	2.73	.94	−.13	1.48	.83	.56
Skiing	.70	−.98	−1.10	−.86	−.36	.85	.37	.63	.03	.19	.78	.31
Snowmobiling	−.07	−.50	−.48	−.08	−.37	.33	.85	.20	.14	1.52	.97	.44
Local winter	.66	−1.49	−1.18	−.80	.15	5.13	1.34	.82	.06	1.84	−.32	1.05

Source: New York State Office of Parks and Recreation (1978).

[a] The control group selected was the highest income, age, and density group. This column gives the estimated base annual participation rate for this group. The other columns give the amounts to be added to this amount to obtain participation rates for any other income, age, and density group. Negative participation rates are possible and should be interpreted as zero.

among people with higher incomes. Here fishing is the sole exception. Finally, participation rates for many activities are higher in rural rather than urban regions, though this relationship is not as strong as the other two. Court games are clearly more popular in urban areas, while biking, golf, and skiing are popular among exurban residents, and tennis is most popular among suburban residents.

To estimate the frequency of participation for an individual in any particular age, income, and population density category, the base participation rate is simply added to the coefficients associated with the individual's categories. (That is, Equation 8.2 is used.) For example, a young adult in the middle-income category residing in a rural region could be expected to go swimming 2.45 − .73 + 9.83 + 8.51 = 20.06 times during the year.

Total participation by the entire population may be derived by first multiplying the group-specific annual participation frequencies by the number of people in the group and then summing across groups. The result is a measure of total annual demand, and the units of measurement are *annual activity days*. One thousand annual activity days could imply 1,000 individuals each swimming 1 day per year, 500 individuals each swimming 2 days per year, 10 individuals each swimming 100 days per year, and so on. Annual activity days is a useful measure of the total demand for facilities.

Forecasts of the number of people in each age/income/density category in the year 2000, combined with the assumption of constant coefficients (which in turn is equivalent to assuming constant preferences for recreation activities) yield a forecast of recreation demand for the year 2000.

Table 8.6 summarizes the statewide demand analysis. The aging of the population results in large participation increases for activities such as golf and tennis. Similarly, activities that are very youth-oriented, such as court and field games, will witness declines in demand. Income projections for the state show that relatively more people will be in the upper-income categories by the year 2000. This leads to forecasts of rapid growth for high-income activities such as skiing and snowmobiling, and only modest increases for activities such as fishing, which are characterized by larger participation rates among those with lower incomes.

Demand Distribution

To examine geographical variation in the demand for recreation, we must distribute the demand generated by regions of residence to regions where the recreation will take place. Annual activity days (A_i) generated by a population within a region are distributed to destination regions by adopting an origin-constrained gravity model similar to the one discussed previously in the application to transportation planning:

$$T_{ij} = A_i \frac{C_j d_{ij}^{-\beta}}{\sum_k C_k d_{ik}^{-\beta}} \tag{8.3}$$

where T_{ij} is the number of recreation trips originating in i and terminating in j, A_i is the number of activity days generated by the population residing in region i, and d_{ij} is the distance between the centroids of region i and region j. C_j denotes the capacity of region j for the particular activity being examined and is taken as the measure of destination attractiveness. The parameter β reflects individuals' unwillingness to travel to participate in recreation. Its value is relatively high for activities such as court and field games, and relatively low for activities such as camping and skiing. As suggested previously, the parameter may be estimated from observed data by minimizing the

8.3 Recreation Facilities Planning 253

TABLE 8.6 Statewide Summary of Recreational Demand and Supply

Activity	Total Demand (000 Activity Days) 1975	Total Demand (000 Activity Days) 2000	Growth, 1975–2000 Millions of Activity Days	Growth, 1975–2000 Percent Change	Activity Days per Capita 1975	Activity Days per Capita 2000	Days of Use per Unit of Public Capacity 1975	Days of Use per Unit of Public Capacity 2000
Summer Activities								
Swimming	228,963	283,435	54.5	23.8%	12.5	14.6	63	78
Biking	130,913	135,224	4.3	3.3	7.2	7.0	42	44
Court games	116,139	111,741	-4.4	-3.7	6.4	5.7	132	127
Camping	43,769	46,836	3.1	7.0	2.4	2.4	74	79
Tennis	57,782	82,688	24.9	43.2	3.2	4.2	86	124
Picnicking	56,556	61,490	4.9	8.8	3.1	3.2	47	51
Golf	7,724	19,041	11.3	146.7	0.4	1.0	63	156
Fishing	43,593	44,626	1.0	2.4	2.4	2.3	62	63
Hiking	45,224	48,000	2.8	6.1	2.5	2.5	53	56
Boating	37,237	47,716	10.5	28.2	2.0	2.5	51	65
Field games	59,208	58,404	-.8	-1.3	3.2	3.0	102	100
Winter Activities								
Skiing	7,464	16,812	9.3	25.2	0.4	0.9	27	62
Snowmobiling	3,042	4,793	1.8	53.3	0.2	0.2	15	23
Local winter	34,041	46,003	12.0	35.1	1.9	2.4	46	64

Source: New York State Office of Parks and Recreation (1978).

sum of squared differences between observed and predicted trip flows, where the predicted flows are given by Equation 8.3.

This model is also sometimes known as Huff's (1963) relative attractiveness model. Essentially, the model states that the likelihood of going to a particular destination is dependent upon (1) the attractiveness of the destination and (2) the distance to it, both relative to the attractiveness and distances to all potential destinations.

Matching Forecasts of Demand with Supply

Ultimately, we would like to match demand with supply. One such attempt is shown in the final two columns of Table 8.6, where days of use per unit of public capacity is shown. To arrive at this measure, total demand must be split into demand occurring at public facilities and demand occurring at private facilities. Unfortunately, no such data that would facilitate this allocation of total demand are available. Instead, the Delphi method (see, e.g., Linstone and Turoff, 1975), which employs a consensus of "expert opinions," was used to estimate these proportions. Since supply is already divided by type of operator in the supply inventory, no such assumption is necessary on the supply side.

The measure of capacity utilization shown in Table 8.6 is inadequate for a number of reasons. Since different activities have different season lengths, it should not be assumed that those activities with high days of use per unit of public capacity should necessarily receive the highest priority in funding decisions. For example, court games have a high number of days of use per unit capacity, but they also have a longer season than many of the other activities. Likewise, skiing has only a small number of days of use per unit capacity, but its season is also relatively short.

The degree to which peaking in demand takes place should also be accounted for. Thus some activities such as swimming are concentrated on a few hot weekend days during the summer, while other activities such as court games and biking have their demand more evenly distributed throughout the week and the season. Ideally we do not want to plan facilities for the periods of heaviest use. (Otherwise the facility would be too big and would be underutilized the majority of the time.) Nor do we want to plan facilities for periods of lowest use, such as weekday mornings. (Otherwise the facilities would be overcrowded too often.) Instead, we plan facilities for the "design day," which might best be thought of in this context as an average weekend day (not including, for example, holiday weekends, where usage tends to be abnormally high). Thus we wish to match design day supply with design day demand. Accounting for the degree of peaking in demand and the lengths of seasons results in Table 8.7. Note that those activities with a high degree of peaking in demand have a relatively fewer number of design days in their seasons, reflecting the fact that a higher percentage of their demand is concentrated on a few average weekend days. Note also that for high-peaking activities, the number of design days in a season is found by dividing the length of season by 7. Similarly, factors of 5 and 4 are used for activities that have medium and low amounts of peaking, respectively. These factors were chosen by the Delphi method, and no special significance should be attached to their particular values. Accounting for peaking and length of season leads to capacity utilization figures in Table 8.7 that are much more comparable across activities than those in Table 8.6.

Capacity utilization is computed by first dividing design day demand by design day supply. Design day supply equals daily capacity multiplied by the number of design days in a season. Design day demand equals annual demand multiplied by 0.4, which is a Delphi estimate of the fraction of total demand that occurs on design days. The final step is to account for the different perceptions of crowdedness that exist in

TABLE 8.7 Design Day Computation

Activity	Length of Season (Days)	Number of Design Days in Season	Percent Capacity Utilization on Design Day (Statewide) 1975	2000
Swimming	90	14	96%	107%
Biking	200	50	58	59
Court games	270	67	88	86
Camping	180	36	90	93
Tennis	180	45	87	105
Picknicking	120	18	102	106
Golfing	150	30	92	144
Fishing	180	45	74	75
Hiking	240	60	61	62
Boating	150	30	84	95
Field games	270	67	77	77
Skiing	90	14	89	133
Local winter	110	28	62	72
Snowmobiling	75	15	63	78

Source: New York State Office of Parks and Recreation (1978).

rural and urban areas. Users of facilities in urban areas are relatively less sensitive to crowding, since they are accustomed to high densities. Likewise, users in rural areas prefer lower densities, and are relatively more sensitive to crowding. By taking the square root of the demand/supply ratio, high capacity utilization figures (over 100 percent) become somewhat smaller, and low capacity utilization figures (under 100 percent) become larger. This achieves the desired accounting of perceived crowding.

The result of this analysis is a forecast of design day capacity utilization for each recreation activity. Figures 8.8 and 8.9, for example, show the increased demand expected on golf courses; this is primarily a consequence of the rapid aging of the population.

Finally, the deficiencies that occur on design days may be translated into projected facilities needs (Table 8.8). Design day deficiencies are calculated by subtracting design day supply from design day demand for those regions where demand exceeds supply, and summing across regions to derive total deficiencies. Deficiencies are translated into facility needs by dividing design day deficiencies by the daily capacity of a facility. Note that this table also allocates responsibility for providing supply to various providers (public and nonpublic) of recreation facilities.

8.4 SITE LOCATION

Until this point in the chapter, we have focused on the generation and distribution of demand for facilities for those situations where facilities already exist. However, we may also use the same concepts to evaluate the demand that may be expected at potential new locations.

For example, suppose we have two residential zones and two grocery stores (Figure 8.10). The owner of a supermarket chain is contemplating two potential sites:

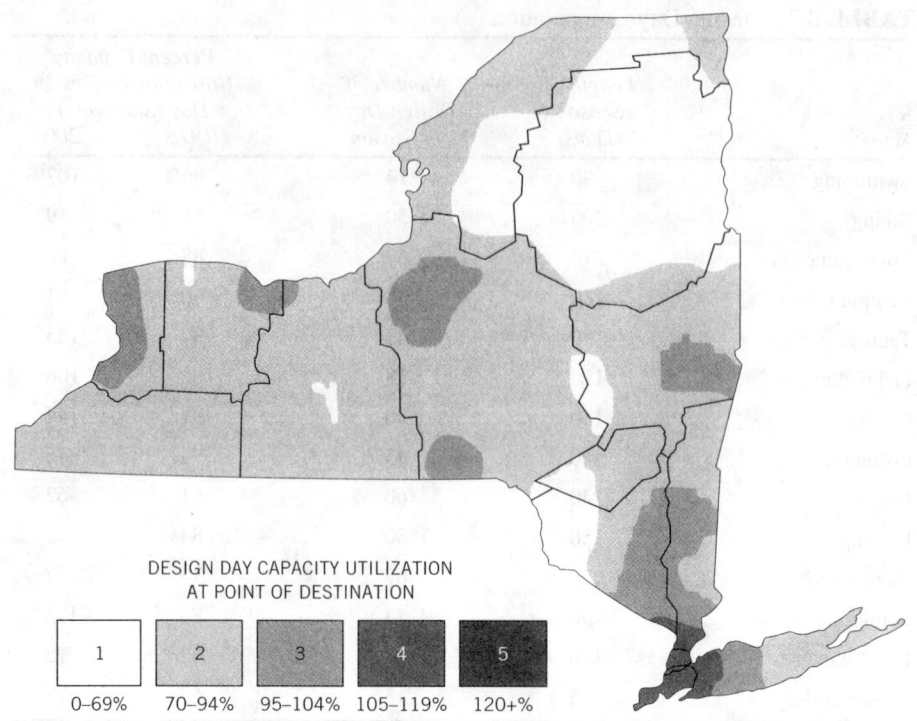

FIGURE 8.8 Design Day Capacity Utilization at Point of Destination: Golf in 1975

Source: New York State Office of Parks and Recreation (1978)

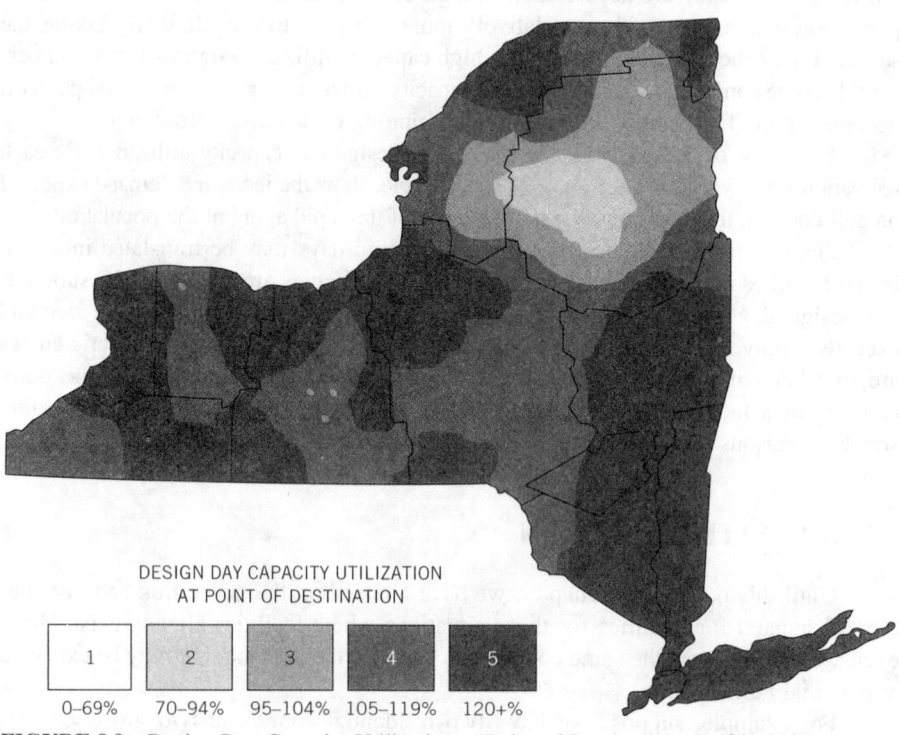

FIGURE 8.9 Design Day Capacity Utilization at Point of Destination: Golf in 2000

Source: New York State Office of Parks and Recreation (1978)

TABLE 8.8 Projected Recreational Deficiencies for the Year 2000

	Design Day Deficiencies (People-Use Capacity)					Facility Needs			
Activity	Total	Nonpublic	Total Public	State	Other Public	Total	Nonpublic	Total Public	State	Other Public
Swimming	1,155,390	433,271	722,119	144,424	577,695	579 pools or beach equiv.	217 pools or beach equiv.	362 pools or beach equiv.	72 pools or beach equiv.	290 pools or beach equiv.
Biking	77,449	—	77,449	7,745	69,704	—	—	—	—	—
Court Games	53,995	8,998	44,997	4,500	40,497	754 basketball 240 handball	126 basketball 40 handball	628 basketball 200 handball	63 basketball 20 handball	565 basketball 180 handball
Camping	29,691	26,390	3,301	2,971	330	9,010 campsites 90 campgrounds	7,995 campsites 180 campgrounds	1,015 campsites 10 campgrounds	915 campsites 9 campgrounds	100 campsites 1 campground
Tennis	163,798	48,334	115,464	11,546	103,918	3,411 courts	1,006 courts	2,405 courts	241 courts	2,164 courts
Picnicking	374,084	36,783	337,301	168,650	168,651	28,774 tables	2,829 tables	25,945 tables	12,972 tables	12,973 tables
Golfing	129,699	77,819	51,880	5,188	46,692	355 courses	213 courses	142 courses	14 courses	128 courses
Fishing	6,392	—	6,392	1,278	5,114	80 access sites	—	80 access sites	16 access sites	64 access sites
Hiking	21,288	—	21,288	8,515	12,773	—	—	—	—	—
Boating	162,326	94,690	67,636	47,100	20,536	373 ramps 9,150 berths	200 ramps 6,650 berths	173 ramps 2,500 berths	138 ramps 500 berths	35 ramps 2,000 berths
Field Games	80,302	20,163	60,139	6,014	54,125	802 fields	201 fields	601 fields	60 fields	541 fields
Skiing	187,266	131,122	56,144	28,072	28,072	80 small ski areas	56 small ski areas	24 small ski areas	12 small ski areas	12 small ski areas
Local Winter Activities	30,672	9,244	21,428	4,286	17,142	77 winter rec. areas	23 winter rec. areas	54 winter rec. areas	11 winter rec. areas	43 winter rec. areas
Snowmobiling	5,616	2,898	2,718	1,903	815	—	—	—	—	—

Source: New York State Office of Parks and Recreation (1978).

PS1 and PS2 in the diagram. Which site is likely to attract more households? We carry out the following steps to respond to this question:

1. Conduct a survey to ascertain current travel patterns from each residential neighborhood.

2. Use the survey results to estimate the "friction of distance" associated with travel to grocery stores. This is done by minimizing the sum of squared deviations between the observed proportions of households going to each store and the predicted proportions. The latter quantities are found from the origin-constrained gravity model:

$$P_{ij} = \frac{S_j d_{ij}^{-\beta}}{\sum_k S_k d_{ik}^{-\beta}}$$

where P_{ij} is the proportion of households in residential zone i that choose store j, S_j is the attractiveness of store j, and d_{ij} represents the distance between zone i and store j.

3. Use the estimated friction of distance value to derive the expected demand at each potential site. Expected demand at a site may be derived by summing the demand received from each of the residential zones:

$$D_j = \sum_i h_i \frac{S_j d_{ij}^{-\beta}}{\sum_k S_k d_{ik}^{-\beta}}$$

where D_j is the number of households visiting store j, and h_i is the number of households in residential zone i.

Suppose that the survey finds that 80 percent of households in residential zone 1 go to store 1 and the other 20 percent patronize store 2. From residential zone 2, 60 percent go to store 1 and the remaining 40 percent go to store 2. The reader may verify that, using the brute-force method to minimize the sum of the squared deviations between observed and predicted proportions, the estimated value of β is 1.153.

The expected demand at each potential site may now be calculated using the estimated value of β and the relative attractiveness equation. Expected demand at potential site 1 is

$$2{,}000 \frac{(20)3^{-1.153}}{(20)3^{-1.153} + (50)5^{-1.153} + (5)2^{-1.153}}$$

$$+ 1{,}000 \frac{(20)8^{-1.153}}{(20)8^{-1.153} + (50)10^{-1.153} + (5)2^{-1.153}} = 1{,}184$$

Similarly, expected demand at potential site 2 is

$$2{,}000 \frac{(20)3^{-1.153}}{(20)3^{-1.153} + (50)5^{-1.153} + (5)2^{-1.153}}$$

$$+ 1{,}000 \frac{(20)1^{-1.153}}{(20)1^{-1.153} + (50)10^{-1.153} + (5)2^{-1.153}} = 1{,}494$$

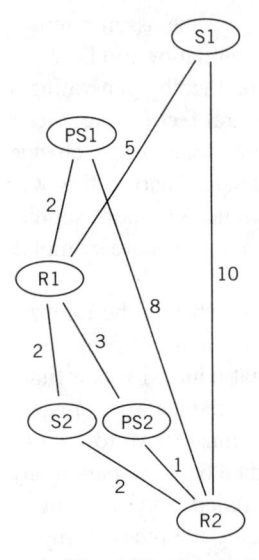

Note: R1 and R2 are residential locations.
S1 and S2 are current stores.
PS1 and PS2 are potential sites.

Numbers indicate travel distances.
Store attractiveness values: S1: 50
S2: 5
PS1: 20
PS2: 20
Households at R1: 2,000
Households at R2: 2,000

FIGURE 8.10 Hypothetical Site Location Problem

Site 2 possesses greater expected demand. Note that we have implicitly assumed that households are homogeneous in their spending habits. Although we have chosen a site that maximizes the number of households expected to patronize the site, an owner would actually be more interested in the amount of revenue derived from those households. If we can make the assumption that households in the two residential zones have similar spending patterns (say in terms of weekly dollars spent on grocery items), then the preceding site analysis is valid.

If on the other hand we have information on how spending behaviors of individuals in the two residential zones differ, we can adjust the analysis accordingly. Suppose, for example, that individuals in residential zone 1 spend $150/week on groceries, while individuals in zone 2 spend $100/week. Then the expected weekly revenue at site 1 is (1,184)($150) = $177,600, while expected weekly revenue at site 2 is (1,494)($100) = $149,400. Because of the higher expenditures by households in residential zone 1, the optimal location has now shifted to site 1, which captures more customers from residential zone 1 than does potential site 2.

8.5 IMPEDIMENTS TO THE PRODUCTION OF IMPROVED FORECASTS

Analyses of the need for public facilities or, even more generally, analyses of the market for consumer goods may all be carried out under a rather general framework that matches forecast demand with anticipated supply. Though the methods often provide suitable forecasts, there exist a number of impediments to the production of more accurate results.

Unfortunately we often have no way of assessing how good the assumption of parameter stationarity actually is. We can sometimes estimate the relationship between trips and the independent variables for a number of recent years, thereby generating a set of parameters for each year where data are available. A general feel for parameter stability may then be obtained by examining the degree to which parameters change over time. It is most common to assume that the most recent set of parameters will remain unchanged over the forecast horizon. It is possible to implement more sophisticated approaches, such as varying parameter regression, but these more complex methods are only rarely employed in most practical situations.

Even more basic than the issue of parameter stationarity is whether the relevant variables have been specified in the regression equation used for demand projection. For example, the projections of recreation demand assumed that demand was a function of age, income, and population density. But clearly other variables, such as gender and car ownership, affect participation as well. The analyst must strive to include those variables that most affect demand, and must strike a balance between parsimony (where there is the danger that not enough variables will be included) and throwing in the kitchen sink (where too many variables are included, clouding the relative importance of each). This often entails performing several analyses to assess the marginal effects of adding assorted variables.

In many cases, we must look for effects of interactions among independent explanatory variables. For example, although we examined the effects of age, income, and population density on recreation demand, the effects of age may vary with income. Thus there may be more of an increase in demand for golf with age among those with higher incomes than among those with lower incomes.

Yet another impediment to accurate forecasts of demand is the usual difficulty in obtaining accurate forecasts of the independent variables. The procedure just described for generating forecasts of demand relies greatly on these assumed future values of the independent variables. If income, population, and car ownership cannot be forecast well, there is clearly little hope for producing accurate forecasts of demand.

What these impediments indicate is the obvious need for monitoring of forecasts. By monitoring parameter stability, measures of the independent variables, and observed demand, the planning process can be a truly continual one. This of course requires that sufficient resources be devoted to the collection and analysis of the necessary data.

8.6 OPTIMAL DEMAND ASSIGNMENT METHODS

Some infrastructure planning and business problems are conceptually different from those discussed previously. In all the examples we have already seen, the aggregate demand for use of a facility is generated by individuals making independent decisions regarding their own trip-making behavior. In this second group of applications, however, geographically dispersed demand is centrally matched up with the supply of services provided at a set of discrete geographic points. Research in the field of *operations research* focuses on developing methods designed to help decision makers find optimal allocations. The operations research literature includes many examples of business and public-sector applications. Scott (1971) provides a good overview of spatial allocation analysis.

Our intention here is to illustrate an optimization technique for a commonly encountered problem in infrastructure planning in which demographics plays a large role—the assignment of pupils to schools so as to minimize the total transportation costs involved. Here we return to the type of application of population analysis described in Chapter 1. In such problems the analyst seeks to find the values for certain variables under the control of a decision maker so as to minimize (or sometimes maximize) the value of an *objective function*. The choices that may be made, however, are restricted by a set of constraints on the values of the control variables that may be feasibly chosen.

Here we present an example of the Transportation Problem of Linear Programming (TPLP). The TPLP is a special case of the more generic class of optimization problems known as linear programs. In linear programming the objective function and each of the constraints is specified as a linear combination of the control variables—meaning that there are, for example, no squared terms or variables multiplied by each other (which would make the problem one in *quadratic* or nonlinear programming). In the TPLP the constraint set assumes a particularly simple structure that makes the solutions amenable to easy computation without the benefit of special computer software packages. Here we take you through the step-by-step calculations of the stepping stone method for solving the TPLP. A fundamental understanding of this simple example should help you understand the outputs from larger, computer-derived programming problem solutions. Such understanding will also lay the basis for future study of optimization methods—should the reader have the need or inclination to become proficient in using these methods.

The Un-unified–Unified School District Problem

Suppose that the hypothetical Norwood–Norfolk Unified School District and the Madrid–Lisbon Unified School District have recently merged. Declining fertility rates during the baby bust period of the early 1970s and continued net out-migration from the region throughout the 1980s and into the 1990s have resulted in fewer schoolchildren and excess school capacity at the new districts' three high schools. At present, pupils in grades 9 through 12 in each of the four towns attend that town's school, except those in Norwood and Norfolk, who attend the Norwood–Norfolk Unified School. Figure 8.11 shows the current assignments.

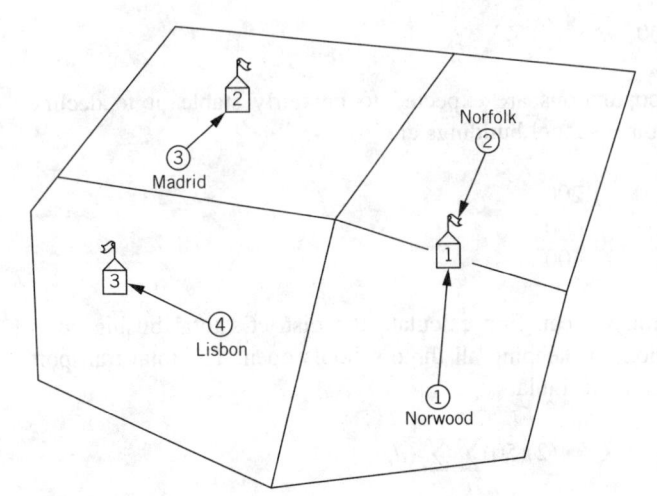

FIGURE 8.11 The Unified–Unified School District Problem: Current Pupil Assignments

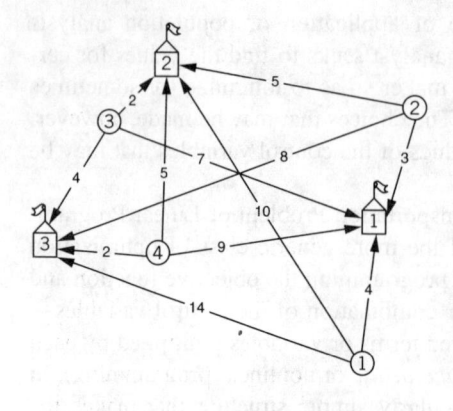

FIGURE 8.12 Road Mileages between Town Population Centroids and High Schools, Norwood–Norfolk–Madrid–Lisbon Unified–Unified School District

Unlike the growth example in the introductory chapter, the school board in this case is contemplating closing an existing school rather than building a new one. A proposal has been made to shut the old Lisbon School building and to rearrange assignments and busing routes to take advantage of the current unused capacity at the Norwood–Norfolk and Madrid Schools.

It is estimated that closing the Lisbon building would save the district $200,000 per year in net operating costs—but would entail some as yet uncalculated additional busing costs. The Unified–Unified Board is far from unified in support for the plan, and no new assignment of pupils to the remaining two schools has yet been devised. How would the geographical analyst of population advise the board?

To generate alternative assignments of pupils to schools and to estimate busing costs for the proposed plan, we must first gather more information. Road mileages between the population centroids of the four towns (see Chapter 2) and each of the three schools are shown in Figure 8.12. Suppose that the estimated annual busing cost next year will be $50 per pupil per mile traveled. Additionally assume that the projected high school populations of the four towns for next year are

Town of Norwood	600
Town of Norfolk	400
Town of Madrid	500
Town of Lisbon	300

Beyond next year these populations are expected to be fairly stable or to decline slightly. Capacities of the three school buildings are

Norwood–Norfolk School	1,200
Madrid School	600
Lisbon School	400

Given this information we can first calculate the district's total busing costs under the "do nothing option" of keeping all three schools open. The total transport cost, C, may be found from the formula

$$C = (2)(50) \sum_i \sum_j (d_{ij} t_{ij})$$

where d_{ij} and t_{ij} are elements of the town-to-school distance matrix, **D**, and the trip distribution matrix (showing current school assignments), **T**:

$$\mathbf{D} = \begin{bmatrix} 4 & 10 & 14 \\ 3 & 5 & 8 \\ 7 & 2 & 4 \\ 9 & 5 & 2 \end{bmatrix}$$

and

$$\mathbf{T} = \begin{bmatrix} 600 & 0 & 0 \\ 400 & 0 & 0 \\ 0 & 500 & 0 \\ 0 & 0 & 300 \end{bmatrix}$$

Current busing costs for the district total $520,000.

To evaluate the additional transportation cost posed by the proposed closing of the Lisbon school, we find a new optimal distribution, **T***, where now the relevant distance matrix is

$$\mathbf{D} = \begin{bmatrix} 4 & 10 \\ 3 & 5 \\ 7 & 2 \\ 9 & 5 \end{bmatrix}$$

The optimal distribution is the (4 × 2) matrix **T*** that minimizes the total busing cost, C^*:

$$C^* = 100 \sum_i \sum_j d_{ij} t_{ij}^* = 100 \min_{\{d_{ij}\}} \sum_i \sum_j d_{ij} t_{ij}$$

This equation is the *objective function* of the transportation problem of linear programming. It says to choose a set of transport flows t_{ij} that results in the lowest possible total cost of travel. Of course, one way to do that would be to choose $t_{ij} = 0$ for all values of i (towns) and j (schools). But we would not want to allow this as a solution because then no one would be going to school! Certain *constraints* must be met by the t_{ij} elements; that is, we are only able to choose values of them within a particular range of allowable values.

For this problem we have three types of constraints: origin, destination, and nonnegativity.

1. *Origin constraints* guarantee that all students in each town must be assigned to a school. The total number of pupils in each town is the number of "origins," O_i:

$$t_{11} + t_{12} = O_1 = 600$$
$$t_{21} + t_{22} = O_2 = 400$$
$$t_{31} + t_{32} = O_3 = 500$$
$$t_{41} + t_{42} = O_4 = 300$$

2. *Destination constraints* state that the capacity of each school cannot be exceeded. The capacity of each school is the number of "destinations," D_j:

$$t_{11} + t_{21} + t_{31} + t_{41} = D_1 = 1{,}200$$
$$t_{12} + t_{22} + t_{32} + t_{42} = D_2 = 600$$

Note that because the total number of pupils in the district and the combined capacity of the two remaining schools are the same, these constraints may be written with = signs rather than $\leq$.

3. Nonnegativity constraints ensure that there are no negative flows of pupils:

$$t_{ij} \geq 0 \text{ for all } i, j$$

Here, and in general in the Transportation Problem of Linear Programming, there are a total of $m + n + mn$ constraints (where m is the number of origin zones and n is the number of destinations). Any set of numbers $\{t_{ij}\}$ that satisfies all these constraints is said to be a feasible solution. The minimization problem then is to select from all possible feasible solutions the one that results in the least total value of the objective function—the minimum busing costs. To solve our school district problem we use the stepping stone method. For more details than can be given in this brief example, see, for instance, Miller (1972).

The Stepping Stone Method for Solving the TPLP

The first step in using the stepping stone method is to set up a tableau as in Figure 8.13. This tableau looks a bit like a bowling scorecard! It summarizes the origin and destination constraints. Above the diagonal lines it notes the various distances that students might be bused. The major section of each box will be used to fill in trial values for the t_{ij} flows.

To decide on the first value to enter, we use the "northwest corner rule," which says to start in the upper left corner, find the smaller of the row and column sums, and use this value for t_{ij}. In this case the origin constraint for town 1 (Norwood) is 600, whereas the column constraint for school 1 (Norwood–Norfolk) is 1,200, so the value we enter (see Figure 8.14) is 600. Because the origin constraint 1 is now totally satisfied, we must also place a zero in the box for element t_{12}.

Then the stepping stone method tells us to step down to the only other adjacent box, that for t_{21}, and follow the same kind of logic to select a value for insertion. At

$i \diagdown j$	1. Norwood – Norfolk	2. Madrid	O_i
1. Norwood	4	10	600
2. Norfolk	3	5	400
3. Madrid	7	2	500
4. Lisbon	9	5	300
D_j	1,200	600	1,800

Note: The rows are for the towns, the columns for the schools. Numbers of students originating in each town (O_i), the school capacities (D_j), and distances between schools (d_{ij}) are shown.

FIGURE 8.13 Tableau for the Stepping Stone Method to Solve the Un-unified–Unified School District Problem

i \ j	1. Norwood–Norfolk	2. Madrid	O_i
1. Norwood	4 / 600	10 / 0	600
2. Norfolk	3 / 400	5 / 0	400
3. Madrid	7 / 200 —→	2 / 300	500
4. Lisbon	9 / 0	5 / 300	300
D_j	1,200	600	1,800

FIGURE 8.14 Initial Feasible Solution to the Un-Unified–Unified School District Problem

Note: Arrows indicate the stepping-stone path taken after using the northwest corner rule.

this second step of the procedure, the origin constraint (for the town of Norfolk) again comes into play before the destination constraint (capacity of Norwood–Norfolk school), so we choose $t_{21} = 400$ and set t_{22} to zero.

The same logic is followed until all the boxes are filled in. As Figure 8.14 shows, at the third step the capacity of the Norwood–Norfolk School was filled with 200 pupils from the town of Madrid, and the 300 other Madrid high schoolers were assigned to Madrid School. Finally, all 300 Lisbon students were assigned to the Madrid School, and we have successfully stepping-stoned our way across the entire tableau.

Notice that there are five positive t_{ij} values in the tableau and three zeros. This is illustrative of a general fact: Any transportation problem solution will have at most $m + n - 1$ nonzero entries; in our case $(4 + 2 - 1) = 5$.

The solution we have obtained with the stepping stone procedure is a *feasible* solution (all the constraints are met), but is it the best we can do in terms of minimizing total busing costs? To determine whether we can do better, we must first calculate C for this solution. Total busing costs under this proposed assignment plan amount to $710,000. But can we find a cheaper feasible solution?

To answer that question, the stepping stone procedure next tells us to compute *opportunity costs* for each of the zero-valued cells. If an opportunity cost is greater than the d_{ij} value for that cell, then it would be beneficial in terms of lowering the overall value of C to enter it into the solution. Of course, if we change the value of any cell in the tableau, a series of compensating changes must be made to certain other cells to ensure that all the constraints continue to be met.

To calculate the opportunity costs, we first need to compute *shadow prices*. There is a shadow price associated with each row or origin constraint of the tableau, which we shall designate U_i ($i = 1, 2, \ldots, m$), and for each column or destination constraint, V_j ($j = 1, 2, \ldots, n$). These may be interpreted as the value (in terms of the objective function) of a marginal slackening of the associated constraint. (See, e.g., Miller, 1972.) They are calculated from the relation

$$V_j - U_i = d_{ij}$$

To find the values for these, all the cells having *nonzero* t_{ij} values are used. To begin, any single shadow price may be set to zero, such as U_1 in Figure 8.15. From here it is a simple matter to determine the rest of the U_i and V_j values. For instance, $V_1 - U_1 = d_{11}$ so $V_1 = 4 - 0 = 4$; $U_2 = V_1 - d_{21} = 4 - 3 = 1$; and so forth. Note that shadow prices may be negative as well as positive.

Once all the shadow prices are determined, the opportunity costs for all the cells having zero-valued t_{ij} are found (Figure 8.16). The formula is

$$k_{ij} = V_j - U_i$$

For example, $k_{41} = V_1 - U_4 = 4 - (-6) = 10$.

Notice that k_{41} is the only one of the opportunity costs shown that is higher than the corresponding actual distance element d_{ij}. This tells us that it would be beneficial in terms of saving busing costs to assign some positive flow of students from the town of Lisbon ($i = 4$) to the Norwood–Norfolk School ($j = 1$).

The major component of the stepping stone procedure now begins. We use (4,1) as the pivot to redistribute flows $\{t_{ij}\}$ so as to improve on our initial feasible solution. The goal is to assign as many of the Lisbon pupils as possible to the Norwood–Norfolk School. Comparing the relevant constraints, we see that we could send all of that town's students to this school since its capacity is 1,200. Thus we would ideally wish to set $t_{41} = 300$; thus $t_{42} = 0$.

There are other repercussions throughout the system, however. We must find 300 pupils previously assigned to Norwood–Norfolk and reassign them to the Madrid School. It would seem that we could subtract some or all of the 300 pupils from any of the other cells in column 1. The stepping stone protocols, however, require that we do this in such a fashion that we make further compensating changes only to positive-(nonzero) valued cells. For instance, if we subtracted 300 from t_{11}, we would also have to add 300 to t_{12} in order to assign all 600 Norwood students—but cell (1,2) is currently a zero-valued one! The same problem exists for cell (2,1). We would have to

i \ j	1. Norwood–Norfolk	2. Madrid	U_i
1. Norwood	4		0
2. Norfolk	3		1
3. Madrid	7	2	−3
4. Lisbon		5	−6
V_j	4	−1	

Note: Shadow prices are computed from the distances (d_{ij}) corresponding to nonzero flows (t_{ij}) in the initial feasible solution in Figure 8.14.

FIGURE 8.15 Shadow Prices (U_i and V_j)

i \ j	1. Norwood–Norfolk	2. Madrid	U_i
1. Norwood		10 / −1	0
2. Norfolk		5 / −2	1
3. Madrid			−3
4. Lisbon	9 / 10		−6
V_j	4	−1	

Note: Opportunity costs are computed from the shadow prices (U_i and V_j) and distances (d_{ij}) corresponding to zero-valued flows (t_{ij}) in the initial feasible solution in Figure 8.14.

FIGURE 8.16 Opportunity Costs (k_{ij})

add to cell (2,2), but it too is zero-valued. Therefore it would seem that we must make our subtraction to cell (3,1).

But troubles again! Since the current value of t_{31} is only 200, the nonnegativity constraint would be violated. There is no such thing as negative students (at least in the Norwood–Norfolk–Madrid–Lisbon Unified–Unified School District—though the authors have encountered a few elsewhere!). We were, in fact, a bit too hasty in trying to assign all 300 Lisbon students to the Norwood–Norfolk School. A more prudent procedure is to first figure out the entire stepping stone path, and then to do the additions and subtractions. The goal is to get back to row 4 and do a negative adjustment in cell (4,2) to offset the positive one we wish to make in (4,1). The most that we can add to (4,1) will be the smallest value of any of the cells to which we will want to make a subtraction.

Using the logic described, we get the new, hopefully improved solution in Figure 8.17. Instead of splitting the Madrid students between the two schools, we are now considering the possibility of assigning Lisbon kids to both schools. As the reader may easily verify, the total busing cost, C, for this new assignment is $690,000. Therefore we have, in fact, lowered the value of the objective function.

Now do we have the best possible solution? Once again we must compute shadow prices and, from them, new opportunity costs (Figure 8.18). Note that now opportunity costs are shown only for the three zero-valued cells in the new solution (Figure 8.17). Comparing these to the corresponding d_{ij} values, we see that in each of the three cases the k_{ij} value is the lower. This means that we have now found the optimum solution and can halt the process. The stepping stone method has found us the trip distribution matrix, **T***, that has the lowest total busing costs. Other problems might require more iterations than were needed here. And for real world analysis, a computer program would likely be needed to solve the relevant TPLP as the analyst would want to have many more and much smaller pupil catchment zones—and thus origin constraints—than in the present example.

268   Chapter 8 • The Role of Population in Infrastructure Planning

i \ j	1. Norwood–Norfolk	2. Madrid	O_i
1. Norwood	4 / 600	10 / 0	600
2. Norfolk	3 / 400	5 / 0	400
3. Madrid	7 / − 0	2 / + 500	500
4. Lisbon	10 / + 200	5 / − 100	300
D_j	1,200	600	1,800

Note: Arrows indicate the stepping-stone path used to make compensating changes after element t_{41} is brought into the solution.

FIGURE 8.17 Second-Round Solution to the Un-unified–Unified School District Problem

Back to the School Board...

Having set up the Transportation Problem of Linear Programming in this fashion, and having found its solution, the geographical analyst of population can provide the decision makers—the members of the Norwood–Norfolk–Madrid–Lisbon Unified–Unified School Board—with useful technical information. But as we discussed in Chapter 1, techniques such as this are not a substitute for judgment.

In terms of simply minimizing the total cost to the district, it would seem that going ahead with the proposed plan of closing the Lisbon school is worthwhile. The annual cost savings of $200,000 would be greater than the additional busing costs. Recall that with all three schools operating and each student going to the high school

i \ j	1. Norwood–Norfolk	2. Madrid	U_i
1. Norwood	4 /	10 / 0	0
2. Norfolk	3 /	5 / −1	1
3. Madrid	7 / 6	2 /	−2
4. Lisbon	9 /	5 /	−5
V_j	4	0	

Note: Because all the opportunity costs are less than the corresponding distances, the optimum solution has been found.

FIGURE 8.18 New Shadow Prices and Opportunity Costs Calculated for the Solution in Figure 8.17

in his or her own town, the district is currently spending $520,000. The assignment plan that we have just worked out for operating only the Norwood–Norfolk and Madrid Schools implies busing costs of $690,000 (an increase of $170,000). Hence net annual savings of $30,000 are possible.

But have we taken into account all possible concerns? With only two schools operating, some students will be spending more time and going longer distances on the district's buses. Of course, some might dismiss this by arguing that the value of a high school student's time is essentially zero (!), but what about the safety implications of having more buses on the roads for longer trips? What about the political ramifications connected with the loss of sense of place some people may experience when the town of Lisbon has no high school to call its own? Politics are often messier than finding some cost-minimizing optimal allocation. Nonetheless, to the extent that we can specify numerically the various concerns relevant to any given problem, the more and more capable are current mathematical programming techniques to provide useful inputs into the decision process.

8.7 THE USE OF FORECASTS IN DECISION MAKING

A number of points should be made regarding the use of forecasts of facility demand. First, there is the issue of forecasts as self-fulfilling prophecies. If a region is forecast to have a need for additional facility capacity, funds will more likely be allocated to it; likewise, forecasts of no additional capacity needs may lead to little or no additional allocations of funds. Hence the planning process may itself cause the forecast to come true. A case can be made for an "indicative" type of planning (Faludi, 1973), where forecasts are produced, not on the basis of the methods just described, but rather on the degree to which the planner or planners would like to see the region developed. This of course inverts the usual direction of causality, and causes us to wonder about the proper role for the forecast methods described in this chapter.

One way out of this dilemma is to change the nature of the planning process. Forecasts of facility demand based on expected changes in population levels and composition should still be produced—they should provide the basis for planning decisions. However, the decisions themselves do not necessarily have to be based upon indexes of relative need. Thus, for example, funding might occur for a facility despite a relatively low index of need if the region was targeted for development. Similarly, a facility in clear need of additional resources might not be funded to the level required to meet demand as a means of controlling growth in a region that was developing too rapidly.

A particular aspect of this jumbling of the directions of relationship between forecast, policy, and reality is well illustrated by the example of water demand forecasting. White (1970) and James and Lee (1971) have argued that the "requirements" approach to forecasting assumes that current use levels are appropriate. This assumption may or may not be desirable. In the context of water demand forecasting, for example, little attention is paid to possible effects on usage of changes in technology, policy, and pricing. Technological change could produce lower amounts of loss and leakage; policies could become more conservation-oriented, and prices could rise. All of these changes would serve to reduce per capita use. Why base the planned provision of water on current use, and therefore risk oversupply, when implementing changes could reduce the amount needed? Also, the forecast itself could undermine conservation efforts. This argument reflects a general shortcoming with the requirements approach outlined in this chapter.

Finally, another reversal of logic sometimes occurs in the use of forecasts by decision makers. Rather than base facility development decisions upon the results of needs and forecasts, there is sometimes a tendency for decisions to be made first. The decision is then justified by building a case that uses the forecast from the best possible perspective. It is easy in these instances for the analyst to become frustrated by the process—why go through the trouble of producing projections if decisions are going to be made without prior examination of relative need? On the bright side, there are just as many instances where the report of projected need is virtually the only "hard" piece of information available. On the occasions where decision makers do wish to rely on such hard information, the role of the forecast becomes paramount to the process.

8.8 SUMMARY

In this chapter, we have seen how important population analysis is in planning for the future needs of individuals. The demographic and socioeconomic composition of the population influences the demand for various goods and services. By accurately projecting the size and composition of the population, we may obtain a good idea of the number, size, and location of facilities that need to be provided.

In the next two chapters, we continue to emphasize the roles of population composition and change in public- and private-sector planning. We do so by shifting our attention away from the individual and toward the household as the unit of analysis.

EXERCISES

1. Suppose that the following journey-to-work pattern characterizes a metropolitan area:

To:	Zone 1	Zone 2	Zone 3	Zone 4
From:				
Zone 1	1,000	800	600	400
Zone 2	900	1,300	800	500
Zone 3	500	800	1,000	300
Zone 4	200	400	600	1,000

Transportation planners wish to forecast the journey-to-work pattern that will characterize the metropolitan area 10 years from now. Projected origin and destination totals for that time are:

	Origin	Destination
Zone 1	3,500	2,800
Zone 2	3,400	3,600
Zone 3	3,000	3,000
Zone 4	2,100	2,600

Use the Cross–Fratar method to derive a projected flow matrix based on the observed commuting pattern, and the projected origin and destination totals.

2. The following matrix of probabilities that trips leaving each of two residential origins (A and B) go to each of two local parks (1 and 2) is observed:

	Park	
Residential Origin	1	2
A	.74	.26
B	.20	.80

Using the following distances between residential origins and parks

	Park	
Residential Origin	1	2
A	3	12
B	10	2

and attractiveness measures for destinations 1 and 2 equal to 1.5 and 1.0, respectively, estimate the parameter β in the origin-constrained version of the gravity model by minimizing the sum of squared deviations between observed and predicted probabilities. For those not using a computer program, try values of $\beta = .2, .3, .4,$ and $.5$. For each value of β, compute the sum of squared errors, and estimate β^* corresponding to the minimum sum of squared errors.

3. Find the probabilities of going from each residential origin to each park, as predicted by the model.

4. Use the value of β found in Exercise 2 to forecast the allocation of flows given a change in the attractiveness measures of destinations 1 and 2 to 2.0 and 1.1, respectively.

5. Assume that the following equation has been estimated for the frequency of monthly shopping trips:

$$S = a + b_1 X_1 + b_2 X_2 + b_3 X_3 + c_1 Y_1 + c_2 Y_2$$

where

$a = 3.65 \quad b_3 = 2.16$
$b_1 = 0.24 \quad c_1 = 0.98$
$b_2 = 0.90 \quad c_2 = 3.47$

and where

$X_1 = 1$ if the individual is in the low-income category
$X_2 = 1$ if the individual is in the middle-income category
$X_3 = 1$ if the individual is in the upper-middle-income category
$Y_1 = 1$ if the individual is in the young adult category
$Y_2 = 1$ if the individual is in the middle-age category

Variables are set to 0 if the respondent is not in the category.

a. How many monthly shopping trips does an individual in the low-income and middle-age categories make per month?

b. Suppose that we have a forecast of the number of people expected in the age/income categories for a particular residential zone:

	Age		
Income	Young Adult	Middle-Age	Elderly
Low	6,500	2,000	3,000
Middle	8,000	4,000	2,000
Upper-Middle	10,000	6,000	3,000
Upper	2,000	8,000	4,000

What is the forecasted number of monthly shopping trips produced by young adults in this residential zone?

6. Assume that a city divided into four zones contains all of its employment and shopping opportunities in zone 4. A survey finds that the three residential zones generate daily shopping trips as follows:

Zone 1:   13,100
Zone 2:   5,000
Zone 3:   15,900

No residential population is contained in zone 4. Current and forecast population and employment data are as follows:

	Current Year (1990)			Projection Year (2000)		
	Zone 1	Zone 2	Zone 3	Zone 1	Zone 2	Zone 3
Population	40,000	10,000	50,000	100,000	20,000	60,000
Households	12,000	4,000	14,000	30,000	10,000	20,000
Household Size						
1–2	5,000	2,000	5,000	10,000	6,000	9,000
3+	7,000	2,000	9,000	20,000	4,000	11,000
Income						
< $20,000	9,000	1,000	10,000	22,000	2,000	12,000
Household Size						
1–2	3,000	600	3,400	6,300	1,100	5,600
3+	6,000	400	6,600	—	—	—
Income						
≥ $20,000	3,000	3,000	4,000	—	—	—
Household Size						
1–2	2,000	1,400	1,600	—	—	—
3+	1,000	1,600	2,400	—	—	—
Total Resident Employees	24,000	4,000	22,000	55,000	12,000	33,000

a. Begin by filling in the blanks in the preceding table. If work and shopping trips are the only types of trips made, find the set of traffic flows, T_{ij}, between each of the zones. Assume that there are no multipurpose trips (e.g., home–work–shop–home), and remember that for every journey-to-work and every journey-to-shop trip, there is a journey-to-home trip.

b. *Trip Generation.* A survey finds that the 34,000 daily shopping trips may be characterized as follows:

Income	Household Size 1–2	Household Size 3+
< $20,000	4,900	15,600
≥ $20,000	5,500	8,000

Using the data on the number of households of different types in the current year, find the trip generation ratios (the ratios of trips to households) in each category. Express the ratios as decimals rounded to the nearest one thousandth (.001). Then, using the projected characteristics of the population, and assuming that the trip generation ratios remain constant, project the total daily number of trips that will originate in each zone.

c. *Trip Distribution.* Assume that in 1980 there is a mall in zone 4 containing 800,000 square feet, and that by the year 2000, there will be a mall in zone 1 containing 600,000 square feet. The distances from residential zones 1, 2, and 3 to zone 4 are 4, 3, and 5, respectively. The distances from zones 1, 2, and 3 to zone 1 are 2, 6, and 8, respectively. Assume that the friction of distance parameter equals 2. Use an origin-constrained gravity model with retail floor space as the attractiveness variable to forecast shopping trips for the year 2000. If the new mall contains 20,000 workers and there are 30,000 new jobs added to zone 4, find the new set of traffic flows (i.e., combined work and shopping trips).

7. A local businesswoman has decided to open a local supermarket and is deciding between two alternative locations. There are two major residential zones (A and B) and two current supermarkets (1 and 2). Store 1 is located 2 miles from A and 8 miles from B. Store 2 is located 1 mile from B and 7 miles from A. Store 1 has 1.5 times the floor space as store 2. Of the sites under consideration, site 3 is 4 miles from A and 5 miles from B; site 4 is located 1 mile from A and 8 miles from B. Assuming (a) that $\beta = 1$, (b) that the new store will have the same retail floor space as store 2, (c) that store attractiveness can be adequately measured by retail floor space, and (d) that the number of households in A and B are 3,000 and 4,000, respectively, find the location that maximizes patronage.

8. *Transportation Problem of Linear Programming.* The management of PFMS, Inc., a personal financial management services firm, has announced plans to close down the company's aging office building in the central business district of Megalopolis and transfer all 500 of its employees to two new suburban sites. PFMS officials hope to attract additional business by moving closer to the company's client markets in wealthier suburban areas. The new offices and the numbers of employees that will work out of each are

Eastgate Office Park	200
Northern Lights Business Center	300

In accordance with the Megalopolis Travel Reduction Ordinance, PFMS management wishes to assign employees to their new work locations so as to minimize their commuting time. A breakdown of PFMS employment by resident zone looks like this:

Central neighborhoods   50
Eastern suburbs   250
Sin Vista subdivision   200

The following table shows approximate one-way travel times (in minutes) from each of the three residential zones to each of the two new office sites:

	Eastgate	Northern Lights
Central	20	10
Eastern	30	50
Sin Vista	30	40

a. Set up a tableau for the Transportation Problem of Linear Programming (TPLP) and use the stepping stone method to find the optimal (travel-time–minimizing) assignment of workers from the three residential zones to the two new offices.

b. What will be the average amount of time spent on daily (one-way) commuting by PFMS employees if the workers are, in fact, assigned using the TPLP solution?

> There's nothing I wouldn't do if
> you would be my POSSLQ.
>
> —Charles Osgood

CHAPTER 9
HOUSEHOLD DEMOGRAPHY

9.1 Introduction
9.2 A Brief Look at Households, Location, and Relocation in the United States
9.3 Geographic Concepts Relevant to the Study of Households
9.4 Demographic and Geographic Analysis of Households
9.5 Summary

9.1 INTRODUCTION

Until now, we have focused upon the study of populations as collections of *individuals*. However, almost all individuals in a population are organized into households (individuals living in institutions constitute an exception), so it is appropriate that attention also be given to the study of the household unit. Such study is particularly appropriate in light of dramatic changes in recent decades in the number, size distribution, and composition of households. During the 1970s, the U.S. population increased 11.4 percent, while the number of households grew by 27 percent. This more rapid relative growth of households continued during the 1980s, though at a less brisk pace. The number of households in the United States grew by 14.4 percent during the 1980s, while the population grew by 9.8 percent. The difference between population and household growth rates is attributable to several causes. Nest-leaving baby boomers, high divorce rates, and low levels of childbearing are major factors.

We must recognize that substantial differences between population and household growth do indeed exist, and that they have significant consequences for planning both public- and private-sector activities. Demand for many consumer goods is driven not by the number of individuals in a population, but rather by the number of households. Cars, appliances, and furniture are examples of products whose demand is tied closely to the number of households. In both the public and private sectors, accurate estimates and forecasts of household characteristics are clearly needed to plan for housing, but they are also needed for the quasi-public sectors and other government agencies. Demands for electricity, gas, and water are all closely related to the size distribution and number of households. One of our aims in this chapter is to examine some basic methods for analyzing and projecting the number, size, and composition of households.

There are also strong interactions between household characteristics, location, and relocation. Single females are more likely to live in metropolitan areas than are married females. Married couples with children relocate less frequently than households comprised of single individuals. When households with children do move, they are more likely to make short, intrametropolitan moves, while households without children are more likely to make longer, intermetropolitan moves. Despite their importance, these geographic effects are often ignored in demographic analyses of population change.

The adoption of a perspective oriented toward households rather than individuals is particularly appropriate for studies that focus on urban residential structure and demographics for marketing and advertising (which we discuss in the next chapter). In this chapter we stress the interactions between geography, demography, and the organization of individuals into households.

In the next section we present empirical evidence to support the contention that location and relocation are important elements of any study of households. We then review geographic concepts that provide an underpinning for the study of local housing markets. Following this, we review methods for housing analysis, moving from demographic methods to approaches frequently employed by geographers.

9.2   A BRIEF LOOK AT HOUSEHOLDS, LOCATION, AND RELOCATION IN THE UNITED STATES

As just noted, changes in household composition in the United States have been quite dramatic over the past several decades. In this section, we examine current demographic and geographic characteristics of households as well as changes in the recent past. The intent is not to provide a comprehensive portrait of U.S. households, but rather to highlight the interactions between households, location, and relocation. Sweet and Bumpass (1987) provide an extensive, relatively recent description of American families and households, though they do not treat geographical aspects in any detail.

Table 9.1 and Figure 9.1 depict changes in the household size distribution during the 1980s. Households of 1 and 2 individuals grew by 24 and 17 percent, respectively. Consequently, in 1990, over 56 percent of all households in the country had only 1 or

TABLE 9.1   Household Size, 1980 and 1990

Household Size	Number of Households 1980	%	Number of Households 1990	%
1	18,202	22.6	22,580	24.6
2	25,133	31.2	29,454	32.6
3	13,958	17.4	15,970	17.4
4	12,409	15.4	13,860	15.1
5	6,335	7.9	6,189	6.7
6+	4,430	5.5	3,894	4.2
Total	80,467	100.0	91,947	100.0

Sources: Decennial data from the U.S. Bureau of the Census and calculations by the authors. Number of households is in thousands.

9.2 A Brief Look at Households, Location, and Relocation in the United States

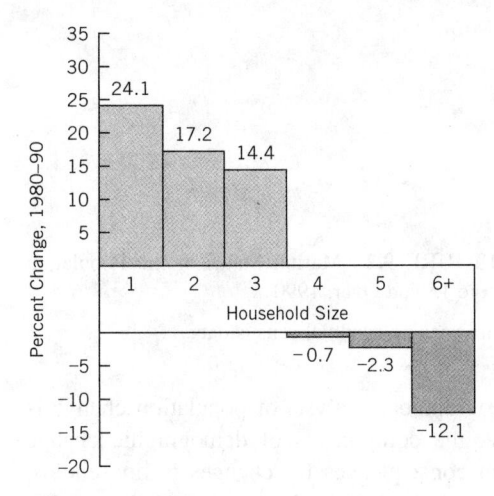

FIGURE 9.1 Percent Changes in Household Sizes, 1980–1990

Source: Decennial census data.

2 individuals. The perhaps stereotypical image of households as families with children is clearly erroneous. Households of 3 or more individuals are not even in the majority. In fact, as Figure 9.2 demonstrates, the fraction of all individuals over age 15 who are "married with spouse present" is only slightly over one-half.

Table 9.2 shows the dramatic increase during the 1980s in the proportion of households classified as either nonfamily or "other family." The former category contains individuals who are either living alone or living with unrelated individuals. The latter category is comprised chiefly of single parent households. The rapid increase in the number of nonfamily households has even prompted the Bureau of the Census to coin a new term, *POSSLQs,* meaning "persons of opposite sex sharing living quarters."

Superimposed on these changes in household size, household composition, and family structures are spatial patterns in the locations of households of different types. For example, Figure 9.3 shows that household type is strikingly different in central cities than elsewhere. Only 43.1 percent of central city households consisted of married couples, while the corresponding percentages elsewhere were over 60 percent. The amount and type of relocation also vary with householder type. Figure 9.4 reveals that "other family" householders have a mobility rate over 50 percent greater than the rate for householders living with their spouse. The figure also depicts differences in destination choice among central city householders who move within metropolitan areas. Householders living with spouses in the central city are much more likely to choose the balance of the metropolitan area (the suburbs) when they move in comparison with "other family" householders, who are more likely to stay within the central city.

TABLE 9.2 Household Types, 1980 and 1990

Household Type	Number of Households 1980	%	Number of Households 1990	%	Percent Change 1980–90
Nonfamily	21,274,967	26.4	27,329,463	29.9	28.9
Family					
Married couple	48,987,847	60.9	50,708,322	55.2	3.5
Other family	10,197,303	12.7	13,809,625	15.0	35.4

Source: Decennial data from the U.S. Bureau of the Census and calculations by the authors.

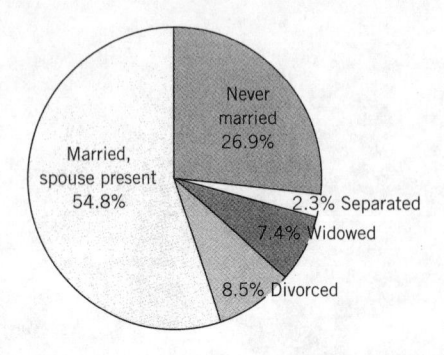

FIGURE 9.2 Marital Status of the Population Age 15 and Over, 1990

Source: Decennial census data.

A focus upon households in the geographical analysis of population change is thus warranted for several reasons. First, the components of demographic change (births, deaths, and migration) have direct consequences for changes in households, since changes in the size of a household and changes in the number of households occur only through births, deaths, and the migrations of individuals. The components of population change also have indirect as well as direct effects on household size and composition. Whenever there is a large birth cohort, there will be rapid household growth 20 years hence as the young adults leave their parental nests. Second, many movers are "tied" movers. Children and spouses usually move together with the householder, yet rarely do migration models and projection methods account for this.

Next we review concepts that provide a geographic basis for analyzing households and housing markets.

9.3 GEOGRAPHIC CONCEPTS RELEVANT TO THE STUDY OF HOUSEHOLDS

Gober (1990) summarizes three approaches to studying the geographic distribution of households: (1) the neoclassical trade-off model, (2) the ecological model, and (3) the life cycle model.

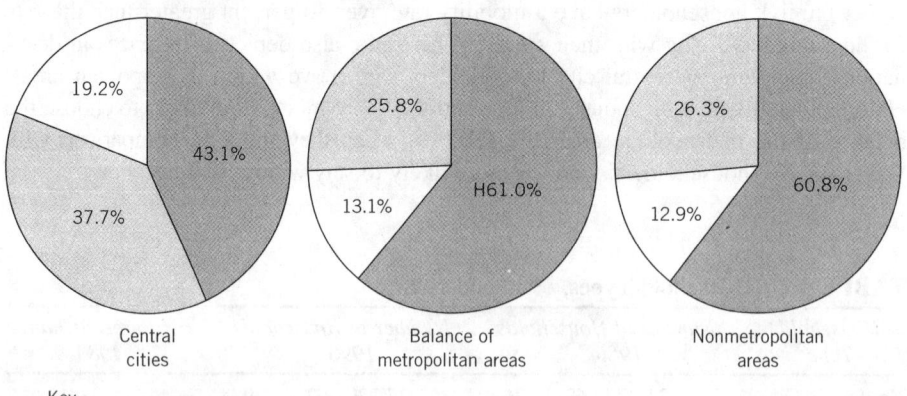

Key:
- ■ : Householder, spouse present
- ▨ : Nonfamily householder
- □ : Other family householder

FIGURE 9.3 Household Composition in Metropolitan and Nonmetropolitan Areas, 1990

Source: Decennial census data.

```
        Central                          Central
         city                              city
         .657                              .747
    Balance of MSA   other MSAs      Balance of MSA   other MSAs
        .313            .030             .218            .035

    Mobility rate: 13.7%            Mobility rate: 21.9%

          (a)                              (b)
      Householder,                    Other family
     spouse present                   householder
```

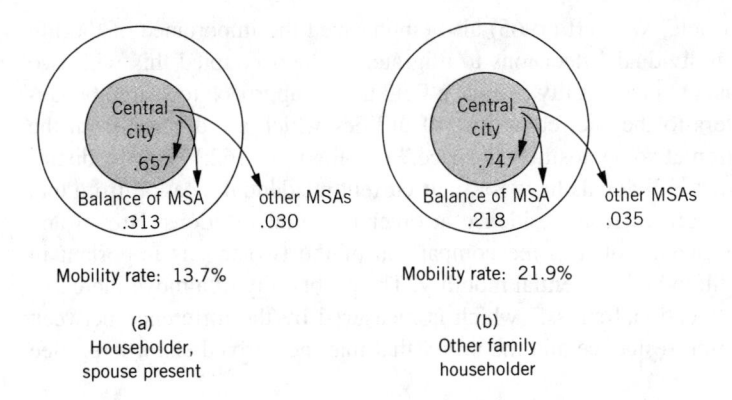

Note: The figure depicts destination choice probabilities for those central city residents staying within metropolitan regions.

FIGURE 9.4 Effects of Householder Type on Mobility and Destination Choice of Central City Residents, 1990–1991

Source: U.S. Bureau of the Census (1992b)

The first category of models was originally developed under the banner of the "new urban economics." (See, e.g., Richardson, 1977.) These models, pioneered by Alonso (1964), are based upon the assumption that there is a trade-off between residential space and commuting costs. Individuals wanting additional land locate farther from the central business district, and are willing to incur additional commuting costs.

The ecological approach was originally developed by sociologists who borrowed concepts from ecology to describe urban residential land use (Park, Burgess, and McKenzie, 1925). New immigrants were viewed as "invading" existing residential areas immediately outside the center of the city. Stages of "competition" and "succession" followed, with the new immigrants becoming the dominant group in the zone, and the previous occupants moving farther away from the city center. This work stimulated much additional effort aimed at understanding where households of various types live within urban areas. Indeed, the foundations of urban geography are built largely upon the related works of Hoyt (1939) and Harris and Ullman (1945) in addition to the pioneering efforts of Park, Burgess, and McKenzie. The quantitative revolution that swept geography and the social sciences took up this subject under the heading of "factorial ecology" in the 1960s. (See Chapter 10.)

Of particular relevance to a demographic view of residential location and mobility is the *life cycle* approach. Indeed, the life cycle is a central notion to the geographical analysis of households. Rossi (1955), in describing why families move, argued that there are key events during periods in individuals' lives that are often associated with the decision to move. Leaving the parental home, getting married, and having children were, at the time, events that occurred to just about everyone, and those events were often associated with mobility that could be viewed as a response to changing housing needs. Since Rossi's work appeared, others (e.g., Speare, Goldstein, and Frey, 1975; Nijkamp, van Wissen, and Rima, 1993) have stressed the relationship between mobility and life cycle events. More recently, the life cycle framework has been modified and generalized to account for the complexity, variety, and dynamism of contemporary living arrangements. Clausen (1986) provides a good description of this new emphasis upon the "life course."

In a seminal article, Wolpert (1965) also emphasized the importance of the life cycle in explaining individuals' decisions to migrate. Wolpert coupled this with two other important ideas: (1) place utility and (2) a field theory approach to search behavior. Place utility refers to the "net composite of utilities which are derived from the individual's integration at some position in space." (Wolpert, p. 162.) Wolpert distinguished this utility that individuals have for their current residential location from that utility that might be derived from residence at other potential locations. Brown and Moore (1970) emphasized that it is the comparison of the two that is important in determining the likelihood of residential mobility. The probability of a move increases with the degree of "locational stress," which is measured by the difference between the utility of the current residence and the utility that may be derived elsewhere. (See Figure 9.5.)

"A field theory approach to search behavior" is a fancy term for an elementary, yet meaningful idea—individuals operate within only a small portion of their entire urban environment. This "action space" of the individual is the subregion from which information about the housing market is obtained, so it would be incorrect to assume that the potential choice set of the individual corresponds to the entire metropolitan area.

What are the implications of all this for neighborhood change? Though we have emphasized the close relationship between mobility and household change, Gober (1986) notes that it is important to distinguish between what occurs for individuals and what occurs in neighborhoods. Gober, McHugh, and Reid (1991) find it useful to define two classes of individuals: unstable stayers (family households that have relatively low residential mobility and a relatively high degree of household change) and stable movers (nonfamily households with a relatively high degree of residential mobility and relatively fewer changes in household structure). One finding that emerges is that although mobility and household change are directly related for individuals, high rates of housing turnover in an area do not necessarily imply changes in household structure for that area. Thus Gober (1986) states that "the linkage between mobility and household change for individuals does not hold for places." Much work remains to be done on the topic of neighborhood change, and on the linkages between the dynamics of neighborhood change and individual relocation decisions. Huff (1979) has also indicated the importance of these linkages, and he has suggested how models of individual choice among housing vacancies may be tied to planning models of neighborhood change.

Though much of the preceding discussion focuses on demand for housing, a view from the housing supply side is also sometimes taken. Housing vacancies are created and filled by residential movement. Chains of opportunity (White, 1970) occur as household A moves to vacancy 1, thereby creating vacancy 2; household B moves to vacancy 2, creating vacancy 3; household C moves to vacancy 3, and so on. Next we explore the characteristics and details of such chains.

Vacancy Chains

White (1970), in an influential book on the structure of organizations, proposed a creative and useful view of job opportunities within an organization. Rather than focus on the movement of individuals between jobs, White chose to focus on the vacancies themselves. When a newly created job is filled by an employed person, another vacancy will occur elsewhere (assuming that the position formerly occupied by the person taking the new job is not eliminated). In this manner, "chains of opportunity" are created. A vacancy created and filled in one part of an organization therefore sends off a chain reaction that has repercussions throughout other parts of the organization.

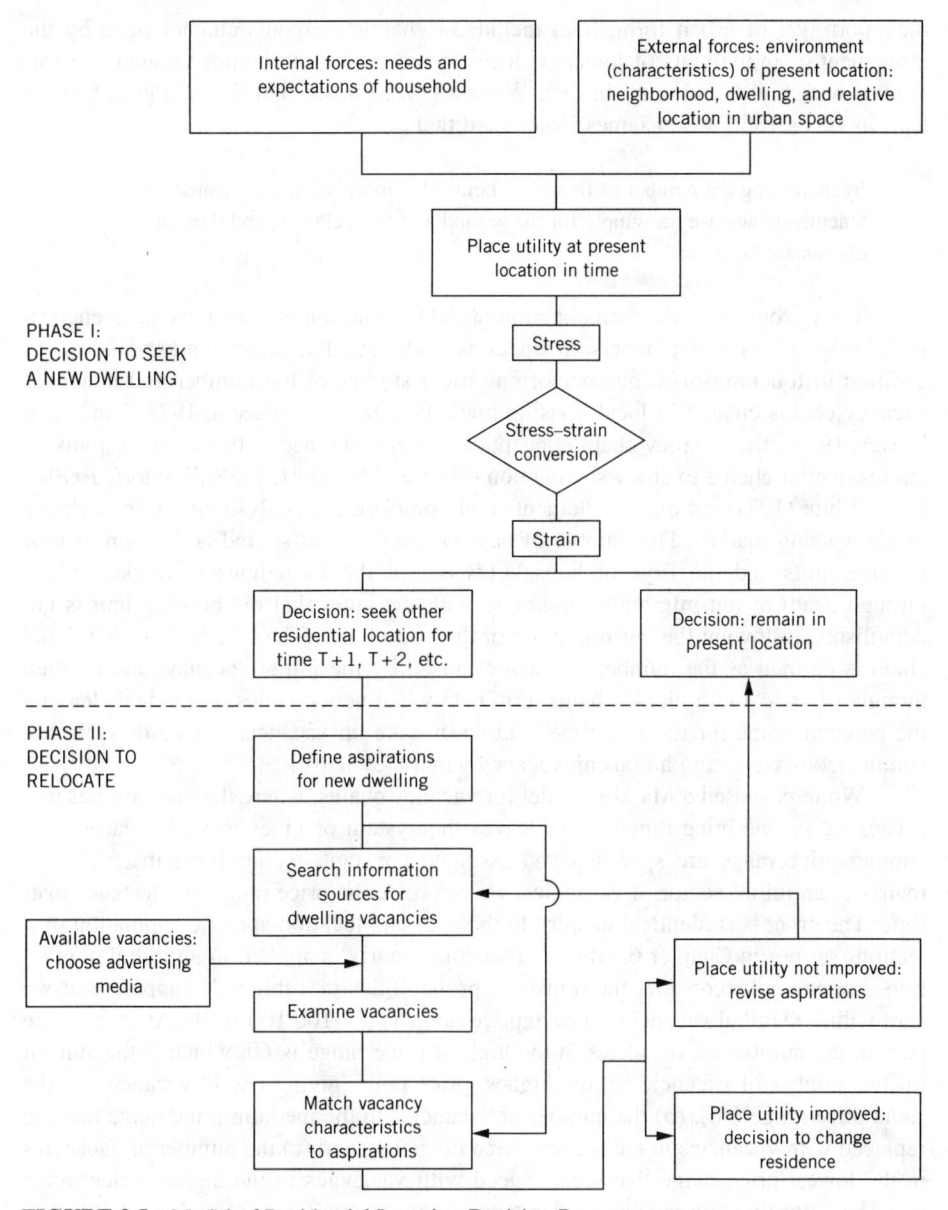

9.3 Geographic Concepts Relevant to the Study of Households

FIGURE 9.5 Model of Residential Location Decision Process
Source: Brown and Moore (1970)

Geographers have used the notion of vacancy chains to model and understand the nature of regional labor markets. (See, e.g., MacKinnon, 1975; MacKinnon and Rogerson, 1980.)

Of course, there is no reason that the idea of a vacancy chain be limited to labor markets. In his book, White suggested that vacancies in housing markets be viewed from a similar perspective. Indeed, the concept of ripple effects in housing markets being caused by the movement of the higher social classes has been around for a long time. The classic works of Park, Burgess and McKenzie (1925) and Hoyt (1939), in their portrayal of urban form, both included dynamic elements characterized by the movement of individuals of lower socioeconomic status into homes vacated by individuals with higher incomes. In fact, Watson (1973) noted that the idea goes back at least as far as 1866, when James Hole noted that

> by increasing the number of first class homes for mechanics, the vacated tenements increase the supply for the second and third classes, and thus all classes are benefitted.

The introduction and formalization of White's notion of vacancy chains encouraged geographers and planners to undertake additional housing market studies. In addition to touching off a number of empirical studies of the number and length of vacancy chains created in local housing markets (See, e.g., Watson, 1973; Sands and Bower, 1974), the vacancy chain concept has been combined with models of individual residential choice to study segregation (Huff and Waldorf, 1988; Waldorf, 1990).

White (1971) set out the elements that comprise the analysis of vacancy chains in the housing market. The number of new vacancy chains started is the sum of new housing units and the flow of households out of the local housing market, either through death or out-migration (assuming with the latter that the housing unit is not demolished following the out-migration of the household). The length of an individual chain is defined as the number of moves caused by the initial vacancy, and is often thought of as a "multiplier." Chains terminate when new couples, individuals leaving the parental home for the first time, and so on take up vacancies, since they are not simultaneously creating a housing vacancy when they leave.

White proposed a Markov model for vacancy chains, where the probabilities that a vacancy in one price range either leaves the system or gives way to a vacancy in another price range are specified and assumed constant. By applying the transition matrix to an initial vector of vacancies, we can trace the price range of vacancies over time. The process is identical in spirit to the age- and location-specific population projections set out in Chapter 6, where a transition matrix is applied to an initial population. For example, consider the transition probabilities in Table 9.3. Suppose that we start with 100 initial vacancies in each price range ($\mathbf{v}$ = [100 100 100]). After one time period, the number of vacancies in the highest price range is 60, which is the sum of (a) the number of vacancies in the highest price range giving rise to vacancies in the same price range (46), (b) the number of vacancies in the medium price range that are replaced with vacancies in the highest price range (8), and (c) the number of vacancies in the lowest price range that are replaced with vacancies in the highest price range (6). The latter two numbers are understandably quite small, since few households will move from the high price range to lower price ranges, causing vacancies to move in the opposite direction. If we form a matrix, $\mathbf{Q}$, with three rows and columns by taking the first three columns of Table 9.3, we can generalize this procedure to find the distribution of vacancies across price ranges after one period of time:

$$\mathbf{v}_{t+1} = \mathbf{v}_t \, \mathbf{Q}$$

For our example,

$$[100\ 100\ 100] \begin{bmatrix} .46 & .33 & .03 \\ .08 & .44 & .11 \\ .06 & .20 & .22 \end{bmatrix} = [60\ 97\ 36]$$

TABLE 9.3 Destination Probabilities for Housing Vacancies in Various Price Ranges

	Destination			
Origin	High Price	Medium Price	Low Price	Outside the Region
High Price	.46	.33	.03	.18
Medium Price	.08	.44	.11	.37
Low Price	.06	.20	.22	.53

Source: White (1971). Reprinted by permission of the *Journal of the American Institute of Planners.*

Similarly, after two time periods,

$$\mathbf{v}_{t+2} = \mathbf{v}_{t+1} \mathbf{Q} = \mathbf{v}_t \mathbf{Q} \mathbf{Q} = \mathbf{v}_t \mathbf{Q}^2$$

The reader should verify that in the example here,

$$\mathbf{v}_{t+2} = [37.52 \quad 69.68 \quad 20.39]$$

Note that the total number of vacancies is declining over time; this is due to the fact that some of the original vacancies leave the system during each period. A plausible projection model for price-specific housing vacancies would add the number of new vacancy chains started to each projection period. Thus if 20 new chains were started in each price range each period, $\mathbf{v}_{t+1}$ would be modified to reflect this:

$$\mathbf{v}_{t+1} = [60 \quad 97 \quad 36] + [20 \quad 20 \quad 20] = [80 \quad 117 \quad 56]$$

What about the length of vacancy chains? One interpretation of an element in row i and column j of the matrix $\mathbf{Q}$ raised to the power k is the average number of times a vacancy starting in price range i is in price range j after k time periods. Thus the total number of times a vacancy starting in price range i will reappear in price range j (on average) is the sum of the row i, column j, elements in the matrices $\mathbf{Q}$, $\mathbf{Q}^2$, $\mathbf{Q}^3$, $\mathbf{Q}^4$ and so on. In the preceding example, we have the result reported by White:

$$\mathbf{Q} + \mathbf{Q}^2 + \mathbf{Q}^3 + \mathbf{Q}^4 + \ldots = \begin{bmatrix} 1.08 & 1.32 & 0.25 \\ 0.35 & 1.10 & 0.29 \\ 0.24 & 0.62 & 1.36 \end{bmatrix}$$

For instance, the element in row 2 and column 1 may be interpreted to mean that a vacancy starting in the medium price range will reappear on average 0.35 times in the high price range before the chain is terminated. Similarly, a chain starting in the medium price range will reappear 1.1 times in that same price range, and 0.29 times in the lowest price range. The total length of the chain is the sum of these three numbers (.35 + 1.1 + .29) plus 1 since we also must account for the initial vacancy. So we can find the average length of chains simply by summing the rows of the matrix $\mathbf{I} + \mathbf{Q} + \mathbf{Q}^2 + \mathbf{Q}^3 + \mathbf{Q}^4 + \ldots$ where $\mathbf{I}$, the *identity matrix,* is a matrix with ones on the diagonal (to account for the initial vacancy) and zeros occupying the off-diagonal elements. In the present case with three price ranges,

$$\mathbf{I} = \begin{bmatrix} 1 & 0 & 0 \\ 0 & 1 & 0 \\ 0 & 0 & 1 \end{bmatrix}$$

In theory, the calculation requires an infinite sum of the powers of the **Q** matrix. In practice, the elements of the powers of the **Q** matrix rapidly become smaller as the power of the matrix increases. The higher powers of **Q** may therefore be ignored, so two- or three-digit accuracy may be achieved quickly without extensive calculation. Faster methods of calculation involving additional knowledge of matrix algebra are also convenient.[1]

The multiplier effects of vacancy chains are important, because they indicate the potential benefits from investments in new housing in various price ranges. It can be argued that investments in higher-priced housing will be more beneficial, since the multiplier effects will typically be greater. (That is, the chains created will typically be longer.) Thus White (1971) states that "policy choices should maximize this leverage, either on average, or with some groups given extra weight." The idea is a sort of "trickle down" theory of housing—the opportunity generated in the lower price ranges will be greater if chains are initiated in the higher price range. A key variable often overlooked, however, is the time lag involved. (These multiplier effects do not occur overnight, so we must understand not only the magnitude of the trickle down effect, but also its timing.)

Finally, we should be aware that applications of vacancy chains to the housing market can focus on variables other than, or in addition to, price ranges. Applications to intraurban migration use disaggregation by subregions within the local housing market. (See, e.g., Huff, 1979.) This requires data on vacancy transitions from one subregion to another. An example is given in the exercises at the end of this chapter.

9.4 DEMOGRAPHIC AND GEOGRAPHIC ANALYSIS OF HOUSEHOLDS

The concepts and ideas discussed in Section 9.3 have proven valuable in obtaining a geographic perspective on household change, intraurban migration, and the nature of local housing markets. Armed with these fundamentals, we have advanced our understanding considerably. Planners, however, must augment this knowledge with household projection methods. This subsection examines some alternative methods for the demographic and geographic analysis of households. Desirable are methods derived from the concepts just discussed. However, demographers and geographers have in general found this difficult. The challenge of extending the traditional and principal methods of population analysis to the study of households has led to what is now an active and exciting subfield within demographic analysis.

There are several alternative techniques for analyzing changes in the number, size, and composition of families and households. The methods typically make use of current

---

[1] The infinite sum of the matrices $\mathbf{I} + \mathbf{Q} + \mathbf{Q}^2 + \mathbf{Q}^3 + \ldots$ equals the inverse of the matrix $\mathbf{I} - \mathbf{Q}$ and is denoted by $(\mathbf{I} - \mathbf{Q})^{-1}$. Finding the length of vacancy chains may be facilitated by calculating this matrix inverse.

or recent demographic rates to project household or family variables forward in time. Often the rates are assumed constant, though simple time trends are sometimes incorporated. The techniques range from elementary methods of extrapolation to generalizations of life table methodology to complex simulation models that attempt to capture the processes of household formation and dissolution. The subsections that follow provide the basic information required to use and appreciate these multifarious approaches.

Householders and Markers

Much of the demographic analysis of households relies on identifying an individual within the household to whom the relationship of other household members may be noted. Brass (1983), for example, uses the eldest female in a household as a *marker*; projections of household dynamics may then be carried out by following the changes in the number of markers (due, e.g., to death, marriage, divorce).

More common is the designation of a householder. This term, coined by the U.S. Bureau of the Census (see, e.g., U.S. Bureau of the Census, 1986) refers to

> the person (or one of the persons) in whose name the housing unit is
> owned or rented (maintained) or, if there is no such person, any adult
> member, excluding roomers, boarders, or paid employees. If the household
> is owned or rented jointly by a married couple, the householder may be
> either the husband or the wife. (pp. 6–7).

The term *householder* replaced the more antiquated term *head of household*. Prior to this definition, in married-couple households, the husband was designated as the head of household.

This definitional change, which allowed for more flexibility in designating the householder, was clearly warranted in view of changes in social norms. An unfortunate effect of the change, however, is a lack of comparability in statistical data series. Consequently some analysts still use the old definition. The U.S. Bureau of the Census (1986) itself made the decision to use the old definition in its projections:

> For purposes of historical comparability in the preparation of householder
> proportions to be used in making the projections, the husband in married-
> couple households was viewed as the householder for the years 1980
> through 1985. (p. 7)

While in the past it was common to use the distribution of householders disaggregated by sex, sex disaggregation of householders is now less meaningful due to the ambiguity inherent in assigning householders to married-couple households.

In the following subsection, we describe variants of a common method of household projection known as the *headship rate method*. The headship rate is defined simply as the proportion of individuals in a particular category (usually an age–sex category, but sometimes further disaggregated by other variables such as marital status or family type) who are heads of households. In place of the *headship rate*, it is now more appropriate to use the *householder proportions* or *householder rates*.

Householder Rate Method

By far the most common procedure used to generate household projections is the householder rate method. Though perhaps obvious, it should be noted that the projected number of householders is equivalent to the projected number of households, since there is exactly one householder per household. To derive the projected number of households in a particular category, a projected householder rate is simply multiplied by the projected population in that category.

The method has the advantage of being simple and operational. Data requirements are limited to the following: (1) data on population by age and sex for a "base" year, (2) data on the number of households in the base year, disaggregated by age and sex of the householder, and (3) projections of the population by age and sex. Because sex disaggregation is now less meaningful, it is not even necessary to have the population and household data disaggregated by sex.

Some assumption must also be made about the future trajectory of householder rates—they are normally assumed to remain constant or follow a linear trend, though the U.S. Bureau of the Census has recently adopted the slightly more sophisticated approach of using time series methods to extrapolate recent changes in householder rates.

The householder rate method has the additional advantage that it is readily generalizable to other variables related to household and family status. The U.S. Bureau of the Census (1963, 1981, 1986), for example, has routinely prepared projections of households by family type (e.g., married couple, nonfamily with female householder) in this manner.

A Detailed Description of the Method

In this subsection we follow, with minor modifications, the notation and explanation of Linke (1988) to describe the steps in producing a household forecast via the householder rate method.

First, age–sex-specific householder rates, v_y, are calculated:

$$v_y = H_y / B_y^P$$

where H_y is the number of householders in age–sex category y, and B_y^P is the number of people in households in age–sex category y. The next step is to apply the householder rates to an age–sex disaggregated projection of the population residing in households. Typically we have access to an age–sex disaggregated projection of the resident population; this must be first modified by multiplying the projected population in each age–sex group by the proportion of that group who will reside in households. These proportions are most often obtained by assuming that the proportions observed in some recent year will remain constant. Thus the proportion of the population residing in households, f_y, is given by

$$f_y = B_y^P / B_y$$

where B_y is the total population in age–sex group y. The astute reader will note that the quantities f_y are identical to the term $(1 - GQ)$, which is the proportion of the population not residing in group quarters, used in the housing-unit method of making population estimates described in Chapter 5.

The number of households with a householder in age–sex group y in projection year t, $H_{y,t}$, is then determined from

$$H_{y,t} = v_{y,t} B_{y,t}^P = v_{y,t} f_{y,t} B_{y,t}$$

Note that in this equation, both the householder rate and the proportion of population residing in households have been given a t subscript. This is to indicate that in the general case, both householder rates and the quantities f_y may be functions of time. Although it is frequently assumed that householder rates are constant, it is also common to fit a linear trend through a set of recent householder rates. Likewise, trends in the proportion of population residing in households may also be accounted for, though this is done less often in practice.

The total number of households is obtained by adding the quantities $H_{y,t}$ over all age–sex groups. Average household size is given by

$$\text{PPH} = \frac{\sum_y B^P_{y,t}}{\sum_y H_{y,t}}$$

An Example

Table 9.4 gives the total U.S. resident population, the population in households, and the number of householders, by age and sex of householder, in 1980. The age-specific householder rate in column 4 is found by dividing the entries in column 3 by the entries in column 2. Column 5 gives the fraction of the population residing in households (f). It is calculated by dividing the entries in column 2 by those in column 1.

The projected resident population of the United States for 2000 is shown by age and sex in column 1 of Table 9.5. The projected population in households (column 2) is derived by multiplying column 1 by column 5 of Table 9.4. Finally, column 3 depicts the projected number of householders in each age–sex category; it is obtained by multiplying column 2 in Table 9.5 by column 4 in Table 9.4.

We should emphasize that this approach to household projection is based solely on projected changes in the age–sex composition of the population. Thus the average number of people per household should decline (alternatively, the average householder rate should increase) simply because in 2000 the baby boom generation will be placed squarely in those age groups that have high householder rates. A more sophisticated analysis of household change would account for the potential changes in householder rates that might occur between 1980 and 2000. For example, age–sex-specific householder rates would be expected to increase if the trend toward an increasing number of individuals living alone and toward fewer married couple households continues.

The householder rate method is useful due to its simplicity and modest data requirements. However, the lack of attention given to the actual processes of household formation, change, and dissolution decreases the confidence that we may place in the projections. Before discussing other methods that explicitly address the determinants of household change, we first turn to a modification of the householder rate method. The household membership rate method, suggested by Linke, is attractive because it avoids the use of the concept of householder or head of household.

The Household Membership Rate Method

Linke (1988) describes a method used for household projections in Germany that does not use the concept of household head or require the designation of a householder. But it retains the concept embodied in the householder rate method of elementary rates that are either are assumed constant or are extrapolated in some simple way.

TABLE 9.4 Data Required for Household Projection

Age	Resident Population (1)	Population in Households (2)	House-holders (3)	House-holder Rate (4)	Proportion of Population in Households (5)
Male:					
<15	26,217,310	26,156,091	—	—	.9976
15–19	10,755,409	9,961,169	349,625	.0351	.9262
20–24	10,663,231	9,619,120	4,185,697	.4351	.9021
25–29	9,705,107	9,405,867	6,997,264	.7439	.9692
30–34	8,676,796	8,599,632	7,354,314	.8552	.9911
35–39	6,861,509	6,762,933	6,053,088	.8950	.9856
40–44	5,708,210	5,639,928	5,128,712	.9094	.9880
45–49	5,388,249	5,287,773	4,848,683	.9170	.9814
50–54	5,620,670	5,549,429	5,095,828	.9183	.9873
55–59	5,481,863	5,431,117	5,008,349	.9222	.9907
60–64	4,669,892	4,632,635	4,274,530	.9227	.9920
65–69	3,902,955	3,813,514	3,506,468	.9195	.9771
70–74	2,853,547	2,786,362	2,541,341	.9121	.9765
75–79	1,847,661	1,763,374	1,578,630	.8952	.9544
80–84	1,019,227	931,329	806,144	.8656	.9138
85+	681,525	554,356	436,366	.7872	.8134
Total		106,894,629	58,165,039		
Female:					
<15	25,073,029	24,978,670	—	—	.9962
15–19	10,412,715	9,784,773	258,440	.0264	.9397
20–24	10,655,473	10,127,673	1,827,314	.1804	.9505
25–29	9,815,812	9,728,979	2,186,339	.2247	.9912
30–34	8,884,124	8,914,652	1,937,634	.2174	1.0034*
35–39	7,103,793	7,078,642	1,518,123	.2145	.9965
40–44	5,961,198	5,931,067	1,264,130	.2131	.9950
45–49	5,701,506	5,648,957	1,221,733	.2163	.9907
50–54	6,089,362	6,058,315	1,424,157	.2351	.9949
55–59	6,133,391	6,105,923	1,639,026	.2684	.9955
60–64	5,417,729	5,380,938	1,763,369	.3277	.9932
65–69	4,879,526	4,805,679	1,990,300	.4141	.9847
70–74	3,944,577	3,840,453	1,938,865	.5049	.9736
75–79	2,946,061	2,766,169	1,627,203	.5883	.9389
80–84	1,915,806	1,659,216	1,053,383	.6349	.8661
85+	1,558,542	1,102,647	652,372	.5916	.7075
Total		113,912,753	22,302,388		

Population/Household: 2.744

Sources: Columns 1 through 3 are taken from 1980 decennial census data.
*The fraction of the population in households exceeds 1 because the number of persons in households (column 2) is based upon sample data. Fractions exceeding 1 should be set equal to 1 when deriving the projected population in households.

TABLE 9.5 Household and Population Projection, United States, 2000

Age	Projected Resident Population (thousands) (1)	Projected Population in (thousands) (2)	Projected Households (thousands) (3)
Male:			
<15	28,241	28,173	—
15–19	9,769	9,048	318
20–24	8,889	8,019	3,489
25–29	8,917	8,642	6,429
30–34	9,745	9,658	8,260
35–39	10,928	10,771	9,640
40–44	11,017	10,885	9,899
45–49	9,828	9,645	8,844
50–54	8,468	8,361	7,677
55–59	6,478	6,418	5,918
60–64	5,079	5,038	4,649
65–69	4,382	4,282	3,937
70–74	3,860	3,769	3,438
75–79	2,971	2,835	2,538
80–84	1,739	1,589	1,375
85+	880	716	563
Total	131,191	127,849	76,974
Female:			
<15	27,811	27,705	—
15–19	9,342	8,779	232
20–24	8,711	8,280	1,494
25–29	8,819	8,741	1,964
30–34	9,668	9,701	2,109
35–39	10,892	10,853	2,328
40–44	11,074	11,018	2,348
45–49	10,057	9,964	2,155
50–54	8,870	8,825	2,074
55–59	6,981	6,950	1,866
60–64	5,620	5,582	1,829
65–69	5,109	5,032	2,084
70–74	4,892	4,763	2,405
75–79	4,311	4,048	2,381
80–84	2,996	2,595	1,647
85+	1,923	1,360	805
Total	137,076	134,196	27,720

Population/household: 2.503.

Source: Column 1 is taken from the middle series national-level projection found in U.S. Bureau of the Census (1989a).

First, household membership rates are defined for each age group and household size:

$$r_y^i = \frac{P_y^i}{B_y}$$

Here r_y^i is the household membership rate for age group y, and household size i; P_y^i is the population in households of size i in age group y; and B_y is the total number of individuals in households in age group y. Note that the household membership rates sum to 1 when they are added across household sizes:

$$\sum_i r_y^i = \sum_i \frac{P_y^i}{B_y} = 1$$

With an age-disaggregated population projection (consisting of the values B_y^*, where we use asterisks to denote projected values), the projected number of people in households of size i in age group y, P_y^{i*}, is

$$P_y^{i*} = r_y^i B_y^* \tag{9.1}$$

where the household membership rate has been assumed constant. Alternatively, household membership rates could be extrapolated on the basis of recent trends under the condition that the extrapolated rates are adjusted, if necessary, to sum to 1.

The projected number of households of size i is

$$H^i = \frac{1}{i} \sum_y P_y^i$$

As Murphy (1991) points out, there are circumstances where the method will not produce accurate results. To illustrate, he cites the extreme case where all individuals age 30 to 34 live with a spouse and two children in households of size four. If fertility falls, there will be many more three-person households, yet the household membership rate method will continue to project all individuals age 30 to 34 in households of size four. Linke, however, compares the approach with the householder rate method using actual data, and finds little difference between the two.

Table 9.6 gives hypothetical data on the population in various household sizes and age groups for the county of Utopia. To project the number of households, by household size, household membership rates are first derived by dividing each entry in the table by its row sum (i.e., the B_y values). These rates, along with the projected age-specific populations also given in Table 9.6, are then used in Equation 9.1 to derive the projected population, by age and household size. These projections are displayed in Table 9.7, along with the projected number of households, by size. The small increase in average household size may be attributed to the anticipated large increase in young adults, who are relatively more likely to reside in larger households.

The most obvious limitation with these methods is that although the effects of changing age composition are captured, there is no modeling or accounting of the processes (e.g., marriage, nest leaving) that lead to changes in household composition. Next, we discuss microsimulation, a way to handle these and other complexities associated with modeling household change.

TABLE 9.6 Hypothetical Population Data for Utopia, by Household Size and Age Group

Age	\multicolumn{4}{c}{Household Size}	Population (B_y)	Projected Population (B_y^*)
	1	2	3	4		
15–24	1,000	3,000	2,400	2,100	8,500	12,000
25–34	1,100	3,100	2,700	2,000	8,900	9,400
35–34	1,200	3,200	2,700	1,800	8,900	9,800
45–54	1,100	3,400	2,400	1,600	8,500	9,800
55–64	1,000	3,200	2,900	1,400	8,000	9,200
65+	1,500	3,000	2,100	1,400	8,000	7,700
Total population	6,900	19,100	14,700	10,300	50,800	57,900
Number of households	6,900	9,550	4,900	2,575		

Total households: 23,925.
Average household size: 50,800 / 23,925 = 2.123.

Microsimulation

One means to model changes in household size, location, and composition is via *microsimulation*. This approach simulates the hypothetical history of the individual units of observation. As applied to the study of households, microsimulation provides a means for tracking the demographic and geographic fate of individuals and households. Microsimulation is particularly appealing in situations where the dynamics are complex, and where the problems associated with standard analytical methods are considerable.

Examples of microsimulation in the field of population geography include the work of Clarke (1986) and Wang (1992). Clarke, for example, uses microsimulation to generate 5- and 10-year projections of the number of different types of households in Yorkshire and Humberside. By detailing all of the processes that lead to household change (e.g., fertility, mortality, migration, marriage, divorce, nest leaving), Clarke produces detailed projections for household types. For example, his projections reveal that the number of households comprised of 1 man and 1 woman will increase by 2.7 percent over the projection period, while the number of households with 1 man, 1 woman, and 1 child will decline by 1 percent. This is illustrative of the level of detail that is achievable with microsimulation. Though Clarke uses his model for projection, the use of microsimulation in "what if" sensitivity analyses should not be overlooked. Microsimulation facilitates the answering of such "what if" questions as "if divorce rates fall, how will the number of 3-person households change?"

A household simulation is carried out by determining the likelihood of a change from one demographic and/or geographic category to another. Random numbers are then chosen from a uniform distribution to determine whether the change actually takes place during the time period. For example, suppose for individuals in the central city, the probability of leaving the parental home at age 18 is 0.4. To simulate whether an individual of this age residing in the central city actually leaves home, we would like to flip a coin whose probability of landing heads is 0.4 and whose probability of landing tails is 0.6. Then, if the coin is heads, the hypothetical individual moves;

292 Chapter 9 • Household Demography

TABLE 9.7 Projected Population Data for Utopia, by Household Size and Age Group

Age	Household Size 1	2	3	4	Projected Population (B_y^*)
15–24	1,412	4,235	3,388	2,965	12,000
25–34	1,162	3,274	2,852	2,112	9,400
35–44	1,321	3,524	2,973	1,982	9,800
45–54	1,268	3,920	2,767	1,845	9,800
55–64	1,150	3,680	2,760	1,610	9,200
65+	1,444	2,887	2,021	1,347	7,700
Total population	7,757	21,520	16,761	11,860	57,900
Number of households	7,757	10,760	5,587	2,965	

Total households: 27,069
Average household size: 57,900 / 27,069 = 2.139

---

otherwise the individual remains in the parental home. Since coins do not actually have the desired probabilities (and since flipping coins many times would be time-consuming), we use any standard random number generator to do the flipping for us. With a random number chosen between 0 and 1, the individual will move if the random number is less than 0.4 and will remain in the parental home otherwise. If the individual moves, the household size must be decreased by 1.

Other probabilities could then be used to determine the destination of the individual. Suppose, for example, that the probability that such a mover chooses to live alone is 0.7, and the probability that the individual chooses to live with another individual who is already living alone is 0.2, and the probability of moving in with a group is 0.1. If the new random number is less than 0.7, the mover sets up a new household, and the number of 1-person households is increased by 1. If the new random number is between 0.7 and 0.9, the number of households does not change, but the number of two-person households is increased by 1. If the random number is greater than 0.9, the number of households does not change, and the group size is increased by one.

The actual procedure associated with microsimulation is therefore quite elementary. It is a simple matter of choosing random numbers, using them to determine whether particular demographic events occur, and updating the records to account for any change.

Difficulties and Advantages of Microsimulation

It should also be apparent that although the number of steps in the procedure is small, it can be quite complicated to actually carry out the simulation accurately. Indeed, a good deal of thought is required to construct a useful microsimulation model.

An important initial step is the specification of an initial population. The initial hypothetical population is often specified on the basis of probabilities derived from empirical observation. Thus if the number of married couple households in the actual population were twice the number of other family households, this relationship would be retained in the construction of the initial simulated population. If, in the observed population, 60 percent of all 18-year-olds lived in married couple families, and 40 percent lived in other families, then random numbers would be chosen to assign 18-year-olds to households of each type.

It is often difficult to acquire accurate data on the likelihoods of transitions between categories. For example, while there may be good national data on the probability of leaving the parental home at a given age, it is less likely that such information will be available for subregions (such as central cities). It is common to have to make some leap of faith to obtain probabilities at the level of disaggregation desired. One common, though not necessarily desirable, method is to assume that the national probabilities also hold at the subnational level of interest. In our present example, it might be necessary to assume that the probabilities of leaving home that were available for the nation would also apply to central cities. Such a leap is indeed a big one, and making it in some ways detracts from the benefits that microsimulation provides in modeling complex situations.

There should also be no necessary expectation that the probabilities, however accurate they may be, will remain the same over time. Unlike the Markov model discussed in Chapter 6, with microsimulation it is not necessary to assume that the probabilities remain constant over time. However, the added flexibility that microsimulation gives us in allowing us to specify what probabilities to apply during each time period brings with it the difficult task of specifying what those future probabilities will in fact be.

We have therefore seen that the ability of a simulation to capture the complex structure of a population and its change over time is highly constrained by the nature of the data available to construct (a) the initial simulated population and (b) the probabilities of making a transition from one category to another.

In household microsimulations, disaggregation along a few interesting dimensions can quickly lead to a rapid proliferation of the number of probabilities that must be specified, and to complex accounting. To return to the preceding example, we would also likely want to disaggregate the population by sex. If nest-leaving behavior varies significantly by sex, separate sets of age-disaggregated probabilities of leaving the parental home would be needed for each sex. For the individual in the example who is leaving the parental home, it would be desirable to know not only the probability of moving into a household with another individual, but also whether the individual in the destination household is male or female. This in turn might be determined as the outcome of a separate marriage model that could be embedded within the simulation.

The length of the time period in microsimulation is entirely arbitrary, but in most demographic applications, it is convenient to choose 1 or 5 years for the length. This is due to the nature of demographic data, and is analogous to the popularity of 1- and 5-year survival rates and population projections.

Note that a distinguishing feature of microsimulations is that a different "answer" is obtained each time the microsimulation is run. This is an attractive feature, since if the microsimulation is run many times, some idea of the stochastic variation in the variables of interest will be obtained.

Matching Housing Demand with Housing Supply

As we saw in Chapter 8, one reason that planners are interested in projecting demand is to match it with available or future supply, to assess needs, and to establish planning priorities. In this subsection, we illustrate techniques for matching supply and demand in local housing markets.

TABLE 9.8 Households, by Household Type and Housing Type, in Flatland

	Apartment	Townhouse or Condo	Single-Family Dwelling	Total
Family				
Married couple	100	200	900	1,200
No spouse present	500	200	200	900
Nonfamily				
Unrelated individuals	400	300	200	900
Living alone	500	200	100	800
Total households	1,500	900	1,400	3,800

Gober (1986) indicates the importance of understanding how well, or how poorly, supply and demand match one another, noting that matches between housing preferences and housing supply are not always perfect. She refers specifically to the 1980s as a period when increases in the number of single-family dwellings coincided with a decline in the number of families that were traditional occupants of such dwellings (namely, married couples with children). In such situations adjustments must be made, and Gober cites evidence that indicates increases in the number of nonfamily households that occupied single-family dwellings during the 1980s. It is therefore of considerable interest to know how severe the adjustments in matching might be for given numbers of household and housing types.

Suppose that we have available information on the numbers of specific household types, by housing type, for the Flatland metropolitan area (Table 9.8). By dividing each entry in Table 9.8 by its row sum, we have a summary of what fraction of each household type lives in each housing type (Table 9.9).

With a projection of household types in hand (possibly using one of the methods suggested earlier in this chapter), we can (a) project the amount of housing of each type that will be required, assuming that the proportions in Table 9.10 reflect the unchanging housing preferences of each family type, and (b) project the changes in the proportions of each household type occupying housing types that would occur if specific types of housing supply were provided.

To illustrate both possibilities, we begin by assuming that the analyst has arrived at the projections in Table 9.10. By multiplying the elements of Table 9.9 by the cor-

TABLE 9.9 Fractions of Flatlander Households by Housing Type and Household Type

	Apartment	Townhouse or Condo	Single-Family Dwelling
Family			
Married couple	.083	.166	.750
No spouse present	.556	.222	.222
Nonfamily			
Unrelated individuals	.444	.333	.222
Living alone	.625	.250	.125

TABLE 9.10 Projected Number of Households for Flatland, by Household Type

Family	
Married couple	1,200
No spouse present	1,200
Nonfamily	
Unrelated individuals	1,000
Living alone	1,000

responding household type data in Table 9.10, we have a projection of household types by housing type (Table 9.11). For example, we project 125 nonfamily households comprised of an individual living alone in a single-family dwelling by multiplying the projected number of nonfamily households characterized by those living alone (1,000) by the proportion of such households who live in single-family dwellings (.125).

This is essentially the same procedure that was used to project trip generation in Tables 8.1 and 8.2; base year rates or proportions are held constant and are then used to multiply a projected population to arrive at the desired information.

In the example, the total number of dwellings needs to increase from 3,800 to 4,400 to accommodate the expected growth. Since a disproportionate amount of the growth in households is expected to be nonfamily households, and nonfamily households have a relatively greater preference for apartments, the growth in the stock of apartments would have to be relatively greater than that for other dwelling types.

Alternatively, we could use the Cross–Fratar method described in Chapter 8 to match the household-type projection with a projection of dwelling types. Suppose that a projection of dwelling types, based possibly on the area's zoning ordinance, the amount of vacant land, and other factors, yields a projection of 1,600, 1,400, and 1,400, for apartment, condo/townhome, and single-family dwelling types, respectively. Exercise 8 at the end of this chapter asks the reader to verify that, using the iterative, Cross–Fratar method, the projected matrix matching households and housing types is that given in Table 9.12. Dividing the elements of Table 9.12 by the respective row sums results in Table 9.13.

Note how the "preferences" among households for different types of housing have changed. For example, since growth in the number of apartments is projected to

TABLE 9.11 Projection of Household Types by Housing Type

	Apartment	Townhouse or Condo	Single-Family Dwelling	Total
Family				
Married couple	100	200	900	1,200
No spouse present	667	267	267	1,200
Nonfamily				
Unrelated individuals	444	333	222	1,000
Living alone	625	250	125	1,000
Total	1,836	1,050	1,514	4,400

TABLE 9.12 Projected Matrix Matching Household Types and Housing Types

	Apartment	Townhouse or Condo	Single-Family Dwelling	Total
Family				
Married couple	89	271	840	1,200
No spouse present	590	362	248	1,200
Nonfamily				
Unrelated individuals	373	430	197	1,000
Living alone	548	337	115	1,000
Total households	1,600	1,400	1,400	4,400

be small, those living alone must choose from among other housing types. Implicit in the Cross–Fratar method is the assumption that people will adjust their housing choices as little as possible, but that they will nevertheless adjust them. Other responses are clearly possible. Those living alone might, for example, migrate out of the local housing market if adequate housing is not available. If interest rates drop, more households may find and choose single-family dwellings. The strength of the Cross–Fratar approach is in providing an idea of the change in the current matching of households and housing that would be required under given projected levels of supply and demand. It is particularly useful in assessing potential future difficulties in meeting the housing needs of the local population.

9.5 SUMMARY

In this chapter, we have emphasized the importance of the household as a unit of analysis. Though still in their infancy, the fields of housing demography and family demography are advancing rapidly. (See, e.g., Myers, 1990; Keilman, 1988; Bongaarts, Burch, and Wachter, 1987.) It is noteworthy that geographers have contributed to this development. Historically, the fields of demographic analysis and population geography have emphasized the individual, yet it is the household unit that is most relevant in many arenas, ranging from the study of residential mobility to the demographics for marketing and advertising. In the next chapter, we build upon our discussion of households to focus on both residential distribution and applied demographics.

TABLE 9.13 Projected Fractions of Household Types Residing in Given Housing Types

	Apartment	Townhouse or Condo	Single-Family Dwelling
Family			
Married Couple	.074	.226	.700
No spouse present	.492	.302	.207
Nonfamily			
Unrelated individuals	.373	.430	.197
Living alone	.548	.337	.115

EXERCISES

Suppose that the following population and household data are available from a recent census:

Age	Population in Households	Householders
<15	5,000	—
15–24	4,000	800
25–34	3,000	1,000
35–44	3,000	1,400
45–54	2,000	1,000
55–64	2,000	1,000
65+	3,000	1,800

The projected population 20 years hence is:

Age	Population
<15	8,000
15–24	5,000
25–34	3,500
35–44	3,500
45–54	2,500
55–64	2,500
65+	2,500

Use the preceding data to answer Exercises 1 through 4.

1. Compute householder rates by age using the census data.

2. Assuming constant householder rates, prepare a projection of householders by age.

3. What is the average number of people per household in the census year? In the projection year?

4. By what percentage will the population in households grow over the projection period? Compare this with the projected percentage increase in households. Explain how the shifting age composition of the population leads to these differences.

5. A survey of a regional housing market reveals that there are currently 230, 110, and 50 vacancies, respectively, in the suburban, central city, and rural subregions. Past surveys have indicated that vacancies are likely to "migrate" according to the following probabilities:

	Destination			
Origin	Suburbs	Central City	Rural	Outside Region
Suburbs	.5	.3	.1	.1
Central city	.1	.5	.2	.2
Rural	.1	.1	.4	.4

Local planning authorities assume that during each period 5, 10, and 25 new vacancies will be created in the suburban, central city, and rural regions, respectively.

a. Project the number of vacancies in each subregion for each of the next two periods.

b. Find the average length of chains for new vacancies created in each of the three subregions.

6. A local utility company is projecting future residential electricity use. The analysts decide to base their forecast on a projection of the number of households, disaggregated by household size. They begin by collecting the following data:

Household Size	<15	15–34	35–54	55+
1	0	6,000	2,000	8,000
2	1,000	6,000	6,000	3,000
3	4,000	2,000	3,000	0
4	5,000	1,000	3,000	0
Total	10,000	15,000	14,000	11,000

Find the household membership rates, and use them with the following age-disaggregated population projection to project the number of households, by household size. What is the projected average household size?

Age	Population Projection
<15	18,000
15–34	20,000
35–54	28,000
55+	23,000

7. A household currently is comprised of a mother, father, and 12-year-old child. A university researcher estimates the following probabilities of events during the next 5-year period:

    Probability of divorce: 0.1

    Probability of a birth, given no divorce: 0.1

    Probability that the child leaves home: 0.3

Use a set of random numbers (e.g., the last two digits from any column of telephone numbers) to determine the composition of the household 5 years from now. Show your work.

8. Verify the entries in Table 9.12 by using the Cross–Fratar method. Use the projected number of households, by type, from Table 9.11, and the projected number of housing types given in the last row of Table 9.12.

> When the Star-Belly Sneetches had frankfurter roasts
> Or picnics or parties or marshmallow toasts,
> They never invited the Plain-Belly Sneetches.
> They left them out cold, in the dark of the beaches.
> They kept them away. Never let them come near
> And that's how they treated them year after year.
>
> —*Dr. Seuss,* The Sneetches

CHAPTER 10
DEMOGRAPHICS*

10.1 Introduction
10.2. The Measurement of Diversity and Segregation
10.3 Factorial Ecology
10.4 Cluster Analysis for Defining Socially Homogeneous Areas
10.5 Life-Style Clustering for Market Segmentation and Targeting
10.6 Summary

10.1 INTRODUCTION

In Chapter 2 we examined elementary methods for exposing the distribution and composition of populations. We concluded with the concept of geographic association—the areal covariation of one demographic characteristic with another. Chapter 9 just highlighted the importance of the household as a basic building block for population analysis. In recent years there has been an explosion of interest in the areal covariation of different types of households. And, according to *Webster's Ninth New Collegiate Dictionary*, a new noun, *demographics*, entered the American lexicon around 1966. The word refers to the whole realm of statistical characteristics of a particular population—especially as used to identify markets. The root of the word, the adjective *demographic*, was purportedly coined much earlier—in 1855—by French political economist Achille Guillard (Francese and Piirto, 1990, p. 45).

---

*Chapter co-authored with Brian Sommers, Ph.D. candidate in geography and regional development, University of Arizona; and in association with Robert Nunley, professor of geography, University of Kansas, and director, Kansas Geographic Bureau.

A surge of new interest in the geographical analysis of population has occurred as a result of the discovery of its usefulness in marketing products to households of consumers. As described by the editors of *American Demographics* (the flagship publication of the booming demographics industry),

> Demographics provide the basic picture of the consumer and psychographic information adds color and depth to that picture. Used together, attribute information can predict what consumers are likely to want and buy in the future. (Francese and Piirto, 1990, p. 45)

Psychographics is similarly a word of new coinage, coming into existence at about the same time as demographics, and used to describe an approach to marketing based on techniques developed in social psychology. Psychographics emphasizes the concept of *life-style* and is based on the belief that persons living in households sharing similar demographic characteristics are likely to have fairly similar sets of beliefs, values, and—most importantly for business applications—patterns of purchase and consumption. Thus by doing survey research or by probing for the tastes and preferences of "focus groups" (panels of individuals with known demographic characteristics), marketers can make inferences about where the most likely consumers for their products are to be found simply by identifying geographical areas containing households exhibiting the appropriate demographics.

Applied research in psychographics and demographics has pushed many companies to think much more explicitly than in the past about *market segmentation* (developing products designed to appeal to consumers with a particular mix of attributes rather than to the mass market) and about *target marketing* (focusing advertising efforts on certain groups of consumers, many times defined by where they live). Common aphorisms used nowadays by proponents of these forms of marketing are "you are where you live" and "you are what your neighbors eat." In an influential book called *The Clustering of America*, Weiss attributes the new emphasis on geographical locale and life-style that has come to permeate current American culture to creation in 1962 by the U.S. Postal Service of the ZIP (Zone Improvement Plan). ZIP codes allowed marketers, by examining census data compiled or estimated for ZIP code areas, to target mass mailings to sections of urban areas where their customers were most likely to live (1988, p. xi).

Indeed, Weiss and others have argued that during the past couple of decades American society has been fundamentally altered by the increasing tendency in not only the business world, but also in popular thought, politics, and the law to focus on groups of the overall population rather than on the average or typical American. The concept of *diversity*, always an important aspect of American society, has assumed increased importance. Diversity may now be one of the defining ideas of our current period of history. For a scholarly treatment of the geographical manifestations of this broad societal tendency in terms of residential differentiation and neighborhood change, see White (1987).

In this chapter we do not seek to provide a cookbook of "how to" instructions for doing target marketing and target segmentation. These techniques belong much more properly in the realm of marketing than geography. Many recent books have appeared that can guide the interested geographical analyst of population in carrying out such studies. In addition to the two excellent general references already cited (Weiss, 1988; Francese and Piirto, 1990), "how to" treatments are given in Ambry

(1990) and Nichols (1990). Crispell (1990) includes a wealth of pragmatic advice about when to use demographic information and lots of ideas about sources of data for business applications. Lazer (1987) details the implications of many recent U.S. demographic trends for current and future marketing strategies.

Our goal in this chapter, rather, is to provide a firm, substantive knowledge about the geographical basis for using demographic information in these fashions. We begin in Section 10.2 by describing useful approaches to actually measuring the amount of diversity or *segregation* inherent in a population. In using the term *segregation*, we intend a broader usage than that associated only with the residential distribution of racial groups. Rather, we mean the areal differentiation of a population according to any single characteristic.

Next in Section 10.3 we explore a body of geographical literature called *factorial ecology* which seeks to expose the social geography of urban areas by examining the correlations across a whole nexus of demographic, economic, and housing variables. A recent application of the technique to study "postindustrial" demographic change in the Cleveland metropolitan area is presented.

Then in Section 10.4 we develop a case study of the use of *cluster analysis* in defining areal units to maximize the homogeneity of demographic characteristics within them.

Finally, Section 10.5 contains a treatment of the application of the Claritas Corporation's PRIZM Lifestyle Clusters to the marketing of *Grit* magazine—an actual demographics project completed by the Kansas Geographic Bureau.

10.2 THE MEASUREMENT OF DIVERSITY AND SEGREGATION

To begin our discussion of demographics we examine the related concepts of (a) measuring the amount of heterogeneity of an area's population (i.e., its diversity) and (b) gauging the extent to which persons with different characteristics are geographically clustered (i.e., the degree of segregation present across a set of subareas). These concepts are inherent in the factorial ecology and cluster analytic methods that we shall subsequently study. Both factorial ecology and cluster analysis are used to partition a region into a set of subareas on the basis of maximizing the internal demographic homogeneity of the subareas, that is, finding units that are demographically distinctive. Note that the concepts of diversity and segregation are specialized notions about population composition and distribution, and thus measurement issues have already been anticipated by some of the material in Chapter 2. Here we give a comparative perspective on the different measures and discuss their use in practical applications—in the public sector as well as in the business world.

Diversity Measures

A diversity index is a single number that informs us about the relative amount of heterogeneity of a population. The various types of indexes have been defined so that higher values indicate that the subgroups of the population are equally present, whereas lower values indicate the predominance of a single subgroup. These measures are most useful for making comparisons—whether across different areas or in studying how the composition of population changes over time.

Entropy Index

The *entropy index* (a widely used statistical measure of uniformity in a distribution) was encountered earlier in Chapter 4 where we used it to measure the degree of spatial focusing in a system of migration flows. It is computed as

$$H = -\sum_{k=1}^{n}\left[(P_k/P)\ln(P_k/P)\right]$$

where now n is the number of subgroups present in the population, P_k is the population of the kth such group, and P is the total population. The maximum value of the entropy index, $\ln n$, is reached when all groups are equally present. Entropy approaches its minimum limit of zero when there is but a single member in each of $n-1$ of the subgroups. It is often useful to "normalize" the index (that is, to adjust H by dividing by its maximum) so that all values will fall on the range 0 to 1:

$$H^* = H/\ln n$$

In practical applications a diversity index may be most useful when there are more than two groups to be considered, because when there are only two, the percentage breakdown is all that is necessary to make the diversity level obvious. As an example of a practical application, consider its use by public officials concerned about racial segregation in an urban area's housing market. The normalized entropy index could be used to compare the distribution of white, black, and American Indian populations across a set of census tracts, and to produce a map showing the overall level of diversity for each. Suppose the population of tract 1 is 60 percent white, 30 percent black, and 10 percent American Indian:

$$\begin{aligned}H^* &= [(-0.60 \ln 0.60) + (-0.30 \ln 0.30) + (-0.10 \ln 0.10)]/\ln 3 \\ &= (0.3065 + 0.3612 + 0.2303)/1.0986 \\ &= 0.817\end{aligned}$$

whereas tract 2 is 65 percent white, 20 percent black, and 15 percent American Indian:

$$\begin{aligned}H^* &= [(-0.65 \ln 0.65) + (-0.20 \ln 0.20) + (-0.15 \ln 0.15)]/\ln 3 \\ &= (0.2800 + 0.3219 + 0.2846)/1.0986 \\ &= 0.887\end{aligned}$$

Notice that according to the entropy index, the diversity of tract 2 is somewhat higher than tract 1, *despite the fact that it has a higher proportion in the majority group*. This is because the two minority groups are found in more equal percentages in tract 2 than in tract 1.

The Interaction Index

The *interaction index* is the principal alternative measure of diversity available to the geographical analyst of population. Called in biology the Simpson index, it is commonly applied in studies of the species diversity of ecosystems. Think of this measure as the probability that any two members selected at random from a population will belong to different groups. It is computed as

$$S = 1 - \sum_{k=1}^{n}(P_k/P)^2$$

The minimum value of the index is zero (when all in the population are members of the same subgroup). The maximum value (reached when all subgroups are equally represented) depends on the total number of subgroups: $S^{max} = (n - 1)/n$. If the population is bifurcated into only two subgroups, the maximum is 0.50; as the number of groups increases, the maximum value approaches 1.0.

For our previous example, for tract 1,

$$S = 1 - [(0.60)^2 + (0.30)^2 + (0.10)^2] = 0.540$$

and for tract 2,

$$S = 1 - [(0.65)^2 + (0.20)^2 + (0.15)^2] = 0.515$$

Notice that according to the interaction index, tract 1 has higher diversity than tract 2, the opposite result to that found from the entropy index! The probability that any two persons selected at random from this tract come from different groups is higher than the similarly computed probability for the other tract. Because there are three groups in this example, the maximum index value is 2/3, which is the probability of encountering a person from one of the other two groups if all three groups were equally represented.

Business Applications of Diversity Indexes

There are important (although, we suspect, largely untapped to date) business uses for diversity measures. Beyond such obvious cases as for proving that a firm's work force complies with Equal Employment Opportunity regulations, there are important marketing applications in which the subgroups used to compute the indexes would not be restricted to racial ones. These measures could be very profitably applied to evaluate the likely effectiveness of, for instance, geographically targeted advertising campaigns.

As we detail in later sections of this chapter on clustering and life-style analysis, typical practice has been to assign geographic subunits (such as ZIP codes or census tracts) to particular market or life-style clusters based on the *predominant* (or average) characteristics of persons or households within them. It is quite likely, however, that among ZIPs or tracts receiving similar designations there will be wide variation in the actual homogeneity of these characteristics. The fundamental tenet of marketing demographics would suggest that the most effective areas for targeting should, in principle, be those having the *lowest* values on a diversity index (if those consumers exhibiting the "right stuff" in terms of demographic characteristics are, in fact, the majority group). The exercises for this chapter include such a business application.

Notice that we have now begun to talk about a spatial dimension of the distribution of population among subgroups. Whereas some of the most pertinent applications of diversity measures are to compare the composition of populations in different geographic units, if the intent is to quantify in a single number the dispersion of subgroups across geographic areas, it is a *segregation* measure that the analyst should compute.

Segregation Measures

The geographic counterpart of an (aspatial) diversity index is a measure of segregation. Segregation indexes reflect the extent to which the various subgroups of a total population are clustered within certain geographic subareas. White (1986) notes that, while there is no clear consensus about which is the best segregation measure to use for all applications, there are a number of desirable features for such an index. It should not be distorted by the overall minority group percentage of the population, nor by the total size of the population. It should be standardized so that it ranges from a minimum value of zero, which is found when each geographic subarea has an equal share of each of the subgroups' total population, to a maximum value of 1, which is obtained only when there is no mixing of subgroups in any of the geographic subareas. There are also a number of other technical criteria on which various measures can be evaluated for specific uses. (See White, 1986.)

Index of Dissimilarity

One of the most commonly employed segregation measures in population research (after receiving high marks in Duncan and Duncan, 1955) is the *index of dissimilarity*. Structurally this is the same measure as the index of concentration (or Hoover index) we presented in Chapter 2 to examine the extent to which total population is clustered across land area. In that application the index was easily interpretable as the fraction of population that would have to be moved from one of the study region's subareas to another to achieve uniform density across all subareas. In the case of segregation applications one big advantage of the measure is the similar ease with which it can be explained to nontechnical audiences. The index value is simply the fraction of either of the two subgroups' population that would have to be moved to obtain an equal percentage breakdown of the subgroups in all areas.

The index of dissimilarity is computed as

$$D = \tfrac{1}{2} \sum_{i=1}^{r} \left| \left(P_{ig} / P_g \right) - \left(P_{ih} / P_h \right) \right|$$

where i references one of the r total subareas and g and h the two demographic subgroups of the population. Note that the population of each group in each subarea is divided by the total population of that group across all subareas, and not by the subarea's total population. The index simply takes the sum of the absolute differences of subgroup shares across areas. Complete desegregation would mean there were the same fractions of majority and minority group populations in each subarea as in the entire study area's population. In such a case $D = 0$. Complete segregation would occur if there were no commingling of subgroup members in any of the subareas, whence $D = 1$.

An Application to School Attendance Zone Planning

The index of dissimilarity does have its limitations. To see these, we take the practical geographic problem of trying to balance the racial composition of a district's schools through setting the boundaries of school attendance zones or catchment areas. Examine the distributions in Table 10.1 of Anglo and Hispanic students across four high school catchment areas. Under neither plan A nor plan B is there complete desegregation. There are higher shares of the Hispanic population going to South and West than to North and East High Schools under plan A. Under plan B there is even more clustering of the Hispanic population in a single school, and, absent any other information about such things as the related busing costs and the splitting of functional neighborhoods (considerations we discussed in the Chapter 1 case study), plan B clearly seems to be less desirable.

The index of dissimilarity, however, yields an identical value for each plan:

$$D^A = \tfrac{1}{2} \left(|0.375 - 0.250| + |0.375 - 0.250| + |0.125 - 0.250| \right.$$
$$\left. + |0.125 - 0.250| \right) = .25$$
$$D^B = \tfrac{1}{2} \left(|0.500 - 0.250| + |0.250 - 0.250| + |0.250 - 0.250| \right.$$
$$\left. + |0.000 - 0.250| \right) = .25$$

TABLE 10.1 School Catchment Area Populations for Segregation Index Calculations

School	Busing Plan A Hispanic	Busing Plan A Anglo	Busing Plan B Hispanic	Busing Plan B Anglo
South High School	750	2,000	1,000	2,000
West High School	750	2,000	500	2,000
North High School	250	2,000	500	2,000
East High School	250	2,000	0	2,000
Total students	2,000	8,000	2,000	8,000

In both cases a quarter of one of the group's total populations would have to be reassigned to different schools to achieve the overall districtwide ratio of 1 Hispanic to 4 Anglos in each of the four schools. In plan A, 250 Hispanics from South High and 250 from West High (a total of 500 or 25 percent of the 2,000 total Hispanic students in the district) could be reassigned to bring the Hispanic student bodies of all schools to 500, thus balancing the ratios. In plan B, the same total number of Hispanic students would be reassigned, but all would move from South to East High. Alternatively, larger numbers of Anglo students could be reassigned from the plan A starting point to achieve the 1:4 ratios in each case (though capacity problems at South and West High might result). The general point about the dissimilarity index is that it is insensitive to any redistribution of a group's members between subareas that are either both already above or both below the mean percentage representation for that group. Only moving group members between overrepresented and underrepresented subareas will lower the value of this index.

The Gini Index of Diversity

A measure that avoids the problem we have just illustrated with the index of dissimilarity is the *Gini index*. Although it has been used infrequently in demographic applications, as was mentioned in Chapter 2 the Gini index is the standard measure used by economists for studies of income inequality. It is hard to explain to a lay audience, but (as our discussion of Figure 2.4 pointed out) it has a geometric interpretation as the ratio of (a) the area between the corresponding Lorenz curve and the 45-degree line to (b) the total triangular area lying under the diagonal.

The Gini index requires more involved computations because it is based on pairwise comparison of one of the two groups' population share in each subarea (P_{ih} / P_i) to that group's share in every other subarea (P_{jh} / P_j). The complete formula is

$$G = \frac{1}{2P^2(P_h/P)\left[1-(P_h/P)\right]} \sum_{i=1}^{r} \sum_{j=1}^{r} \left[P_i P_j \left|(P_{ih}/P_i) - (P_{jh}/P_j)\right|\right]$$

For the high school catchment area problem we previously discussed (Table 10.1), the Gini index is $G = 0.250$ for plan A and 0.375 for plan B. Unlike the index of dissimilarity, the Gini index clearly identifies the latter as the more segregated.

The Exposure Index

Much more commonly used in actual school desegregation and other demographic studies than the Gini index is the *exposure index*. The basic concept of the exposure index is to define the probability that a member of a specific subgroup comes into contact with a member of another specific subgroup in any random "exposure" to a member of the general population of his or her own subarea of residence. If the specified groups are the same, the calculation yields an *intragroup* exposure index. For group h,

$$E_{hh} = \sum_{i=1}^{r} \left[(P_{ih}/P_h)(P_{ih}/P_i) \right]$$

The *intergroup* index value for the exposure of group h to another group g is

$$E_{hg} = \sum_{i=1}^{r} \left[(P_{ih}/P_h)(P_{ig}/P_i) \right]$$

The exposure index concept is easily generalizable to more than two groups, a flexibility not inherent to either the index of dissimilarity or the Gini index. When there are just two subgroups, $E_{hg} = 1 - E_{hh}$.

Let us compute the intra- and intergroup exposure index values for the two alternative busing plans in Table 10.1.

The intragroup exposure index value for the Hispanic population under plan A is

$$E_{hh}^{A} = [(750 / 2,000) (750 / 2,750)] + [(750 / 2,000) (750 / 2,750)]$$
$$+ [(250 / 2,000) (250 / 2,250)] + [(250 / 2,000) (250 / 2,250)]$$
$$= 0.2323$$

and the intergroup exposure of Hispanics to Anglos under this same plan is

$$E_{hg}^{A} = [(750 / 2,000) (2,000 / 2,750)] + [(750 / 2,000) (2,000 / 2,750)]$$
$$+ [(250 / 2,000) (2,000 / 2,250)] + [(250 / 2,000) (2,000 / 2,250)]$$
$$= 0.7677$$

Note that these two probabilities do, in fact, sum to 1. The remaining exposure indexes for plan A are $E_{gg}^{A} = 0.8081$ (Anglo intragroup) and $E_{gh}^{A} = 0.1919$ (Anglos to Hispanics). The corresponding exposure index values for plan B are $E_{hh}^{B} = 0.2667$; $E_{hg}^{B} = 0.7333$; $E_{gg}^{B} = 0.8167$; and $E_{gh}^{B} = 0.1833$. This measure, like the Gini index, indicates that plan B is the more segregated of the two.

One final technical observation about the exposure index: whereas the maximum value obtainable for the intragroup exposure probability is 1 (when all subgroups are completely segregated in separate subareas), the minimum value depends on the relative fraction of the overall population that belongs to the group. A simple numerical example illustrates this minimum value. Suppose a total population of 1,000 has equal representation of both minority and majority population in each of two subareas. If the minority population overall accounts for 20 percent of the total,

$$E_{hh} = [(100 / 200) (100 / 500)] + [(100 / 200) (100 / 500)] = 0.20$$

In fact, the minimum value of the intragroup exposure index is just equal to the overall fraction of the population accounted for by the group.

The Entropy Index of Segregation

The final segregation measure that we shall describe is the counterpart to the entropy index of diversity that we discussed earlier. Theil and Finizza (1971) used an *entropy index* of racial mixing in the Chicago public school system computed as

$$H^* = (H - H') / H$$

where the H term is the entropy measure of the diversity of the entire study area's population and H' is a weighted average of the diversity of each individual subunit's population. These individual terms are calculated by the formulas

$$H = -\sum_{k=1}^{n}\left[(P_k/P)\ln(P_k/P)\right]$$

$$H' = -\sum_{i=1}^{r}\left\{(P_i/P)\sum_{k=1}^{n}\left[(P_{ik}/P_i)\ln(P_{ik}/P_i)\right]\right\}$$

For our high school catchment area example, the entropy diversity index for the entire school district, H, is 0.5004. The weighted average diversities for the four high schools are $H' = 0.4792$ under plan A and 0.4624 under B. From these the overall entropy index values (H^*) are derived: 0.0423 and 0.0822 for the two respective busing plans. The minimum value of this index is zero, indicating in our example that all schools have exactly the same mix of Hispanic and Anglo students as the district as a whole. The maximum value of 1 would occur if each of the four schools had students from a single ethnic group (yielding $H' = 0$). The higher entropy index value for plan B than plan A once again discloses that this is the more segregated of the two busing plans.

"Cracking and Packing": Diversity and Segregation in Political Redistricting

The measures described in this section have obvious utility for real world applications to political redistricting similar to the example we have given for use in school planning. The various segregation indexes can be used to evaluate the fairness of alternative proposed plans for carving up a region into political units.

As Morrison and Clark (1992) point out, however, the legal and data measurement issues connected with these public policy issues can be thorny. In the case of political redistricting the courts have consistently insisted that electoral districts be drawn up with almost exactly the same *total* population in each. This is the famous "one person, one vote" requirement. At the same time, a legal requirement (of Section 2 of the Voting Rights Act) has been that districts be drawn so as not to dilute the voting strength of minority populations, although as this book was going to press a Supreme Court decision had just been handed down that would seem to question some recently evolved practices devised by state legislatures under the mandates of federal officials in creating "minority majority" districts.

Two practices that have sometimes been used in the past to minimize the number of minority group representatives are given scrutiny when districts are drawn up today. *Cracking* means to draw boundaries such that the majority group dominates all districts—such plans would have values close to zero on a segregation index. On the other hand, *packing* involves placing a disproportionate percentage of minorities in a few districts so that overall minority representation may become substantially less than suggested by the relative proportion of minorities in the population at large. A high value (close to 1) on a segregation index would be found for such a redistricting plan.

Because of this legal consideration with respect to recognized minority subpopulations, in drawing up districting plans acceptable to the courts some districts have recently been created that are "segregated" in the sense that minority subpopulations are given a reasonable chance to elect a member from their own group. What may appear to be a "minority majority" district, however, is not always what it seems.

Case Study: A "Minority Majority" District?

A case in point (see Morrison and Clark, 1992) is a so-called Hispanic district drawn up in Los Angeles following the release of initial 1980 census data. The total population of the district was 1,392,114, of which 951,961 enumerated residents were "of Spanish origin." However not everyone is eligible to vote. Persons under voting age and non-U.S. citizens made up a more substantial proportion of the Hispanic population of the district than did non-Hispanics. Furthermore, at the time any districting plan is drawn up, only limited amounts of detail on population characteristics have been released by the U.S. Bureau of the Census. In this case it was only at the county level that a cross-tabulation by age, ethnicity, and citizenship could be obtained.

Using county-level ratios, the estimated population in this district of Hispanic persons of voting age who were also U.S. citizens was only 298,460, a bare majority (50.1 percent) of persons eligible to register to vote. Later on, when a special census tabulation was carried out, it was found that the actual enumerated population of Hispanic voting-age citizens in the district accounted for just 46.9 percent of the population eligible to register to vote. Thus, this so-called Hispanic district was, in fact, 53.1 percent *non*-Hispanic from the perspective of "potential voting strength."

10.3 FACTORIAL ECOLOGY

The segregation indexes we have examined measure the extent to which a *single demographic characteristic* is geographically clustered. In this section and the next we examine approaches to identifying areal units that are homogeneous with respect to *multiple* characteristics. A fairly standard methodology called *factorial ecology* has been developed that is widely applied in geographic research to identify distinctive social areas.

The theoretical underpinnings for numerous urban geography studies of neighborhood differentiation were provided by sociologists Shevky and Bell (1955). Their research took off from a study of the social areas of Los Angeles (Shevky and Williams, 1949) and posited three "constructs" to help explain the geographic patterns. These constructs were originally termed *social rank*, *urbanization*, and *segregation*, but in more recent work the comparable underlying dimensions of urban social ecology are commonly known by the labels *socioeconomic status*, *family status* (or *stage in life cycle*), and *ethnic status*. A distinction between the early *social area analyses* and the more recent *factorial ecology* studies is that the former imposes or presupposes the three indexes or constructs set forth by Shevky, Williams, and Bell, whereas in the latter the statistical method of *factor analysis* is used on a wider set of variables to extract a unique set of composite indexes (called factors) for the particular study area and time period. (See Berry and Kasarda, 1977, pp. 122–24.) Despite these technical differences, however, in many factorial ecology studies three of the major factors extracted are essentially the same three predicted by the theories of the social area analysts.

The usefulness of factorial ecology for many practical applications is that it produces a set of composite indexes—each of which measures the joint influence of a group of interrelated variables. The method can uncover underlying relationships not immediately apparent in a large data set. The actual determination of what these revealed patterns mean in terms of describing or delimiting uniform social areas rests on the ability of the user (Cadwallader, 1985, p. 141). As an example of the use of factorial ecology to uncover shifting demographic patterns in an urban area, we present a case study of Cleveland, Ohio.

Case Study: Postindustrial Demographic Change in Cleveland

Sommers (1993) makes extensive use of factorial ecology in examining the shifting neighborhood demographics of the Cleveland area from 1960 to 1990—a period when Cleveland, like many major metropolitan areas in the traditional American Manufacturing Belt, was undergoing intense metamorphoses as a result of job restructuring, significant net out-migration of population, and the coming of age of the large post–World War II baby boom generation. Here we present selected results for 1990 and compare them to 1980 to show how the geographical analyst of population can use factor analysis to gain insights into the shifting demographics of an urban area. First, however, we take the reader on a step-by-step tour of how a factorial ecology is accomplished.

The Data Base

Much of the work in carrying out a factorial ecology is in compiling the data base. With modern, user-friendly interactive software, the statistical manipulations are easily accomplished, as, for instance, in the Sommers study with the Statistical Package for the Social Sciences (SPSS).

The two data sets compiled for this analysis consisted of 1980 and 1990 decennial census data for a portion of the overall Cleveland Metropolitan Area including the City of Cleveland and inner suburbs. The geographic units used were block groups, rather than census tracts which have been employed in most previous studies. By using block groups (for which the same data are available as for tracts—though not always in as easily used media) the analyst can gain a finer-grain picture of geographic patterns. The Cleveland study used 1,123 populated block groups for 1990 data.

Variable Selection

Some 150 variables representing a broad range of social, economic, and housing characteristics were first extracted from the census standardized tape files (STF). From these, the total list of variables used in the analysis was pared down to 28. Chosen were characteristics measured in a similar fashion on both censuses. Also of concern was to select, *a priori*, variables measuring unique aspects and not redundant ones. Table 10.2 gives the roster of selected variables.

The selection of variables is a crucial step in performing a factorial ecology. The analyst can make one or more underlying dimensions seem particularly significant simply by including a large number of highly related variables measuring essentially the same phenomenon. Critics have sometimes argued that factorial ecology studies have told us more about the structure of the questions asked on the U.S. census than about the real social ecology of our urban areas! In defense, we would note that census content is to a great measure a reflection of characteristics that our society values—thus census variables inherently measure attributes with meaning to our culture.

One final point about variable selection: academic honesty and integrity count! Once a factor analysis is conducted, we could repeat the procedures, adding in additional

TABLE 10.2 The 28 Census Variables Used in the Cleveland Factorial Ecology Case Study

Descriptor	Description
Income	Median household income
Poverty	Population below the poverty limit (percentage of total persons)
House Value	Median housing value (in thousands of dollars)
Rent	Median contract rent
Owner Occ	Owner-occupied housing units (percentage of total housing units)
Renter Occ	Renter-occupied housing units (percentage of total housing units)
Vacant Units	Vacant housing units (percentage of total housing units)
Youths	Persons under 18 years of age (percentage of total population)
Elderly	Persons 65 years and older (percentage of total population)
Female Hshldr	Households with female householder (percentage of total households)
Fertility	Average number of children per household
PPH	Persons per housing unit
Rooms	Average number of rooms per housing unit
Whites	White persons (percentage of total population)
Blacks	Black persons (percentage of total population)
Hispanics	Persons of Hispanic descent (percentage of total population)
Asians	Persons of Asian descent (percentage of total population)
Grad Degrees	Persons with graduate or professional degrees (percentage of total population 25 years and older)
Dropouts	Persons with no high school diploma (percentage of population 25 years and older)
Managerial Emp	Persons employed in management, professional, technical, or administrative positions (percentage of total employment)
Service Emp	Persons employed in service positions (percentage of total employment)
Laborer Emp	Persons employed as operators and laborers (percentage of total employment)
CC Emp	Persons working in the central city (percentage of total employed persons)
Metro Migrants	Households living outside MSA 5 years earlier (percentage of total households)
Interstate Migrants	Households living out of state 5 years earlier (percentage of total households)
Nonmovers	Households living in the same housing unit as 5 years earlier
Natives	Native-born citizens (percentage of total population)
New Housing	Housing units built since 1960 (percentage of total housing units)

variables to reinforce those relationships already identified by the factor analysis. Although such actions might increase the amount of variation in the data set explained by the factors, they do not necessarily contribute to the goal of using the technique as an exploratory tool to uncover new relationships among population characteristics.

The Correlation Matrix and Factor Extraction

The first step in the statistical massaging of the data is to compute a *correlation matrix* showing the level of statistical association between each variable and each of the other variables. Although for space reasons we do not report all 378 such values, it should be noted that most of these *r* values are in the low or middle range. Table 10.3 reports the 10 highest correlations for 1990.

The complete correlation matrix is used as the basis for extracting factors. Table 10.4 shows intermediate statistics as produced by the SPSS software in extracting the factors. Factors are composite indexes made up from linear combinations (transformations) of the original variables. To account for all the variation encompassed by the original variables, an equal number of factors as variables is required. However (as can be seen in this table), the factors are chosen in such a way that the first group of them are compact, summary representations of many of the original variables.

The statistics shown in the table may be interpreted as follows:

Eigenvalue. This measure shows the actual variance in the data set accounted for by each factor. Note that in this case, the first factor is quite dominant, accounting for more than twice as much variance as the second factor.

Percentage and cumulative percentage of total variance. These are computed directly from the eigenvalues and simply standardize the raw values into percentage terms for easier interpretation. Note that by the time the third factor is extracted, more than half the original variance in the entire set has been accounted for, illustrating the ability of the first few factors to serve as summary, composite indexes fairly representative of the entire data base.

The Factor Matrix, Factor Loadings, Factor Retention, and Rotation

After all the factors are extracted, an initial *factor matrix* is produced. This matrix contains the *factor loadings* that represent the strength of the statistical relationship between

TABLE 10.3 Highest Correlations (*r*) between Variables in the 1990 Cleveland Factor Analysis

Positive correlations	
.866	Metro Migrants and Interstate Migrants
.862	Fertility and Youths
.815	House Value and Income
.728	PPH and Youths
.724	Rent and Income
.724	Income and Grad Degrees
.721	PPH and Income
.720	House Value and Grad Degrees
Negative (inverse) correlations	
−.990	Whites and Blacks
−.967	Owner Occ and Renter Occ
−.751	Managerial Emp and Dropouts
−.722	Rent and Dropouts
−.720	Managerial Emp and Laborer Emp

Source: Sommers (1993).

TABLE 10.4 Principal Component Factors and Intermediate Statistics for Cleveland Case Study, 1990

Factor	Eigenvalue	Percentage of Variation	Cumulative Percentage
1	8.3599	29.9	29.9
2	4.1549	14.8	44.7
3	3.0351	10.8	55.5
4	2.2260	8.0	63.5
5	1.2025	4.3	67.8
6	.9888	3.5	71.3
7	.9213	3.3	74.6
8	.8753	3.1	77.7
9	.7772	2.8	80.5
10	.7604	2.7	83.2
11	.6295	2.2	85.5
12	.5906	2.1	87.6
13	.5500	2.0	89.5
14	.4380	1.6	91.1
15	.3970	1.4	92.5
16	.3851	1.4	93.9
17	.3219	1.1	95.0
18	.2896	1.0	96.1
19	.2417	0.9	96.9
20	.1914	0.7	97.6
21	.1737	0.6	98.2
22	.1489	0.5	98.8
23	.1136	0.4	99.2
24	.0873	0.3	99.5
25	.0739	0.3	99.8
26	.0630	0.2	100.0
27	.0030	0.0	100.0
28	.0005	0.0	100.0

Source: Sommers (1993).

each of the original variables and each factor. Such a table is generally shown in an appendix, but not much commented on in the full report of a factorial ecology study, because usual practice is to retain only some of the original factors for further analysis. The retained factors are then *rotated* in some fashion to ease their interpretation.

In this case study a common rule of thumb was used in determining that only factors with eigenvalues of 1.0 or greater would be retained, meaning only the top 5 (refer back to Table 10.4), and the frequently chosen Varimax (variance maximization) rotation technique was employed. This rotation scheme means that the original 5 factors are modified (rotated around in the 28-dimensional space of the data set!) in such a fashion as to produce factor loadings that approach either 1.0 or 0, thereby lessening the number of variables with difficult-to-interpret, midlevel loadings. Geometrically speaking, this cleaner factor structure means that the computer attempts to orient the factors to the greatest extent possible with the best-fit regression lines of the original variables in the data set. Whereas the original, unrotated factors were chosen to group variables that are highly correlated with each other, the procedure does not necessarily guarantee that the variables will be represented by only one factor. Rotation thus produces a "more unique" set of factors.

Interpreting the Rotated Factor Loading Matrix

The *rotated factor matrix* for 1990 is reproduced in Table 10.5. The reported loadings show the strength of the relationship between each variable and the rotated factor. Values approaching 1 or −1 indicate variables that are closely related to the factor (that "load highly on it"). Note the column labeled *Communality*. This measure is the proportion of variation in each variable accounted for by *all* the factors. At the previous stage of the analysis when the initial factors were extracted, all the communalities equaled 1; there were the same number of factors as there were variables, so each of the data values for each block group on each of the original variables could be expressed exactly as a linear combination of the 28 factors. Now, with only 5 retained factors, some of the variance of the original variables is unexplained by linear combinations of factors.

Much of the art of factorial ecology rests on interpreting the factor loadings. The SPSS program conveniently groups the variables according to their highest loadings on the factors. These values are shown in boldface type in Table 10.5. Despite the goal of the Varimax rotation scheme to eliminate as much as possible variables loading on more than one factor, some significant "secondary" loadings exist. Any such secondary factor score greater than 0.4 is distinguished in the table by being typeset in boldface italic numerals.

We now interpret the 5 rotated factors. Traditionally each factor is given a name based on the group of original variables that load highly on it.

Factor 1. Socioeconomic Status. The first factor extracted in this case study, as in many previous factorial ecology studies, is characterized by high positive loadings by variables measuring income, house value, the size of housing units (rooms), grad degrees, and high-status occupations (managerial emp). Equally, variables indicative of poverty, low educational achievement (dropouts), and low-status occupations (laborer emp) have strong negative loadings.

Factor 2. Family Life Cycle. Included here are some of the prime age and household composition variables normally loading on the family status or stage in life cycle dimension typical of many other factorial ecologies. Significant positive loadings are found for percentage of children (youths) and family size (PPH). Elderly percentage has a high negative loading. There is a fairly high positive loading for the fertility variable. On this factor are also loadings for new housing units (those built since 1960), and a strong secondary loading for the size of housing units (rooms). This is likely due to movement of young families of baby boomers into newer, larger homes in more peripheral locations.

Factor 3. Mobility. This factor represents a mobility index with high positive loadings indicating high percentages of in-movers (metro and interstate migrants) and high percentages of renters (renter occ). Conversely, high negative loadings are found for household stability (non-movers) and percentage of home owners. This is not one

TABLE 10.5 Varimax Rotated Factor Matrix and Final Communalities, 1990

Variable	Factor 1 Socioeconomic Status	Factor 2 Life Cycle	Factor 3 Mobility	Factor 4 Race	Factor 5 Asian Immigration	Communality
Income	**.860**	.101	−.107	−.201	.109	.81
Managerial Emp	**.831**	−.174	.108	−.149	−.036	.76
House Value	**.829**	−.088	.009	−.126	.134	.73
Dropouts	**−.814**	.156	−.008	.245	.126	.76
Rent	**.811**	−.034	−.064	−.236	−.004	.72
Grad Degrees	**.801**	.004	.244	−.006	.212	.75
Rooms	**.674**	*.505*	−.276	−.106	.128	.81
Laborer Emp	**−.672**	.282	−.088	−.074	.075	.55
CC Emp	**−.463**	.187	.339	.374	.136	.52
Vacant Units	**−.450**	.111	.277	*.423*	.084	.48
Female Hshldr	**.446**	−.171	−.369	−.143	−.067	.39
Hispanics	**−.395**	.331	.278	−.351	.291	.55
PPH	.119	**.864**	−.296	.133	.034	.87
Youths	−.248	**.844**	−.085	.192	−.092	.83
Fertility	−.264	**.812**	−.008	.130	−.095	.76
Elderly	.055	**−.730**	.356	−.036	.143	.68
New Housing	−.011	**−.493**	−.010	.369	.159	.41
Metro Migrants	.243	−.061	**.814**	−.138	.139	.76
Nonmovers	.132	.055	**−.805**	.032	−.039	.67
Interstate Migrants	.241	.044	**.759**	−.113	.225	.70
Renter Occ	*−.538*	−.207	**.608**	.336	−.018	.82
Owner Occ	*.585*	.140	**−.601**	*−.405*	−.006	.89
Blacks	−.166	.164	−.087	**.891**	−.216	.90
Whites	.216	−.205	.040	**−.879**	.134	.88
Service Emp	*−.405*	−.040	−.014	**.536**	−.048	.46
Asians	.062	−.071	.216	−.034	**.751**	.62
Natives	−.091	.180	−.090	.365	**−.704**	.68
Poverty	.082	.101	−.007	−.323	**−.339**	.24

Note: A number in **boldface** type indicates a variable's primary loading; those in *italic boldface* indicate secondary loadings greater than .400.

Source: Sommers (1993).

of the original three constructs of social area analysis, although the social area analyst's "urbanization" concept and the typical family status dimension of many factorial ecology studies subsume it. In these Cleveland results typical life cycle variables are evidently split out onto this factor and factor 2.

Factor 4. Race and Ethnicity. This is the third of the three traditional factorial ecology factors. A positive loading is associated with high percentage black and a negative one with high percentage white. Certain employment and housing variables also have strong loadings on this variable because the black areas of Cleveland tend to be more impoverished than predominantly white sections (but remember that a separate socioeconomic status factor has already been described). Service employment loads on this factor.

Factor 5. Asian Immigration. A high positive loading is found for the population of Asian descent and a significant negative one for the native-born population percentage. There is a modest (.21) positive loading of the variable measuring the percentage of the population holding graduate or professional degrees as well as a primary negative loading for percentage of the population below the poverty limit. The influx of Asian student population at U.S. universities may be reflected here; all the major four-year institutions of higher education in metropolitan Cleveland are included in the study area. Note that the extraction of a separate factor for Asians suggests that the racial geography of Cleveland is more complex than a segregated white-versus-"minority" dichotomy of neighborhoods. Note, too, that race and ethnicity do not simply load on the socioeconomic status factor—income segregation and the clustering of the American population by ethnicity are related, but inherently different underlying dimensions of our cities' social geographies.

Analyzing the Temporal Stability of Factor Structure

The 1980 factor structure is quite similar to that for 1990, although there are a number of differences worth noting. Table 10.6 summarizes the factors extracted and retained for both periods. The first factor, socioeconomic status, is the same for both years in the sense that the same variables load on it. The highest positive loading in 1980 was for the percentage of the population 25 years and older without a high school diploma. Another interesting change with respect to this factor is the loading for female heads of household (VAR79). For 1980 this variable loads on a separate factor 5 along with the percentage of housing built since 1960 and the renter-occupied percentage. But for 1990 the correlation is with higher-income variables. It is highly correlated with the population 65 years and older (reflecting the widowed population). The change from 1980 to 1990 in how it loads could be a reflection of growing divorce rates among the middle and upper classes, or simply of an increasingly elderly population base.

Another notable change is the switch in position of the race and ethnicity and mobility factors. In 1990 the owner/renter occupancy variables loaded on the mobility factor whereas in 1980 they were merged with socioeconomic status indicators on factor 1. On the other hand, the 1980 race and ethnicity factor included the Hispanic population—a variable that in 1990 was encompassed by the socioeconomic factor.

TABLE 10.6 Temporal Stability of Factors Extracted for the Cleveland Case Study, 1980 and 1990

1980 Factors	1990 Factors
1. Socioeconomic Status	1. Socioeconomic Status
2. Life Cycle	2. Life Cycle
3. Race and Ethnicity	3. Mobility
4. Family Life Cycle	4. Race
5. New Housing / Female Householders	5. Asian Immigration
6. Asian Population	

316   Chapter 10 • Demographics

One further difference is worth noting. The 1990 Asian population factor reflects relatively well-educated immigrants, whereas the 1980 factor is a unique one with only the indigenous Asian population variable (VAR77) loading significantly.

Computation and Interpretation of Factor Scores

The final step of a factorial ecology—and for the geographical analyst of population, truly the key part of the analysis—is the computation and mapping of *factor scores*. These are simply the numerical values that an areal unit rates on the composite indexes represented by the factors. Thus the factor scores on factor 1 for the Cleveland case study provide us with a useful index of the relative economic status of all the 1,123 block groups included in the study area, those on factor 2 are a summary indicator of the stage of life cycle of the population living within each block group, and so forth. These have considerable utility in target marketing, political campaigning, and other applied settings where it is extremely useful to know in what parts of town do peoples' characteristics suggest that they may be most susceptible to a pitch for the particular product being peddled.

To compute the factor scores, the analyst uses the factor score coefficient matrix produced as part of the output of the statistical package used for the factor analysis. Table 10.7 reports these coefficients for the first three factors in the 1990 Cleveland analysis. As can be seen, there is a coefficient for each variable's contribution to each factor. To obtain the actual factor score for any block group in the Cleveland study area, the *actual data value* for that block group for each of the variables is multiplied by the appropriate coefficient. Thus there is a factor score for every areal unit in the study on each factor.

Obviously for the Cleveland case study the matrix of factor scores (1,123 × 5) is too bulky to include in this book! Maps, though, are an efficient, useful way to examine these. After describing how the scores are calculated, we will conclude our case study by briefly examining the social geography of Cleveland as portrayed by selected factor score maps.

To illustrate the computation of factor scores, consider four different block groups (BGs): BG 4 of tract 1038, BG 2 of tract 1182, BG 3 of tract 1834.01, and BG 1 of tract 1928. Their values on each of the variables as tabulated from 1990 census returns are shown in Table 10.8. Note that these are diametrically opposite types of areas in Cleveland, as registered, for instance, by the data values for VAR138 (median income). BG 4 of tract 1038 has the lowest median income in the entire study area, $7,335. Income is also low for BG 2 of tract 1182 ($15,057), whereas it is quite high for BG 3 of tract 1834.01 in affluent Shaker Heights ($73,780) and BG 1 of tract 1928 in the community of Bratenahl—located on Lake Erie ($61,364).

To compute the respective factor scores for these block groups on the socioeconomic status factor, the coefficients from column 1 of Table 10.7 are multiplied by the data values in Table 10.8—but first the data values must be standardized into *z* scores. This requires subtracting the variable's mean value ($\bar{X}$) over all the included block groups and dividing by its standard deviation (σ):

$$z_i = (X_i - \bar{X})/\sigma$$

For instance, since the mean and standard deviations of median income are $25,788.41 and $16,925.65, the respective median-income z scores of the four block groups are -1.1, -0.6, $+2.8$, and $+2.1$. Note that z scores lying outside the range -1 to $+1$ indicate extreme values of the variable.

10.3 Factorial Ecology

TABLE 10.7 Factor Score Coefficients for Three Selected Factors, 1990

Variable	Factor 1 Socioeconomic Status	Factor 3 Mobility	Factor 4 Race and Ethnicity
Income	.13834	−.00445	.03797
Poverty	−.02135	.07087	−.16141
House Value	.14215	.02144	.06951
Rent	.12452	.02195	.00490
Owner Occ	.03881	−.15845	−.08393
Renter Occ	−.03928	.16652	.05921
Vacant Units	−.02010	.05142	.12123
Youths	−.00625	−.00613	.02316
Elderly	−.03003	−.15863	.03309
Female Hshldr	.04840	−.09361	.00055
Fertility	−.01417	.02002	−.00621
PPH	.04691	−.07695	.06542
Rooms	.11587	−.06340	.05348
Whites	−.07274	.01933	−.31968
Blacks	.08182	−.02160	.32253
Hispanics	−.10333	.04857	−.17285
Asians	.00664	−.04161	.08348
Grad Degrees	.16379	.08622	.11789
Dropouts	−.12588	−.06148	.01320
Managerial Emp	.14423	.07871	.03789
Service Emp	−.01288	−.03073	.16785
Laborer Emp	.13799	−.06187	−.10857
CC Emp	−.02465	.06696	.10288
Metro Migrants	.06437	.26359	−.02954
Interstate Migrants	.06628	.23459	−.00904
Non-movers	−.00577	−.25605	.04445
Natives	.03714	.06959	.05169
New Housing	.02941	−.05196	.18219

Source: Sommers (1993).

The *contribution* of any single variable to the overall factor score for a block group is simply its z value multiplied by the corresponding factor score coefficient. Thus the contribution of median income to the socioeconomic status scores for BG 4 of tract 1038 is $-.15 = (-1.1) \times (.1383)$. To obtain the final factor score for any block group requires summing up the contributions of all 28 included variables. Note, however, that certain variables have much more of an influence than others. As we see by comparing Table 10.5 to Table 10.7, the variables that have the highest loadings are the most significant in fixing the values of the factor scores via the factor score coefficients.

TABLE 10.8 Actual 1990 Data Values for Four Selected Block Groups and Variable Means and Standard Deviations for All 1,102 Included Block Groups

	Block Group					
Variable	1038–4	1182–2	1834.01–3	1928–1	Mean	Standard Deviation
Income	$7,335	$15,057	$73,780	$61,364	$25,788.41	$16,925.65
Poverty	<0.5%	<0.5%	<0.5%	<0.5%	.0046	.0079%
House Value (× 1,000)	$15.30	$26.30	$130.70	$128.20	$54.38	$41.22
Rent	$207	$256	$725	$400	$314.94	$131.84
Owner Occ	45%	52%	93%	72%	53.09%	26.88%
Renter Occ	45%	42%	5%	20%	38.52%	22.66%
Vacant Units	12%	7%	0%	9%	8.12%	8.01%
Youths	36%	30%	27%	11%	25.10%	7.61%
Elderly	9%	21%	10%	34%	18.29%	8.83%
Female Hshldr	32%	63%	33%	58%	58.35%	12.01%
Fertility	.302	.238	.205	.071	.2044	.0876
PPH	2.88	3.05	3.03	1.93	2.52	0.42
Rooms	5.09	6.51	8.80	6.37	5.69	1.08
Whites	77%	0%	74%	92%	61.15%	40.26%
Blacks	3%	100%	26%	6%	35.15%	41.57%
Hispanics	29%	0%	0%	1%	3.54%	7.31%
Asians	2%	0%	0%	1%	1.15%	2.82%
Grad Degrees	0%	3%	44%	28%	6.20%	10.24%
Dropouts	66%	62%	7%	7%	33.89%	19.10%
Managerial Emp	15%	57%	88%	88%	54.13%	19.51%
Service Emp	15%	21%	8%	5%	16.70%	11.79%
Laborer Emp	51%	23%	0%	4%	18.94%	13.43%
CC Emp	67%	77%	42%	58%	57.56%	19.76%
Metro Migrants	0%	2%	26%	7%	8.21%	9.07%
Interstate Migrants	0%	2%	24%	5%	5.89%	6.97%
Nonmovers	58%	73%	39%	62%	60.39%	15.11%
Natives	100%	100%	92%	92%	93.95%	6.88%
New Housing	.101	.171	.000	.466	.1618	.1898

Source: Sommers (1993).

The factor score coefficients are derived such that factor scores will also be in z score units. That means that the scores on each factor are standardized by the mean and standard deviation of the distribution of scores for all the block groups. Table 10.9 shows the factor scores for our 4 selected block groups on each of the 5 factors. Thus

TABLE 10.9 Factor Scores for the Four Selected Cleveland Block Groups, 1990

Factor	Tract 1038, BG 4	Tract 1182, BG 2	Tract 1834.01, BG 3	Tract 1928, BG 1
1. Socioeconomic Status	−2.01	−0.15	+3.03	+1.58
2. Life Cycle	+1.50	+0.55	+1.59	−1.51
3. Mobility	−0.17	−0.90	+1.71	−0.12
4. Race	−1.18	+1.42	+0.21	+0.31
5. Asian Immigration	+0.82	+0.05	+0.29	+0.71

Source: Sommers (1993).

the −2.01 and −0.15 scores of BG 4 of tract 1038 and BG 2 of tract 1182 indicate that these tracts are extremely low and just below average, respectively, in socioeconomic status among all the block groups of the study area. Though both are below average in socioeconomic status, the two tracts differ significantly in other ways. BG 4 of tract 1038 on the inner west side near the Central Business District has a low minority population and thus a negative score on factor 4 (−1.78). BG 2 of tract 1182 in the east side Glenville neighborhood, on the other hand, has 100 percent black population and a high positive score (+1.42) on the race and ethnicity dimension. Note the very high positive socioeconomic status factor scores and average race and ethnicity scores for the other two, wealthy block groups.

Mapping Factor Scores and Analyzing the Geographic Patterns of Urban Social Structure

Figures 10.1 through 10.3 are maps of the 1990 factor scores for all the included block groups in the study area on 3 of the 5 retained factors. Many previous factorial ecologies have found a tendency for the three usual underlying dimensions of urban social structure, when mapped in this fashion, to suggest three classic urban models.

Geographically, socioeconomic status in U.S. cities often follows a sectoral pattern such as suggested by Hoyt (1939). Family status and stage in life cycle, on the other hand, are most likely to express themselves spatially in a concentric zone pattern similar to the concentric zone theory of Burgess (Park, Burgess, and McKenzie, 1925). And the ethnic status dimension most often follows a multiple-nuclei pattern much like the zones in Harris and Ullman's (1945) model. For a fuller discussion of these models and the patterns typical of the underlying dimensions of urban social geography, consult an urban geography text such as Hartshorn (1992).

The maps of the Cleveland factor scores do seem to be typified by these traditional findings. On the other hand, specific land use patterns, transport routes, historical circumstances, and physical amenities in the region (for example, the Lake Erie shoreline) exert unique influences. For instance, the highest socioeconomic status areas lie along the lakefront and in a wide eastern suburban wedge, with only scattered pockets of dark shading appearing elsewhere in Figure 10.1. The race and ethnicity factor scores, when mapped, disclose a clear area of overwhelmingly minority population clustered in a single broad inner city eastern zone (Figure 10.2). The pattern for the mobility factor scores (Figure 10.3) is more complex. As we earlier mentioned, this is a somewhat atypical factor, and its spatial pattern does not seem to follow particularly the concentric zone model suggested by more classical family status factors.

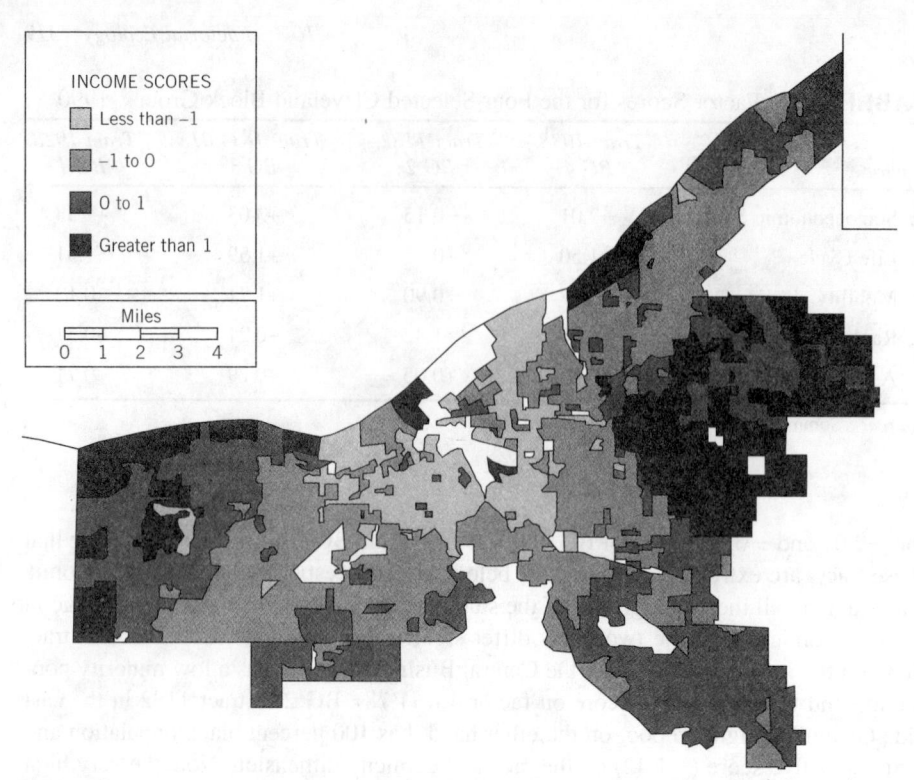

FIGURE 10.1 Factor Scores for Factor 1 (Socioeconomic Status), Cleveland Case Study, 1990

Source: Sommers (1993).

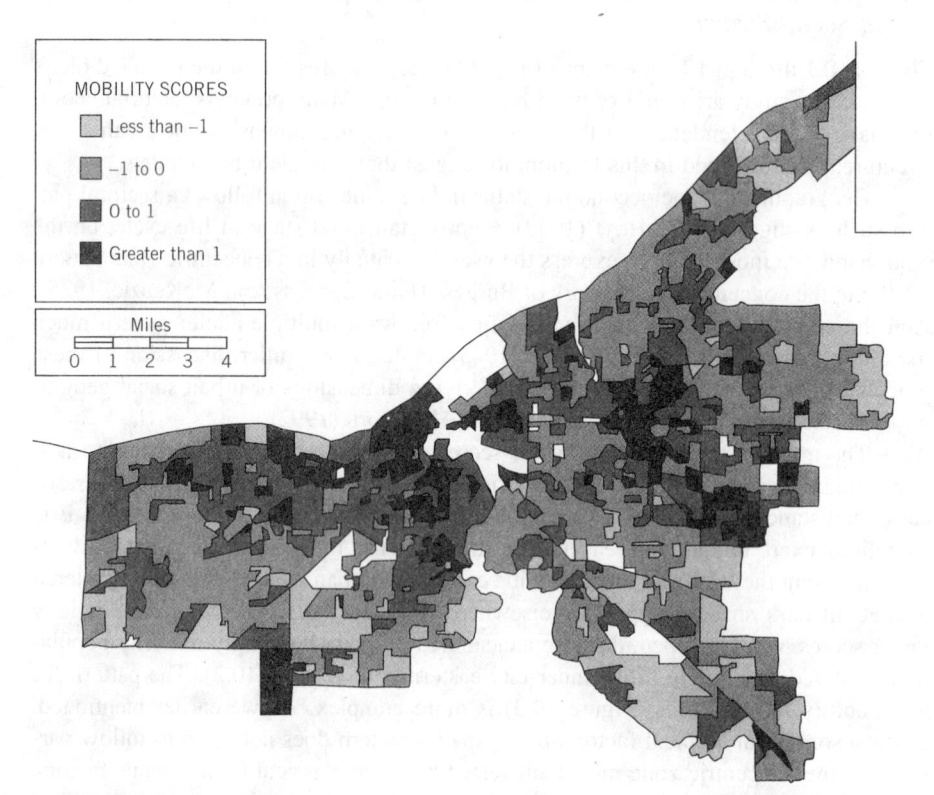

FIGURE 10.2 Factor Scores for Factor 4 (Race and Ethnicity), Cleveland Case Study, 1990

Source: Sommers (1993).

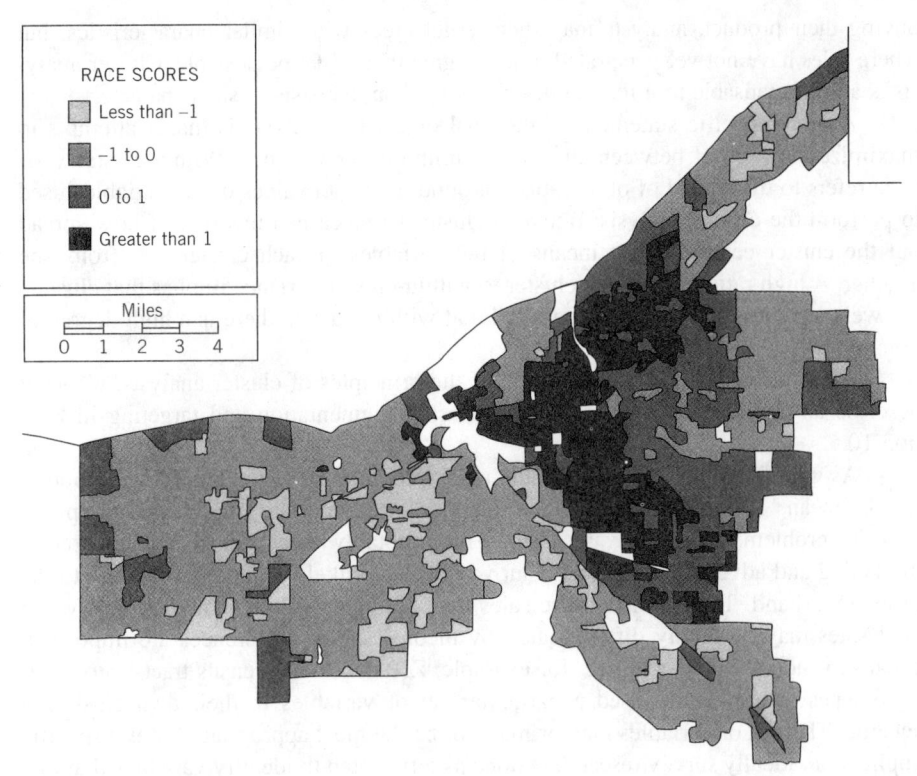

FIGURE 10.3 Factor Scores for Factor 3 (Mobility), Cleveland Case Study, 1990
Source: Sommers (1993).

Sommers (1993) examines the geographic patterns of all 5 factors in full detail, focusing particularly on the role played by large-scale natural amenities in ameliorating the broad, regionwide forces of economic and demographic change on the Cleveland area's housing markets.

Whereas a large body of academic work in geography has used the technique of factorial ecology, and much of the basic theory of demographics used today in business applications can trace its pedigree to the early social area analysts, perhaps more widely used than factor analysis in applied situations is *cluster analysis*. We now examine how cluster analysis works in seeking to identify socially homogeneous subunits of a broader study area.

10.4 CLUSTER ANALYSIS FOR DEFINING SOCIALLY HOMOGENEOUS AREAS

The objective of cluster analysis is to identify explicitly those observations (in our case, areal units) that have more in common (in terms of their values on prespecified variables) with one another than they do with other observations. This classification of small regions on the basis of their characteristics is in fact precisely the aim of market researchers in carrying out market segmentation and target marketing studies. By first identifying the general demographic and socioeconomic characteristics of those buying their product, analysts may then target areas with similar characteristics, but where sales have not yet penetrated to the degree that might be possible. Cluster analysis is an indispensable tool for classifying regions on the basis of such characteristics.

A more specific statement of the goal of cluster analysis is that it attempts to maximize the ratio of between-cluster to within-cluster variance. Within-cluster variance refers to the spread of observations around the mean values of the variables used to perform the cluster analysis. Between-cluster variance is a measure of how spread out the cluster centroids (the means of the variables in each cluster) are from one another. A high ratio of between-cluster to within-cluster variance implies that clusters are well separated from one another, and that within clusters there is a high degree of homogeneity.

After our discussion here of some of the principles of cluster analysis, we shall describe an application of clustering for market segmentation and targeting in Section 10.5.

As with factor analysis, one of the most important steps in classifying a population by means of cluster analysis is the selection of appropriate variables. For applications to problems in the private sector—most notably those associated with target marketing and advertising—the relevant variables are likely to be quite specific to the problem at hand. Thus compact disc sales may be highly differentiated by age, while boat sales may be highly differentiated by income. These differences are important, because vendors who categorize, for example, ZIP codes and census tracts into "lifestyle types" have clearly used a particular set of variables in their own clustering scheme. That set of variables may or may not be the most appropriate for any specific application. Ideally survey research is done as a first step to identify variables that distinguish likely adopters of the product being marketed.

In our illustrative example here, we assume that a hypothetical product is related to household consumption, and that variables indicating (1) the degree of home ownership, (2) family size, and (3) the proportion of persons in nonfamily households are the most relevant in differentiating those households buying the product from those who do not. Figure 10.4 depicts a portion of Palo Alto, California, and Table 10.10 provides the data for the three variables across the eight census tracts that we wish to cluster.

Methods of classifying the areal units (in this case, census tracts) into clusters fall into two general types: *hierarchical agglomerative methods* and *nonhierarchical, iterative partitioning methods* (Aldenderfer and Blashfield, 1984). Hierarchical agglomerative methods are based on a measure of similarity (or dissimilarity) between "cases." In the geographic applications that we are interested in, the term *case* refers to an areal unit.

Similarity Measures

Perhaps the most common measure of the similarity between two cases i and j is the square root of the sum of squared differences between the values of the individual variable values for the two cases. It is analogous to the way that the Euclidean distance between two points is calculated:

$$d_{ij} = \sqrt{\sum_{k=1}^{p} \left(x_{ik} - x_{jk}\right)^2}$$

where x_{ik} is the value of case i on variable k, and p is the number of variables.

10.4 Cluster Analysis for Defining Socially Homogeneous Areas

Another measure of the similarity between cases is the correlation coefficient

$$r_{ij} = \frac{\sum_{k=1}^{p}\left(x_{ik}-\overline{x}_i\right)\left(x_{jk}-\overline{x}_j\right)}{\sum_{k=1}^{p}\left(x_{ik}-\overline{x}_i\right)^2 \sum_{k=1}^{p}\left(x_{jk}-\overline{x}_j\right)^2}$$

where $\overline{x}_i$ is the mean of all of the values for the variables associated with case i. That is,

$$\overline{x}_i = \frac{1}{p}\sum_{k=1}^{p}x_{ik}$$

Table 10.11 depicts the distance matrix and Table 10.12 portrays the correlation matrix associated with the data in Table 10.10.

Agglomerative Clustering

Agglomerative forms of clustering are based on rules for adding cases to clusters. A particularly easy form of hierarchical agglomeration to illustrate and understand is known as *complete linkage*. In this form of agglomeration, the two most similar cases

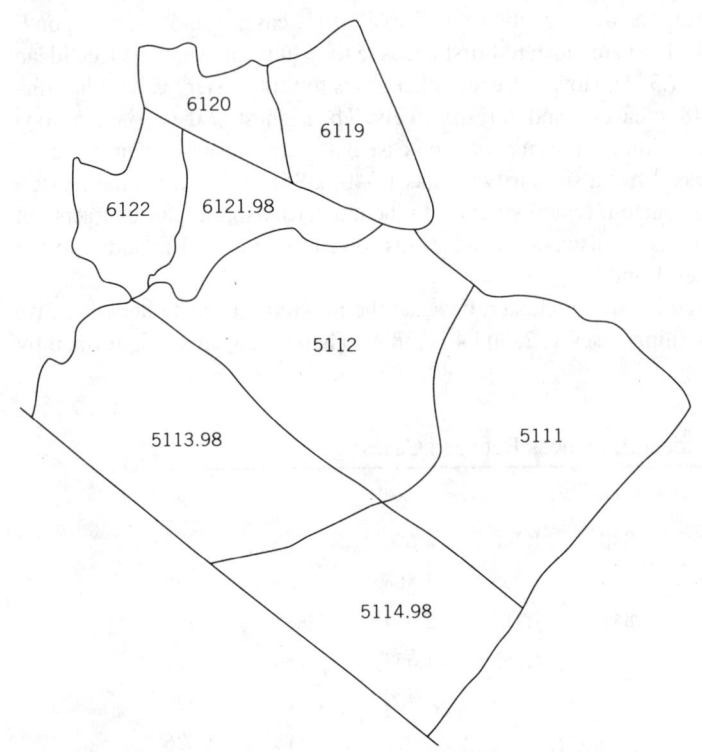

FIGURE 10.4 Palo Alto, California, Census Tracts Used for the Cluster Analysis Example

TABLE 10.10 Data Values for Palo Alto Cluster Analysis Example

Case	Census Tract	Percentage Owner Occupied	Persons Per Family	Proportion of Population in Nonfamily Households
1	5111	.885	2.88	.129
2	5112	.767	2.91	.179
3	5113.98	.294	2.55	.483
4	5114.98	.732	2.82	.201
5	6119	.584	4.09	.071
6	6120	.574	4.41	.080
7	6121.98	.151	3.16	.303
8	6122	.567	2.78	.303

are clustered first. Next, either (a) two other cases are clustered together or (b) a case is added to the existing cluster. Which of the two options is followed depends on which has the highest measure of similarity. The similarity of a case to a cluster is measured by the minimum level of similarity between that case and individual members of the cluster. The candidate case to possibly be added to the preexisting cluster is the case that has the highest similarity with the cluster.

To illustrate, using the distance matrix in Table 10.11, cases 1 and 2 (corresponding to tracts 5111 and 5112) are clustered first because they have the smallest Euclidean distance between them (.334). How close are other cases to this cluster? Case 3 has distances of .596 and .548 to cases 1 and 2, respectively. The highest of these (i.e., .596) is used as a measure of minimal similarity between case 3 and the cluster containing cases 1 and 2. Likewise, case 4 has a similarity of max (.340, .385) = .385 with the cluster. No other tract has a distance measure so close to the cluster. Also, no pair of tracts not already in the solution has a distance measure this low. Hence case 4 is added to the cluster containing cases 1 and 2.

Next, case 8 is added to the cluster, because the maximal distance between case 8 and the cluster containing cases 1, 2, and 4 is .3849. Then a new cluster is formed by

TABLE 10.11 Euclidean Distances Between Cases

	2	3	4	5	6	7	8
1	.334	.596	.340	1.296	1.600	.809	.385
2		.548	.385	1.265	1.568	.728	.380
3			.841	1.713	2.004	.885	.721
4				1.353	1.658	.750	.371
5					.409	1.315	1.626
6						1.315	1.626
7							.727

TABLE 10.12 Correlation Between Cases

	2	3	4	5	6	7	8
1	.9980	.9411	.9971	.9885	.9864	.9510	.9853
2		.9605	.9999	.9961	.9948	.9687	.9941
3			.9642	.9814	.9838	.9995	.9851
4				.9972	.9961	.9719	.9954
5					.9999	.9868	.9998
6						.9889	1.0000
7							.9899

combining cases 5 and 6; the distance between them is .409. Figure 10.5 shows the completed "tree." Cases 5 and 6 form their own cluster, and the remaining tracts form another cluster. The latter cluster is comprised of 4 tracts that form a fairly tight subcluster (cases 1, 2, 4, and 8) and 2 others (cases 3 and 7) that are more isolated). Returning to the original data in Table 10.10, we see that cases 5 and 6 represent tracts that have a high number of persons per family and few nonfamily households. We

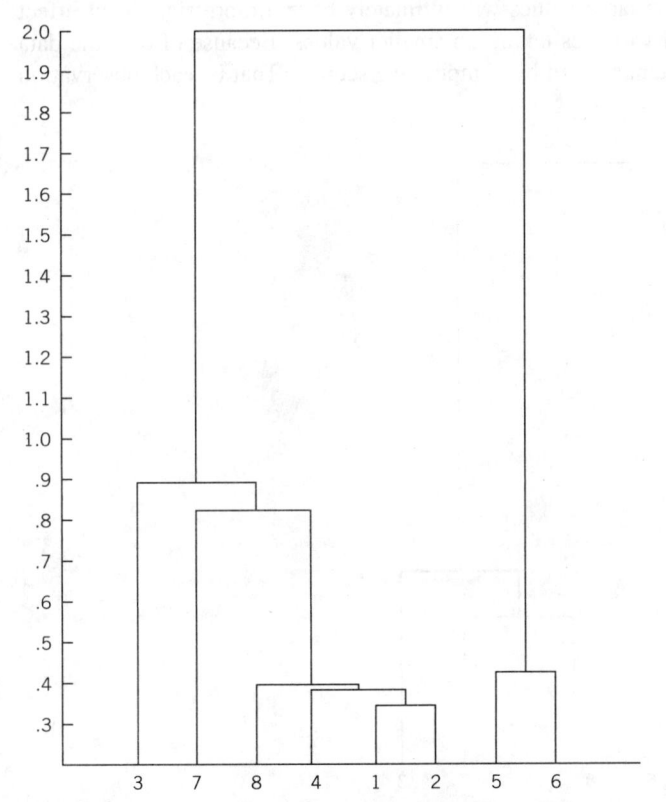

FIGURE 10.5 Complete Linkage Clusters Using the Distance Measure

might label this a "families with children" category. Cases 1, 2, 4, and 8 are characterized by a high degree of homeownership and relatively small family sizes—"small-family homeowners." Case 3 is unique because of its high percentage of nonfamily households, while case 7 is unique because of the high proportion of renters.

Figure 10.6 represents the outcome of a similar complete linkage analysis of the data, this time using the entries of the correlation matrix as measures of similarity. Note the differences between this figure and the previous one. The particular similarity measure that is chosen can influence the outcome. In Figure 10.6, two clusters again emerge from the analysis. One cluster consists of cases 1, 2, and 4—tracts with high percentages of home ownership, small percentages of nonfamily households, and small families. In the other cluster, cases 3 and 7 again are somewhat peripheral to the principal members of the cluster (cases 5, 6, and 8), the latter being characterized by medium levels of home ownership.

We wish to emphasize here that "complete linkage" is merely one of many alternative sets of rules for aggregating clusters into cases. Ward's (1963) method is one of the most commonly used hierarchical cluster methods (though its detailed description is beyond the scope of this book). The objective of Ward's method at each stage is to merge those two observations or clusters that yield the smallest increase in the within-cluster sum of squares. (See, e.g., Anderberg, 1973.) Each set of clustering methods has its own advantages and drawbacks. Anderberg, Aldenderfer and Blashfield, and others discuss these alternatives in detail.

Distance measures are affected by the size of the values associated with the variables. Variables taking on large values will ultimately have a more significant effect on the solution than will variables taking on smaller values. Because of this, the data are almost always first standardized by computing z scores. (That is, each observation

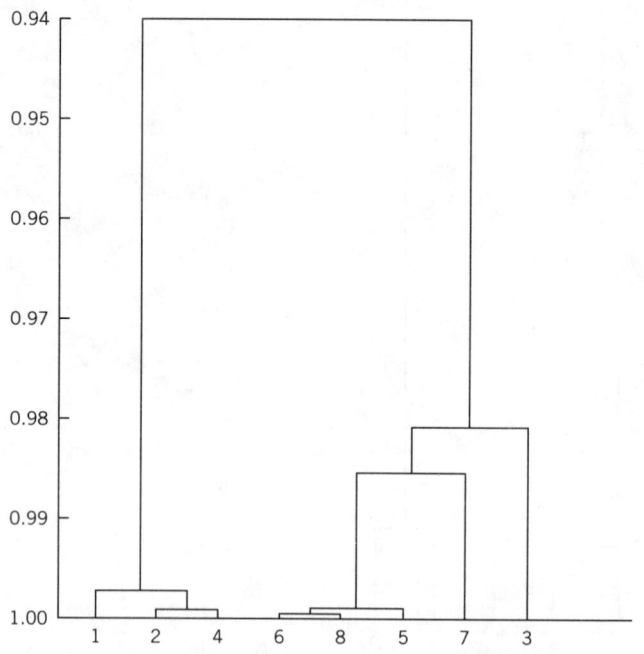

FIGURE 10.6 Complete Linkage Clusters Using the Correlation Measure

TABLE 10.13   Distance Between Cases and Cluster Centroids: First Iteration

	{1, 2, 4}	{3, 5, 6, 7, 8}
1	.010	.486
2	.002	.354
3	**.452**	.794
4	.008	.424
5	1.542	.532
6	2.429	1.071
7	.516	.140
8	**.078**	.403

has its mean subtracted from it, and the result is divided by the standard deviation of the variable; see our discussion of the calculation of z scores in Section 10.3.)

Iterative Clustering

Iterative methods for clustering start with an initial partition of the data into a prespecified number of clusters. There are no definitive guidelines with respect to how the data should be initially divided into the prespecified number of clusters. However, it is desirable to start with a "good" initial partition. Such a partition could be arrived at either by inspection of the data or by a hierarchical agglomerative method. This approach is intuitively more likely to lead to a "good" solution than is a totally arbitrary initial assignment of cases to clusters, since the latter would clearly lead to more iterations, and the likelihood of getting "stuck" with a solution having less well defined clusters.

After choosing an initial partition the centroids of each cluster are calculated. The centroid for a cluster is defined simply as the set of means (one mean for each variable) for the cases in the cluster. The mean for each variable equals the average value of that variable for all cases in the cluster.

Next, all cases are assigned to their nearest centroid, where nearness is measured, for example, by Euclidean distance. If any changes in assignment have occurred, new centroids are recomputed, and cases are assigned to the new centroid that they are closest to. This iterative procedure continues until no reassignments are made.

If we take as a starting partition the two-cluster solution {1, 2, 4}, {3, 5, 6, 7, 8} found using the correlation measure, the centroids of the respective clusters are {.795, 2.87, .170} and {.434, 3.40, .248}, where the three ordered elements refer to the mean (a) proportion of householders that are homeowners, (b) average family size, and (c) fraction of population in nonfamily households. Table 10.13 shows the Euclidean distances from each case to each cluster centroid. Cases 3 and 8 are closer to cluster {1, 2, 4} than they are to their present cluster, so they are reassigned. The recomputed centroids of clusters {1, 2, 3, 4, 8} and {5, 6, 7} are {.649, 2.79, .259} and {.436, 3.89, .151}, respectively. Distances from each case to these centroids are shown in Table 10.14. This time, case 7 is reassigned, and the new clusters become {1, 2, 3, 4, 7, 8} and {5, 6}. The reader may wish to verify that no reassignment occurs after this itera-

TABLE 10.14 Distance Between Cases and Cluster Centroids: Second Iteration

	{1, 2, 3, 4, 8}	{5, 6, 7}
1	.081	1.22
2	.035	1.06
3	.233	1.92
4	.011	1.23
5	1.73	.070
6	2.67	.298
7	**.389**	.632
8	.009	1.265

tion. Note also that this solution is the same as the two-cluster solution we found previously using complete linkage and the distance measure of similarity.

10.5 LIFE-STYLE CLUSTERING FOR MARKET SEGMENTATION AND TARGETING

One of the earliest and most successful applications of demographic and psychographic concepts to market segmentation and targeting is the PRIZM system developed by the Claritas Corporation. Their 40–life-style–cluster classification scheme has received considerable exposure and forms the basis for subscribers to the Claritas data base to carry out a large variety of specific marketing and marketing-related studies. Table 10.15 shows the catchy, descriptive nicknames assigned to each of the 40 clusters as well as the 12 broader social groups to which the clusters can themselves be assigned.

An important observation to make at the outset in examining how such a system is used in practice is to note that these PRIZM clusters are not really groupings of individual consumers sharing the same characteristics. They are actually classifications of the entire set of households found in any geographical unit according to the predominant characteristics found in that area. Implicit in regionalizing the nation into zones of 40 different types is that variations *within* the units are less important than the variations among units belonging to different types. (Recall our earlier suggestion of how diversity measures can be useful in evaluating geographic life-style clusters.)

As Table 10.15 shows, Claritas provides users of its system with classifications of neighborhoods into the 40 clusters at three different levels of geography. What are called *block clusters* are classifications of census block groups and enumeration districts; these units have an average of 340 households. For the version of the model based on the 1980 census, there were some 254,000 block clusters classified. *Tract clusters* are built up from census tracts in metropolitan areas and from minor civil divisions elsewhere. There were approximately 68,000 of these in the version of the PRIZM system we are describing, averaging 1,270 households. Finally, some 37,000 *ZIP clusters* are based on 5-digit ZIP codes, with an average of 2,320 households in each. As the final three columns of Table 10.15 show, the overall percentage of households falling into any of the 40 neighborhood life-style clusters or the 12 social groups depends on which of the three levels of geography is used.

Although the exact variables and procedures employed to develop the Claritas clusters are proprietary, some of the demographic and other characteristics that distinguish the 40 different types of areas are (1) density, (2) type of area (city, suburban, town, farming area), (3) whether a dominant ethnic group or mixture of ethnicities is represented, (4) typical structure of families, (5) relative importance of adult age groups, (6) average level of education, (7) most common type of employment, (8) predominant housing type, and (9) overall social rank (based on a weighted composite of socioeconomic variables like the SES index derived from a factor analysis).

The listing of 40 clusters in Table 10.15 is sorted by the overall socioeconomic status of the areas, from "Blue Blood Estates" (the wealthiest cluster) to "Public Assistance" (the poorest). The letter in the social group code indicates the type of area (S for suburban, U for urban, T for town, R for rural). Densities correspond fairly closely to these. Twenty-eight of the clusters have a dominant ethnic group. The 12 clusters with a mixture of groups range from the relatively high status "Two More Rungs," "Urban Gold Coast," and "Bohemian Mix," to the low socioeconomic-status "Share Croppers," "Hard Scrabble," and "Heavy Industry" clusters.

Family type, age, education, employment, and housing type further discriminate between the clusters included in each of the 12 social groups. For example, Cluster 12, "Towns and Gowns," is associated with university areas. Single people and couples with no children predominate; the adult age groups 18 to 24 and 25 to 34 are most common; residents tend to have a college education or a graduate degree; employment is white collar and single-unit housing is typical. By contrast, Cluster 17, "New Homesteaders" (which is ranked just above "Towns and Gowns" according to overall socioeconomic status) encompasses family-oriented areas with similar-aged adults but with less overall education and with a mix of blue and white collar employment.

Figure 10.7 shows a map of Midtown Manhattan with residential areas classified according to the block cluster level of geographic detail. In all, eight different cluster types are represented, although "Urban Gold Coasters" predominate, with more than half the total households falling into areas so designated.

Case Study: The Geography of *Grit Magazine* Subscription

To see in a bit more detail how such a cluster analysis of areas can be used for marketing purposes, we move from the urban landscape of Midtown Manhattan to consider the rural landscape of subscribers to *Grit Magazine*.

The Market Segmentation Phase of Analysis

Figure 10.8 shows a map of U.S. 5-digit ZIP code areas broken down into 5 quintiles based on the frequency of subscribers to *Grit* found within each. The correspondence between *penetration rates* (the percentage of persons who subscribe) for this conservatively themed magazine and the ZIP cluster designations is the first step toward developing a hypothetical target marketing scheme. This step is an application of *market segmentation analysis* (which can also be done somewhat more precisely by survey techniques). The idea is to figure out what characteristics of people distinguish likely adopters of the product (in this case, subscribers to *Grit*) from those who are most likely resistant to appeals to purchase it.

TABLE 10.15 The PRIZM Neighborhood Life Style Cluster System

The 12 Social Groups		The 40 Lifestyle Clusters		Percent of U.S. Households		
Code	Descriptive Title	Number	Nickname	Block Clusters	Tract Clusters	ZIP Clusters
S1	Educated, Affluent Executives & Professionals in Elite Metro Suburbs	28	Blue Blood Estates	1.12	0.84	0.64
		8	Money & Brains	0.94	0.99	1.14
		5	Furs & Station Wagons	3.16	2.87	2.44
S2	Pre & Post-Child Families & Singles in Upscale, White-Collar Suburbs	7	Pools & Patios	3.41	3.66	3.28
		25	Two More Rungs	0.74	0.85	1.03
		20	Young Influentials	2.85	2.62	3.02
S3	Upper-Middle, Child-Raising Families in Outlying, Owner-Occupied Suburbs	24	Young Suburbia	5.33	5.45	5.64
		30	Blue-Chip Blues	6.00	6.13	5.19
U1	Educated, White-Collar Singles & Couples in Upscale, Urban Areas	21	Urban Gold Coast	0.47	0.49	0.45
		37	Bohemian Mix	1.14	1.16	0.81
		31	Black Enterprise	0.76	0.75	1.21
		23	New Beginnings	4.30	5.12	4.77
T1	Educated, Young, Mobile Families in Exurban Satellites & Boom Towns	1	God's Country	2.70	2.37	2.97
		17	New Homesteaders	4.15	4.76	5.08
		12	Towns & Gowns	1.17	1.39	2.18
S4	Middle-Class, Post-Child Families in Aging Suburbs & Retirement Areas	27	Levittown, U.S.A.	3.05	3.29	4.51
		39	Gray Power	2.90	2.04	2.26
		2	Rank & File	1.42	1.37	1.07
T2	Mid-Scale, Child-Raising, Blue-Collar Families in Remote Suburbs & Towns	40	Blue-Collar Nursery	2.24	2.41	1.70
		16	Middle America	3.19	3.54	4.76
		29	Coalburg & Corntown	1.96	2.02	2.55

10.5 Life-Style Clustering for Market Segmentation and Targeting

U2	Mid-Scale Families, Singles & Elders in Dense, Urban Row & Hi-Rise Areas	3	New Melting Pot	0.91	0.96	1.33
		36	Old Yankee Rows	1.60	1.84	1.80
		14	Emergent Minorities	1.73	1.78	2.07
		26	Single City Blues	3.34	2.75	2.08
R1	Rural Towns & Villages Amidst Farms & Ranches Across Agrarian America	19	Shotguns & Pickups	1.87	1.84	2.53
		34	Agri-Business	2.13	2.62	4.28
		35	Grain Belt	1.27	1.23	1.43
T3	Mixed Gentry & Blue-Collar Labor in Low-Mid Rustic, Mill & Factory Towns	33	Golden Ponds	5.24	5.03	3.06
		22	Mines & Mills	2.84	2.60	1.85
		13	Norma Rae–Ville	2.32	2.75	2.95
		18	Smalltown Downtown	2.46	2.76	1.95
R2	Landowners, Migrants & Rustics in Poor Rural Towns, Farms & Uplands	10	Back-Country Folks	3.42	3.16	4.29
		38	Share Croppers	4.00	3.78	3.65
		15	Tobacco Roads	1.22	1.08	0.96
		6	Hard Scrabble	1.51	1.05	1.03
U3	Mixed, Unskilled Service & Labor in Aging, Urban Row & Hi-Rise Areas	4	Heavy Industry	2.75	2.52	1.95
		11	Downtown Dixie-Style	3.39	3.51	2.30
		9	Hispanix Mix	1.88	1.94	1.52
		32	Public Assistance	3.12	2.67	2.30

Source: Claritas, L. P. Used by permission of Kansas Geographic Bureau.

332 Chapter 10 • Demographics

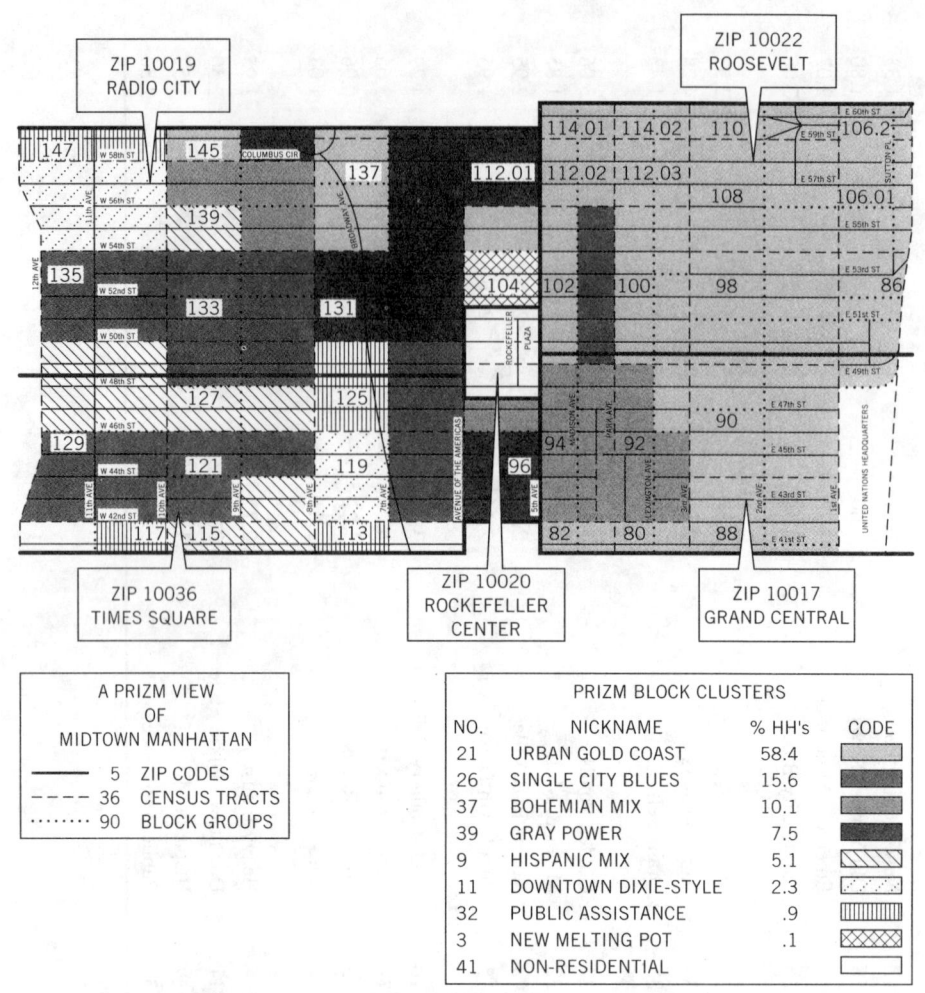

FIGURE 10.7 A PRIZM View of Midtown Manhattan

Source: Claritas, L. P. Adapted from a color original; used by permission of Kansas Geographic Bureau.

A descriptive measure useful in these kinds of analyses is the *penetration index*. It is computed as 100 times the proportion of adopters of a product in a specific area divided by the average probability that any person in the entire reference region (typically the whole nation) adopts the product. Thus a penetration rate of 200 for an area means consumers there are twice as likely to buy the product as are consumers nationally, a rate of 100 means purchases are made at national frequencies, and so forth.

One Claritas cluster type, "Shotguns and Pickups," turns out to be especially significant in describing the geographic patterns of *Grit* subscribership. "Shotguns and Pickups" households subscribe to the magazine, on average, at rates three times higher than the national average, and persons in this cluster are found in high percentages in the areas of the nation where *Grit* is most avidly read. The heartland of the *Grit* world consists of the states of Pennsylvania, Ohio, Indiana, Michigan, and Illinois, which have very high percentages of neighborhoods classified as belonging in this cluster. In

Ohio, 13.5 percent of all *Grit* subscribers live in "Shotguns and Pickups" ZIP codes, where the penetration rate is two-and-a-half-times the national average.

We have noted the importance of the "Shotguns and Pickups" cluster in describing the overall geography of *Grit* readership. The impact of a cluster is actually made up of a combination of its penetration rate and the overall prevalence of that cluster in an area's population. For Ohio, for instance, the "Middle America" ZIP cluster (which nationally has a somewhat lower penetration rate) turns out to be even more important in terms of absolute numbers of *Grit* subscribers, accounting for 16 percent of the approximately 16,000 *Grit* subscriber base in the state. On the other hand, still another life-style cluster, "Grain Belt," makes up a vastly smaller percentage of the total count, but households in these ZIP codes subscribe to the magazine at rates over four times the national average.

By contrast to these conservative, middle-class, rural and town clusters, the lowest penetration rates and the smallest overall numbers of subscribers for *Grit* in Ohio are found in two urban clusters: "Downtown Dixie Style" and "Young Influentials."

The Target Marketing Phase of Analysis

After analyzing the current distribution of penetration rates according to the life-style geography of PRIZM, the marketer can decide to target advertising campaigns on similarly classified areas of the country where subscriptions appear to be lagging. By "lag-

GritMap: Subscribers to Grit by 5-Digit Zip Codes

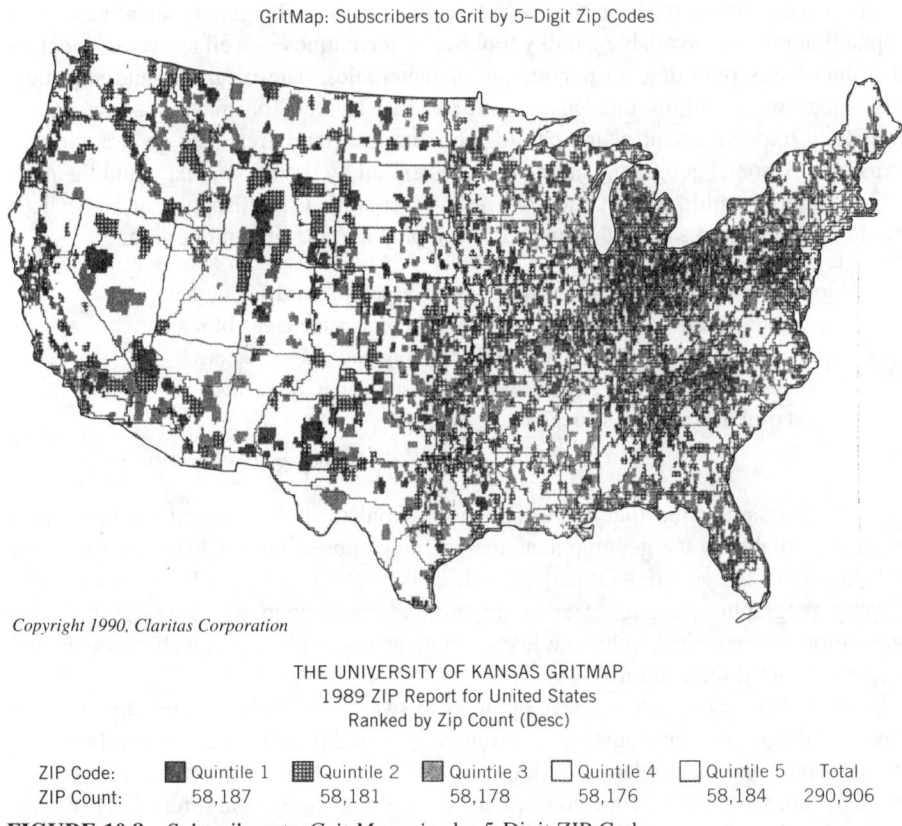

Copyright 1990, Claritas Corporation

THE UNIVERSITY OF KANSAS GRITMAP
1989 ZIP Report for United States
Ranked by Zip Count (Desc)

ZIP Code:	Quintile 1	Quintile 2	Quintile 3	Quintile 4	Quintile 5	Total
ZIP Count:	58,187	58,181	58,178	58,176	58,184	290,906

FIGURE 10.8 Subscribers to *Grit Magazine* by 5-Digit ZIP Code

Source: Kansas Geographic Bureau. Used by permission.

ging," we mean that actual sales are less than potential sales based on the characteristics of the people found there and such persons' national-scale consumption preferences.

As an example, consider the case of three local television markets in Kansas in which the "Heartland" cluster group—made up of the "Shotguns and Pickups," "Grain Belt," and "Agribusiness" clusters—accounts for a significant share of the overall household base. The Kansas City ADI (area of dominant interest) includes six counties in which *Grit* has a penetration index value over 300. (Recall that this means that households subscribe to it at rates more than three times the national norm.) In both the Topeka and Wichita ADIs, however, index values tend to be much lower, despite the similar life-style classification of the households. Within the Topeka ADI the highest county-level index is around 240; of the 68 counties making up the Wichita ADI, only two have an index of 300 or greater.

Based on this analysis of PRIZM clusters in the three ADIs, a marketing consultant's suggestion might be to focus new media advertising on the Topeka and Wichita markets to profit from some of the unrealized potential subscriber base located there. At a nationwide scale, it would be found that parts of the South are similarly markets that could be better exploited than at present with respect to sales of *Grit*.

10.6 SUMMARY

Whether for increasing business profits or for achieving a public policy goal (such as reducing segregation in a public school district), the geographical analyst of population now has available a bulky tool bag of techniques as well as access to extensive data bases reporting on population characteristics. The measures and statistical techniques presented in this chapter are useful devices for making sense of the extremely intricate social geography of the American land area. The geographic patterns of U.S. social geography are now characterized by their complexity and the nesting of patterns within other patterns. The late economist, philosopher, and poet Ken Boulding had this to say on the subject of regional complexity and nesting:

> Most regions have subregions within them. The subregions have further
> subregions. A universe, indeed, can be thought of as a hierarchy of regions.
> Some patterns and parameters run through all regions and subregions.
> Some may be confined to a single elementary subregion. Some may be
> spread over a number of subregions.
>
> —K. E. Boulding (1985, p. 19)

At the same time that more and more sophisticated statistical methods have become available to the geographical analyst, new appreciation is being given to that old standby of the geographer's art, mapping. To sort out the complexity of the overlapping geographic patterns of demographic and socioeconomic characteristics of a population, no statistical technique is ever quite as compelling as a well-designed and properly focused thematic map.

One of the most exciting developments to sweep the field of geography in recent times is the power of the computer to bring into the hands of the practitioner the quick, easy mapping and manipulation of data bases. In the next (and final substantive) chapter of the book we describe the concept and uses of a geographical information system (GIS). We show how a GIS can be useful for population analysis, and why this technology that welds together data base management and computer cartography has been receiving ever burgeoning interest in the academic, business, and public sectors.

EXERCISES

1. *Entropy diversity index.* Suppose your friend, a personal investment counselor, has designed a new system especially for managing the financial affairs of upscale, family-oriented clients. She is thinking of opening an office in either of two well-to-do suburban communities—West Snootford or Silver Spoon—on opposite sides of your metro area. They have equal population bases and both are home to small private colleges. A marketing consultant told her they are equally good locations because their Claritas ZIP cluster classifications are both "Furs & Station Wagons." This cluster is in the top social group and has a predominance of married couples with children. Your friend, however, intuitively feels that these communities have some differences in their demographics and has asked you to look into the matter. You obtain tract-level population figures and the tract cluster classifications:

West Snootford ZIP Code		
Tract 1021	2,000	Blue Chip Blues
Tract 1022.01	8,000	Towns and Gowns
Tract 1022.02	4,000	Furs and Station Wagons
Tract 1034	8,000	Furs and Station Wagons

Silver Spoon ZIP Code		
Tract 1382.04	7,000	Blue Chip Blues
Tract 1383	5,000	Furs and Station Wagons
Tract 1384	4,000	Towns and Gowns
Tract 1391.01	6,000	Furs and Station Wagons

Compute the entropy diversity index values for these two communities. What do the index values suggest about considering the communities as equally good "Furs & Station Wagon" territory? Comment on how you would advise your friend about which community to choose.

2. *Interaction diversity index.* Calculate the interaction diversity index values for the two communities in Exercise 1.

3. *Index of dissimilarity.* Which metropolitan area is more segregated?

	Rogersonburgh		Planeville		
	Whites	Blacks		Anglos	Hispanics
Central city	450,000	200,000	Central City	150,000	125,000
North and west suburbs	150,000	5,000	North and east suburbs	300,000	100,000
South and east suburbs	150,000	45,000	South and west suburbs	150,000	175,000
Total	750,000	250,000	Total	600,000	400,000

Calculate the index of dissimilarity for each and interpret the values.

4. *Gini index.* Compute the Gini index for the two metro areas in Exercise 3. Note: The calculations can become quite tedious if done by hand. If you have the expertise, write a short computer program to solve for the values. If not, consider skipping on to the next exercise!

5. *Exposure indexes.* Calculate the intragroup and intergroup exposure index values for both the majority and the minority populations of each of the two metro areas in Exercise 3. Comment on these values vis-à-vis what you found for the index of dissimilarity and the Gini index.

6. *Entropy index of segregation.* Again repeat the analysis for Exercise 3, this time computing values of entropy. Again comment on what this measure tells us relative to the previous three indexes. Which of the two metro areas really is more segregated?

7. *Factorial ecology.* Calculate (a) mobility and (b) race and ethnicity factor scores for one or more of the Cleveland census tracts for which actual data values are given in Table 10.8. Note that you must first convert the actual values into z scores using the means and standard deviations shown. You will also be using the factor score coefficients in Table 10.7. Compare the result(s) to the actual factor score(s) in Table 10.9. How good was your intuition in predicting the result?

8. *Cluster analysis.* Use the complete linkage form of hierarchical agglomeration and a Euclidean measure of distance to cluster the following census tract data:

Tract	Median Income	Median Age	Mean Family Size
1	35	35	2.2
2	50	45	2.1
3	27	34	3.6
4	29	36	3.4
5	44	48	2.3
6	23	36	3.2

9. Use the solution to Exercise 8 to form an initial partition of the data into two clusters, and determine whether the iterative method would reassign any cases to new clusters, based on the Euclidean distance of cases to cluster centroids.

10. *Market segmentation and penetration rates.* Suppose you run a marketing firm, Psycho Demons, Inc., which has been retained by a corporation that is currently test marketing a new cosmetic product in a metropolitan area with 100,000 households. Your firm has pioneered the use of a three-cluster demographic life-style segmentation of consumers. According to your patented technique, all U.S. households can be classified as either "American Dreamers," "Career Careeners," or "Life Zonkers" based on the characteristics of their householders. In the test market metro area (selected for its representativeness with respect to the nation as a whole) your method indicates that 50 percent of the households are American Dreamers, 30 percent are Career Careeners, and 20 percent are Life Zonkers. In interviews you have conducted with a random sample of 50 total consumers, you have found that of 30 adult respondents from American Dreamer households, 6 have made purchases of the new cosmetic. Among 10 Career Careener householders, 5 have bought the product, whereas among 10 Life

Zonker householders, there was only one purchase. Using the metro area as representative of the nation as a whole, compute penetration index values for each of the three life-style clusters.

11. *Tarket marketing.* Suppose the cosmetics company in Exercise 10 wants to introduce the new product next month into either the Rogersonburgh or Planeville metro market. The current household demographics in these two markets according to the patented Psycho Demons, Inc. clusters look like this:

	Rogersonburgh	Planeville
American Dreamers	50,000	150,000
Career Careeners	50,000	25,000
Life Zonkers	100,000	25,000
Total households	200,000	200,000

Using the test market results as a basis for comparison, which of the two markets would you advise the company to enter? What is the expected overall penetration index for each metro area? Is the firm making a good business decision?

> No State office has found the volume and complexity of its operations sufficiently large to justify the acquisition of a computer for vital statistics work alone.
>
> —Shryock and Siegel, (1976), p. 46

CHAPTER 11
DEMOGRAPHIC AND GEOGRAPHIC INFORMATION SYSTEMS FOR POPULATION ANALYSIS

11.1 Introduction
11.2 A Brief Historical Account of Early Population Mapping
11.3 TIGER: A Digital Cartographic Database
11.4 Geographic Information Systems
11.5 Demographic and Geographic Information Systems for Population Analysis
11.6 Some Capabilities of Demographic and Geographic Information Systems Relevant to Population Analysis
11.7 Limitations of GIS Used for Population Analysis
11.8 Prospects
11.9 Summary

11.1 INTRODUCTION

The amount of data that population analysts work with is often voluminous. It is a nontrivial matter to decide how such massive amounts of information should be stored, retrieved, displayed, and analyzed. Demographic information systems and geographic information systems that automate these tasks have facilitated our ability to perform the jobs of storage, retrieval, and display of spatially referenced data. While these emergent technologies have increased our capability enormously, there is room for improvement, and the future promises further substantial advances. In this chapter, we review some of these developments and indicate both the realized progress and the future potential that demographic and geographic information systems have in aiding and encouraging population analysis.

Before we do so, a caveat is in order. The chapter's opening quote from Shryock and Siegel, while certainly amusing in retrospect, indicates the pace of change in the application of computer technology to demographic applications. It is likely that some of what we cover in this chapter will also seem amusing 15 years from now—if not sooner! We do however feel that it is important to try to capture both some of the underlying principles and some of the current trends in this area.

In their discussion of desktop marketing systems, Thomas and Kirchner (1991) suggest a classification into two categories: market analysis systems and geographic information systems. As the name suggests, geographic information systems focus more on locational variables; market analysis systems focus more on "demographics" as discussed in the previous chapter—that is, demographic information, socioeconomic characteristics, and behavioral variables that describe the life-styles of populations residing in specific regions. Because we wish to emphasize that many of the so-called market analysis systems may also be used for public-sector planning problems, we prefer to use the term *demographic information systems* (DIS) for these systems. Later in the chapter we distinguish between DIS and GIS in detail.

But first we review some methods to store and display population data. We then describe the digital cartographic database (TIGER) of the U.S. Bureau of the Census as well as the role that GIS plays in making the database useful. Following a brief outline of the essential features of geographic information systems, we describe both the characteristics of GIS and DIS as used in population analysis, and the differences between the two. Next we give examples of basic geographic operations that these systems perform, and we illustrate simple algorithms that are often used for these tasks. We conclude the chapter with (1) a discussion of limitations of these systems and (2) a look to the future, which holds the promise for the integration of many of the types of analysis described in this book with GIS and DIS. Also relevant to the material in this chapter is Appendix B, which describes the geographical entities or building blocks used by the U.S. Bureau of the Census for tabulating and reporting population data, as well as the hierarchical relationships that exist among them.

11.2 A BRIEF HISTORICAL ACCOUNT OF EARLY POPULATION MAPPING

Perhaps surprisingly, the history of the thematic mapping of population data dates back no further than the early to mid-19th century. One of the earliest known population maps is an 1828 map of population densities in Prussia, using the technique of graduated shading (Palsky, 1991). Robinson (1955, 1982) notes that during the period 1835–55, the majority of techniques now used to represent population data came into existence. Robinson pays particular attention to the 1837 maps of Henry Drury Harness, which were produced to accompany the Second Report of the Irish Railway Commissioners. These maps contained a number of cartographic innovations. The maps employ flow lines to represent the magnitudes of passenger flow (with the width of the line related to the size of the flow), graduated circles to represent the population of Irish cities, and four classes of shading to represent different levels of population density. Figures 11.1 and 11.2 show two of the maps.

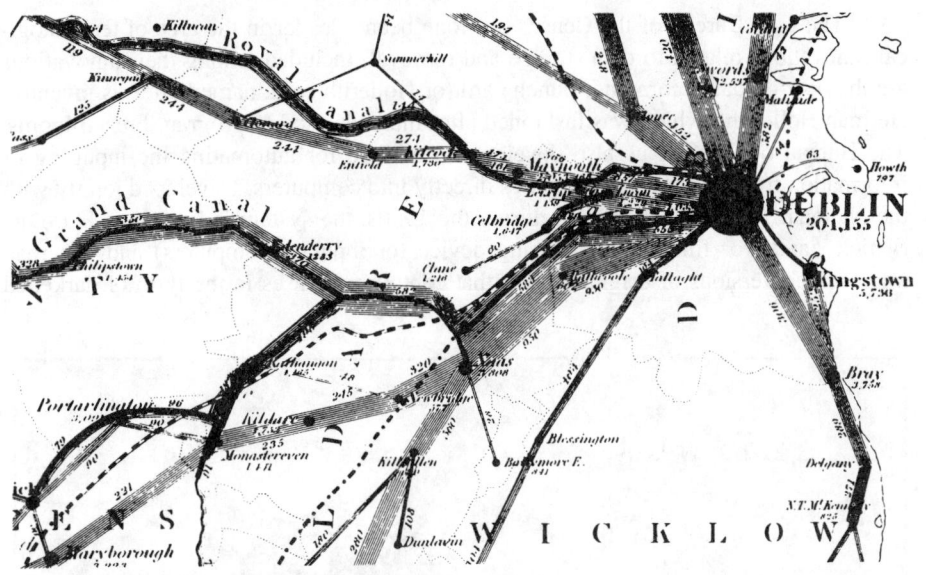

FIGURE 11.1 Harness's Passenger Conveyance Map
Source: Robinson (1982). Reprinted by permission of University of Chicago Press.

Hargreaves (1961) notes that the first use of dot maps to represent population is apparently a map of New Zealand made in 1863, where each symbol represents 100 Maoris. Hargreaves makes the interesting point that the development of what are now regarded as more advanced techniques (choropleth mapping, with regions shaded in relation to their values; flow maps; and proportional circle maps) actually preceded the development of the dot map, which is now considered an effective, yet simple way to portray the population. However, according to Monmonier and Schnell (as noted in Chapter 2) and Robinson (1982), the first dot map of population was produced in France in 1830 (Figure 11.3). The gap of over three decades between these two early population dot maps illustrates the fact that this innovation did not diffuse rapidly.

Cartograms, where regions are reproportioned to reflect not their geographic size, but rather the relative magnitude of the data being portrayed, have also been used to represent population distribution. Perhaps the earliest use of a cartogram to represent population in the United States is the map of Raisz (1934) (Figure 11.4). The early history of cartograms is somewhat more difficult to trace, but early work evidently appeared in France in the 1870s and 1880s, and in Germany in 1902 (Hunter and Young, 1968; Palsky, 1991).

11.3 TIGER: A DIGITAL CARTOGRAPHIC DATABASE

Because population data are now collected and displayed for a large number of very small geographic units, the effort involved in producing both base maps and thematic population maps would be overwhelming if it were not automated. Indeed, for the 1980 Census, the U.S. Bureau of the Census employed literally thousands of employees to work on the task of providing maps for census enumerators. For the 1990 Census, the process of producing maps was automated for the first time, and the work was carried out by a few dozen people.

342 Chapter 11 • Demographic and Geographic Information Systems

The U.S. Bureau of the Census has long been a leader in the area of technological innovations related to data storage and retrieval. Included among their innovations are the now-outdated computer punch card (or Hollerith card, named after its inventor, Herman Hollerith), which was fashioned after methods used to program French looms. The Bureau of the Census also developed a method for automating the inputting of information from census questionnaires directly into computers. Developed jointly with the National Bureau of Standards during the 1950s, the system is known as FOSDIC (which stands for film optical sensing device for input to computers) and can read microfilmed versions of original forms that contain responses in the form of darkened

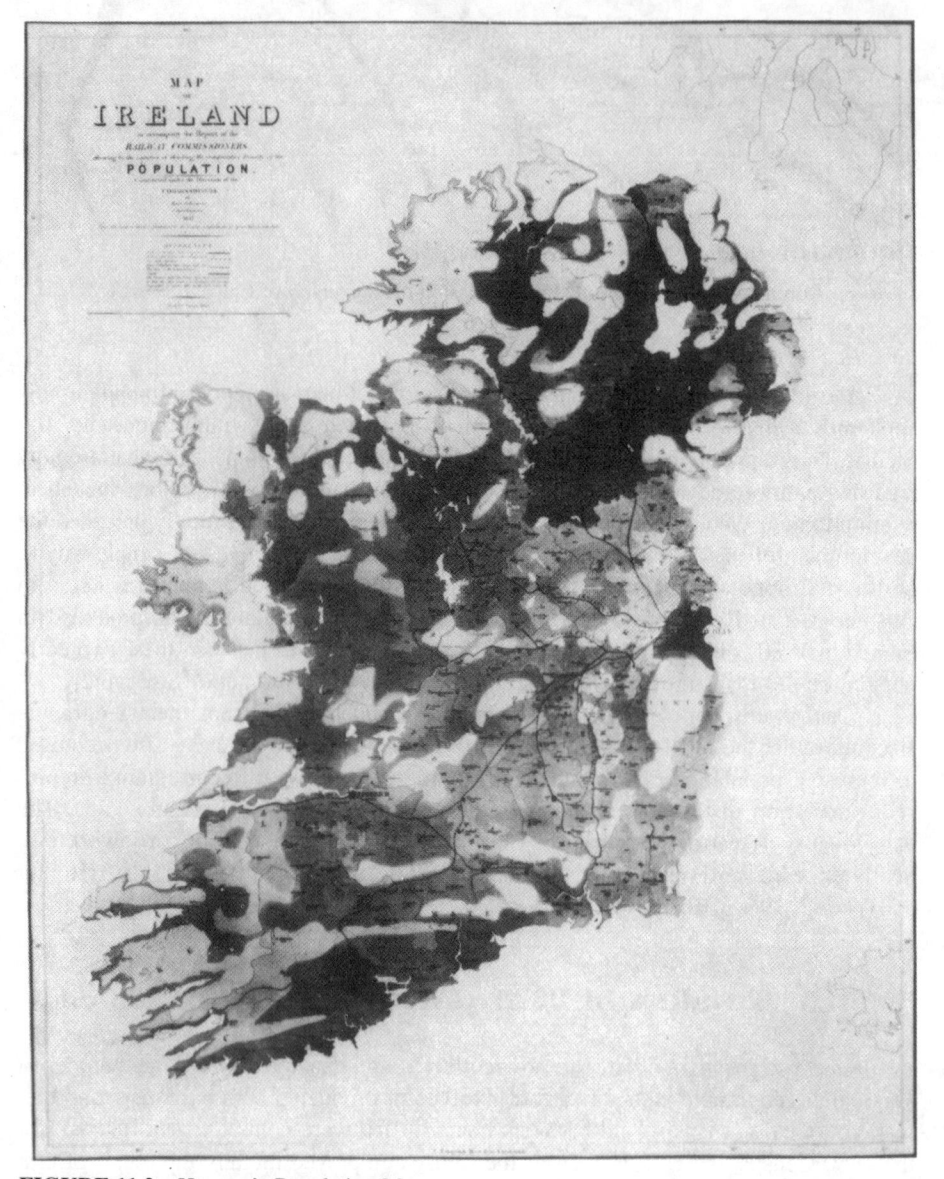

FIGURE 11.2 Harness's Population Map

Source: Robinson (1982). Reprinted by permission of University of Chicago Press.

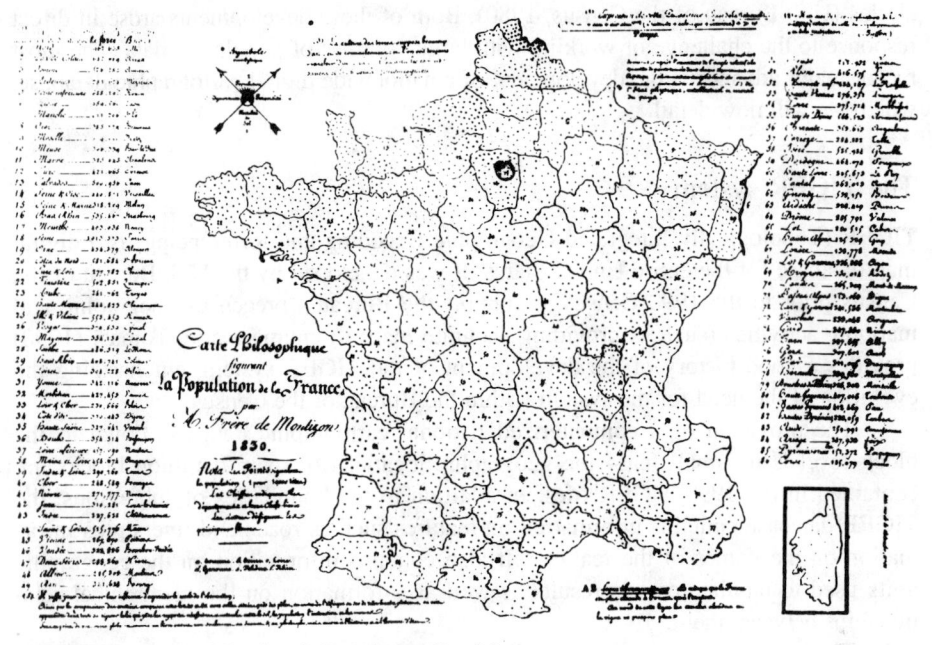

FIGURE 11.3 De Montizon's Population Dot Map

Source: Robinson (1982). Reprinted by permission of Bibliotheque Nationale.

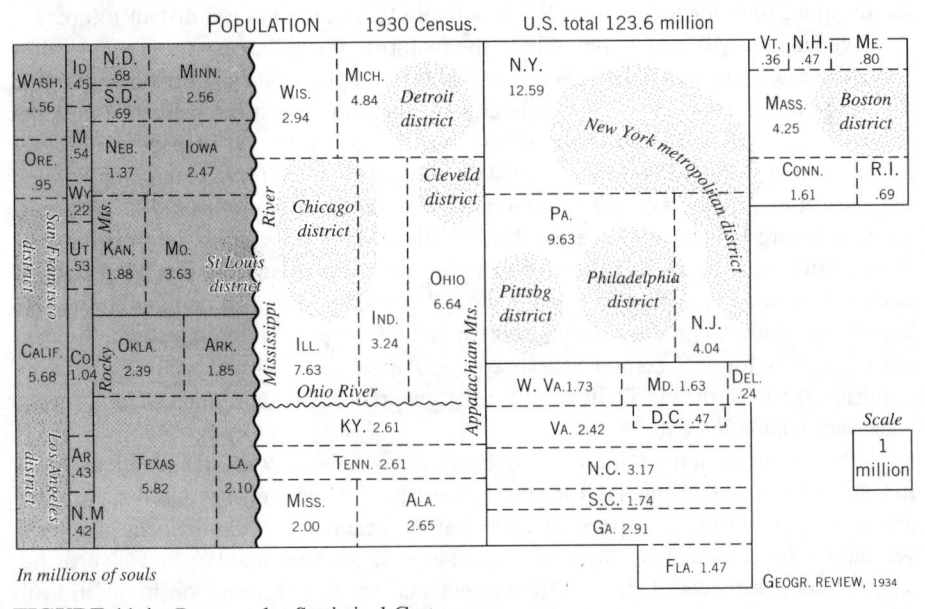

FIGURE 11.4 Rectangular Statistical Cartograms

Source: Raisz (1934). Courtesy of the American Geographical Society.

circles (U.S. Bureau of the Census, 1980). Both of these developments arose in direct response to the challenge of working with large volumes of population data. The most recent major advance is the development of a nationwide digital cartographic database, which we will now detail.

The TIGER system

TIGER (Topologically Integrated Geographic Encoding and Referencing) is a digital map database for the entire United States. It was developed by the U.S. Bureau of the Census prior to the 1990 Census to support the Bureau's precensus geographic and mapping activities (such as providing accurate maps to enumerators). Tomasi (1990) provides a good history of the development of the TIGER system, emphasizing the evolution of the need for the system within the Bureau of the Census.

The provision of a computerized national cartographic database down to the block level is no small feat—the geographic locations of over 7 million blocks are contained in the 1990 TIGER files. In addition to the boundaries of subregions, the TIGER database contains information on features such as roads and rivers plus information on the names of the features. It also contains information on the geographic units used to tabulate census results as well as information on the hierarchical relationships between them.

Through the process of "geocoding" the addresses of housing units, the system automatically assigns households and housing units to the proper block, census tract, county, and so on, allowing enormous improvements in the efficiency of the tabulation process. Address geocoding works well primarily in the nation's urban areas, where the traditional form of address (house number and street name) prevails. Though geocoding is more difficult in rural areas because of the lack of appropriate address types, many parts of the country are now assigning house numbers and street names to rural areas. One reason for doing so is to improve the emergency 911 system by lowering the response time for incoming calls. It is likely that in the not too distant future, address geocoding will also be possible in the majority of rural areas.

We must note that TIGER is *not* a GIS. It is simply a database containing digital cartographic information (e.g., coordinates of intersections, street network topology, feature names). The Bureau of the Census itself recognizes that perhaps the biggest misperception of TIGER is that it "is just one big map—load it into your computer and see a street map pop up on your monitor" (U.S. Bureau of the Census, 1990b). Additional software is needed to use the TIGER files. Many geographic information systems permit input of TIGER files for mapping purposes. Also note that TIGER itself does not contain population or housing data. Population data must be input to the GIS before, for example, producing a choropleth map of the population of a county by census tract. This form of computer-assisted population mapping and analysis therefore requires (a) a GIS, (b) the TIGER file for the geographic region of interest, and (c) the necessary population data.

The combination of these three elements permits a wide range of analyses. Indeed, applications of the TIGER system have progressed well beyond their original use as support for census operations. Current applications include thematic mapping, vehicle routing, political redistricting, market area delineation, and targeted advertising.

Good overviews of the TIGER system that provide information in addition to that presented here include Carbaugh and Marx (1990), Marx (1986, 1990), and U.S. Bureau of the Census (1990b).

11.4 GEOGRAPHIC INFORMATION SYSTEMS

For TIGER and other geographic databases to be useful, a GIS is a necessary prerequisite.

According to Star and Estes (1990), "[a] GIS is an information system that is designed to work with data referenced by spatial or geographic coordinates. . . . [it] is both a database system with specific capabilities for spatially referenced data, as well as a set of operations for working with the data." Star and Estes describe five essential steps in using a GIS:

1. *Data acquisition.* Though self-explanatory, this step is one of the most critical, since (a) deciding upon the appropriate data and obtaining them are often difficult and (b) the output of any analysis depends heavily upon data quality.

2. *Preprocessing.* This refers to getting the data into the format necessary for input to the GIS. Manual preprocessing may involve taking information off of maps or air photos; automated preprocessing may involve, for example, conversion of data from one format to another through the use of a computer program.

3. *Data management.* The data management functions of GIS include a focus on database structure, capabilities for creating and accessing the database, and input/output management. As Star and Estes note, these functions, if designed well, make the system much easier to use.

4. *Manipulation and analysis.* This is what users view as the "core" of a GIS—namely the "geoprocessing" capabilities such as handling queries (e.g., finding the number of customers within a census tract and computing the areas of polygons) and carrying out operations such as address geocoding.

5. *Product generation.* Geographic information systems produce output that may be obtained in the form of printed reports, color displays, or computer files. All geographic information systems contain product generation capabilities.

Geographic information systems contain all of these functions, and increasingly they include automated methods of data acquisition and preprocessing. Next we elaborate on the role of demographic and geographic information systems in population analysis.

11.5 DEMOGRAPHIC AND GEOGRAPHIC INFORMATION SYSTEMS FOR POPULATION ANALYSIS

A growing number of private vendors are "packaging" the three necessary elements described in Section 11.3: a digital map, demographic data, and a GIS capable of handling (storing, displaying, retrieving, and analyzing) the information. These systems are remarkable in their ability to quickly find and display information for any level of geography. What used to require hours of poring through printed reports or magnetic data tapes can now be done in minutes on a desktop personal computer. Most systems allow users to specify whether the data are to be retrieved for either a predefined or user-defined geographic area. The data may be displayed in the form of a map or a report. Demographic and geographic information systems have been applied to an incredible variety of problems requiring demographic analysis, including site location, finding markets, advertising, and improving the efficiency of vehicle routing. *American Demographics* (1992) published a review of GIS that explores the interface of GIS and population analysis; it contains many examples of these and other applications.

As indicated in the introduction to this chapter, these systems may be broadly classified into two categories. Demographic information systems, such as PCensus, CONQUEST, INFOMARK, and the Claritas system described in the previous chapter (see Table 11.1) are centered around demographic information and population databases, but also possess options that are linked to geographic databases and GIS capabilities. The systems almost always have a very good user interface. They are small enough (in terms of storage and computational requirements) to fit on personal computers. Because the user interfaces are well designed, these packaged systems are extremely easy (and fun!) to use.

Table 11.2 provides an example of the type of output that is possible using a DIS—in this case PCensus. In the example, areal interpolation methods have been used to produce demographic and socioeconomic information on the population within given radii of the corner of Cornwall Avenue and Alabama Street in Bellingham, Washington. Such summaries are highly worthwhile in evaluating alternative locations, assessing the strength of current locations, and planning ad campaigns. As just noted, additional output options allow such information to be summarized in map form.

Despite their attraction, demographic information systems have some notable limitations. Shultz and Regan (1991) note that the data necessary to make the systems operational are costly and are often proprietary. Furthermore, in some cases we cannot digitize geographic features for input into the systems, nor can we import data that have already been digitized.

The second category has at its core a GIS, resulting in a system that has a much more flexible architecture that permits the user to make full use of all of the system's GIS capabilities. As Exter (1992) notes, "Trading your market research workstation for a GIS is like putting away the kitchen knives and turning on the food processor." Examples include ARC/INFO and GISPlus. (See Table 11.3.) Mapping packages such

TABLE 11.1 Demographic Information Systems

CACI Marketing Systems 1110 N. Glebe Road Arlington, VA 22201 1-800-292-CACI	**Equifax** National Decision Systems 5375 Mira Sorrento Place, Suite 400 San Diego, CA 92121 1-800-866-6510
Claritas/NPDC 201 N. Union St. Alexandria, VA 22314 1-800-284-4868	**PCensus-USA** Tetrad Computer Applications Ltd. 3873 Airport Way Box 9754
Donnelley Marketing Information Services	Bellingham, WA 98227-9754 1-800-663-1334
Strategic Mapping Inc. 70 Seaview Ave. 5th Floor, P.O. Box 120058 Stamford, CT 06912-0058 1-800-866-2255	

11.5 Demographic and Geographic Information Systems for Population Analysis

TABLE 11.2 Typical Output From a DIS Using 1990 Census Information

	Cornwall Ave. & Alabama St. 0–1 mi		Cornwall Ave. & Alabama St. 1–2 mi		Cornwall Ave. & Alabama St. 0–2 mi	
Total Population	11,838		13,819		25,657	
Population/square mile	4,380		2,280		2,927	
Persons living in households	11,255	95%	13,253	96%	24,508	96%
Persons in group quarters	583	5%	566	4%	1,149	4%
Male	5,645	48%	6,736	49%	12,381	48%
Female	6,193	52%	7,083	51%	13,276	52%
Average age	37.1		35.0		36.0	
White	10,990	93%	12,810	93%	23,800	93%
Black	107	.90%	90	.65%	197	.77%
American Indian/Eskimo/Aleut	277	2%	353	3%	630	2%
Asian/Pacific Islander	240	2%	439	3%	679	3%
Other race	224	2%	127	.92%	351	1%
Hispanic origin (any race)	367	3%	380	3%	747	3%
Persons 25+ years	7,716		8,651		16,367	
By Educational Attainment:						
Less than complete high school	1,570	20%	1,697	20%	3,267	20%
High school graduate	2,100	27%	2,776	32%	4,876	30%
Some college/college degree	4,046	52%	4,178	48%	8,224	50%
Persons 16+ years	9,300		11,085		20,385	
By Age:						
Under 5 years	805	7%	936	7%	1,741	7%
5 to 17 years	1,975	17%	2,083	15%	4,058	16%
18 to 24 years	1,342	11%	2,149	16%	3,491	14%
25 to 34 years	2,112	18%	2,811	20%	4,923	19%
35 to 44 years	2,037	17%	2,002	14%	4,039	16%
45 to 54 years	800	7%	998	7%	1,798	7%
55 to 64 years	661	6%	970	7%	1,631	6%
65 to 69 years	379	3%	579	4%	958	4%

TABLE 11.2 (continued)

	Cornwall Ave. & Alabama St. 0–1 mi		Cornwall Ave. & Alabama St. 1–2 mi		Cornwall Ave. & Alabama St. 0–2 mi	
70 to 74 years	475	4%	355	3%	830	3%
75 years and over	1,252	11%	936	7%	2,188	9%
Mean age	37.1		35.0		36.0	
Total Households (% base)	4,936		5,784		10,720	
Family households	2,754	56%	3,351	58%	6,105	57%
Married-couple households	2,098	43%	2,700	47%	4,798	45%
With own children < 18	1,033	21%	1,192	21%	2,225	21%
Other households	656	13%	651	11%	1,307	12%
Nonfamily households	2,182	44%	2,433	42%	4,615	43%
Householder living alone	1,680	34%	1,767	31%	3,447	32%
Persons per household	2.3		2.3		2.3	
Average household income ($)	25,934		28,856		27,511	
Average family income ($)	31,723		34,995		33,519	
Married couple families ($)	34,778		38,549		36,900	
Other families ($)	21,947		20,252		21,103	
Average nonfamily household ($)	17,688		19,211		18,491	
Percent of persons below the poverty level	16.9		13.9		15.3	
Total Housing Units	5,056		6,103		11,159	

Source: PCENSUS-USA.

as Atlas Pro and MapInfo also have much to offer the population analyst. With these systems, files can be imported and exported, users can add their own data, and high-level programming languages permit users to add their own procedures for manipulating and analyzing the data. Census TIGER files can be imported directly, avoiding the higher costs of proprietary population data. Figure 11.5 gives an example of output from one of these systems.

Still, there are also barriers to the use of these more sophisticated systems:

> First, the cost of the software and all of the necessary support equipment (from high powered computers to digitizers and plotters) is prohibitively expensive for many users. Second, this software is somewhat cumbersome and difficult to learn. In summary, these software packages do not permit users to quickly and inexpensively begin geographical analyses. (Shultz and Regan, 1991, p. 25)

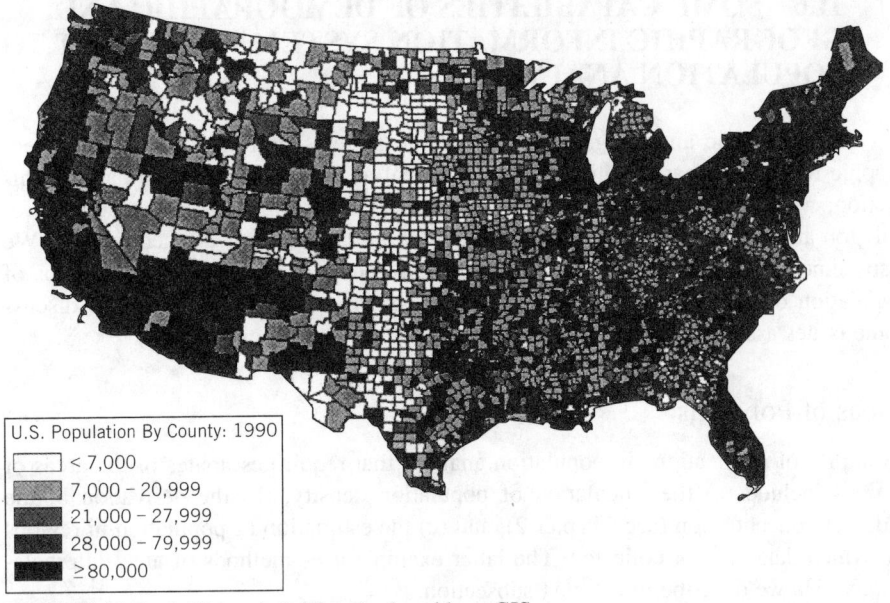

FIGURE 11.5 A Choropleth Map Produced by a GIS

Despite these barriers, the outlook is very promising for the emergence of systems capable of combining the best features of demographic and geographic information systems for population analysis and mapping. Indeed, we must note that the distinction between the two categories is already beginning to be blurred as (a) demographic information systems become more flexible in their input and output options, and include an increasing variety of geographical operations and (b) as geographic information systems become more user friendly.

In the next section, we explore some methods used by these systems to perform elementary geographic operations in applications to population analysis.

TABLE 11.3 Geographic Information Systems for Demographic Analysis

Atlas	**GISPlus**
Strategic Mapping, Inc.	Caliper Corporation
3135 Kifer Road	1172 Beacon Street
Santa Clara, Calif. 95051	Newton, Mass. 02161
Geographic Data Technology	**MapInfo Corp.**
13 Dartmouth College Highway	Hendrick Hudson Building
Lyme, N.H. 03768-9713	200 Broadway
1-800-331-7881	Troy, N.Y. 12180-9981
ARC/INFO	1-800-327-8627
Environmental Systems Research Institute	
380 New York Street	
Redlands, Calif. 92373	

11.6 SOME CAPABILITIES OF DEMOGRAPHIC AND GEOGRAPHIC INFORMATION SYSTEMS RELEVANT TO POPULATION ANALYSIS

Demographic and geographic information systems perform several basic geographic operations that simplify the geographical analysis of population data. In this section, we examine two of them: (1) elementary algorithms for the calculation of polygon areas and (2) the estimation of population data for user-defined regions. We also demonstrate how the latter may be generalized to draw more accurate maps of population density (a technique known as *dasymetric mapping*). Finally we discuss some issues associated with population mapping and geographic scale.

Areas of Polygons

Examples of applications in population analysis that require estimates of the areas of regions include (a) the calculation of population density, (b) the derivation of the index of concentration (see Chapter 2), and (c) the estimation of population in regions for which data are not collected. The latter example uses methods of areal interpolation, which we describe in the next subsection.

Geographic information systems contain algorithms for computing the area of polygons. One of the most common methods involves summing the areas of the trapezoids formed by dropping straight vertical lines from each vertex of the polygon to the horizontal axis (Figure 11.6). First, the areas in (a) are added, and then the areas in (b) are subtracted. Recalling from Chapter 2 that the area of a trapezoid equals the dis-

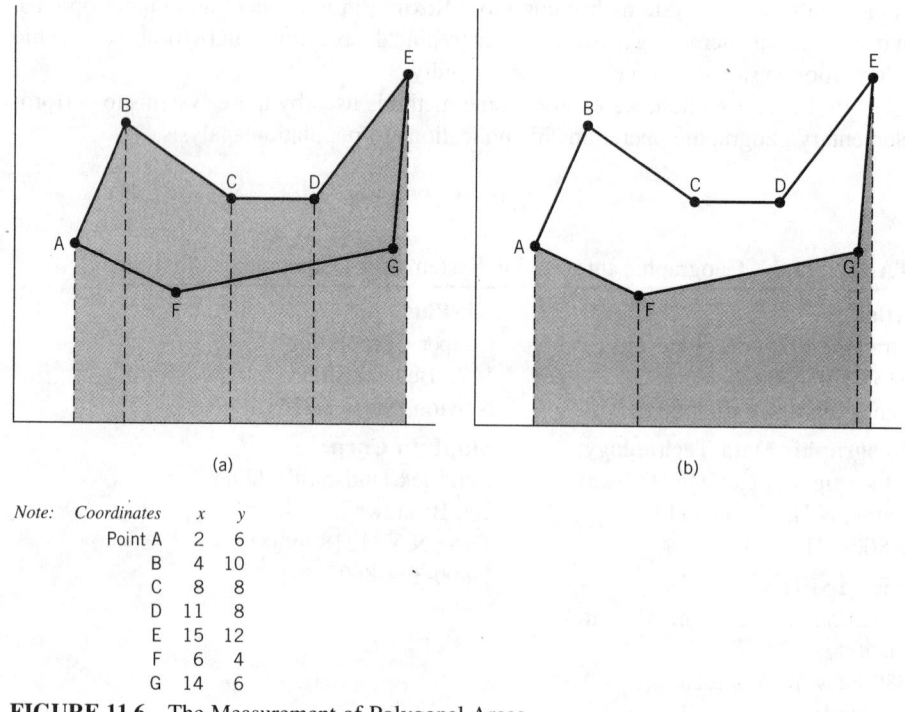

	(a)	(b)

Note:
Coordinates	x	y
Point A	2	6
B	4	10
C	8	8
D	11	8
E	15	12
F	6	4
G	14	6

FIGURE 11.6 The Measurement of Polygonal Areas

tance between the parallel sides multiplied by one-half the sum of the lengths of the two parallel sides, areas of the trapezoids in Figure 11.6 are calculated using

$$(x_{i+1} - x_i)(y_i + y_{i+1}) / 2$$

where the x and y coordinates of successive polygon vertices are used.

To find the area of the polygon in Figure 11.6, we begin by calculating the area of the trapezoid formed by dropping vertical lines from vertices A and B to the horizontal axis. The area of that trapezoid is $(4 - 2)(10 + 6) / 2 = 16$. Areas of the other trapezoids in panel (a) are found in a similar manner; the total of the shaded area in panel (a) is $16 + 36 + 24 + 40 = 116$. From this total of 116, we must subtract the area formed by the three trapezoids in panel (b). For example, the area of the trapezoid formed by dropping vertical lines from vertices A and F is $(6 - 2)(4 + 6) / 2 = 20$. The total shaded area in (b) is $20 + 40 + 9 = 69$, implying that the area of the polygon is $116 - 69 = 47$.

A general formula for the area is

$$A = \sum_{i=0}^{n-1} (x_{i+1} - x_i)(y_i + y_{i+1}) / 2$$

where the n vertices are sequentially labeled, clockwise, from 0 to $(n - 1)$, and where $y_n = y_0$ and $x_n = x_0$.

Areal Interpolation

One of the most common applications of demographic information systems is to find information for user-defined geographic territories. Often analysts need estimates for a subregion for which data are not collected. Some type of interpolation or extrapolation is called for to transform the data from a set of administrative regions to a set of "target" regions defined by the analyst. For example, a store owner may wish to know how many households lie within 2 miles of the store. Or a planner may wish to know the number of young children in a 10-block by 10-block square that is centered upon a playground. Since inevitably data will not be available for such "custom-defined" subregions, we need methods that provide estimates using what data are available. This is one common task performed by GIS, and, as just noted, virtually all demographic information systems (where GIS, a digital map, and demographic data are combined) provide this capability. The methods for producing such estimates are called methods of "areal interpolation." We now examine some of the simpler approaches.

Suppose that we wish to estimate the total population of school districts A, B, and C in Figure 11.7. The population data, however, instead of being available by school district, are tabulated by census tract. (Census tracts are indicated by numbers in the figure.) Perhaps the most straightforward way to obtain population estimates by school district is to assume, first, that population is distributed uniformly within each of the census tracts. For example, the population of district A may be determined by forming the product of (a) the proportion of area in a given census tract occupied by district A and (b) the population of that census tract. These products are formed for each census tract that intersects with district A, and are summed to derive the population estimate. For example, consider Table 11.4, which contains data to accompany the figure. An estimate of the population of district A is

$$1{,}000 \left(\frac{.4}{2}\right) + 2{,}000 \left(\frac{.6}{3}\right) + 1{,}500 \left(\frac{.2}{2}\right) + 2{,}500 \left(\frac{.1}{3}\right) = 833$$

(rounding the answer to the nearest whole number).

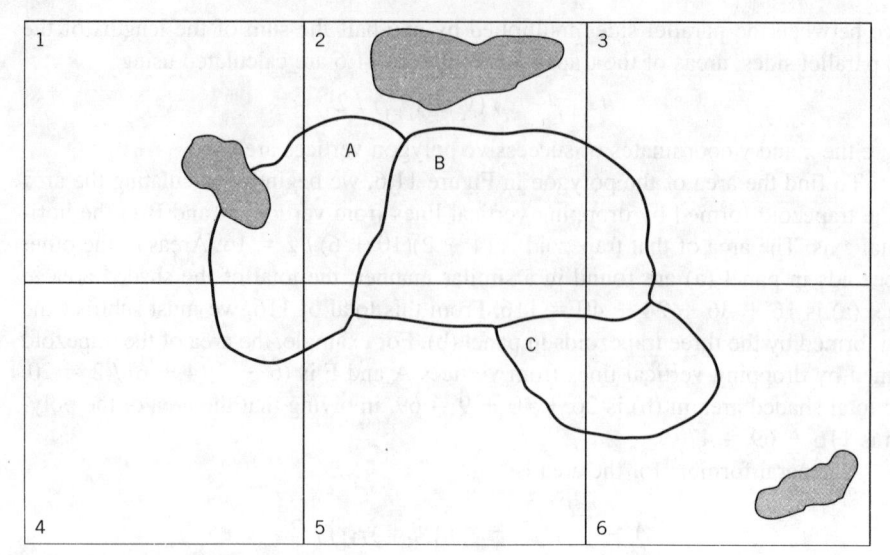

FIGURE 11.7 Hypothetical School District System

This elementary method of areal interpolation has been discussed by Markoff and Shapiro (1973) and by Goodchild and Lam (1980).

Dasymetric Mapping

Flowerdew, Green, and Kehris (1990) discuss more sophisticated approaches that avoid the often questionable assumption that population is distributed uniformly within the regions for which data are available. The simplest form of *dasymetric mapping* (Wright, 1936), for example, allows us to incorporate the knowledge that population must equal zero in water bodies, parks (assuming that we are only estimating the nonvagrant population!), and other nonresidential land uses. Though the term *dasymetric mapping* may appear somewhat daunting, the technique itself is not complicated. The nonresidential areas to be eliminated from consideration are simply subtracted from the areal measurements, and the areal weighting method is applied as before. For example, consider the additional data in Table 11.5. The population of district A may now be estimated as

$$1,000\left(\frac{0.4 - 0.1}{2.0 - 0.2}\right) + 2,000\left(\frac{0.6}{3.0 - 0.6}\right) + 1,500\left(\frac{.2}{2}\right) + 2,500\left(\frac{.1}{3}\right) = 900$$

This new population estimate for district A is higher than the previous one. This is because we have now accounted for the large lake in census tract 2. Our initial estimate assumed that the population was evenly distributed over the 3 sq mi constituting the tract. With our new estimate, we have assigned a higher percentage of the tract's population to district A, thereby recognizing that the population resides in a more limited region within the tract.

An extension of this idea allows the population analyst to produce more detailed maps of population density. Wright (1936), for example, describes how we can employ general dasymetric mapping techniques to produce detailed population density maps. First, the study region is divided into two subregions—a sparsely populated region and a more densely populated one. Suppose that a region has 100 persons per square mile.

11.6 Some Capabilities of Demographic and Geographic Information Systems

TABLE 11.4 Hypothetical Data for Areal Interpolation Problem Depicted in Figure 11.7

Subregion	Population	Area (sq mi)
1	1,000	2
2	2,000	3
3	3,000	3
4	1,500	2
5	2,500	3
6	3,000	3
A	?	
B	?	
C	?	

Amount of areal overlap (sq mi) between	
A and 1:	0.4
A and 2:	0.6
A and 4:	0.2
A and 5:	0.1
B and 2:	1.2
B and 3:	0.2
B and 5:	0.2
B and 6:	0.1
C and 5:	0.4
C and 6:	0.3

Seven-tenths of the region is sparsely populated, and the remaining three-tenths is densely populated. If we estimate that the sparsely populated region has 20 persons per square mile, we may solve for the density of the densely populated subregion since we know that the population density for the entire region is 100/sq mi. That is,

(20 persons/sq mi × .7) + (? persons/sq mi × .3) = 100 persons/sq mi

Solving for the density of the more densely populated subregion yields

$$D_d = [D_t - D_s (A_s)] / A_d \qquad (11.1)$$

TABLE 11.5 Additional Data for the Areal Interpolation Problem

Subregion	Lake Area
1	0.2 sq mi
2	0.6 sq mi
6	0.4 sq mi

Note: Area of overlap between lake in subregion 1 and district A: 0.1 sq mi.

where D refers to density, and A refers to area of the subregion (expressed as a fraction of the total area of the region). The s, d, and t subscripts refer to the sparse and dense subregions and the total region, respectively. In our example, $D_d = 267$ persons/sq mi. The densely populated region may then be further subdivided and the process repeated, if desired. Clearly the key to accurate dasymetric mapping as described here is accurate information on the density of the sparsely populated subregions. The needed information might be obtained from aerial photos or from topographic maps depicting residential locations.

It may seem fairly unusual and arbitrary to begin estimation of the population density of the densely settled subregion by simply assuming a population density for the sparsely settled subregion. However, the estimates of the population density in the more densely settled subregion are relatively insensitive to errors in the initial assumption made for the density of the sparse subregion. To see this, suppose that the figure of 20 persons per square mile was 100 percent too high; that is, the actual figure used should have been 10 persons/sq mi. Using $D_s = 10$ persons/sq mi in Equation 11.1 yields $D_d = 310$ persons/sq mi, which implies that the original estimate of $D_d = 267$ persons/sq mi was just 14 percent too low. Similarly, if the actual density for the sparse region was $D_s = 40$ persons/sq mi (implying that our value of $D_s = 20$ persons/sq mi was 50 percent too low), then Equation 11.1 yields $D_d = 240$ persons/sq mi, implying that our original estimate for D_d was 11 percent too high.

A good application of this form of mapping is to the mapping of population density on the fringe of metropolitan areas, where there is often a large amount of vacant land. As Figure 11.8 demonstrates, the technique produces a map that conveys more of a visual impression of the geographical distribution of population than does, for example, a dot map, where the dots are more arbitrarily placed.

Choropleth Maps and Geographic Scale

A common method for depicting a population characteristic across a set of regions for which census data are available is the choropleth map, on which regions are shaded according to the category into which the variable of interest falls. As Bracken (1992) notes, however, such displays may be highly misleading due to the artificiality of the regional boundaries of the administrative units for which data are collected:

> Such mapping is now very readily produced by many PC-based GIS and this availability makes all the more urgent a careful assessment of the "quality of the message" that is provided. Such a display can be highly misleading, mainly because the "fixed" area base has no relationship to the distribution of the phenomenon in question; the apparent distributional properties of the topic in question will be distorted by the imposed boundaries; and all areas must be populated, which is clearly a distortion at any scale.

One solution Bracken proposed is to produce, for example, population counts for a square grid having a fine level of resolution. The hypothetical advantage of a square grid is that the geometry of the grid is essentially independent of the underlying distribution. Yet simply changing the shape of the region will not always produce accurate displays—square grids are just as artificial, if not more so, than administrative regions. The real key is to produce accurate data at a fine enough resolution that a reliable depiction at the scale desired emerges. For an accurate portrayal at a given scale, data should be preferably collected (or at least interpolated) and displayed at a smaller scale, so that the depiction in subregions of concern is not affected by possible heterogeneity

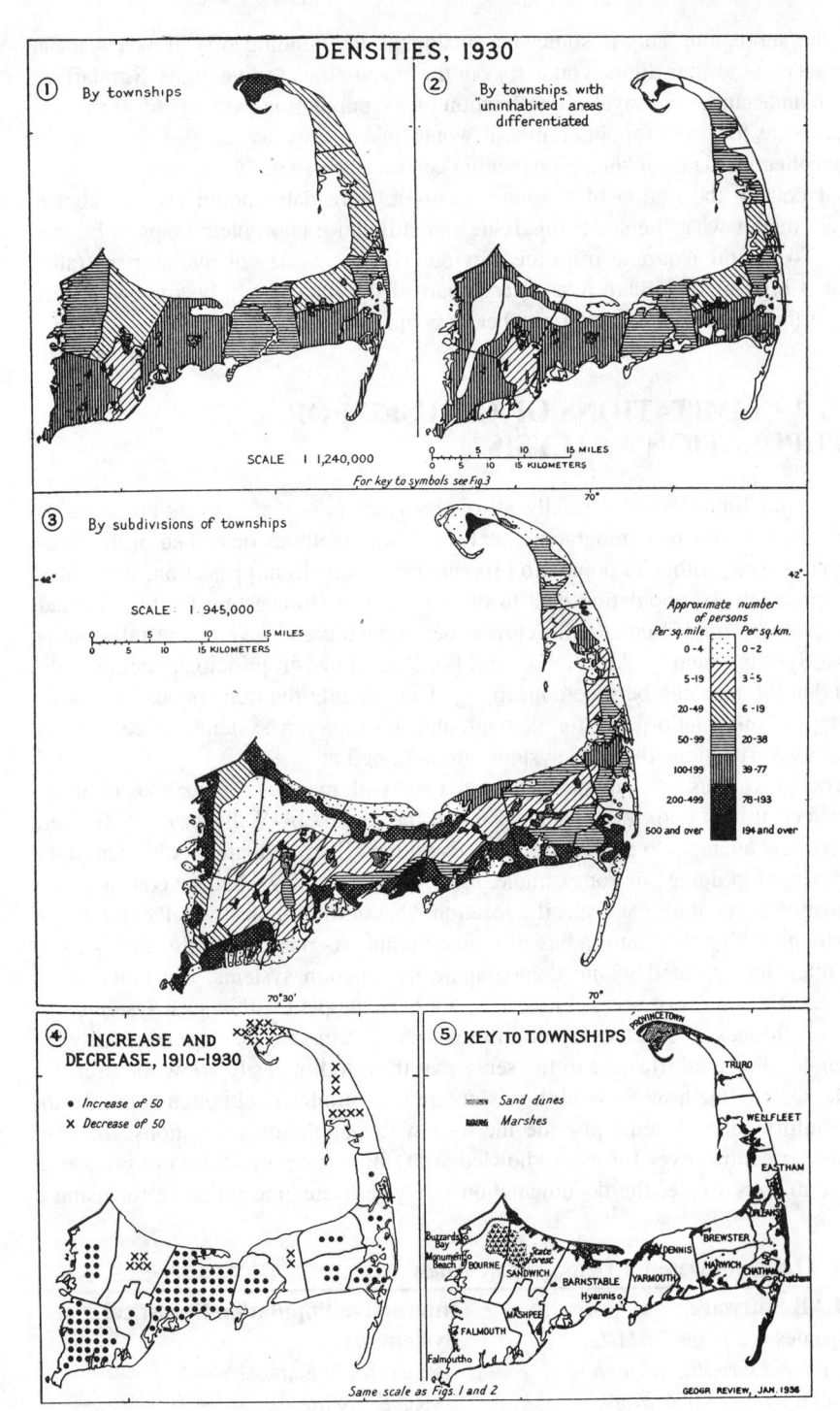

FIGURE 11.8 Population Density Maps for Cape Cod

Source: Wright (1936). Courtesy of the American Geographical Society.

within the subregion. This is somewhat analogous to rounding off—if you want an answer accurate to four digits, you carry out the calculations to five digits. Similarly, if you wish an accurate portrayal of the fraction of the population over age 65 at the census tract scale for a metropolitan area, it would make sense to actually construct the map by collecting data for subregions within census tracts (e.g., block groups).

Of course, the notion of absolute versus relative data should also be kept in mind. No matter what the scale, this issue can still cause choropleth maps to be misleading. To use the example from the previous chapter, a map of market penetration showing a high market share over a large part of the map would be somewhat misleading if there were only a small number of people in that area.

11.7 LIMITATIONS OF GIS USED FOR POPULATION ANALYSIS

A major limitation of virtually all of the systems is that there is no provision made for most forms of demographic analysis. Many methods described in this book that relate directly either to population (population estimation, projection, life tables) or to applications of population data to other problems (forecasting facility demand) cannot easily be implemented with current demographic and geographic information systems. Systems such as ARC/INFO and GISPlus allow, in principal, such an integration, but the task can be time-consuming. Thus despite the tremendous market for demographic information and for demographic information systems, there are still many analysis functions that GIS systems are not good at.

Though various software packages can assist with many of the methods of analysis described in this book (see Table 11.6 and the review by Willekens, 1993), there have been few attempts to embed the methods within GIS. Yet there would seem to be much potential in doing so. For example, in Chapter 2, we described several measures of population accessibility at a specific location. One of the simplest was the number of people living within a certain radius of a given point. As mentioned, the ability to do this is often incorporated within demographic information systems. But other, more sophisticated measures of accessibility have not been adopted within such systems.

Though packaged demographic information systems are user friendly for what they *can* do, they are inflexible in the sense that they do not easily allow the user, for example, to redefine how accessibility is computed. Similarly, although some demographic information systems provide the user with population projections, the user often has no control over (or even knowledge of) how the projections are produced. To make matters worse, the documentation rarely indicates the rationale for using a

TABLE 11.6 Software for Demographic Analysis

DEM-LAB Software Accompanies the book *DEM-LAB: Teaching Demography through Computers*, by Vivian Z. Klaff. (Englewood Cliffs, NJ: Prentice-Hall, 1992).	**Interactive Population Statistical System** (IPSS) (for Macintosh) PSRC Software Bowling Green State University Bowling Green, Ohio 43403
FIVFIV/SINSIN The Population Council One Dag Hammarskjold Plaza New York, N.Y. 10017	**VANPRO** Office of Population Studies University of San Carlos Cebu City 6000 Philippines

particular algorithm and, in fact, often does not even indicate which algorithm has been employed! The user of packaged demographic information systems is therefore often limited by the constraints of the system, and usually cannot program it to do anything else. In other words, what the systems do, they do well, but users often want to do things that systems cannot do. Along with the user-friendliness of such systems often comes a transparent, black-box mentality, where the user is assumed to simply accept what appears on the screen.

The more sophisticated and flexible systems would allow programming modules to produce life tables, population estimates, and population projections, but to date, the impediments associated with the combination of knowledge required in demography, geography, and programming, as well as the steep learning curve, have apparently dissuaded people from trying.

Examples of the use of GIS and population data include market area delineation, location analysis, and political redistricting. In all of these applications, GIS, when loaded with the appropriate geographic and demographic information, can efficiently provide and display information on alternatives that the analyst wishes to consider. What they typically do not do is, for example, to suggest a neutral, bipartisan redistricting plan, or find a new store or branch location that is in some sense optimal (as discussed in Chapter 8), or draw in market area boundaries from consideration of observed travel behavior.[1] The systems are thus passive, rather than prescriptive, in how they approach these problems.

These limitations should by no means be viewed as fatal flaws. Current systems have been invaluable to countless businesses, industries, and planners. Also, the entire field is changing extremely rapidly, and progress aimed at removing some of these impediments will undoubtedly be made between the time these words are written and the time the book is published (which should not be taken as a comment on publication lags!). We again emphasize that there is a largely untapped potential for these systems to embed within them the methods of analysis such as those covered in this book.

11.8 PROSPECTS

There are several ways in which demographic analyses may ultimately be made more viable within GIS. One avenue is to pursue the development of population analysis modules using elemental building blocks or "atoms" (Densham and Goodchild, 1989; Densham, 1992). Atoms in this sense are defined as very basic, algorithmic operations that are used repetitively in geographical analysis. Thus one atom may add a column of numbers; another may compute the great circle distance between two geographic coordinates. If the appropriate atoms are available, the creation and modification of routines is greatly simplified. This is extremely important in many applications to population analysis, since users are often in a position where they need the program to do essentially what it already does, but with some small modification to the algorithm. Having the appropriate atomic programming elements at your disposal greatly facilitates program modification.

A somewhat related approach is the provision of a higher-level set of routines to perform common types of demographic and spatial analysis. An example of this kind of effort is the collection of spatial analysis routines being assembled in the United Kingdom (Dixon, Openshaw, and Wymer 1987). Ultimately there may be a collection of spatial analysis routines that facilitate geographical analysis of population much as

---

[1] An exception is the work of Densham, who has embedded location models within GIS.

SPSS and SAS now facilitate statistical analysis in the social sciences. Several researchers have recently emphasized the potential of integrating the methods of spatial analysis and GIS (e.g., Anselin and Getis, 1992; Fotheringham and Rogerson, 1993a, and Rogerson and Fotheringham, 1993). Indeed, this integration has been the subject of recent research initiatives in both the United States (Fotheringham and Rogerson, 1993b) and Britain (Haining and Wise, 1991).

We can also combine programs designed explicitly for population analysis with a GIS, demographic data, and geographic data. This alternative may be further subdivided into a loose coupling or strong coupling approach. (See, e.g., Goodchild, 1991.) "Loose coupling" implies that the demographic software is not fully integrated with the GIS—data and results flow between the demographic software and the GIS. This approach allows us to take full advantage of existing demographic software, some of which is quite sophisticated. However, it requires a concerted effort at developing a good input–output interface, since the link between the demographic software and the GIS is so important. A good example of this approach is the work of Lolonis et al. (1992) on population projections at the block level.

"Strong coupling" occurs when the analysis and the GIS are fully integrated, and there is no passing of data into and out of the GIS to carry out the spatial analysis. Densham (1993) argues against the use of loose coupling and for the use of strong coupling in applications to location problems, since with loose coupling, the specialized software that resides outside of the GIS inevitably cannot take full advantage of the locational data and functions that reside within the GIS.

While prospects for the merger of population methods and models with DIS and GIS technology are excellent, it is vital for population analysts to become part of the development process. Otherwise, desktop (or laptop) demographic information systems and geographic information systems devoted to population applications are more likely to contain inferior algorithms for such tasks as population estimation and projection. The point was made earlier that it is difficult for population analysts to know what algorithms are used in present systems, and, if they do know, they may often wish that other ones had been used in their place. Future generations of systems not only need to be equipped with clearer documentation of the methods; they also require input from experts to ensure that proper methods are being used, and there must be sufficient flexibility so that methods may be replaced with alternative methods defined by the user.

The advent of demographic and geographic information systems, and the potential for integrating them with routines capable of population analysis, makes this an exciting time for both applied and basic population studies. These new tools will be increasingly used for demographic research in marketing applications, and in the efforts of local planning agencies. In addition, they are helping to change the way that demography and population geography are taught (e.g., by further emphasizing the importance of location). Demographic and geographic information systems will help to facilitate research in population much as SAS and SPSS have facilitated research in the social sciences. Finally, the systems should ultimately help to promote the proper use of the methods of population analysis.

11.9 SUMMARY

Both the field of GIS and its applications to population analysis are growing rapidly, and it is virtually impossible to do justice to the topic. In this chapter, we have chosen to provide an overview of the principal concepts and applications. We have focused on the relationship between GIS and the geographical analysis of population, and have given virtually no attention to the nitty-gritty details of actually using GIS for population analysis. Readers interested in the latter topic are encouraged to consult the references cited in this chapter. However, there is no substitute for actually obtaining one of these systems and to begin "learning by doing."

EXERCISES

1. Using Figure 11.7 and Table 11.4, estimate the populations of districts B and C, assuming that the population of each census tract is spread uniformly throughout the tract.

2. Repeat Exercise 1, this time making the dasymetric mapping assumption that the population of each tract is spread uniformly over the land area within each tract.

3. A planner wishes to map the population density along a lakefront shown in the figure below. Because of zoning regulations, no houses are permitted within 0.1 mi of the shoreline. Further inland, a steep, mountainous region has limited population density to approximately 30 persons/sq mi. Three census tracts comprise the region of interest. The solid lines in the figure represent census tract boundaries. Compute the population density for the remaining subregions in each tract using the following data.

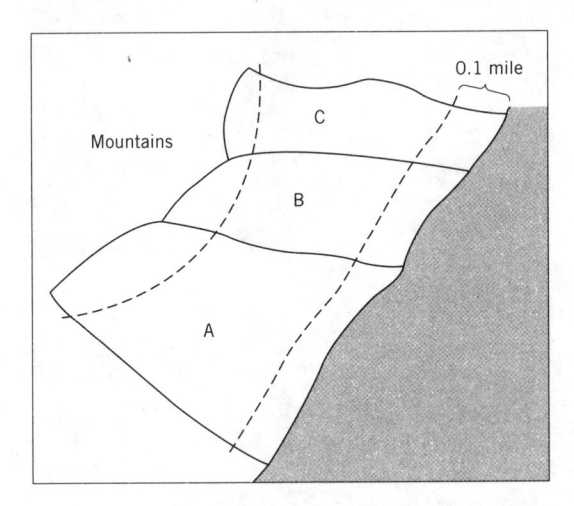

Census Tract	Population	Miles of Shoreline	Area (Sq Mi)	Area of Mountainous Region (Sq Mi)
A	600	4	8	1
B	500	2	4	0.5
C	400	1	2	0.25

4. A regional analyst has been asked to form recreation management areas for a three-county region. As part of her report, she must include the area of each region. After overlaying a square grid (with grid units marked in miles), she finds that the vertices of one polygonal area are at points (2,2), (3,5), (7,5), (9,4), (13,3), and (6,1). Vertices are given in clockwise order. Find the area.

Think globally; act locally.

CHAPTER 12
CONCLUSIONS

12.1 Importance of the Geographical Analysis of Population
12.2 Scale
12.3 Long-Term Perspectives on Population Change
12.4 Geographic Perspectives on Future Population Change

12.1 IMPORTANCE OF THE GEOGRAPHICAL ANALYSIS OF POPULATION

Throughout this book one of our primary objectives has been to provide a geographic perspective on population analysis. The circumstances that require the analysis of population data are of course numerous and varied. Certainly the majority of public- and private-sector planning efforts benefit from a careful study of population characteristics and change. Both public-sector planning by government agencies and the market-driven planning activities of businesses and industries require population-related information as the most basic of all inputs. A geographic perspective on population is particularly appropriate in many of these applications, either because of the importance of inherently spatial attributes of the application, such as location and accessibility, or because of the importance of migration as a component of population change.

In this final chapter, we wish to emphasize a few important issues associated with applying the methods we have described to planning problems. These relate primarily to the issues of scale and long-term perspectives on population change.

12.2 SCALE

Scale-Dependence in Choice of Method

We must bear in mind that we have provided a geographical perspective on population analysis through methods, examples, and applications at particular geographic scales. Different methods are often employed at different scales of analysis. This point may be illustrated by using population projections as an example. The earth is a closed system with respect to population change, so we do not need to consider migration in projections of global population change (barring alien immigration from extraterrestrial sources!). This is quite different from the situation experienced by many local communities where migration is the key component in determining population change. Similarly, the methods used for estimating the current population of a town often use symptomatic variables, whereas the methods used for estimating the current population of the country do not. The methods used at one scale do not automatically generalize to other scales, and data available at one scale are not always obtainable at other scales. A careful consideration of the underlying assumptions at each scale of analysis must be made prior to proceeding.

Different Questions for Different Scales

We must also recognize that by shifting the scale of analysis, we may be in the position of being able to ask an entirely different set of questions. We have focused upon scales intermediate between the individual level and a global scale of analysis. But an entirely different analysis—for example, that of the demographic behavior of individuals—is also possible. At the individual level, the modeling of human behavior related to fertility and migration decisions occupies a vast part of the population literature. (See, e.g., Wolpert, 1965; Brown and Moore, 1970.) The reader should be aware that we have barely touched upon this literature in this book, and that an analysis of the demographic behavior of individuals would aim to answer different questions using different approaches.

The Modifiable Areal Unit Problem

The importance of scale may also be illustrated in another way. Specifically, at the core of what is known as the *modifiable areal unit problem* is the fact that results of statistical analyses almost always depend on the spatial scale of the analysis. A variable that is significant in a regression analysis at one scale may be insignificant at another scale. Point patterns that appear random at one scale may appear clustered at another. In panel (a) of Figure 12.1, quadrats have a distribution of points per quadrat that is consistent with a random pattern. In panel (b), where the point pattern is the same, but where the scale of analysis has changed (since the quadrat size is bigger), there is a definite clustering of points into one particular quadrat.

It is interesting to note that among the earliest papers to note this scale-dependence was one by Gehlke and Biehl (1934), showing that correlation coefficients tended to increase with the level of aggregation of census tracts. More recently, Openshaw (1984)

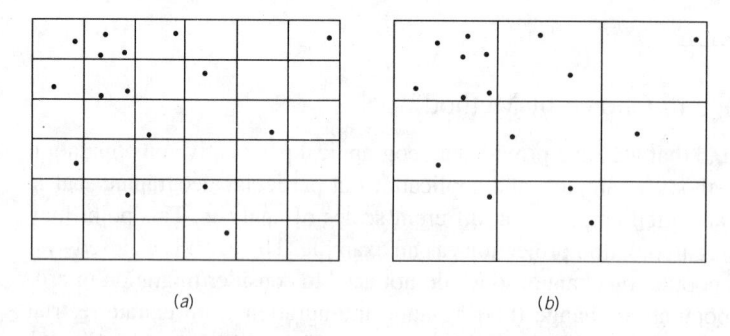

FIGURE 12.1 Effects of Scale on Point Pattern Interpretation

and Fotheringham and Wong (1991) have called renewed attention to the substantially different conclusions that may be reached when data are aggregated to different scales.

Positive Benefits from Studying Phenomena at Alternative Spatial Scales

Since different questions can be asked if the scale of analysis is shifted, since the choice of technique is often scale-dependent, and since statistical results may be a function of the spatial scale of the units of analysis, we might think that analyses at a given scale should always be interpreted completely independently of analyses at other scales. This is not always the case. There is often an important, positive interaction between phenomena studied at alternative scales of analysis. Examples of the advantageous consequences that this interaction may produce include (1) the understanding of a phenomenon at one particular scale benefiting from an examination of that same phenomenon at another scale and (2) methods typically employed at one scale being generalized for use at other scales.

An illustration of the former is provided through an examination of school enrollment projections. Following the departure of the baby boom cohort from elementary schools during the 1970s, enrollment declines led to the closing of a large number of elementary schools. In many cases, there were significant lags from the time enrollments had declined substantially to the time that a school closing decision was made. Delays were due to typical lags associated with the planning and decision making process, as well as to the fact that the difficult decision to close a school is often carried out in a highly-charged atmosphere. Consequently, decisions to close schools were often not made until the mid-1980s.

By using projection methods specific to local areas, the majority of local districts failed to foresee that the baby boom would have an "echo."[1] Figure 12.2 displays the emergence of the baby boom echo cohort using U.S. population data for 1990. Kindergarten enrollments in many districts began to rise at just about the time that schools were being closed. Consequently, many elementary schools witnessed a severe shortage of space during the early 1990s. Options including grade reconfiguration (changing the grade levels assigned to elementary, middle, and high schools), addition of temporary space, and new school construction were discussed in districts throughout the country, such as in the Arizona district that we examined in our Chapter 1 case study. Though the issue is complicated by increases in the space requirements brought about by the proliferation of mandated school programs, certainly many problems might have been avoided if the local enrollment projections had been made in the context of the national trends that showed a substantial rise in children of elementary school age during the early 1990s. There are many such circumstances where local planning efforts would benefit both from a removal of the "blinders" created by focusing too much on the local scale, and from the additional information conveyed by trends on a larger scale. The situation is summarized nicely by Goldscheider (1971),

---

[1] The baby bust may be approximately defined as the relatively small cohort born between 1965 and 1978. The baby boom echo may be approximately defined as beginning in 1979 and ending in the mid-1990s. The baby bust echo refers to the relatively small, yet-to-be-born cohort that will follow the baby boom echo cohort. The baby bust echo cohort will begin to reach elementary school age around the beginning of the next century.

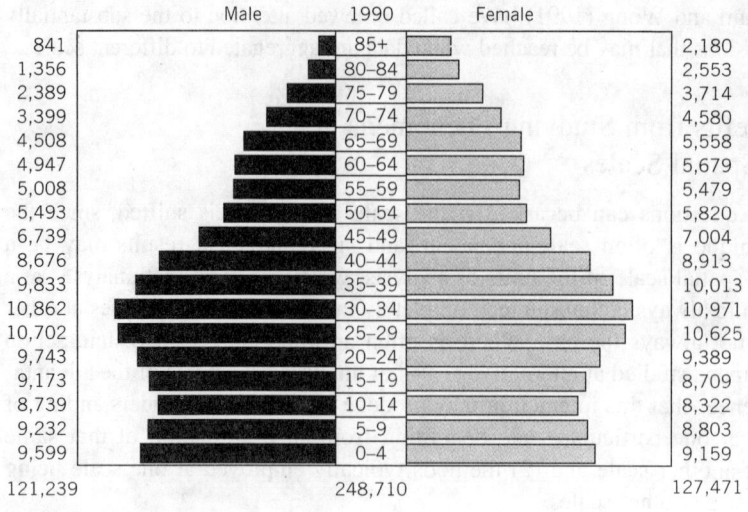

Male		1990		Female
841		85+		2,180
1,356		80–84		2,553
2,389		75–79		3,714
3,399		70–74		4,580
4,508		65–69		5,558
4,947		60–64		5,679
5,008		55–59		5,479
5,493		50–54		5,820
6,739		45–49		7,004
8,676		40–44		8,913
9,833		35–39		10,013
10,862		30–34		10,971
10,702		25–29		10,625
9,743		20–24		9,389
9,173		15–19		8,709
8,739		10–14		8,322
9,232		5–9		8,803
9,599		0–4		9,159
121,239		248,710		127,471

Note: Numbers in thousands.

FIGURE 12.2 U.S. Population Pyramid, 1990

Source: U.S. Bureau of the Census. Decennial census data.

who emphasizes both the importance of separating analyses at different scales and the importance of reintegrating them:

> [T]he failure to separate analysis levels results in a host of logical and methodological problems. Ultimately we are required to integrate macro- and micro-explanations. This integration is not achieved by mixing levels between the data and interpretation; synthesis can only result from preliminary separation of micro- and macro-analysis and their eventual reintegration. (pp. 39–40)

Although results and techniques may often differ at different scales of analysis, there are also situations where the results and techniques employed at a given scale may generate ideas for analysis at other scales. For example, Rogerson (1992) notes that

> [A]pplications of GIS to local scales include private sector geodemographic market research. It is easy to imagine that these desktop systems that now make it so easy for the user to gather data and produce reports about the demographic characteristics, lifestyles, and purchasing behavior of the American population will soon be capable of aiding the marketing, location, and advertising decisions of multinational firms.

Results at a Given Scale May Hide Significant Spatial Variation

One of the best and most vivid examples of the importance of spatial disaggregation in population studies concerns global population growth. The annual growth rate of the world's population currently stands at approximately 1.71 percent. But annual growth is approximately 2.1 percent in developing regions (implying a doubling time of 35 years), while it is only 0.5 percent in the developed regions of the world (corresponding to a doubling time of 87.5 years). To focus solely on world population growth in this instance is misleading for several reasons. First, a fixed-rate projection made in 1987 (when the world's population was 5 billion) using the global annual growth rate of 1.71 percent implies

$$5(1.0171)^{63} = 14.55 \text{ billion}$$

people in 2050. In contrast, a projection of global population in 2050 based on the separate projections of developed and developing regions yields

$$3.8(1.021)^{63} + 1.2\,(1.005)^{63} = 15.71 \text{ billion}$$

Now some might argue that this is a difference of "only" 1.16 billion people. Still, we certainly do not wish to plan for a world where the needs of these "extra" people are not accounted for, so the percentage difference between the two projections of 6 or 7 percent is in that sense significant. Also hidden by the projection that simply uses the world population growth rate is the expectation that the proportion of the world's population living in the developing regions will grow from 3.8 / 5 = 76 percent in 1987, to 90 percent by 2050.

Thus when significant spatial heterogeneity characterizes an analysis of population change, two significant consequences are (1) overall population change will be underestimated by focusing on the growth rate of the region as a whole and (2) the population of the rapidly growing region will become an increasingly larger fraction of the total population. The significance of both of these consequences increases with the length of the projection period.

These points are well taken not only for global population growth, but for the growth of metropolitan areas and other spatial units as well. Disaggregation of the region of interest into a greater number of subregions will also have a tendency to produce projections that differ even more substantially from the projection produced by using the region as a whole. In our previous example we simply used the developing and developed regions of the world; had we used country-specific growth rates, the discrepancies between the two approaches would have been even more apparent.

But is it always desirable to disaggregate as far as possible and then aggregate the answers when the analyst is interested in some variable at the aggregate level? The answer is "not necessarily." Spatial disaggregation does not yield benefits if (a) the quality of the data at finer scales is unreliable or (b) the processes at the finer scale are sufficiently unstable that we have little faith that the underlying assumptions (e.g., that a particular rate will remain stationary) will hold. A good example is again provided by school enrollment projections. Lolonis et al. (1992) describe a GIS-based analysis of the variation of grade progression ratios with spatial scale. It is clear from their results that it would make little sense to use grade progression ratios at too small a spatial scale—say at the block or block group level. At that level of geographic detail, we simply cannot plausibly assume that the grade progression ratios will be stable enough to use in projections, since they are based on such a small number of students. If, for example, there were 5 pupils in grade 4 in a given block last year and, due to net in-migration, there were 10 pupils in grade 5 this year, the grade progression ratio (associated with moving from grade 4 to grade 5) for that block would be 10 / 5 or 2. This would imply that, if there were 6 children in grade 4 this year, there would be 12 children in grade 5 next year. It is not plausible to use the grade progression ratio of 2 for projections in this example; the net in-migration of a small number of children has produced a grade progression ratio that is unlikely to hold in future years. Rather than make enrollment projections by block that are based on unstable block-specific grade progression ratios, it is more desirable to use grade progression ratios calculated for larger spatial units, since the assumption of a constant grade progression ratio will then be more tenable.

12.3 LONG-TERM PERSPECTIVES ON POPULATION CHANGE

Just as it is important to consider other spatial scales of analysis in particular studies, so too it is important to consider other time scales. It is enlightening to know, for example, that the general stability of the annual mobility rate in the United States over the past decade is a continuation of the general stability that has prevailed for over four decades.

In addition to being placed in a spatial context, studies should also be placed in a temporal context. With respect to school enrollment projections, not only do we wish to have district projections supplemented by current enrollment trends in the county, state, and country, it is also desirable to place the projections in the context of both past changes and potential future changes. Such a perspective leads to a greater realization that the relatively small cohort known as the "baby bust echo" will be reaching elementary school age before too long. By not overbuilding and by not overreacting to the current space shortage, the current planning effort can be made more efficient.

Similarly, by adopting a long-term perspective, we can more readily appreciate and anticipate mobility trends and changes in the markets for housing, labor, and so on that are at least in part determined by the age composition of the population. The tightness in housing and labor markets during the late 1970s and early 1980s was due in part to the large size of the baby boom cohort then in their twenties; a repeat of such conditions due to demographic supply pressure is not likely until the baby boom echo cohort reaches its twenties during the first decade of the next century. (And even then, it is likely to occur with diminished force.) Markets for consumer products are also strongly affected by the age composition of the population. Merrill Lynch, for example, established a mutual fund called Fund for Tomorrow whose objective is investing in companies that sell products oriented toward demographic segments of the population that are expected to experience significant growth.

The Interaction Between Rates and Composition

Much of our discussion of population projection methods in Chapter 6 emphasized methods that relied on the assumption of fixed demographic rates or probabilities to produce projections. This is a common practice among demographers, geographers, and planners. By assuming that birth, death, and migration rates remain the same, it is quite straightforward to project the changing age composition of the population. As we noted at the end of that chapter, the more challenging task is to anticipate how the rates themselves may possibly change over time.

Easterlin (1980) has argued that demographic rates, as well as other socioeconomic measures, are functions of the relative sizes of cohorts. He maintains that among the characteristics of large cohorts are lower fertility rates, more postponement of marriage and childbearing, higher rates of divorce, increased rates of female labor force participation, and higher crime rates when members of the cohort are in their young adult years. The argument rests largely on a simple appeal to the laws of supply and demand. Large groups of young adults face relatively more competition for a limited number of opportunities in housing and labor markets. As a direct result of these difficulties, the need for two incomes becomes greater, the likelihood of additional stress borne by individuals is increased, and the ability to comfortably afford the time and monetary costs associated with raising children is reduced.

Figure 12.3 depicts the fairly strong positive relationship that has persisted over the past 60 years between the total fertility rate and the Easterlin ratio, which is defined as the ratio of the size of the male population age 35 to 64 to the size of the male population age 15 to 34. The higher the ratio, the smaller the relative size of the young adult cohort, and, according to Easterlin's thesis, the higher will be fertility rates.

Though this brief description of Easterlin's thesis is highly simplified, it illustrates the important point that rates and composition are intertwined. The rates of childbearing and divorce are themselves seen to be dependent on the age composition

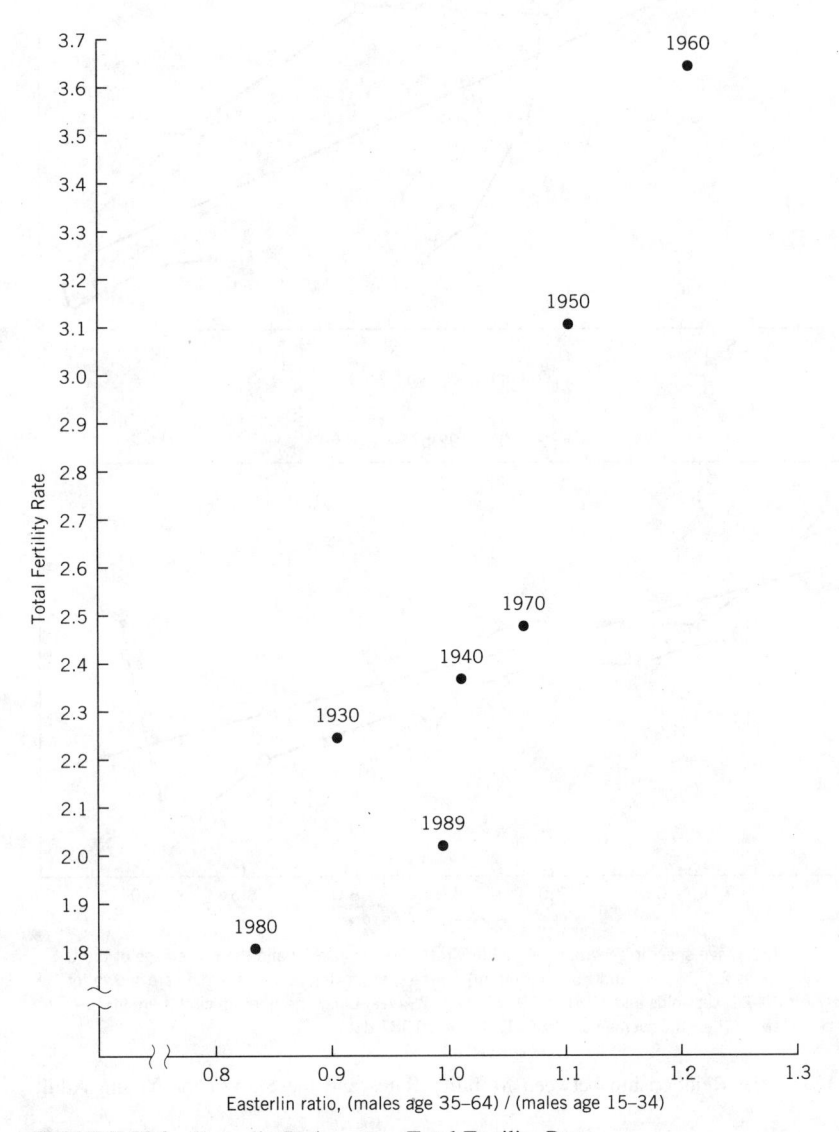

FIGURE 12.3 Easterlin Ratio versus Total Fertility Rate

Source: Graphed by the authors from data in U.S. Bureau of the Census (1975, 1992).

368 Chapter 12 • Conclusions

of the population. Though we have seen that many projection methods assume that future population composition can be determined from rates, it is also true that rates are determined, in part, by population composition.

In Chapter 1, we noted that the ideas of Easterlin can also be viewed from a geographical perspective. We suggested that migration and mobility are also affected by generation size. Plane and Rogerson (1991) provide evidence for a negative correlation between mobility rates and the relative size of the young adult cohort. This

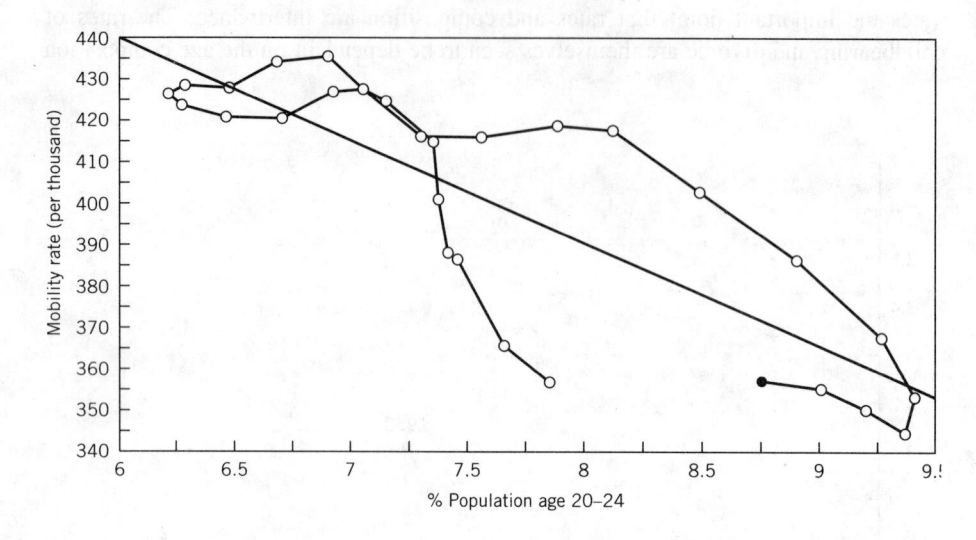

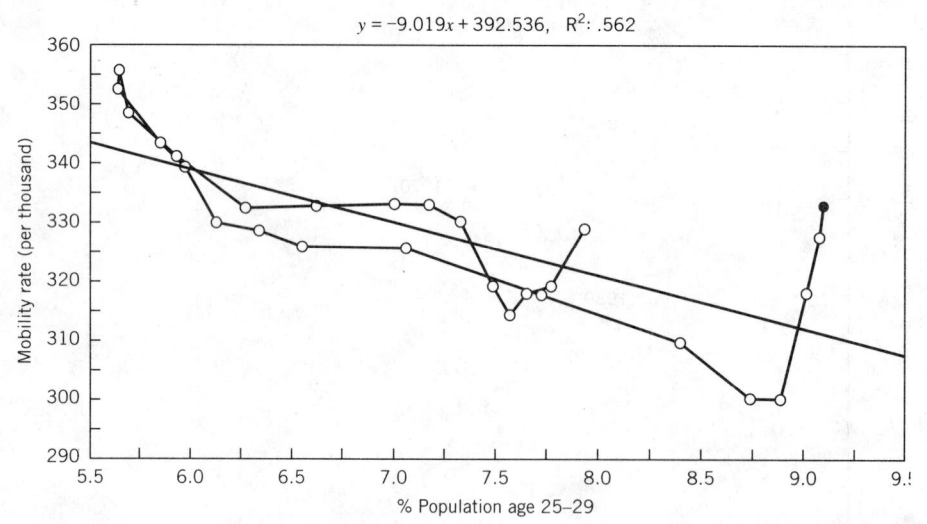

$y = -9.019x + 392.536, \quad R^2: .562$

The relationship between age-specific geographical mobility rates (per thousand) and the percentage of the U.S. population in age groups 20–24 (top) and 25–29 (bottom). Three-year moving averages of data are shown for 1949 to 1987 for 20–24-year-olds and 1951 to 1987 for 25–29-year-olds. The filled-in circles are for the most recent observations (i.e., the average of 1985, 1986, and 1987 data).

FIGURE 12.4 The Relationship between Mobility Rates and the Size of the Young Adult Cohort

Source: Plane and Rogerson (1991). Reprinted by permission. Calculated from data in U.S. Department of Commerce, Bureau of the Census, *Current Population Reports,* Series P-20 and P-25, various numbers.

implies that large generations have low rates of mobility during their young adult years, and that small cohorts have higher rates of mobility. Figure 12.4 depicts the negative correlation between the mobility rate and the relative size of the 20-to-24 and 25-to-29 cohorts. At work in determining this relationship is the tightness of housing markets and the simultaneous difficulty in moving that occurs when a large, young adult cohort is competing for a relatively fixed number of housing vacancies. The demographic effectiveness of migration flows is also greater for larger cohorts. Thus during the 1970s, the migration of the baby boom cohort was sharply focused on the relatively small number of destinations where job prospects were the brightest.

These relationships have several implications for future mobility and migration. First, the prospects for sustaining the extremely high 1970s magnitude of Frostbelt-to-Sunbelt migration throughout the 1990s seem dim, given the current decline in the number of young adults now reaching their migration-prone years. Though the magnitude of such flows remained fairly high during the early years of 1990s, as we noted in Chapter 4, the efficiency of these flows had already begun to decline by the late 1980s. Also possible is a return to flow patterns exhibiting a high degree of demographic efficiency when the baby boom cohort begins to retire to a limited number of favored destinations shortly after 2010. The high efficiency of that flow pattern may be reinforced by the creation of additional jobs at those locations brought about by the influx of the elderly.

12.4 GEOGRAPHIC PERSPECTIVES ON FUTURE POPULATION CHANGE

It is apparent that issues relating to population change will continue to be of central importance at virtually all geographic scales for a long time.

At the global scale, the dichotomy between the developed regions that have entered a period of slow population growth and even decline and the rapidly growing developing regions has already sparked a lively debate about the future of the world's population. For the past several decades many have been sounding the alarm that the current rate of global population growth is too rapid, and that the consequent increased burden upon resources and the environment is too great. (See, for example, Ehrlich, 1990; Fornos, 1991.) At the same time, some economists (e.g., Simon, 1982; Kasun, 1991) have argued that this concern is misplaced, and that increased productivity is a positive byproduct of a larger population. Still others have suggested that we should focus more attention on the likelihood of population decline (Teitelbaum, 1985; Wattenberg, 1987).

Not only the debate itself is of interest; also noteworthy is the recognition that such debates are inherently difficult to resolve because of differing interpretations. As Keyfitz (1991, p. 16) notes,

> The great obstacle to explanation and forecasting is that so much depends on how people look at their situations. If they can anticipate prospective crowding and its discomforts, and accordingly limit their numbers as Malthus urged, that gives rise to one demographic condition; if they see themselves as needing children for their support in their individual old age, or collectively as warriors, that leads to something different.[2]

---

[2]Reprinted by permission from *Population Index*.

Keyfitz adds that just as two tribes in similar situations may exhibit different fertility patterns because of differing interpretations of their situation, economists and biologists also have different perspectives on population.

Though insights to such difficult questions are not readily forthcoming, global-scale studies of population are rapidly gaining in relevance. The increased international migration precipitated by recent and widespread changes in political borders as well as the globalization of the world economy suggest that geographic perspectives on world population change will be increasingly significant.

The major demographic changes that will take place in the United States over the next several decades also stand to derive substantial benefit from geographic analysis. The aging of the baby boom cohort is clearly the major demographic force now operating in the United States. Although on the surface the aging of this cohort is a fairly aspatial phenomenon, since the cohort is not particularly well differentiated by location, there are several important and potentially significant geographic dimensions associated with it. During the early decades of the next century, we can anticipate substantial increases in elderly migration focused on a small number of favored destinations. This will have direct effects on population growth (including potential in-migration from other age groups to meet increased labor demand associated with providing services to the elderly, as just indicated), the location and provision of health care facilities, and so on.

Morrison (1992) has pointed out that perhaps even more important than elderly migration is "aging in place." Through the non-events of deciding to remain at a current residence, aging individuals can create geographic concentrations in the distribution of the elderly population. These concentrations are made even more pronounced by the out-migration of young adults to other locations. Localities characterized by relatively old age structures that have resulted from aging in place have indeed already become so profuse that they have been given their own acronym, *NORC*, for "naturally occurring retirement community."

Rogerson, Weng, and Lin (1993) emphasize yet another geographic issue of particular relevance to an aging population—the spatial separation between parents and their adult children. Sociologists have convincingly demonstrated (e.g., Crimmins and Ingegneri, 1990) the important roles that distance and accessibility play in determining the frequency of intergenerational contact and support. The mobility and migration of young adults is of primary importance in determining the proximity between parents and their adult children. The relative location of family members and potential changes in the degree of geographic separation will become increasingly important as members of the baby boom generation look to their children for support during their elderly years. Even now the issue is important, since the parents of baby boom members are entering their retirement years in increasing numbers, and will require increasing amounts of support and care from their children. Ever increasing numbers of people are also finding themselves "squeezed" between the responsibilities of caring for parents and children simultaneously. (See, e.g., Menken, 1985.) In addition, the rise in the number of two-worker households has made it even harder to balance the responsibilities of work and childrearing. Consequently, more than ever before, intergenerational support runs in two directions. Grandparents watch the kids for their offspring in two-worker households, and adult children provide care for their aging parents.

Finally, geographic perspectives on demographic analysis at the local level seem destined to flourish. In many ways, population analysis at the local scale is more difficult than at any other scale. The quality and availability of demographic data are typically inadequate for the task at hand. In addition, migration is likely to be an important component of population change, and, as we have noted, it is notoriously difficult to measure, model, and project. Still, encouraging developments point toward an increased sophistication of local demographic analysis. First and foremost is the development and continued improvement of geographic information systems. GIS only provides the tools to facilitate the geographic analysis of population; by itself, it cannot provide the data and the methods of analysis, to say nothing of allowing us to improve our understanding of population processes. However, the widespread adoption of the technology by government agencies is stimulating planners and local officials to take additional steps toward acquiring and sharing data, as well as making them more aware of the potential for valuable geographic and demographic analyses. The need for analysts trained with a geographic perspective is now growing along with this recognition of the importance of geography in planning and business.

We are confident that the geographic analysis of population will continue to grow in importance during coming decades, in response to both technological improvements and the unfolding demographic events that occur at all spatial scales.

APPENDIX A

ADDRESSES FOR U.S. ORGANIZATIONS OF INTEREST

American Demographics, Inc.
127 State Street
Ithaca, NY 14850

American Marketing Association
250 South Wacker Drive
Chicago, IL 60606

American Planning Association
1313 East 60th Street
Chicago, IL 60637-2891

Association of American Geographers
1710 16th Street, NW
Washington, DC 20009-3198

Population Association of America
1722 N Street, NW
Washington, DC 20036

Population Reference Bureau
1875 Connecticut Avenue, NW
Washington, DC 20009

Regional Science Association International
1 Observatory
University of Illinois at Urbana-Champaign
901 South Mathews
Urbana, IL 61801-3681

U.S. Bureau of the Census
Data Users Services Division
Washington, DC 20233

Zero Population Growth
1400 16th Street, NW
Washington, DC 20036

APPENDIX B

GEOGRAPHICAL SUBUNITS AND HIERARCHICAL RELATIONSHIPS

In this appendix, we summarize the various geographical units often used to report population data in the United States, as well as the hierarchical relationships between them.

Geographical reporting units may be of many types: political, administrative, postal, or related to marketing and media.

The largest subnational units for which census data are tabulated are census Regions and Divisions. The four census Regions and nine census Divisions are comprised of states (Figure B.1). Counties constitute the political subdivision that most, but not all, states are broken into. (See the following Bureau of the Census description.) In 1990 there were 3,248 counties (or statistically equivalent entities). Counties are further subdivided into minor civil divisions (MCDs) or census county divisions (CCDs). Data were tabulated for 30,386 MCDs and 5,581 CCDs in 1990. Counties are also disaggregated into census tracts and block numbering areas. The latest census had 50,690 of the former, and 11,586 of the latter. As we zoom our geographic microscope down, we encounter the "block group." In metropolitan areas block groups contain, on average, about 1,000 people. Approximately four block groups typically constitute a census tract. Finally, the smallest geographical unit for which data are reported is the block. A rough rule of thumb for metropolitan areas is that a census block generally contains about 100 persons, so there are typically about 10 blocks per block group. For the 1990 Census, the Bureau of the Census divided the United States into 229,192 block groups and 7,017,427 blocks!

Perhaps the best description of census geography comes from the census itself. The following detailed description comes from *Maps and More*, a 1992 Bureau of the Census publication. In addition to the material reprinted here, the publication contains much information on other aspects of census geography, such as the TIGER system and TIGER products, map products, and the use of 1990 Census maps. We strongly urge interested readers to study this publication.

Geographic Entities

The Census Bureau's geographic units are identified as either legal/administrative or statistical entities. Legal entities have their own governments that exercise jurisdiction within legally defined boundaries; administrative entities normally do not have their own governments, but do have boundaries established by officially prescribed laws or regulations to implement specific programs or operations. Statistical entities are geographic units delineated on the basis of criteria prepared by the Census Bureau or some other Federal agency for the purpose of data collection and presentation. The different types of entities are described below.

Legal/Administrative Geographic Entities

United States, States, the District of Columbia, Puerto Rico, and the Outlying Areas. The outlying areas are: American Samoa, Guam, the Commonwealth of the Northern Mariana Islands, the Republic of Palau, and the Virgin Islands of the United States.

Counties. Counties are the primary legal and administrative subdivisions of States. Also treated as county equivalents are boroughs and census areas in Alaska; parishes in Louisiana; independent cities in Maryland, Missouri, Nevada, and Virginia; the part of Yellowstone National Park in Montana; municipios in Puerto Rico; and a variety of entities in the outlying areas. The District of Columbia and Guam do not have primary governmental subdivisions; therefore, the entire area of each serves as the statistical equivalent of a county.

Minor Civil Divisions (MCD's). These are legally defined county subdivisions, such as the towns and townships found in many States. In some States, incorporated places are not part of any MCD, and the Census Bureau treats them as the statistical equivalents of MCD's as well as reporting them as places. In other States, incorporated places are subordinate to the MCD(s) in which they are located, and in some States the pattern is mixed.

Sub-MCD's. These subdivisions of MCD's are recognized only in the decennial census of Puerto Rico, and are called subbarrios.

Incorporated Places. These are entities incorporated under the laws of each State. For hierarchical presentations, the Census Bureau generally treats places or parts of places as subdivisions of MCD's or census county divisions (CCD's); in a few States, incorporated places also serve as MCD equivalents. They are known as cities, towns, boroughs, or villages, except for the towns in the New England States, New York, and Wisconsin and the boroughs in New York, which are treated as MCD's, and the boroughs in Alaska, which serve as county equivalents.

Consolidated Cities. For the 1990 census, the Census Bureau recognized six cities that have consolidated their governmental functions with a county or MCD, but continue to contain governmentally active incorporated places within and as part of those cities. These cities are: Butte–Silver Bow, MT; Columbus, GA; Indianapolis, IN; Jacksonville, FL; Milford, CT; and Nashville–Davidson, TN.

American Indian Reservations and Trust Lands. Reservations are areas with boundaries established by treaty, statute, and/or executive or court order. The reservations and their boundaries were identified to the Census Bureau by the Bureau of Indian Affairs and State governments. The boundaries of reservations may cross State, county, county subdivision, and place boundaries. Trust lands are properties held in trust outside a reservation for American Indians; they are identified for the Census Bureau by the Bureau of Indian Affairs.

Alaska Native Regional Corporations (ANRC's). The 12 ANRC's conduct business for profit for Alaska Natives with a common heritage and common interest. Each comprises an area with boundaries legally established by the Secretary of the Interior in cooperation with Alaska Natives under the terms of the Alaska Native Claims Settlement Act.

Congressional Districts (CD's). Congressional districts are the 435 areas from which people are elected to the U.S. House of Representatives. The CD's are reapportioned among the States after every decennial census.

Voting Districts. Voting districts, including election districts, precincts, legislative districts, and wards, are defined by State and local governments for the purpose of conducting elections. The boundaries of some voting districts for which the Census Bureau presents data are approximate and may not necessarily represent legally defined locations.

ZIP Codes. These administrative entities, identified by five-digit codes, are established by the U.S. Postal Service to expedite mail delivery. They do not have specific boundaries and frequently cross State, county, metropolitan area, and city boundaries.

Statistical Geographic Entities

Regions and Divisions. The Census Bureau has divided the United States into four regions: Northeast, South, Midwest, and West. The regions are divided into nine divisions, with each region having two or more divisions.

Metropolitan Areas (MA's). These areas are designated and defined by the Federal Office of Management and Budget (OMB) following a set of published standards. To meet the needs of users, the standards provide for the creation of three types of areas: metropolitan statistical areas (MSA's), consolidated metropolitan statistical areas (CMSA's), and primary metropolitan statistical areas (PMSA's). Collectively, these three types are designated "metropolitan areas" (MA's).

MA's are defined in terms of entire counties, except in the six New England States where cities and towns are used. A MA must contain either a city with a population of at least 50,000 or an urbanized area delineated by the Census Bureau; in the latter case, the MA must consist of one or more counties containing a population of at least 100,000 (75,000 in cities and towns in New England). The OMB also has established New England county metropolitan areas (NECMA's) to provide county-based metropolitan areas in New England. The Census Bureau does not prepare separate data tabulations for the NECMA's, but users may do so by aggregating county data.

A MSA with a population of one million or more may be divided into component areas called PMSA's. A PMSA consists of one

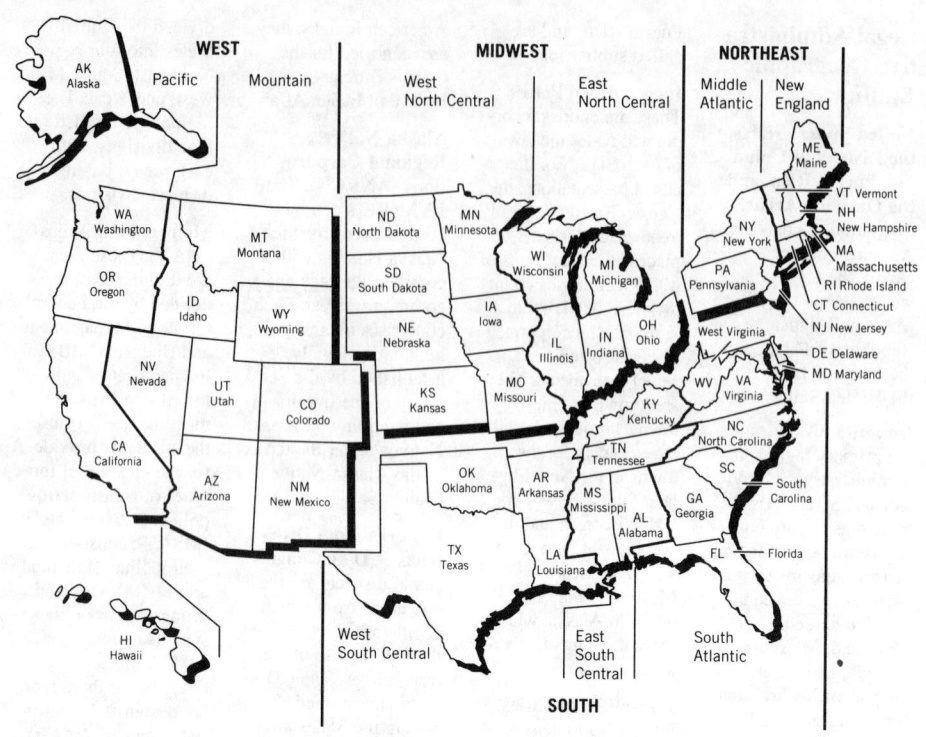

Figure B.1 U.S. Bureau of the Census Regions and Divisions

Source: U.S. Bureau of the Census.

or more counties (cities and towns in New England) that demonstrate, based on specific standards, strong internal economic and social links separate from its ties to other portions of the MSA. A MSA is redesignated as a CMSA if the OMB establishes PMSA's within the MSA. Every MSA and CMSA—but not every PMSA—has at least one core place, which is called a "central city."

Until June 30, 1983, MA's were referred to as standard metropolitan statistical areas (SMSA's) and standard consolidated statistical areas (SCSA's).

Urbanized Areas (UA's). A UA comprises one or more places and the adjacent densely settled surrounding territory (urban fringe) that together have a minimum population of 50,000. The urban fringe generally consists of contiguous territory having a density of at least 1,000 people per square mile. A UA may exclude low-density territory in one or more "extended cities"—incorporated places that contain substantial territory with a population density of fewer than 100 people per square mile.

One or more central places function as the primary centers of each UA. Each place that is a central city of a MA and lies within the UA also is a central place of the UA. If the UA does not contain a central city, its central place(s) is determined by population size. The term "central place" was instituted for the 1990 census primarily to avoid confusion with the MA "central city."

Alaska Native Village Statistical Areas (ANVSA's). Alaska Native villages (ANV's), consisting of tribes, bands, clans, villages, communities, or associations, were established pursuant to the Alaska Native Claims Settlement Act. Because ANV's do not have legally designated boundaries, the Census Bureau has established ANVSA's in cooperation with officials of each participating ANRC. ANVSA's are located within ANRC's and do not cross ANRC boundaries. The ANVSA's replace, for purposes of data presentation, the ANV's recognized in the 1980 census.

Tribal Jurisdiction Statistical Areas (TJSA's). The TJSA's were delineated by officials of federally recognized tribes without reservations in Oklahoma. They define for purposes of data presentation, areas that contain population over which the tribes have jurisdiction. The TJSA's replace the single "Historic Areas of Oklahoma" recognized in the 1980 census.

Tribal Designated Statistical Areas (TDSA's). The TDSA's were delineated by officials of federally and State-recognized tribes without reservations outside Oklahoma. They define, for purposes of data presentation, areas that contain population over which the tribes have jurisdiction.

Census Subareas. These areas are subdivisions of boroughs and census areas, the county equivalents in Alaska. They were delineated cooperatively by the State of Alaska and the Census Bureau to serve as the statistical equivalents of MCD's.

Census County Divisions (CCD's). These county subdivisions are delineated by the Census Bureau in cooperation with State and local officials in 21 States. In these States, MCD's are not legally established, do not have governmental or administrative functions, have frequently changing boundaries, are not generally known to the public, and/or are not suitable for reporting subcounty statistics. The CCD's have no legal function.

Unorganized Territories (UT's). Nine of the States with MCD's recognized by the Census Bureau contain one or more counties with territory that is not included in a recognized MCD. Such territory is treated as one or more UT's—MCD equivalents—for statistical purposes.

Census Designated Places (CDP's). These entities are designed to recognize significant population concentrations that are not in incorporated places, but have characteristics similar to such places, including community identity, high population density, and commercial development.

The population size criteria for CDP's for the 1980 census required at least 5,000 people when the CDP was located in an urbanized area having a central city of at least 50,000 people, and 1,000 people in most other areas. For the 1990 census, the minimum population size for a CDP was changed to 2,500 people if the CDP was located within a 1980 urbanized area. Elsewhere, settlements still must have at least 1,000 people to qualify with the following exceptions: Hawaii and the outlying areas—minimum of 300 people; Alaska—minimum of 25 people; American Indian reservations—minimum of 250 people; and "zonas urbanas" in Puerto Rico—no minimum population. Changes in population distribution, new incorporations, and alterations in the boundaries of incorporated places affect the inventory, names, and boundaries of CDP's from census to census.

Census Tracts. Census tracts are small, relatively permanent areas delineated to cover entire counties, primarily those in metropolitan areas. They are designed by local census statistical areas committees to be relatively homogeneous with respect to population characteristics, economic status, and living conditions at the time they are established. Census tracts average 4,000 people, but generally range from fewer than 2,500 to more than 8,000.

Block Numbering Areas (BNA's). BNA's are statistical subdivisions created for grouping and numbering blocks in counties for which census tracts have not been established. BNA's are delineated by State agencies and the Census Bureau, using guidelines similar to those used in the delineation of census tracts.

Block Groups (BG's). BG's are combinations of census blocks within a census tract or BNA that share the same first digit in their identifying numbers. For example, BG 3 within a particular census tract comprises all blocks numbered between 301 and 399. A census tract or BNA contains a maximum of nine BG's.

Blocks. These are the smallest geographic units for which the Census Bureau tabulates data. Many census blocks correspond to individual city blocks bounded by streets, but census blocks in rural areas may include many square miles and may have some boundaries that are not streets. The entire United States and its territories were divided into blocks for the first time for the 1990 census.

Hierarchical Relationships

The Census Bureau organizes geographic entities into hierarchies for tabulating and reporting statistics. The entities included in these hierarchies range from census blocks (the smallest and most numerous type of entity) to the United States. Each step up in a hierarchy has fewer units.

The relationships among the most common units of census geography and the way they overlap are shown in the accompanying Figure B.2. States are combined to form regions and divisions, and are subdivided into counties and statistically equivalent entities. These are further subdivided into legal entities (MCD's) or statistical ones (CCD's, UT's). (ZIP Codes generally do not fit into a geographic structure or hierarchy. They are administrative units, established by the U.S. Postal Service for purposes of mail delivery, that generally do not respect legal/administrative boundaries, often do not have specific boundaries, and

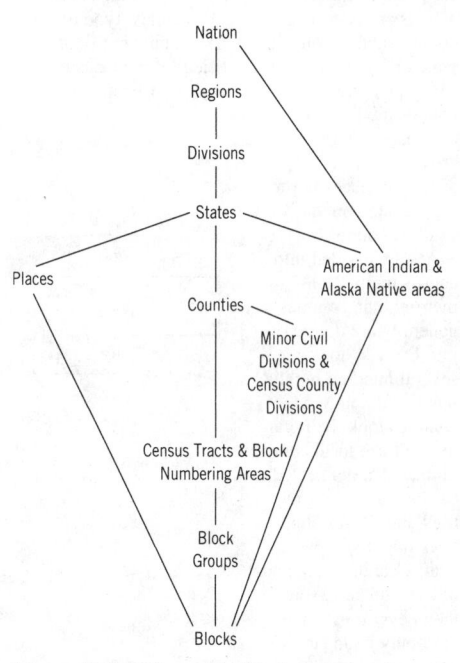

Figure B.2 Hierarchical Relationships Among U.S. Bureau of the Census Geographic Units

serve continually changing areas.)

Although most American Indian entities may cross State boundaries, some tribes are recognized only by the States in which they are located and their defined areas do not cross State boundaries. All Alaska Native entities are located in Alaska. The encompassing term for these entities is American Indian/Alaska Native areas (AI/ANA's). In eight States, every incorporated place is part of the MCD(s) in which it is located; for example, a village located within and legally part of a township. In nine States, all incorporated places are independent of the adjacent MCD's, and the Census Bureau treats them as the statistical equivalents of MCD's. In 12 States, the pattern is mixed. All places in CCD States are part of the CCD's in which they are located. CDP's always are part of the county subdivisions in which they are located. Many places cross county subdivision boundaries and some cross county boundaries, but places never cross State boundaries.

Most counties in MA's are divided into census tracts. Nonmetropolitan counties generally are divided into BNA's; however, several hundred highly populated non-MA counties took the initiative to form local committees that delineated census tracts. Census tract and BNA boundaries may cross place, county subdivision, and other boundaries, but they never cross State or county boundaries. The Census Bureau provides data both for whole census tracts and BNA's, and for each portion created when such an area is split by the boundary of a higher-level geographic entity; for example, the portion inside a place and the portion outside a place. In publications and some data files, the split census tract/BNA data are provided only for places with a population of 10,000 or more.

Both census tracts and BNA's are further divided into BG's. BG's never cross census tract or BNA boundaries (and therefore never cross State or county boundaries), but may cross the boundaries of other entities. The Census Bureau provides data both for whole BG's and for each portion of a BG that is split by the boundary of a higher-level geographic entity; for example, the portion of a BG inside a place and the portion outside.

Census blocks are the smallest type of geographic entity identified by the Census Bureau. Census block numbers are assigned within each census tract and BNA. All blocks sharing the same first digit in their identifying numbers constitute a BG.

The United States and its territories also are divided into urban and rural categories. This classification cuts across other hierarchies: for example, there generally is both urban and rural territory within each MA and within the nonmetropolitan portion of each State. "Urban" comprises all territory, population, and housing units in UA's and in non-UA places (both incorporated places and CDP's) of 2,500 or more people; however, it excludes persons living in the rural portions of "extended cities." This latter concept was applied by the Census Bureau in the 1990 census not only to incorporated places within UA's, but, for the first time, to other incorporated places that contained large areas of sparsely settled territory.

Territory, population, and housing units not classified as urban constitute "rural." Rural population and housing units are subdivided into "rural farm" and "rural nonfarm." In the 1990 census, "rural farm" includes all rural households and housing units on farms from which $1,000 or more of agricultural products were sold in 1989; the remaining population and housing units are termed "rural nonfarm."

Machine-readable data files, especially from the decennial census of population and housing, present data for the lowest-common-denominator areas within a variety of geographic hierarchies; e.g., for county subdivisions (MCD's/CCD's) within county within State. See the later section on "Geographic Reporting in Data Products."

Census Regions and Divisions

Decennial Census Small-Area Geography

Source: Bureau of the Census. 1992. *Maps and More*. Washington, DC: U.S. Government Printing Office.

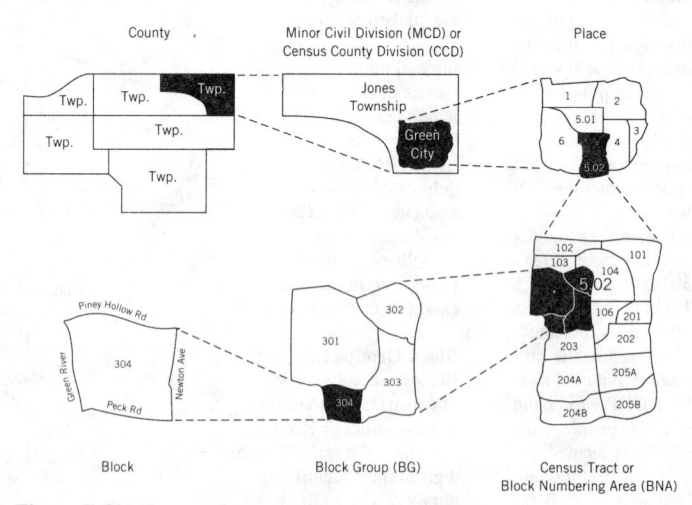

Figure B.3 Decennial Census Small-Area Geography

APPENDIX C
LOGARITHMS

There are two kinds of logarithms that are commonly used: base 10 and base $e = 2.718\ldots$ (natural logs). Base 10 logs are indicated by the notation "log" and base e (natural) logs are indicated by the notation "ln."

You can think of logarithms as the exponent to which the base (usually either e or 10) must be raised to give you the number that you are taking the log of.

Examples

$\log 10 = 1 \qquad \log 100 = 2 \qquad \log 1{,}000 = 3 \qquad \ln 10 = 2.303$

Thus, in the second example, 2 is the exponent to which the base (10) must be raised to give you 100. Likewise, for the other examples,

$$10^1 = 100 \qquad 10^3 = 1{,}000 \qquad 2.718^{2.303} = 10$$

Rules

(1) $\log x^y = y \log x$.
(2) $\log xy = \log x + \log y$
(3) If $y = \ln x$, then $x = e^y$ and if $y = \log x$, then $x = 10^y$

When using logarithms in population examples it makes no difference whether you use natural logs or base 10 logs, though it is much easier to use natural logarithms when working with the exponential growth model.

Where does the number e come from? We have already indicated in Chapter 3 that exponential growth results when a continuous growth rate is used. Consider a periodic growth rate r. At the end of one period, the original amount of money or population has grown to $1+r$ times its original size. Now suppose that the period is cut into n small intervals, and during each interval, a per-interval growth rate of $1+(r/n)$ is applied. At the end of the period, the original amount has grown to $[1+(r/n)]^n$ times is original size. If n is a large number, the quantity $[1+(r/n)]^n$ is approximately equal to the quantity e^r. Suppose that $1 was invested at a period growth rate of $r = 1$ (corresponding to 100 percent growth per period). At the end of the period, the investment would be worth $2. But if that same growth rate were applied continuously, the initial investment would be worth $e^1 = \$2.72$.

APPENDIX D
MATRIX ALGEBRA

Matrix algebra provides a convenient, compact way to express sets of linear equations. In addition, matrices have various mathematical properties that prove useful in population analysis. For example, as we saw in Chapter 6, growth matrices with row sums equal to 1 will always yield long-run equilibrium populations that are independent of initial conditions.

A matrix is simply a collection of elements—symbols or numbers—arranged as a set of rows and columns. A matrix with m rows and n columns is said to be of order $m \times n$. For example, the matrix

$$\begin{bmatrix} 2 & 3 \\ 6 & 7 \\ 3 & 4 \end{bmatrix}$$

is of order 3×2, and the matrix

$$\begin{bmatrix} a & b & c \\ d & e & f \end{bmatrix}$$

is 2×3.

A matrix consisting of one row is called a *row vector*, and a matrix with just one column is called, not surprisingly, a *column vector*. A matrix with one row and one column is termed a *scalar*.

Regarding notation, the usual convention is to represent matrices in boldface type, with uppercase letters for matrices and lowercase letters for vectors. Elements within matrices are denoted by lowercase letters (in italics, and not boldfaced), with two subscripts, the first representing the row, and the second representing the column. For example, a_{34} represents the element in the third row and fourth column of the matrix **A**.

Matrices may be subject to arithmetic operations. Here we describe matrix addition, subtraction, and multiplication.

Two matrices **A** and **B** may be added together to form a new matrix **C** only if **A** and **B** have the same order. Addition then proceeds element by element. That is,

$$c_{ij} = a_{ij} + b_{ij}$$

For example,

$$\begin{bmatrix} 3 & 6 \\ 2 & 10 \\ 3 & 68 \end{bmatrix} + \begin{bmatrix} 2 & 10 \\ 2 & 10 \\ 8 & 11 \end{bmatrix} = \begin{bmatrix} 5 & 16 \\ 4 & 20 \\ 11 & 79 \end{bmatrix}$$

Matrix subtraction also requires matrices of the same order. To subtract a matrix **B** from a matrix **A**, element-by-element subtraction is carried out:

$$\begin{bmatrix} 2 & 3 & 5 & 9 \\ 1 & 4 & 7 & 2 \end{bmatrix} - \begin{bmatrix} 1 & 6 & 10 & 4 \\ 0 & 0 & 3 & 5 \end{bmatrix} = \begin{bmatrix} 1 & -3 & -5 & 5 \\ 1 & 4 & 4 & -3 \end{bmatrix}$$

To postmultiply **A** by **B** requires that the number of columns in **A** equal the number of rows in **B**. That is, if **A** is $m \times n$, **B** must be $n \times p$. The resulting matrix, say **C**, will have order $m \times p$; that is, it will have the same number of rows as **A** and the same number of columns as **B**. The element in row i and column j of **C** is found by summing the products formed by pairing an element of row i in **A** with the corresponding element in column j of **B**. That is,

$$c_{ij} = \sum_{k=1}^{n} a_{ik} b_{kj} \tag{A.1}$$

For example, let

$$\mathbf{A} = \begin{bmatrix} 3 & 5 & 6 \\ 2 & 4 & 6 \end{bmatrix} \quad \mathbf{B} = \begin{bmatrix} 6 & 3 \\ 2 & 3 \\ 4 & 5 \end{bmatrix}$$

When **A** is postmultiplied by **B**, the result is a matrix **C** with two rows and two columns:

$$\mathbf{C} = \begin{bmatrix} 52 & 54 \\ 44 & 48 \end{bmatrix}$$

Each element in **C** is derived by using Equation A.1. For instance, the element in the second row and first column of **C** is found by summing the products of the second row of **A** and the first column of **B**:

$$44 = (2 \times 6) + (4 \times 2) + (6 \times 4)$$

A more detailed treatment of matrix algebra may be found in, for instance, Miller (1972).

REFERENCES

Aldenderfer, M. S., and Blashfield, R. K. 1984. *Cluster analysis.* Beverly Hills, CA: Sage.

Anderberg, M. 1973. *Cluster analysis for applications.* New York: Academic Press.

Alonso, W. 1964. *Location and land use: toward a general theory of land rent.* Cambridge, MA: Harvard University Press.

Ambry, M. 1990. *The almanac of consumer markets: a demographic guide to finding today's complex and hard-to-reach customers.* Chicago: Probus.

American Demographics. 1992. *American geography.* Desk Reference Series, no. 2. Ithaca, NY: American Demographics.

Anselin, L., and Getis, A. 1992. Spatial statistical analysis and geographic information systems. *Annals of Regional Science* 26: 19–33.

Arizona Department of Economic Security. 1985. *A demographic guide to Arizona.* Phoenix: State of Arizona Department of Economic Security, Population Statistics Unit.

Asimov, I. 1977. *Asimov on numbers.* Garden City, NY: Doubleday.

Bartholomew, D. J., and Forbes, A. F. 1979. *Statistical techniques for manpower planning.* New York: Wiley.

Beale, C. L. 1969. The relation of gross outmigration rates to net migration. Paper presented at the annual meeting of the Population Association of America, Atlantic City, NJ, April 10–12. U.S. Department of Agriculture.

Beaumont, P. M. 1984. *ECESIS: an interregional economic–demographic model of the United States.* Ph.D. dissertation in economics. Philadelphia: University of Pennsylvania.

Beaumont, P. M. 1989. *ECESIS: an interregional economic–demographic model of the United States.* New York: Garland.

Beaumont, P. M., Isserman, A. M., McMillen, D., Plane, D. A., and Rogerson, P. A. 1986. The ECESIS economic–demographic model of the United States. In *Population change and the economy: social science theories and models.* A. M. Isserman (ed.), pp. 203–38. Boston: Kluwer–Nijhoff.

Berry, B. J. L. 1988. Migration reversals in perspective: the long-wave evidence. *International Regional Science Review* 11: 245–51.

Berry, B. J. L., and Kasarda, J. D. 1977. *Contemporary urban ecology.* New York: Macmillan.

Blau, P. M., and Duncan, O. D. 1967. *The American occupational structure.* New York: John Wiley.

Blumen, I., Kogan, M., and McCarthy, P. J. 1955. *The industrial mobility of labor as a probability process.* Cornell Studies of Industrial and Labor Relations, Number 6. Ithaca, NY: Cornell University.

Bongaarts, J., Burch, T., and Wachter, K. (eds.). 1987. *Family demography: methods and their applications.* Oxford, England: Clarendon.

Borts, G. H., and Stein, J. L. 1964. *Economic growth in a free market.* New York: Columbia University Press.

Boulding, K. E. 1985. Regions of time. *Papers of the Regional Science Association* 57: 19–32.

Boyce, D. 1984. Urban-transportation network equilibrium and design models: recent achievements and future prospects. *Environment and Planning A* 16: 1445–74.

Bracken, I. 1993. A surface model approach to the representation of population-related social indicators. In *Research directions in GIS and spatial analysis.* A. S. Fotheringham and P. Rogerson (eds.), pp. 247–59. London: Taylor and Francis.

Brass, W. 1983. The formal demography of the family: an overview of the proximate determinants. In *Proceedings of the British Society for Population Studies Conference.* Office of Population Censuses and Surveys, Occasional Paper no. 31.

Brown, L. A., and Lawson, V. A. 1989. Polarization reversal, migration related shifts in human resource profiles, and spatial growth policies: a Venezuelan study. *International Regional Science Review* 12: 165–88.

Brown, L. A., and Moore, E. G. 1970. The intra-urban migration process: a perspective. *Geografiska Annaler* 52B: 1–13.

Cadwallader, M. 1985. *Analytical urban geography: spatial patterns and theories.* Englewood Cliffs, NJ: Prentice-Hall.

Carbaugh, L. W., and Marx, R. W. 1990. The TIGER system: a Census Bureau innovation serving data analysis. *Government Information Quarterly* 7: 285–306.

Champion, A. G. (ed.). 1989. Counterurbanization: the changing pace and nature of population deconcentration. London: Edward Arnold.

Clark, C. 1951. Urban population densities. *Journal of the Royal Statistical Society A* 114: 490–96.

Clark, W. A. V. 1980. Residential mobility and neighborhood change: some implications for racial residential segregation. *Urban Geography* 1: 95–117.

Clark, W. A. V. 1983. Structures for research on the dynamics of residential mobility. In *Evolving geographical structures.* D. Griffith and A. Lea (eds.), pp. 325–57. The Hague: Martinus Nijhoff.

Clark, W. A. V. 1986. *Human migration.* Beverly Hills: Sage.

Clark, W. A. V., and Burt, J. E. 1980. The impact of workplace on residential location. *Annals of the Association of American Geographers* 70: 59–67.

Clark, W. A. V., and Huff, J. O. 1977. Some empirical tests of duration-of-stay effects in intra-urban migration. *Environment and Planning A* 9: 1357–74.

Clarke, J. I. 1972. *Population geography* (2nd edition). Oxford, England: Pergamon.

Clarke, M. 1986. Demographic processes and household dynamics: a microsimulation approach. In *Population structures and models.* R. Woods and P. Rees (eds.), pp. 245–72. Boston: Allen & Unwin.

Clausen, J. A. 1986. *The life course: a sociological perspective.* Englewood Cliffs, NJ: Prentice-Hall.

Clayton, C. 1977. The structure of interstate and interregional migration 1965–70. *Annals of Regional Science* 11: 109–22.

Clayton, C. 1982. Hierarchically organized migration fields: the application of higher order factor analysis to population migration tables. *Annals of Regional Science* 16: 11–20.

Coale, A. J., and Demeny, P. 1966. *Regional model life tables and stable populations.* Princeton: Princeton University Press.

Coale, A. J., and Trussell, T. J. 1974. Model fertility schedules: variations in the age structure of childbearing in human populations. *Population Index* 40: 185–258.

Coffey, W. J. 1981. *Geography: towards a general spatial systems approach.* New York: Methuen.

Collins, L. 1972. *Industrial migration in Ontario: forecasting aspects of industrial activity through Markov chain analysis.* Ottawa: Statistics Canada.

Cordey-Hayes, M. 1975. Migration and the dynamics of multiregional population systems. *Environment and Planning A* 7: 793–814.

Crimmins, E., and Ingegneri, D. 1990. Interaction and living arrangements of older parents and their children. *Research on Aging* 12: 3–35.

Crispell, D. 1990. *The insider's guide to demographic know-how: how to find, analyze, and use information about your customers.* Ithaca, NY: American Demographics Press.

Davies, W. K. D., and Musson, T. C. 1978. Spatial patterns of commuting in South Wales, 1951–1971: a factor analysis definition. *Regional Studies* 12: 353–66.

Densham, P. 1992. Spatial decision support systems. In *Geographical information systems: principles and applications.* D. J. Maguire, M. F. Goodchild, and D. W. Rhind (eds.), pp. 403–12. London: Longman.

Densham, P. 1993. Integrating GIS and spatial modeling: the role of visual interactive modeling in location selection. *Geographical Systems* 1.

Densham, P., and Goodchild, M. 1989. Spatial decision support systems: a research agenda. *Proceedings, GIS/LIS '89,* pp. 707–16.

Dent, B. D. 1990. *Cartography: thematic map design* (2nd edition). Dubuque, IA: William C. Brown.

Dicken, P., and Lloyd, P. E. 1981. *Modern western society: a geographical perspective on work, home, and well-being.* New York: Harper & Row.

Dixon, J., Openshaw, S., and Wymer, C. 1987. A proposal and specification for a geographical analysis subroutine library. Research Report 3. Northern Regional Research Laboratory, Department of Geography, University of Newcastle, Newcastle, U.K.

Dorigo, G., and Tobler, W. 1983. Push–pull migration laws. *Annals of the Association of American Geographers* 73: 1–17.

Draper, N., and Smith, H. 1981. *Applied regression analysis.* New York: Wiley.

Duncan, O. D., and Duncan, B. 1955. A methodological analysis of segregation indexes. *American Sociological Review* 20: 210–17.

Duncan, O. D., Cuzzort, R., and Duncan, B. 1961. *Statistical geography.* Glencoe, IL: Free Press.

Durand, J. D. 1970. World population: trend and prospects. In *Population geography: a reader.* G. J. Demko, H. M. Rose, and G. A. Schnell (eds.). New York: McGraw-Hill.

Easterlin, R. 1980. *Birth and fortune: the impact of numbers on personal welfare.* New York: Basic Books.

Ehrlich, P. 1990. *The population explosion.* New York: Simon & Schuster.

Eldridge, H. 1965. Primary, secondary, and return migration in the United States, 1955–60. *Demography* 2: 444–55.

Ellis, M., Barff, R., and Renard, B. 1993. Migration regions and interstate labor flows by occupation in the United States. *Growth and Change* 24: 166–90.

Engels, R., and Healy, M. K. 1981. Measuring interstate migration flows: an origin–destination network through Internal Revenue Service records. *Environment and Planning A* 13: 1345–60.

Exter, T. 1992. The next step is called GIS. In *American geography.* Desk Reference Series, no. 2. Ithaca, NY: American Demographics.

Faludi, A. 1973. *A reader in planning theory.* Oxford, England: Pergamon.

Feder, G. 1982. On the relation between origin income and migration. *Annals of Regional Science* 16: 46–61.

Feeney, G. 1973. Two models for multiregional population dynamics. *Environment and Planning A* 5: 31–43.

Flowerdew, R., Green, M., and Kehris, E. 1990. Using areal interpolation methods in geographic information systems. *Papers in Regional Science* 70: 303–15.

Foot, D. K., and Milne, W. J. 1989. Multiregional estimation of gross internal migration flows. *International Regional Science Review* 12: 29–43.

Forbes, J. 1989. Migration monitoring and strategic planning. In *Advances in regional demography.* P. Congdon and P. Batey (eds.), pp. 41–57. London: Belhaven Press.

Fornos, W. 1991. Population politics. *Technology Review* 94: 42–51.

Fotheringham, A. S., and Rogerson, P. 1993a. GIS and spatial analytic problems. *International Journal of Geographical Information Systems* 7: 3–19.

Fotheringham, A. S., and Rogerson, P. (eds.). 1993b. *Research directions in GIS and spatial analysis.* London: Taylor & Francis.

Fotheringham, A. S., and Wong, D. 1991. The modifiable areal unit problem in multivariate statistical analysis. *Environment and Planning A* 23: 1025–44.

Francese, P., and Piirto, R. 1990. *Capturing customers: how to target the hottest markets of the '90s.* Ithaca, NY: American Demographics Press.

Fratar, T. 1954. Vehicular trip generation by successive approximations. *Traffic Quarterly* 8: 53–65.

Gaile, G. L., and Willmott, C. J. (eds.). 1989. *Geography in America.* Columbus, OH: Merrill.

Gale, S. 1972. Stochastic stationarity and the analysis of geographic mobility. In P. Adams and F. Helleiner (eds.), *Papers presented to the 22nd International Congress of the International Geographical Union.* Montreal: University of Toronto Press.

Gehlke, C., and Biehl, K. 1934. Certain effects of grouping upon the size of the correlation coefficient in census tract material. *Journal of the American Statistical Association* 29: 169–70.

Gober, P. 1986. How and why Phoenix households changed: 1970–1980. *Annals of the Association of American Geographers* 76: 536–49.

Gober, P. 1990. The urban demographic landscape. In *Housing demography.* D. Myers (ed.), pp. 232–48. Madison: University of Wisconsin Press.

Gober, P., McHugh, K. E., and Reid, N. 1991. Household instability, residential mobility, and neighborhood change. *Annals of the Association of American Geographers* 81: 80–88.

Goldscheider, C. 1971. *Population, modernization, and social structure.* Boston: Little, Brown.

Goldstein, S. 1954. Repeated migration as a factor in high mobility rates. *American Sociological Review* 19: 536–41.

Goodchild, M. F. 1987. A spatial analytical perspective on geographic information systems. *International Journal of Geographical Information Systems* 1: 327–34.

Goodchild, M. F. 1991. Progress on the GIS research agenda. *Proceedings, EGIS '91* (European Geographic Information Systems): 342–50.

Goodchild, M. F., and Lam, N. S.-N. 1980. Areal interpolation: a variant of the traditional spatial problem. *Geo-processing* 1: 297–312.

Graves, P. E. 1980. Migration and climate. *Journal of Regional Science* 20: 227–37.

Graves, P. E. 1983. Migration with a composite amenity: the role of rents. *Journal of Regional Science* 23: 541–46.

Graves, P. E., and Linneman, P. D. 1979. Household migration: theoretical and empirical results. *Journal of Urban Economics* 6: 383–404.

Green, A. 1994. Migration and labor market change in Great Britain: key issues and trends. *Environment and Planning A* 26.

Greenberg, M. 1978. *Local population and employment projection techniques.* New Brunswick, NJ: Center for Urban Policy Research.

Greenwood, M. J. 1969. An analysis of the determinants of geographic labor mobility in the United States. *The Review of Economics and Statistics* 51: 189–94.

Greenwood, M. J. 1970. Lagged response in the decision to migrate. *Journal of Regional Science* 10: 375–84.

Greenwood, M. J. 1975. Research on internal migration in the United States: a survey. *Journal of Economic Literature* 13: 397–433.

Greenwood, M. J. 1985. Human migration: theory, models, and empirical studies. *Journal of Regional Science* 25: 521–44.

Greenwood, M. J., Mueser, P. R., Plane, D. A., and Schlottmann, A. M. 1991. New directions in migration research: perspectives from some North American regional science disciplines. *Annals of Regional Science* 25: 237–70.

Griffith, D. A., and Amrhein, C. G. 1991. *Statistical analysis for geographers.* Englewood Cliffs, NJ: Prentice Hall.

Grinold, R. C., and Marshall, K. T. 1977. *Manpower planning models.* New York: North–Holland.

Hägerstrand, T. 1970. What about people in regional science? *Papers of the Regional Science Association* 24: 7–21.

Haining, R. P., and Wise, S. M. 1991. GIS and spatial analysis: report on the Sheffield Workshop. Regional Research Laboratory Initiative Discussion Paper 11, Department of Town and Regional Planning, University of Sheffield, England.

Hansen, W. 1959. How accessibility shapes land use. *Journal of the American Institute of Planners* 25: 72–77.

Hargreaves, R. P. 1961. The first use of the dot technique in cartography. *The Professional Geographer* 13, no. 5: 37–39.

Harris, C. D., and Ullman, E. L. 1945. The nature of cities. *Annals of the American Academy of Political Science* 242: 7–17.

Hartshorn, T. A. 1992. *Interpreting the city: an urban geography.* New York: John Wiley.

Haworth, J. M., and Vincent, P. J. 1979. The stochastic disturbance specification and its implications for log-linear regression. *Environment and Planning A* 11: 781–90.

Hole, J. 1985. *The homes of the working classes with suggestions for their improvement.* New York: Garland.

Hoover, E. M., and Giarratani, F. 1984. *An introduction to regional economics* (3rd edition). New York: Alfred A. Knopf.

Hoyt, H. 1939. *The structure and growth of residential neighborhoods in American cities.* Washington, DC: Federal Housing Administration.

Hua, C., and Porell, F. 1979. A critical review of the development of the gravity model. *International Regional Science Review* 4: 97–126.

Huff, D. L. 1963. A probabilistic analysis of shopping center trade areas. *Land Economics* 39: 81–90.

Huff, J. O. 1979. Residential mobility patterns and population redistribution within the city. *Geographical Analysis* 11: 133–48.

Huff, J. O., and Waldorf, B. 1988. A predictive model of residential mobility and residential segregation. *Papers of the Regional Science Association* 65: 59–77.

Hoover, E. M. 1941. Interstate redistribution of population, 1850–1940. *Journal of Economic History* 1: 199–205.

Hunter, J. M., and Young, J. C. 1968. A technique for the construction of quantitative cartograms by physical accretion models. *The Professional Geographer* 20: 402–07.

***International encyclopedia of population.* 1982.** (Compiled by Center for Population and Family Health, Columbia University; J. A. Ross, editor-in-chief.) New York: Free Press.

Ishikawa, Y. 1992. The 1970s migration turnaround in Japan revisited: a shift-share approach. *Papers in Regional Science* 71: 153–74.

Isserman, A. M. 1977. The accuracy of population projections for subcounty regions. *Journal of the American Institute of Planners* 43: 247–59.

Isserman, A. M. (ed.). 1986. *Population change and the economy: social science theories and models.* Boston: Kluwer–Nijhoff.

Isserman, A. 1993. The right people, the right rates. *Journal of the American Planning Association* 59: 45–64.

Isserman, A. M., Plane, D., and McMillen, D. 1982. Internal migration in the United States: a review of federal data. *Review of Public Data Use* 10: 285–311.

Isserman, A. M., Plane, D. A., Rogerson, P. A., and Beaumont, P. M. 1985. Forecasting interstate migration with limited data: a demographic-economic approach. *Journal of the American Statistical Association* 80: 277–85.

James, L. D., and Lee, R. 1971. *Economics of water resources planning.* New York: McGraw-Hill.

Jones, H. R. 1981. *A population geography.* New York: Harper & Row.

Kalish, S. 1993. Pregnancy rates fall while numbers rise. *Population Today,* March 1993. Washington, DC: Population Reference Bureau.

Kasun, J. 1989. Too many people? The myth of excess population. *Economic Affairs* 9, no. 5: 15–18.

Keilman, N. (ed.). 1988. *Modelling household formation and dissolution.* Oxford, England: Clarendon Press.

Keyfitz, N. 1966. How many people have lived on the earth? *Demography* 3: 581–82.

Keyfitz, N. 1981. The limits of population forecasting. *Population and Development Review* 7: 579–593.

Keyfitz, N. 1985. *Applied mathematical demography* (2nd edition). New York: Springer–Verlag.

Keyfitz, N. 1991. Population and development within the ecosphere: one view of the literature. *Population Index* 57: 5–22.

Keyfitz, N., and Flieger, W. 1991. *World population growth and aging: demographic trends in the late twentieth century.* Chicago: University of Chicago Press.

Kinsella, W. P. 1982. *Shoeless Joe.* Boston: Houghton Mifflin.

Kitsul, P., and Philipov, D. 1981. The one-year/five-year migration problem. In *Advances in multiregional mathematical demography.* A. Rogers (ed.), pp. 1–34. Laxenburg, Austria: International Institute for Applied Systems Analysis.

Klaff, V. Z. 1992. *DEM-LAB: Teaching demography through computers.* Englewood Cliffs, NJ: Prentice-Hall.

Kono, S. 1987. The headship rate method for projecting households. In *Family demography.* J. Bongaarts, T. Burch, and K. Wachter (eds.), pp. 287–308. Oxford, England: Clarendon Press.

Kriesberg, E. M., and Vining, D. R., Jr. 1978. On the contribution of out-migration to changes in net migration: a time series confirmation of Beale's cross-sectional results. *Annals of Regional Science* 12: 1–11.

Lansing, J. B., and Mueller, E. 1967. *The geographical mobility of labor.* Ann Arbor: Survey Research Center, University of Michigan.

Lazer, W. 1987. *Handbook of demographics for marketing and advertising: sources and trends on the U.S. consumer.* Lexington, MA: Lexington Books.

Ledent, J. 1993. The influence of the place of birth on interprovincial migration and population redistribution in Indonesia. Paper presented at the Pacific Conference of the Regional Science Association International, Whistler, British Columbia, July 10–14.

Lee, E. S. 1966. A theory of migration. *Demography* 3: 47–57.

Leslie, P. H. 1945. On the use of matrices in certain population mathematics. *Biometrika* 33: 183–212.

Linke, W. 1988. The headship rate approach in modelling households: the case of the Federal Republic of Germany. In *Modelling household formation and dissolution.* N. Keilman, A. Kuijsten, and A. Vossen (eds.), pp. 108–22. Oxford, England: Clarendon Press.

Linstone, H. A., and Turoff, M. 1975. *The Delphi method: techniques and applications.* Reading, MA: Addison-Wesley.

Lolonis, P., Armstrong, M. P., Pavlik, C. E., and Lin, S. 1992. Accuracy of a GIS-based small-area population projection method used in spatial decision support systems. *GIS/LIS '92 Proceedings* 2:473–83. Bethesda, MD: American Society for Photogrammetry and Remote Sensing.

Long, J. F. 1981. *Population deconcentration in the United States.* U.S. Department of Commerce, Bureau of the Census, special demographic analyses, CDS-81-5. Washington, DC: U.S. Government Printing Office.

Long, J. F. 1993. Postcensal population estimates: states, counties, and places. In *Statistical policy working paper on small area estimates.* Federal Committee on Statistical Methodology. Washington, DC.

Long, L. H. 1973. New estimates of migration in the United States. *Journal of the American Statistical Association* 68: 37–43.

Long, L. H. 1988. *Migration and residential mobility in the United States.* New York: Russell Sage Foundation. (Census Monograph Series: The Population of the United States in the 1980s.)

Louviere, J. J., Levin, I. P., Pampel, F. C., and Rushton, G. 1989. Determinants of migration of the elderly in the United States: a model of individual-level migration decision making. In *Regional structural change: experience and prospects in two mature economies.* L. J. Gibson and R. J. Stimson (eds.), pp. 61–77. Peace Dale, RI: Regional Science Research Institute.

Lowry, I. S. 1966. *Migration and metropolitan growth: two analytical models.* San Francisco: Chandler.

MacKinnon, R. 1975. Controlling interregional migration processes of a Markovian type. *Environment and Planning A* 7: 781–92.

MacKinnon, R., and Rogerson, P. 1980. Vacancy chains, information filters, and interregional migration. *Environment and Planning A* 12: 649–58.

Markoff, J., and Shapiro, G. 1973. The linkage of data describing overlapping geographical units. *Historical Methods Newsletter* 7: 34–46.

Marx, R. 1986. The TIGER system: automating the geographic structure of the United States census. *Government Publications Review* 13: 181–201.

Marx, R. (ed.). 1990. *Cartography and geographic information systems* 17, no. 1. Special issue on the Bureau of the Census's TIGER system.

McGinnis, R. 1968. A stochastic model of social mobility. *American Sociological Review* 23: 712–22.

McHugh, K. E. 1990. Seasonal migration as a substitute for, or precursor to, permanent migration. *Research on Aging* 12: 229–49.

McHugh, K. E., and Gober, P. 1992. Short-term dynamics of the U.S. interstate migration system, 1980–1988. *Growth and Change* 23: 428–45.

McHugh, K. E., and Mings, R. C. 1991. On the road again: seasonal migration to a Sunbelt metropolis. *Urban Geography* 12: 1–18.

Menken, J. 1985. Age and fertility: how late can you wait? *Demography* 22: 469–84.

Miller, E. 1973. Is out-migration affected by economic conditions? *Southern Economic Journal* 39: 396–405.

Miller, R. E. 1972. *Modern mathematical methods for economics and business.* New York: Holt, Rinehart & Winston.

Milne, W. J. 1981. Migration in an interregional macroeconometric model of the United States: will net outmigration from the northeast continue? *International Regional Science Review* 6: 71–83.

Mincer, J. 1978. Family migration decisions. *Journal of Political Economy* 86: 749–73.

Mitchneck, B. 1990. *Geographical and economic determinants of interregional migration in the USSR, 1965–1986.* Ph.D. dissertation, Department of Geography, Columbia University.

Monmonier, M., and Schnell, G. A. 1988. *Map appreciation.* Englewood Cliffs, NJ: Prentice Hall.

Morrill, R. L. 1988. Migration regions and population redistribution. *Growth and Change* 19: 43–60.

Morrill, R. L. 1993. Development, diversity, and regions of demographic variability in the U.S. *Annals of the Association of American Geographers* 83: 406–33.

Morrison, P. A. 1973. *Migration from distressed areas: its meaning for regional policy.* Santa Monica, CA: Rand Corporation.

Morrison, P. A. 1977. The functions and dynamics of the migration process. In *Internal migration: a comparative perspective.* A. A. Brown and E. Neuberger (eds.), pp. 61–72. New York: Academic Press.

Morrison, P. A. 1992. Is "aging in place" a blueprint for the future? Invited address at the session on Major Directions in Population Geography, Annual Meeting of the Association of American Geographers, San Diego, CA.

Morrison, P. A., and Clark, W. A. V. 1992. Local redistricting: the demographic context of boundary drawing. *National Civic Review* (Winter–Spring): 57–63.

Mulligan, G. 1991. Equality measures and facility location. *Papers in Regional Science* 70: 345–65.

Murphy, M. 1991. Household modelling and forecasting—dynamic approaches with use of linked census data. *Environment and Planning A* 23: 885–902.

Myers, D. 1990. *Housing demography.* Madison: University of Wisconsin Press.

Myers, D. 1992. *Analysis with local census data: portraits of change.* New York: Academic Press.

Nam, C. B., and Philliber, S. G. 1984. *Population: a basic orientation.* Englewood Cliffs, NJ: Prentice Hall.

Namboodiri, K., and Suchindran, C. M. 1987. *Life table techniques and their applications.* New York: Academic Press.

Nelson, P. 1959. Migration, real income, and information. *Journal of Regional Science* 1: 43–74.

Newling, B. 1969. The spatial variation of urban population densities. *Geographical Review* 59: 242–52.

New York State Office of Parks and Recreation. 1978. People, resources, recreation. New York State Comprehensive Recreation Plan. Albany, NY.

Nichols, J. E. 1990. *By the numbers: using demographics and psychographics for business growth in the '90s.* Chicago: Basic Books.

Nijkamp, P., van Wissen, L., and Rima, A. 1993. A household life cycle model for residential relocation behaviour. *Socio-Economic Planning Sciences* 27: 35–53.

Notestein, F. 1945. Population—the long view. In *Food for the world.* T. W. Schultz (ed.), Chicago: University of Chicago Press, pp. 36–57.

Olsson, G. 1980. *Birds in egg.* London: Pion.

Openshaw, S. 1984. *The modifiable areal unit problem.* Concepts and techniques in modern geography, 38. Norwich, UK: Geo Books.

Palsky, G. 1991. Statistical mapping of population in the 19th century. *Espace, populations, sociétés* 3: 451–58 (in French).

Pant, P. N., and Starbuck, W. H. 1990. Innocents in the forest: forecasting and research methods. *Journal of Management* 16: 433–60.

Park, R. E., Burgess, E. W., and McKenzie, R. D. 1925. *The city.* Chicago: University of Chicago Press. (Reprinted in 1967.)

Pearl, R., and Reed, L. J. 1920. On the rate of growth of population of the United States since 1790 and its mathematical representation. *Proceedings of the National Academy of Sciences* 6: 275–88.

Pearl, R., Reed, L. J., and Kish, J. F. 1940. The logistic curve and the census count of 1940. *Science* 92: 486–88.

Pickles, A. R., Davies, R. B., and Crouchley, R. 1982. Heterogeneity, nonstationarity, and duration-of-stay effects in migration. *Environment and Planning A* 14: 615–22.

Pittenger, D. 1976. *Projecting state and local populations.* Cambridge, MA: Ballinger.

Plane, D. A. 1981. The geography of urban commuting fields: some empirical evidence from New England. *Professional Geographer* 33: 182–88.

Plane, D. A. 1982. An information–theoretic approach to the estimation of migration flows. *Journal of Regional Science* 22: 441–56.

Plane, D. A. 1984. Migration space: doubly constrained gravity model mapping of relative interstate separation. *Annals of the Association of American Geographers* 74: 244–56.

Plane, D. A. 1987. The geographic components of change in a migration system. *Geographical Analysis* 19: 283–99.

Plane, D. A. 1992. Age composition change and the geographical dynamics of interregional migration in the U.S. *Annals of the Association of American Geographers* 82: 64–85.

Plane, D. A. 1993a. Requiem for the fixed-transition-probability migrant. *Geographical Analysis* 25: 211–23.

Plane, D. A. 1993b. Demographic influences on migration. *Regional Studies* 27: 375–83.

Plane, D. A. 1994. The wax and wane of migration patterns in the US in the 1980s: a demographic effectiveness field perspective. *Environment and Planning A* 26.

Plane, D. A., and Isserman, A. M. 1983. U.S. interstate labor force migration: an analysis of trends, net exchanges, and migration subsystems. *Socio-Economic Planning Sciences* 17: 251–66.

Plane, D. A., and Rogerson, P. A. 1985. Economic–demographic models for forecasting interregional migration. *Environment and Planning A* 17: 185–98.

Plane, D. A., and Rogerson, P. A. 1986. Dynamic flow modeling with interregional dependency effects: an application to structural change in the U.S. migration system. *Demography* 23: 91–104.

Plane, D. A., and Rogerson, P. A. 1989. US migration pattern responses to the oil glut and recession of the early 1980s: an application of shift-share and causative matrix techniques. In *Advances in regional demography*. P. Congdon and P. Batey (eds.), pp. 258–80. London: Belhaven Press.

Plane, D. A., and Rogerson, P. A. 1990. The ten commandments of migration research. In *Regional science: retrospect and prospect*. D. E. Boyce, P. Nijkamp, and D. Shefer (eds.), pp. 15–42. New York: Springer–Verlag.

Plane, D. A., and Rogerson, P. A. 1991. Tracking the baby boom, the baby bust, and the echo generations: how age composition regulates U.S. migration. *The Professional Geographer* 43: 416–30.

Plane, D. A., Rogerson, P. A., and Rosen, A. 1984. The cross-regional variation of in-migration and out-migration. *Geographical Analysis* 16: 162–75.

Population Reference Bureau. 1987. *World population data sheet*. Washington, DC.

Population Reference Bureau. 1991. *Population Today* 19, no. 2: 9.

Population Reference Bureau. 1992. *World population data sheet*. Washington, DC.

Population Reference Bureau. 1993. *World population data sheet*. Washington, DC.

Porell, F. W. 1982. Intermetropolitan migration and quality of life. *Journal of Regional Science* 22: 137–58.

Raisz, E. 1934. The rectangular statistical cartogram. *Geographical Review* 24: 292–96.

Ravenstein, E. G. 1885. The laws of migration. *Journal of the Royal Statistical Society* 48: 167–227.

Rees, P., and Wilson, A. 1977. *Spatial population analysis*. London: Edward Arnold.

Richardson, H. W. 1977. *The new urban economics*. London: Pion.

Richardson, H. W. 1980. Polarization reversal in developing countries. *Papers of the Regional Science Association* 45: 67–85.

Rider, R. V., and Badger, G. F. 1943. Family studies in the Eastern Health District. *Human Biology* 15: 101–26.

Ritchey, P. 1976. Explanations of migration. *Annual Review of Sociology* 2: 363–404.

Robinson, A. H. 1955. The 1837 maps of Henry Drury Harness. *The Geographical Journal* 121: 440–50.

Robinson, A. H. 1982. *Early thematic mapping in the history of cartography*. Chicago: University of Chicago Press.

Robinson, A. H., Sale, R. D., Morrison, J. L., and Muehrcke, P. C. 1984. *Elements of cartography* (5th ed.). New York: John Wiley.

Rogers, A. 1967. A regression analysis of interregional migration in California. *The Review of Economics and Statistics* 49: 262–67.

Rogers, A. 1971. *Matrix methods in urban and regional analysis*. San Francisco: Holden–Day.

Rogers, A. 1975. *Introduction to multiregional mathematical demography*. New York: John Wiley.

Rogers, A. 1985. *Regional population projection models.* Beverly Hills, CA: Sage.

Rogers, A. 1990. Requiem for the net migrant. *Geographical Analysis* 22: 283–300.

Rogers, A. 1992. Heterogeneity, spatial population dynamics, and the migration rate. *Environment and Planning A* 24: 775–91.

Rogers, A., and Castro, L. 1986. Migration. In *Migration and settlement: a multiregional comparative study.* A. Rogers and F. J. Willekens (eds.), pp. 157–208. Boston: D. Reidel.

Rogers, A., Raquillet, R., and Castro, L. 1978. Model migration schedules and their applications. *Environment and Planning* A 10: 475–502.

Rogers, A., and Woodward, J. 1991. Assessing state population projections with multiregional demographic models. *Population Research and Policy Review* 10: 1–26.

Rogerson, P. A. 1979. Prediction: a modified Markov chain approach. *Journal of Regional Science* 19: 469–78.

Rogerson, P. A. 1984. New directions in the modelling of interregional migration. *Economic Geography* 60: 111–21.

Rogerson, P. A. 1990a. Buffon's needle and the estimation of migration distances. *Mathematical Population Studies* 2: 229–38.

Rogerson, P. A. 1990b. Migration analysis using data with time intervals of differing widths. *Papers of the Regional Science Association* 68: 97–106.

Rogerson, P. A. 1992. Some issues in global population modeling. Paper presented at the National Center for Geographic Information and Analysis Workshop in Global Modeling and Geographic Information Systems, United Nations, NY.

Rogerson, P. A., and Fotheringham, A. S. 1993. Spatial analysis and GIS: introduction. *Geographical Systems 1.*

Rogerson, P. A., and MacKinnon, R. 1982. An interregional migration model with source and interaction information. *Environment and Planning A* 14: 445–54.

Rogerson, P. A., and Plane, D. A. 1984. Modeling temporal change in flow matrices. *Papers of the Regional Science Association* 54: 148–64.

Rogerson, P. A., and R. Stack. 1987. Demographic changes in the metropolitan areas of the Great Lakes. Occasional Paper 87–2, Great Lakes Program, 212 Engineering West, State University of New York at Buffalo.

Rogerson, P. A., Weng, R., and Lin, G. 1993. The spatial separation of parents and their adult children. *Annals of the Association of American Geographers,* 83: 656–71.

Rossi, P. 1955. *Why families move.* Glencoe, IL: Free Press.

Ruiter, E. 1967. Towards a better understanding of the intervening opportunities model. *Transportation Research* 1: 47–56.

Sands, G., and Bower, L. 1974. *Vacancy chains in the local housing market: an investigation of the public policy implications of housing turnover.* Ithaca, NY: Cornell University, Center for Urban Development Research.

Schmitt, R., and Greene, D. 1978. An alternative derivation of the intervening opportunities model. *Geographical Analysis* 10: 73–77.

Scott, A. J. 1971. *An introduction to spatial allocation analysis.* Washington, DC: Association of American Geographers.

Shaw, R. P. 1975. *Migration theory and fact.* Bibliography series, no. 5. Philadelphia: Regional Science Research Institute.

Sheppard, E. 1980. The ideology of spatial choice. *Papers of the Regional Science Association* 45: 197–213.

Sheppard, E. 1986. Modeling and predicting aggregate flows. In *The geography of urban transport,* S. Hanson (ed.), p. 91–118. New York: Guilford Press.

Shevky, E., and Bell, W. 1955. *Social area analysis: theory, illustrative applications, and computational procedures.* Menlo Park, CA: Stanford University Press.

Shevky, E., and Williams, M. 1949. *The social areas of Los Angeles: analysis and typology.* Los Angeles: University of California Press.

Shultz, S., and Regan, J. 1991. The 1990 census, GIS technology, and rural data needs. *Rural Sociologist* 11: 23–29.

Shryock, H. S. 1959. The efficiency of internal migration in the United States. In *International Population Conference (proceedings) Vienna* 1959, pp. 685–694. Union internationale pour l'étude scientifique de la population. Vienna: Im Selbstverlag.

Shryock, H. S. 1964. *Population mobility within the United States.* Chicago: University of Chicago, Community and Family Study Center. (Chapter 9: "The effectiveness of migration," pp. 285–94.)

Shryock, H. S., and Siegel, J. 1976. *The methods and materials of demography.* New York: Academic Press.

Silvers, A. L. 1977. Probabilistic income-maximizing behavior in regional migration. *International Regional Science Review* 2: 29–40.

Simon, J. 1982. A scheme to promote world economic development with migration. In *Research in population economics,* vol. 4. J. Simon and P. Lindert (eds.). Greenwich, CT: JAI Press.

Sjaastad, L. A. 1962. The costs and returns of human migration. *Journal of Political Economy* 70 (supplement): 80–93.

Slater, P. B. 1976. A hierarchical regionalization of Japanese prefectures using 1972 inter-prefectural migration flows. *Regional Studies* 10: 123–32.

Slater, P. B. 1981. Combinatorial procedures for structuring internal migration and other transaction flows. *Quality and Quantity* 15: 179–202.

Smith, S. K. 1986. Accounting for migration in cohort-component projections of state and local populations. *Demography* 23: 127–35.

Smith, S. K., and Sincich, T. 1990. The relationship between the length of the base period and population forecast errors. *Journal of the American Statistical Association* 85: 367–75.

Smith, T. E. 1978. A cost-efficiency principle of spatial interaction behavior. *Regional Science and Urban Economics* 8: 313–37.

Snickars, F., and Weibull, J. 1977. A minimum information principle: theory and practice. *Regional Science and Urban Economics* 7: 137–68.

Sommers, B. J. 1993. *The impact of natural amenities on residential stability: a case study of post-industrial demographic change in Cleveland, Ohio.* Ph.D. dissertation in Geography and Regional Development. Tucson: University of Arizona.

Sonis, M. 1980. Locational push–pull analysis of migration streams. *Geographical Analysis* 12: 80–97.

Speare, A., Goldstein, S., and Frey, W. 1975. *Residential mobility, migration, and metropolitan change.* Cambridge, MA: Ballinger.

Spilerman, S. 1972. The analysis of mobility processes by the introduction of independent variables into a Markov chain. *American Sociological Review* 37: 277–94.

Star, J., and Star, J. 1990. *Geographic information systems: an introduction.* Englewood Cliffs, NJ: Prentice Hall.

Stillwell, J. C. H. 1978. Interzonal migration: some historical tests of spatial interaction models. *Environment and Planning A* 10: 1187–1200.

Stone, L. O. 1971. On the correlation between metropolitan area in- and out-migration by occupation. *Journal of the American Statistical Association* 66: 693–701.

Stouffer, S. A. 1940. Intervening opportunities: a theory relating mobility and distance. *American Sociological Review* 5: 845–67.

Sweet, J., and Bumpass, L. 1987. American families and households. New York: Russell Sage Foundation.

Taueber, C. 1961. Duration of residence analysis of internal migration in the United States. *Milbank Memorial Fund Quarterly* 29: 116–31.

Teitelbaum, M. 1985. *The fear of population decline.* Orlando: Academic Press.

Theil, H., and Finizza, A. J. 1971. A note on the measurement of racial integration of schools. *Journal of Mathematical Sociology* 1: 187–93.

Thomas, D. S. 1941. *Social and economic aspects of Swedish population movements: 1750–1933.* New York: MacMillan. (Chapter 7: "Population mobility," pp. 299–98.)

Thomas, R., and Kirchner, R. 1991. *Desktop marketing: lessons from America's best.* New York, NY: American Demographics Books.

Tobler, W. 1983. An alternative formulation for spatial-interaction modeling. *Environment and Planning* A 15: 693–704.

Tomasi, S. 1990. Why the nation needs a TIGER system. *Cartography and Geographic Information Systems* 17: 21–26.

Trewartha, G. T. 1953. A case for population geography. *Annals of the Association of American Geographers* 43: 71–97.

U.S. Bureau of the Census. 1963. *Interim revised projections of the number of households and families: 1965 to 1980.* Current Population Reports, Series P-20, no. 123.

U.S. Bureau of the Census. 1975. *Historical statistics of the United States: colonial times to 1970.* Washington, DC: U.S. Government Printing Office.

U.S. Bureau of the Census. 1979. Illustrative projections of state populations by age, race, and sex: 1975 to 2000. *Current Population Reports,* Series P-25, no. 796.

U.S. Bureau of the Census. 1980a. *Census '80: continuing the factfinder tradition.* Washington, DC: U.S. Government Printing Office.

U.S. Bureau of the Census. 1980b. *Census of population. United States summary.*

U.S. Bureau of the Census. 1981. *Projections of the number of households and families, 1979–1995.* Current Population Reports, Series P-25, no. 805.

U.S. Bureau of the Census. 1982. *1980 census of population.* Volume 1, Part 1, US Summary, PC(1)-D.

U.S. Bureau of the Census. 1985. Estimates of the population of Arizona counties and metropolitan areas: July 1, 1981, 1982, and 1983. *Current Population Reports,* Series P-26, no. 83-3-C.

U.S. Bureau of the Census. 1986. Projections of the number of households and families, 1986 to 2000. *Current Population Reports,* Series P-25, no. 986.

U.S. Bureau of the Census. 1988. Estimates of the population of Kentucky counties and metropolitan areas: July 1, 1981, to 1985. *Current Population Reports,* Series P-26, no. 85-KY-C.

U.S. Bureau of the Census. 1989a. State population and household estimates, with age, sex and components of change. *Current Population Reports,* Series P-25, no. 1044.

U.S. Bureau of the Census. 1989b. Projections of the population of the United States by age, sex, and race: 1988–2080. *Current Population Reports,* Series P-25, no. 1018.

U.S. Bureau of the Census. 1989c. Geographical mobility: March 1986 to March 1987. *Current Population Reports,* Series P-20, no. 430.

U.S. Bureau of the Census. 1990a. U.S. population estimates by age, sex, race, and hispanic origin: 1980–1988. *Current Population Reports,* Series P-25, no. 1045.

U.S. Bureau of the Census. 1990b. *TIGER: the coast-to-coast digital map data base.*

U.S. Bureau of the Census. 1990c. State and local agencies preparing population and housing estimates. *Current Population Reports*, Series P-25, no. 1063.

U.S. Bureau of the Census. 1991. Geographical mobility, 1987 to 1990. *Current Population Reports*, Series P-20, no. 456.

U.S. Bureau of the Census. 1992a. *Census of population and housing.* STF 1C.

U.S. Bureau of the Census. 1992b. Geographical mobility: 1990 to 1991. *Current Population Reports*, Series P-20, no. 463.

U.S. Bureau of the Census. 1992c. *Maps and more: your guide to Census Bureau geography.*

U.S. Bureau of the Census. 1993. *1990 selected place of birth and migration statistics for states.* Population Division. Report CPH-L-121.

U.S. Department of Commerce. 1992. *Statistical abstract of the United States.* Washington, DC: U.S. Government Printing Office.

U.S. Department of Health and Human Services. 1983. *Life tables for the United States, 1900–2050.* Actuarial Study no. 89, Social Security Administration, Office of the Actuary.

U.S. Department of Health and Human Services. 1990a. National Center for Health Statistics. *Monthly vital statistics report. Advance report of final natality statistics, 1988.* Volume 39, no. 4.

U.S. Department of Health and Human Services. 1990b. National Center for Health Statistics. *Monthly vital statistics report. Advance report of final mortality statistics, 1988.* Volume 39, no. 7.

van der Tak, J., Haub, C., and Murphy, E. 1979. Our population predicament: a new look. *Population Bulletin* 34, no. 5: 1–48.

Vining, D. R., Jr., and Kontuly, T. 1978. Population dispersal from major metropolitan regions: an international comparison. *International Regional Science Review* 3: 49–73.

Vining, D. R., Jr., and Pallone, R. 1982. Migration between core and peripheral regions: a description and tentative explanation of the patterns in 22 countries. *Geoforum* 13: 339–410.

Vining, D. R., Jr., and Strauss, A. 1977. A demonstration that the current deconcentration of population in the United States is a clean break with the past. *Environment and Planning A* 9: 751–58.

Waldorf, B. 1990. Housing policy impacts on ethnic segregation patterns: evidence from Dusseldorf, West Germany. *Urban Studies* 27: 637–52.

Wang, J. 1992. *A microsimulation model for regional household analysis.* Ph.D. dissertation in Geography. Buffalo: State University of New York.

Ward, J. 1963. Hierarchical grouping to optimize an objective function. *Journal of the American Statistical Association* 58: 236–44.

Watson, C. J. 1973. Household movement in west central Scotland: a study of housing chains and filtering. Occasional Paper 26, Centre for Regional Studies, University of Birmingham.

Wattenberg, B. 1987. *The birth dearth.* New York: Ballantine.

Weidlich, W., and Haag, G. 1986. Stochastic migration theory and migratory phase transitions. In *Transformations through space and time.* D. A. Griffith and R. P. Haining (eds.), pp. 104–17.

Weiss, M. J. 1988. *The clustering of America.* New York: Harper & Row.

Westing, A. H. 1981. A note on how many humans that have ever lived. *Bioscience* 31: 523–24.

White, G. 1970. *Strategies of American water management.* Ann Arbor: University of Michigan Press.

White, H. C. 1970. *Chains of opportunity: systems models of mobility in organizations.* Cambridge, MA: Harvard University Press.

White, H. C. 1971. Multipliers, vacancy chains, and filtering in housing. *Journal of the American Institute of Planners* 37: 88–94.

White, M. J. 1986. Segregation and diversity measures in population distribution. *Population Index* 52: 198–221.

White, M. J. 1987. *American neighborhoods and residential differentiation.* New York: Russell Sage Foundation.

White, S. E. 1980. A philosophical dichotomy in migration research. *Professional Geographer* 32: 6–13.

Wilber, G. W. 1963. Migration expectancy in the United States. *Journal of the American Statistical Association* 58: 444–53.

Willekens, F. 1993. Final report of the demographic software committee. Presented at the IUSSP meeting in Montreal. Liege, Belgium: International Union for the Scientific Study of Population.

Willekens, F., and Rogers, A. 1978. *Spatial population analysis: methods and computer programs.* Report 4–24. Laxenburg, Austria: International Institute for Applied Systems Analysis.

Wilson, A. G. 1970. *Entropy in urban and regional modelling.* London: Pion.

Wilson, A. G. 1974. *Urban and regional models in geography and planning.* New York: John Wiley.

Winchester, H. P. M. 1977. Changing patterns of French internal migration 1891–1968. School of Geography, University of Oxford, Resource Paper Number 17.

Wolpert, J. 1965. Behavioral aspects of the decision to migrate. *Papers of the Regional Science Association* 23: 154–65.

Wright, J. K. 1936. A method of mapping densities of population with Cape Cod as an example. *Geographical Review* 26: 103–10.

Zellner, A. 1962. An efficient method of estimating seemingly unrelated regressions and tests for aggregation bias. *Journal of the American Statistical Association* 57: 348–68.

SOLUTIONS TO SELECTED EXERCISES

Chapter 2

1. $(\bar{x}, \bar{y}) = (3.833\ldots, 3.166\ldots)$
2. $(X, Y) = (5, 4); D^* = 17{,}429.55$.
   Because the population of City 4 is more than the sum of the populations of all the other cities, any location away from City 4 implies more miles of total travel by its residents than the sum of all potential travel distance savings for the residents of all the other cities. In the case of responses to fires, it may be desirable to minimize the number of potential trips that are particularly long ones, rather than to simply minimize the aggregate distance of trips from the station to all residences served.
3. $(\bar{x}, \bar{y}) = (2.75, 2.5)$
   $(X, Y) = (2.998615, 2.99724)$
   $D^* = 20{,}124.62$
5. $H = 37.6$
6. *Planeville:* $D(0) = 2{,}000$ (persons / sq mi); $D(1) = 1{,}846$; $D(2) = 1{,}704$; $D(3) = 1{,}573$; $D(4) = 1{,}452$; $D(5) = 1{,}341$; $D(6) = 1{,}238$; $D(7) = 1{,}142$; $D(8) = 1{,}055$.
   Rogersonburgh: $D(0) = 5{,}000$; $D(1) = 3{,}352$; $D(2) = 2{,}247$; $D(3) = 1{,}506$; $D(4) = 1{,}009$.
   Planeville has lower densities spread over a wider area, so it exhibits more "sprawl." Although Rogersonburgh has substantially more people close to its center than Planeville, the radius of Planeville is twice that of Rogersonburgh, meaning that it covers four times as much area. Outer rings have more land area than inner ones, so Planeville has more total population. (To find an exact population for either city requires some calculus—it equals the area under the "tentlike" three-dimensional surface defined by the density gradient model.)
7. $V_A = 12{,}985$ people / km; $V_B = 10{,}860$ people / km. Site A is thus the more accessible and potentially better location—but the chain's management should also consider some other criteria before making a final decision. Most critically, perhaps, is to take into account the location of competing supermarkets already located in the West End.

9. Overall sex ratio: $100 \times 8{,}339{,}422 / 9{,}218{,}650 = 90.46$

Age	Sex Ratio
0–4	104.47
5–9	104.52
10–14	104.06
15–19	101.10
20–24	94.27
25–29	93.71
30–34	92.43
35–39	90.64
40–44	91.30
45–49	90.59
50–54	88.93
55–59	87.13
60–64	83.42
65–69	75.85
70–74	67.74
75–79	58.78
80–84	50.66
85 and over	42.91

11. $Q_{7.07} = 0.25$; $Q_{7.08} = 0.28$; $Q_8 = 2.01$; $Q_{13.01} = 1.81$; $Q_{13.02} = 0.72$
12. $E_{11} = 12.075$; $E_{12} = 8.925$; $E_{21} = 10.925$; $E_{22} = 8.075$
13. $\chi^2 = 6.32$ with 1 df. Since this value exceeds the critical ones for the .05 and .01 probability levels, we can confidently state that there is a statistically significant relationship between the location quotients for Irish and Portuguese subpopulations. It is a *direct* relationship because the observed values for the low–low and high–high categories are those that exceed the expected values.

Chapter 3

1. The population in 20 years is $13{,}000 \, (1 + .01)^{20} = 15{,}862$.
2. The population in 10 years is $133{,}000 \, (1 - .021)^{10} = 107{,}567$.
3. The annual rate of increase is $-1 + 10^{[\log(18{,}000/13{,}000)] / 10} = .03308$. If natural logarithms are used to find the annual rate of increase, we have $-1 + e^{[\ln(18{,}000/13{,}000)] / 10} = .03308$.
   The continual rate of increase is $\ln(18{,}000 / 13{,}000) / 10 = .03254$.
4. $\log(20{,}000 / 15{,}000) / \log(1 + .015) = 19.32$ years
   or $\ln(20{,}000 / 15{,}000) / \ln(1 + .015) = 19.32$ years
5. $\ln(20{,}000 / 15{,}000) / .015 = 19.18$ years
6. Using Equation 3.6 for a population increasing at a continual rate,
$$P_t = 3P_0 = P_0 e^{rt}$$

Cancelling the common factor P_0 and taking the natural logs of both sides implies that $\ln 3 = rt$, and solving for t yields $t = \ln 3 / r$

7. Using the standard demographic accounting equation, the elderly female population expected on July 1, 1990, is $971{,}800 - 149{,}200 + 153{,}800 - 45{,}000 = 931{,}400$. Adoption of the Rees–Wilson accounting framework yields $d_{FL} = .1568$, $D_{US,FL} = 12{,}550$, and $D_{FL,US} = 4{,}488$. Thus the elderly female population on July 1, 1990, is $971{,}800 - (149{,}200 - 12{,}550 + 4{,}488) + 153{,}800 - 45{,}000 = 939{,}462$. The standard accounting equation produces a 1990 population that is 0.9 percent too low. The difference between the two approaches is smaller for females than it is for males because of the lower female mortality rate.

8. Crude birth rate = $(122{,}561 + 117{,}416) / 17{,}558{,}072 \times 1{,}000 = 13.67$

9. General fertility rate:

$$\text{Births} / (\text{Female pop 15–44}) \times 1000 = (122{,}561 + 117{,}416) / 4{,}087{,}273 \times 1{,}000 = 58.71$$

10. Age-specific fertility rates:

Age	Fertility rate
10–14	0.74
15–19	34.84
20–24	89.05
25–29	108.17
30–34	66.37
35–39	23.24
40–44	4.34
45–49	0.18

11. Total fertility rate = $5 \times$ (sum of age-specific fertility rates) $/ 1{,}000 = 1.6347$

12. Gross reproduction rate = Total fertility rate $\times$ Proportion of births that are female
    $= 1.6347 \times .4893 = .800$

15. Net reproduction rate = .773.

Chapter 4

1.
$$\mathbf{P} = \begin{bmatrix} .9377 & .0095 & .0370 & .0158 \\ .0065 & .9367 & .0339 & .0229 \\ .0098 & .0154 & .9588 & .0160 \\ .0070 & .0169 & .0280 & .9481 \end{bmatrix}$$

Mobility declined for all regions except the South. The most notable off-diagonal decreases were in the probabilities for flows from the Northeast to Midwest and West, Midwest to South and West, and West to South. Probabilities increased notably for Northeast to South flows, for all flows from the South (except to the West), and for West to Northeast flow.

2. Northeast: IMR = 26.4; OMR = 63.4; NMR = −37.0
   Midwest: IMR = 38.9; OMR = 64.1; NMR = −25.3
   South: IMR = 69.9; OMR = 40.6; NMR = +29.3
   West: IMR = 82.2; OMR = 51.1; NMR = +31.1

3a. In-migration: East Java—56.4%, 17.0%, 26.6%
                   West Java—68.3%, 16.2%, 15.6%
    Out-migration: East Java—91.2%, 4.0%, 4.7%
                   West Java—85.8%, 8.0%, 6.5%

For these two provinces there is a much higher proportion of primary migration, and a lower proportion of secondary migration, than in the United States. We can speculate that this may be due to lower overall mobility levels and fewer economically attractive migration destinations. The out-migration return percentages are lower than for the United States. On the other hand, return migrants are a significant fraction of the small number of in-migrants to East Java. Ledent (1993) concluded that there were surprisingly similar characteristics for U.S., Canadian, and Indonesian migration with respect to the structural relationships among primary, secondary, and return migration—despite the very different circumstances within which migration takes place in these countries.

b. East Java: −50.3%, −66.1%, +16.4%, +29.9%.
   West Java: +4.1%, −7.3%, +37.8%, +45.0%.

4a. Northeast: −41.2%; Midwest: −24.5%; South: +26.5%; West: +23.3%.

b.
$$\mathbf{E}(75-80) = \begin{bmatrix} — & +12.7 & +47.1 & +49.7 \\ -12.7 & — & +29.2 & +33.6 \\ -47.1 & -29.2 & — & +1.2 \\ -49.7 & -33.6 & -1.2 & — \end{bmatrix}$$

$$\mathbf{E}(85-90) = \begin{bmatrix} — & -1.8 & +36.5 & +17.1 \\ +1.8 & — & +17.3 & +17.8 \\ -36.5 & -17.3 & — & +2.8 \\ -17.1 & -17.8 & -2.8 & — \end{bmatrix}$$

5. Northeast: IMC = 61%; OMC = 39%. Midwest: IMC = 32%; OMC = 67%.
   South: IMC = 74%; OMC = 26%. West: IMC = 109%; OMC = −9%.

6. Out-migration (row): ENT = 1.3759; in-migration (column): ENT = 1.3072.
   (As predicted, the row entropy is higher, meaning a more uniform distribution.)

7a.
$$\mathbf{P} = \begin{bmatrix} — & .6437 & .3563 \\ .6116 & — & .3884 \\ .4605 & .5395 & — \end{bmatrix}$$

b. $\mathbf{P} = .5395\mathbf{S}_1 + .3563\mathbf{S}_2 + .0721\mathbf{S}_3 + .0321\mathbf{S}_4$
   where

$$\mathbf{S}_1 = \begin{bmatrix} 0 & 1 & 0 \\ 1 & 0 & 0 \\ 0 & 1 & 0 \end{bmatrix}, \mathbf{S}_2 = \begin{bmatrix} 0 & 0 & 1 \\ 0 & 0 & 1 \\ 1 & 0 & 0 \end{bmatrix}, \mathbf{S}_3 = \begin{bmatrix} 0 & 1 & 0 \\ 1 & 0 & 0 \\ 1 & 0 & 0 \end{bmatrix}, \mathbf{S}_4 = \begin{bmatrix} 0 & 1 & 0 \\ 0 & 0 & 1 \\ 1 & 0 & 0 \end{bmatrix}.$$

c. The extremal tendencies are the same as those for 1985–90. The 1985–90 weights on the first two extremal tendencies are smaller than those for 1975–80, implying that the flow pattern in 1985–90 is less focused than it was during 1975–80.

8. $\Delta m_{24} = -238$; $B_{24} = +5$; $U_{24} = -126$; $G_{24} = -117$.

Chapter 5

1. $(0.8 \times 49.173) + (0.2 \times 51.400) = 49.6$ thousand (to the nearest 0.1 thousand)

2. $P_{7/1/91} = P_{4/1/90(\text{census})} e^{1.25r}$. Taking natural logarithms,
$\ln(P_{7/1/91}) = \ln(P_{4/1/90}) + 1.25r$
$r = [\ln(P_{7/1/91} / P_{4/1/90})] / 1.25 = 0.035435$ (approximately) or 3.5435 percent

3. $P_{4/1/90} = P_{4/1/80} e^{10r} = 49{,}173 / e^{0.35435} = 34{,}501$

4. $X_1 = (422 / 11{,}642) / (445 / 11{,}790) = 0.960$
$X_2 = (55 / 526) / (64 / 538) = 0.879$
$X_3 = (2{,}944 / 21{,}005) / (3{,}112 / 20{,}617) = 0.929$
$Y = 0.863$
$P^{92}_{\text{Saltlick}} = (P^{92}_{\text{Deerkill}} / P^{90}_{\text{Deerkill}})(P^{90}_{\text{Saltlick}}) Y$
$= (32{,}101 / 31{,}714)(4{,}805)(0.863) = 4{,}197$

5. After we compute similar, preliminary estimates for all townships of the county, a pro rata adjustment should be performed to ensure that the township populations sum to the county's 1992 population estimate of 32,101.

6. Using 1980 and 1990 numbers, $Y = 0.669$;
$P^{90} = (31{,}714 / 25{,}680)(5{,}766)(0.669) = 4{,}764$; the estimate differs from the census-based figure by 41, or 0.85 percent.

7. (*Note:* The following line numbers are as per Table 4.2; those numbers that are unchanged from the example shown in Table 4.2 are not repeated here.)

1a. Births (4/1/80 to 12/31/84)	4,649
1b. Deaths (4/1/80 to 12/31/84)	3,540
2d. Enrollment on 12/31/84	6,075
2e. Estimated population age 5 to 14 on 12/31/84	6,344
3a. Population age 0.25 to 9.25 on 4/1/80	6,380
(Note: Includes three-fourths of <1-year-olds and three-fourths of 7-to-9 year-olds)	
3c. Estimated population age 5 to 14 on 12/31/84	6,363
4. Net migration of school-age children (2e − 3c)	−19
5. Net migration rate for school-age children	−.00282
7. Net migration rate for all ages (5 × 6)	−.00336
8. Population 4/1/80 + 1/2 Births − 1/2 Deaths	74,261
9. Net migration of persons of all ages (7 × 8)	−250
10. Population on 12/31/84 = 1 + 1a −1b + 9	74,565

8. Census occupied units = 266,102
Census vacancy rate = 0.0411844
Occupancy rate = 0.9588155
Census housing-unit population = 660,602
Census persons per occupied housing unit = 2.4825142

Date	Population	Housing-Unit Inventory
1/1/90	661,890	277,438
4/1/90 (Census)	662,114	277,532
7/1/90	662,292	277,607
10/1/90	662,657	277,760
1/1/91	662,830	277,833
4/1/91	663,080	277,938
7/1/91	663,516	278,121
10/1/91	663,630	278,169
1/1/92	663,744	278,217

Example of calculations: 1/1/90 estimate of housing units is the census count less net units coming on line during 1/90, 2/90, and 3/90; that is, those permitted in 6/89, 7/89, and 8/89 less units demolished during 1/90, 2/90, and 3/90; $34 + 50 + 16 = 100$ permitted units; $100 \times .95$ buildout factor = 95 units; subtract one demolition, so net change of 94 units from 1/1/90 to 4/1/90 or 277,438 total units; $277,438 \times 0.9588155 \times 2.4825142 + 1,512 = 661,890$ estimated population.

9. $\rho = .94667$; $P_1 = 9,653$; $P_2 = 8,476$; $P_3 = 1,903$; $P_4 = 4,170$; $P_5 = 7,804$.

Chapter 6

1a. Solutions using t in years

	Linear	Geometric	Exponential
Slope:	14,239.64	.024975	.024975
Intercept:	−27,396,685	−35.8375	−35.8375

b.

	Linear	Geometric	Exponential
1990 projection:	933,995	1,014,352	1,016,179
2000 projection:	1,076,392	1,299,787	1,304,474

3.

	Northeast	Midwest	South	West
1995	54,083,837	53,941,445	79,137,607	47,147,111
2000	44,119,857	53,231,431	80,335,487	47,650,207

6a. Average grade progression ratios:

Grade 6 to grade 7: 1.066

Grade 7 to grade 8: 1.072

b. Projected enrollment:

	Grade 7	Grade 8
1994	63	72
1995	62	68
1996	61	66

7. Average enrollment/housing-unit ratio: .20

   Projected enrollment:
   1994: 2,260
   1995: 2,320
   1996: 2,380
   1997: 2,400
   1998: 2,420

Chapter 7

1a. $nm_R = -.10$; $nm_M = +.05$; $P_R^1 = 90,000$; $P_M^1 = 210,000$; $P_R^2 = 81,000$; $P_M^2 = 220,500$; $P_*^2 = 301,500$. The overprediction error is 1,500.

b. $P_R^1 = 110,000$; $P_M^1 = 190,000$. For the net migration rate model: $P_R^2 = 121,000$; $P_M^1 = 181,500$; $P_*^2 = 302,500$. For the constant transition probability model: $om_R = p_{RM} = .20$; $om_M = p_{MR} = .05$; $P_R^2 = 119,500$; $P_M^2 = 178,500$; $P_*^2 = 298,000$. Overprediction "error" by the net migration rate model is 4,500. Differential rates of natural increase may have feedback effects that lead to higher errors of overprediction.

2a. $k = 251,189$

b. California: 120,648; Georgia: 9,483; Illinois: 16,603; Minnesota: 9,808; Nevada: 16,467; New York: 13,533

c. California: +15,817; Georgia: −4,129; Illinois: +22,568; Minnesota: +7,542; Nevada: −5,533; New York: +10,220. The positive residuals for the "Snowbird" states (Illinois, Minnesota, and New York) mean that the model significantly underpredicted these flows; Arizona's warmer winter climate drew large numbers of retirees. By contrast, Georgia (which also has warm winters) sent fewer actual migrants than the number predicted by the model. The large positive residual for California may be evidence of recent "Californication," as that state redistributes people out to the rest of the west due to problems of congestion, high housing prices, and so on. Nevada has been growing rapidly, receiving far more migrants from Arizona than it sends back from its currently quite small population base.

3a.
$$\mathbf{P} = \begin{bmatrix} .630 & .255 & .115 \\ .077 & .767 & .156 \\ .022 & .097 & .881 \end{bmatrix} \quad \mathbf{M} = \begin{bmatrix} 63,043 & 25,456 & 11,501 \\ 15,483 & 153,380 & 31,137 \\ 6,544 & 29,129 & 264,327 \end{bmatrix}$$

b.
$$\mathbf{P} = \begin{bmatrix} .915 & .075 & .010 \\ .019 & .954 & .026 \\ .002 & .018 & .980 \end{bmatrix}$$

404 Solutions to Selected Exercises

With the higher β distance deterrence parameter, the model predicts relatively more intrazonal and nonmovement and distributes fewer migrants to the longest-distance destinations.

4a. $L = 10 / 347 = 0.0288184$

b. $(1 - L)^{D_1} = (0.9711816)^{37} = 0.338935$
$p(2) = (1 - L)^{D_1} - (1 - L)^{D_1 + D_2} = 0.338935 - 0.136908 = .202027$

5. $p(1) = .03923; p(2) = .04688; p(3) = .04459; p(4) = .05065$
$M_{11} = 16; M_{12} = 19; M_{13} = 18; M_{14} = 20.$

6.
$$\mathbf{P}_{2000} = \begin{bmatrix} .913 & .011 & .055 & .022 \\ .006 & .929 & .044 & .021 \\ .008 & .010 & .969 & .013 \\ .007 & .017 & .033 & .944 \end{bmatrix}$$

7.
$$\mathbf{P} = \begin{bmatrix} .8168 & .0913 & .0919 \\ .0556 & .8444 & .1000 \\ .0332 & .0594 & .9073 \end{bmatrix}$$

Chapter 8

1.

To:	Zone 1	Zone 2	Zone 3	Zone 4	Total
From:					
Zone 1	1,213	1,012	695	580	3,500
Zone 2	845	1,275	718	562	3,400
Zone 3	566	946	1,082	406	3,000
Zone 4	176	367	505	1,052	2,100
Total	2,800	3,600	3,000	2,600	12,000

2. $\beta = .404$

3. Predicted probabilities:

Origin	Destination 1	2
1	.724	.276
2	.439	.561

4.

Origin	Destination 1	2
1	.761	.239
2	.487	.513

5a. $3.65 + 0.24 + 3.47 = 7.36$

5b. 153,055

7. Site 3 would attract 656 households from origin A and 577 households from origin B. Site 4 would attract 1,585 households from origin A and 381 households from origin B, making it the more attractive site in terms of expected patronage.

8a. $T^{11} = 0$; $T^{12} = 50$; $T^{21} = 200$; $T^{22} = 50$; $T^{31} = 0$; $T^{32} = 200$.

b. 34 minutes (17,000 combined minutes for all 500 employees).

Chapter 9

1.

Age	Householder Rates
15–24	.2000
25–34	.3333
35–44	.4667
45–54	.5000
55–64	.5000
65+	.6000

2.

Age	Projected Householders
15–24	1,000
25–34	1,167
35–44	1,633
45–54	1,250
55–64	1,250
65+	1,500

3. People per household in census year: 22,000 / 7,000 = 3.143
People per household in projection year: 27,500 / 7,800 = 3.526

4. Percentage increase in population: (27,500 − 22,000) / 22,000 × 100% = 25%
Percentage increase in households: (7,800 − 7,000) / 7,000 × 100% = 11.4%

The relatively rapid growth of the younger age groups, coupled with the relatively smaller householder rates for these age groups, causes the average householder rate to decline or, equivalently, causes the average number of people per household to increase.

5a. One time period ahead: [131 129 65] + [5 10 25] = [136 139 90]
Two time periods ahead: [90.9 119.3 77.4] + [5 10 25] = [95.9 129.3 102.4]

5b.
$$I + Q + Q^2 + Q^3 + \ldots = \begin{bmatrix} 2.55 & 1.73 & 1.00 \\ 0.73 & 2.64 & 1.00 \\ 0.55 & 0.73 & 2.00 \end{bmatrix}$$

so that the average lengths of chains for vacancies beginning in suburban, central city, and rural regions are 5.28, 4.37, and 3.28, respectively.

6. Projected number of households:

Household Size	<15	15–34	35–54	55+	Total
1	0	8,000	4,000	16,727	28,727
2	1,800	8,000	12,000	6,273	28,073
3	7,200	2,667	6,000	0	15,867
4	9,000	1,333	6,000	0	16,333
Total	18,000	20,000	28,000	23,000	89,000

Average household size: 2.22

Chapter 10

1. West Snootford: $H^* = .834$; Silver Spoon: $H^* = .929$. The higher value for Silver Spoon indicates that it has a more uniform distribution of households among the three life-style clusters. This suggests that West Snootford might be the better location. Note that there are 12,000 as opposed to 11,000 "Furs & Station Wagons" tract-cluster households in West Snootford. On the other hand, there are also twice as many households in "Towns & Gowns" tract clusters as there are in Silver Spoon. These are the households least likely to become clients, whereas your friend may be able to pick up business from the family-oriented "Blue Chip Blues" households as well as from the "Furs and Station Wagon" set. Maybe she should decide on the basis of which of the two colleges she has stronger ties to!

2. West Snootford: $S = .542$; Silver Spoon: $S = .616$. Note that $S^{max} = .667$. Again, West Snootford is indicated as the less diverse community.

3. Rogersonburgh: $D = .20$; Planeville: $D = .25$. These values can be interpreted to mean that 20 percent of Rogersonburgh's population or 25 percent of Planeville's population would have to be relocated to one of the other sections of the metro area to achieve equal mixing of majority and minority subpopulations in each of the three sections. Thus this measure suggests that Planeville is the more segregated. Is this a surprising result to you, based on your intuition after quickly glancing at the numbers?

4. Rogersonburgh: $G = .232$; Planeville: $G = .375$. Again, Planeville appears to be the more segregated metro area.

5. Rogersonburgh: $E_{hh} = .2883$, $E_{hg} = .7117$, $E_{gg} = .7628$, $E_{gh} = .2372$. Planeville: $E_{hh} = .4401$, $E_{hg} = .5599$, $E_{gg} = .6267$, $E_{gh} = .3733$. There is a higher probability of a member of the minority subpopulation being "exposed" to a member of the majority subgroup in Planeville than in Rogersonburgh. This would suggest that Rogersonburgh is the more segregated community—but note that the overall minority share of population is higher in Planeville. The maximum achievable values for E_{hg} are .7500 and .6000 for Rogersonburgh and Planeville, respectively. Dividing the actual values by their maximums so that they range from 0 to 1, we get $E_{hg}^{adj} = .9489$ for Rogersonburgh and .9332 for Planeville. On this basis Planeville could again be considered the more segregated.

6. Rogersonburgh: $H = .0599$; Planeville: $H = .0509$. The higher value of entropy for Rogersonburgh indicates that it has a more uniform distribution of subgroups

across the three areas. The evidence provided by all the indexes seems to be that, after controlling for its relatively greater minority population, Planeville has the more segregated population distribution.

7. Sample z scores: For the PPH variable: Tract 1038 BG 4, +0.9; Tract 1182 BG 2, +1.3; Tract 1834.01 BG 3, +1.2; Tract 1928 BG 1, −1.4; For the Laborer Emp variable: Tract 1038 BG 4, +2.4; Tract 1182 BG 2, +0.3; Tract 1834.01 BG 3, −1.4; Tract 1928 BG 1, −1.1. Table 10.9 gives the correct factor score values. (*Note:* Your answers may differ slightly due to rounding.)

8. {1 3 4 6}, {2 5}.

9. In this case iteration would not change assignment of tracts to clusters. Distances of cases to cluster centroids:

	Centroid {1, 3, 4, 6}	Centroid {2, 5}
Case 1	43.12	276.25
Case 2	558.31	11.26
Case 3	4.06	558.21
Case 4	0.90	435.69
Case 5	403.45	11.26
Case 6	30.82	687.25

10. Weighting the sample up to the metro area shares of households, we would predict a 27 percent overall adoption percentage. So the relevant penetration index values (based on 20 percent, 50 percent, and 10 percent adoption percentages, respectively, among respondents in the sample) are "American Dreamers" 74; "Career Careeners" 185; "Life Zonkers" 37.

11. Based on the adoption percentages from the test market sample, we would expect to find an equal number of adopters in the two metro areas: 45,000. The overall penetration indexes are the same for both markets: 83. (45,000 / 200,000 = 22.5 percent; 100 × 22.5 / 27 = 83.) The predicted penetration index under 100 suggests that elsewhere in the nation the demographics are more favorable for introduction of this product. The company should hire Psycho Demons, Inc., to conduct a broader study to seek out better potential markets in other metro areas.

Chapter 11

1. B: 1,267   C: 633
2. B: 1,482   C: 679
3. A: 86 persons per sq mi
   B: 147 persons per sq mi
   C: 238 persons per sq mi
4. 26.5 square units

AUTHOR INDEX

Aldenderfer, M. S., 322, 326
Alonso, W., 104
Ambry M., 300–301
Amrhein, C. G., 35, 135, 158
Anderberg, M., 326
Anselin, L., 358
Asimov, I., 2, 4

Badger, G. F. 114
Barff, R., 116, 117
Bartholomew, D. J., 183
Beale, C. L., 102–103, 104, 105(fig.)
Beaumont, P. M., 227
Bell, W., 308
Berry, B. J. L., 219, 308
Biehl, K., 363
Blashfield, R. K., 322, 326
Blau, P. M., 112
Blumen, I., 114
Bongaarts, J., 296
Borts, G. H., 220, 222
Boulding, Ken, 334
Bower, L., 282
Boyce, D. 233
Bracken, I., 354
Brass, W., 285
Brown, L. A., 219, 280, 362
Bumpass, L., 276
Burch, T., 296
Burgess, E. W., 279, 282, 319
Burt, J. E., 24

Cadwallader, M., 309
Carbaugh, L. W., 344
Castro, L., 107
Champion, A. G., 219
Clark, Colin, 27, 38
Clark, W. A. V., 24, 92, 115, 307, 308
Clarke, J. I., 25, 43, 44
Clarke, M., 291
Clausen, J. A., 280
Clayton, C., 116
Coale, A. J., 108
Coffey, W. J., 39
Collins, L., 211
Cordey-Hayes, M., 100, 101(fig.), 104
Crimmins, E., 370

Crispell, D., 301
Crouchley, R., 115
Cuzzort, R., 29

Davies, W.K.D., 115, 116
Demeny, P., 108
De Montizon, F., 343
Densham, P., 357
Dent, B. D., 25
Dicken, P., 24
Dixon, J., 357
Dorigo, G., 204
Draper, N., 249
Duncan, B., 304
Duncan, O. D., 29, 112, 304
Durand, J. D., 4

Easterlin, R., 8
  on demographic rates, 366–369
Ehrlich, P., 369
Eldridge, H., 113
Ellis, M., 116, 117
Engels, R., 95, 140
Exter, T., 346

Feder, G., 102
Feeney, G., migration model of, 215–218, 219
Fibonnacci, Leonardo, population studies by, 2–4
Flieger, W., 162
Flowerdew, R., 352
Foot, D. K., 225
Forbes, A. F., 183
Fornos, W., 369
Fotheringham, A. S., 358, 363
Francese, P., 299, 300
Fratar, T., 243
Frey, W., 279

Gaile, G. L., *Geography in America*, 20, 91
Gale, S., 212
Gehlke, C., 363
Getis, A., 358
Gober, P., 11, 99, 279, 280, 294
Goldscheider, C., 364
Goldstein, S., 114, 279
Goodchild, M., 357, 358
Graves, P. E., 222, 223

Green, M., 352
Greene, D., 206, 207–208
Greenwood, M. J., on migration, 112, 219–220, 222, 223, 224
Griffith, D. A., 35, 135, 158
Grinold, R. C., 183
Guillard, Achille, 299

Haag, G., 100
Hägerstrand, T. 24
Haining, R. P., 358
Hansen, W., 39
Hargreaves, R. P., 341
Harness, Henry Drury, 340, 341, 342
Harris, C. D., 279, 319
Hartshorn, T. A., 319
Healy, M. K., 95, 140
Hole, James, 282
Hollerith, Herman, 342
Hoover, Edgar, 28
Hoyt, 279, 282, 319
Huff, D. L., 115, 254, 280, 282, 284
Hunter, J. M., 341

Ingegneri, D., 370
Isserman, A. M., 116
  on cohort component models, 174–175
  on migration, 95, 169, 176, 225
  on population estimation, 140, 160

James, L. D., 269

Kalish, S., 9
Kasarda, J. D., 308
Kasun, J., 369
Kehris, E., 352
Keilman, N., 296
Keyfitz, N., 4, 85, 162, 166, 369–370
Kish, J. F., 64
Kitsul, P., 95
Klein, Lawrence, 227
Kogan, M., 114
Kontuly, T., 219
Kriesberg, E. M., 103–104

Lansing, J. B., 220, 224
Lawson, V. A., 219
Lazer, W., 301
Lee, Everett, "A theory of
    migration," 112
Lee, R., 269
Lin, G., 370
Linke, W., 286, 287
Linneman, P. D., 222
Linstone, H. A., 254
Lloyd, P. E., 24
Lolonis, P., 358, 365
Long, J. F., 155
Long, L. H., 109, 113
Louviere, J. J., 223
Lowry, I. S., 100
    on in- and out-migration,
        101(fig.), 102, 104
    on migration, 219, 220–222

McCarthy, P. J., 114
McGinnis, R., 114
McHugh, K. E., 11, 24, 99, 280
McKenzie, R. D., 279, 282, 319
MacKinnon, R., 113, 281
McMillen, O., 95, 140
Markoff, J., 352
Marshall, K. T., 183
Marx, R. W., 344
Menken, J., 370
Miller, R. E., 102
Milne, W. J., 224–225
Mincer, J., 10
Mings, R. C., 24
Mitchneck, B., 204
Monmonier, M., 24, 25, 341
Moore, E. G., 280, 362
Morrill, R. L., 146
Morrison, P. A., 102, 189, 190,
    307, 308, 370
Mueller, E., 220, 224
Mulligan, G., 31
Murphy, M., 290
Musson, T. C., 116
Myers, D., 296
    *Analysis with local census data:
        portraits of change,* 21

Namboodiri, K., 80
Nelson, P., 112
Newling, B., 27–28
Newton, Isaac, 197
Nichols, J. E., 301
Nijkamp, P., 279
Notestein, F., 5

Olsson, G., *Eggs in Bird,* 233
Openshaw, S., 357, 363

Pallone, R., 10, 219

Palsky, G., 340, 341
Pant, P. N., 160
Park, R. E., 279, 282, 319
Pearl, R., 64
Philipov, D., 95
Pickles, A. R., 115
Piirto, R., 299, 300
Plane, David A., 11, 12–13, 116,
    140
    on migration, 99, 105, 196, 204,
        205, 211, 218, 219, 220,
        223, 225, 226, 368–369
    on population movement, 24, 95
Porell, F. W., 223

Raisz, E., 341, 344
Raquillet, R., 107
Ravenstein, E. G., 198
    "The Laws of Migration," 196
Reed, L. J., 64
Rees, P., 68, 69, 176
Regan, J., 346
Reid, N., 280
Renard, B., 116, 117
Richardson, H. W., 219, 279
Rider, R. V., 114
Rima, A., 279
Ritchey, P., 106
Robinson, A. H., 25, 340, 341, 342
Rogers, A., 109, 171, 174
    on migration, 96, 107, 169,
        192, 194–195, 219, 220
Rogerson, Peter A., 11, 12–13,
    105, 113, 140, 281, 358, 370
    on areal movement, 93, 95
    on migration, 204, 212, 220,
        223, 225, 226, 368–369
Rosen, A., 105
Ruiter, E., 208

Sands, G., 282
Schmitt, R., 206, 207–208
Schnell, G. A., 24, 25, 341
Scott, A. J., 260
Shapiro, G., 352
Shaw, R. P., 220
Sheppard, E., 233
Shevky E., 308
Shultz, S., 346
Shyrock, H. S., 47, 80, 99, 140
Siegel, J., 47, 80, 140
Silvers, A. L., 221
Simon, J., 369
Sincich, T., 160
Sjaastad, L. A., 220
Slater, P. B., 115
Smith, H., 249
Smith, S. K., 160, 176
Smith, T. E., 197
Snickars, F., 225

Sommers, B. J., 309, 319
Sonis, M., 118
Speare, A., 279
Spilerman, S., 212
Starbuck, W. H., 160
Stein, J. L., 220, 222
Stillwell, J. C. H., 205
Stone, L. O., 102
Stouffer, S. A., 206
Strauss, A., 29, 219
Suchindran, C. M., 80
Sweet, J., 276

Taueber, C., 114
Teitelbaum, M., 369
Thomas, D. S., 99
Tobler, W., 204
Tomasi, S., 343
Trewartha, G. T., 20
Trussell, T. J., 108
Turoff, M., 254

Ullman, E. L., 279, 319

Van Wissen, L., 279
Vining, D. R., 10, 29, 103–104, 219

Wachter, K., 296
Waldorf, B., 282
Wang, J., 291
Watson, C. J., 282
Wattenberg, B., 369
Weibull, J., 225
Weidlich, W., 100
Weiss, M. J., *The Clustering of
    America,* 300
Weng, R., 370
Westing, A. H., 4, 5
White, G., 269
White, H. C., 280, 282, 284
White, M. J., 300
Wilber, G. W., 109, 110
White, S. E., 91
Willekens, F., 109
Williams, M., 308
Willmott, C. J., *Geography in
    America,* 20, 91
Wilson, A. G., 68, 69, 169, 176,
    197
Winchester, H.P.M., 116
Wise, S. M., 358
Wolpert, J., 280, 362
Wong, D., 363
Woodward, J., 171
Wright, J. K., 352, 355
Wymer, C., 357

Young, J. C., 341

Zellner, A., 225

SUBJECT INDEX

AAG, *see* Association of American Geographers
Accessibility, 23
   indexes of, 37–41
Accessibility surface, 37
Accounting
   demographic, 66–68
   migration modeling and, 224–225
   population change and, 68–70
   projections and, 176–177
ADI, *see* Area of dominant interest
Administrative records, 140–141
Afghanistan, 74
Africa, 74
Age, 77, 82, 109
   cohort component models and, 160, 161–163, 175
   dependency ratio and, 42–43, 56, 367–368
   economic participation and, 43–44
   estimation and, 138–139
   migration and, 106, 107–108
   and population pyramids, 46–48
   sex ratio and, 44–45
Age composition, 8, 9, 12–13, 101
Age heaping, 47
Age-specific fertility rates (ASFR), 82–83, 84(fig.), 85
Age-specific mobility rates, 107
Age-specific mortality rates, 71, 77–78
Agglomerative clustering, cases of, 323–327
Aggregate accessibility index, measuring, 38–41
Aggregate unit estimates, 147–148
Agricultural density, 25, 26, 27
Air photos, 143
AMA, *see* American Marketing Association
Amenities, 222–223, 319–320
American Demographics, Inc., 373
American Demographics (Dow Jones), 22, 300

American Manufacturing Belt, 222, 309
American Marketing Association (AMA), 20, 373
American Planning Association (APA), 20, 373
Amherst (N.Y.), 177(table), 178–179, 181–182
Analysis with local census data: portraits of change (Myers), 21
Annals of the AAG, 20
Annual activity days, 252
APA, *see* American Planning Association
ARC/INFO, 346, 356
Areal units, 93–94, 96–97
Area of dominant interest (ADI), 334
Arizona, 48, 145. *See also* Tucson
   Lorenz curve for, 30, 32–33(table)
   migration to, 70, 205(fig.)
   mortality rates in, 72–74
   population estimates for, 141, 150–151
Arizona Department of Economic Security, 134, 195–196
Asia, 94, 315
ASFR, *see* Age-specific fertility rate
Association of American Geographers (AAG), 19, 20, 373
Atlas Pro, 346
At-risk populations, 65, 69, 192
Attenuation parameter, *see* Distance deterrence
Austria, 71

Baby boom, 8–10, 13, 47
Behavioral research, 223–224
Birth rates, 5, 6, 9, 66, 138. *See also* Crude birth rate; Fertility rates
   and cohort component models, 162–163, 164
   falling, 7–8
   and school censuses, 179, 180(table)

Buffalo (N.Y.), 58, 66
Building permits, 143
Bulgaria, 7
Bureau of the Census, *see* U.S. Bureau of the Census
Businesses, 9, 303

Canada, 74, 144
Capacity utilization, 254–255
Cape Cod, 355
Cartograms, 341, 344
Catalina Foothills School District (Tucson), reassignment in, 14–18
Catchment areas, 14(fig.), 16(fig.), 304–305
Causative matrix model, 115
   and migration patterns, 211–214
CBR, *see* Crude birth rate
CDR, *see* Crude death rate
Censuses, 25, 131
   school, 179, 180(table)
   in United States, 21–22, 149–150
Center of gravity, 31. *See also* Population centroid
Centers of population, determining, 31–37
Centroids, 31, 34–35
   accessibility indexes and, 39–41
   and iterative clustering, 327–328
   and migration modeling, 204, 205(fig.)
Chains of opportunity, in housing, 280–284
Change, 158
   accounting of, 68–70
   components of, 65–66
   future population, 369–371
   long-term, 366–369
   world-scale, 2–8
Children, 43–44, 107, 278
China, 5
Choropleth maps, 27, 354–356
Circulation, 92
Civilization, 4–5
Claritas Corporation, 346

Subject Index

PRIZM Lifestyle Clusters, 301, 328–334
Cleveland Metropolitan Area, 157–158
  factorial ecology of, 301, 309–320, 321(fig.), 322(fig.)
Cluster analysis, 301, 304
  life-style, 303, 328–334
  objectives of, 320–323
  types of, 323–328
Clustering of America, The (Weiss), 300
Cohort component models
  interregional, 174–176
  matrix form of, 164–166, 168(table)
  migration and, 169–174, 177
  single-region, 160–164, 166–67
Cohorts, 86. *See also* Cohort component models
  estimation and, 139–140
  and demographic rates, 366–369
  life tables and, 75, 77, 78–79
Commuting, 235–236
Commuting flows, 242, 243–244
Complete linkage, 323–324, 326
Component method II, calculations in, 137–140
Composition, 23
  and demographic rates, 367–369
  and distribution, 41–42, 53
  of households, 276–279
  measurements of, 42–50
Computers, 342–343
Conjoint analysis, 224
Connecticut, 197, 218
CONQUEST, 346
Contingency tables, and geographic association, 50–53
Convergence criterion, 36
Correlation matrix, 310–311
Costs, 148
  school transportation, 263, 268–269
Counterurbanization, 218–219
Cracking, 307
Crete, 5
Cross-Fratar method, 243–244, 295
Crude birth rate (CBR), 81–82, 85–86
Crude death rate (CDR), 70–71
Cumulative inertia, 114–115
Current Population Reports, 140
Current Population Survey, 203

Daily activity bundle, 24
Daily Urban Systems, 24
Dasymetric mapping, 350
  objectives of, 352–354
Death rates, 78, 138, 176. *See also* Mortality rates
  and demographic accounting, 66–67
  determining, 64–65
  measuring, 70–75
  population change accounting and, 68–70
Decision making, 269–270
Delphi method, 254
Demand
  forecasting, 231–232, 259–260
  housing, 293–296
  optimization techniques, 260–261
  recreation, 248–255
  site location and, 255–259
  travel, 232–233
Demand generation, 231
Demographic accounting equation, 66–68, 194
Demographic analysis, 64–65
Demographic effectiveness, 95, 98–99
Demographic information systems, 339–340
  capabilities of, 350–356
  data in, 345–349
  limits of, 356–357
Demographic rates, and cohort sizes, 366–369
Demographics, 2, 22, 299
Demographic transition, 5–6
  logistic curve and, 62–64
Demographic Yearbook (United Nations), 22
Demography, 19
Demography (journal), 19
Denmark, 5
Density, 25, 27–28, 350
Density-dampened DPW models, 219
Dependency ratio, 42–43, 367–368
Design day supply and demand, 254–255, 256(figs.)
Destination constraints, 263–264
Destination population, 215
Destination population weighted (DPW) model, 218–219, 225
Digital cartograpic database, 340. *See also* Topologically Integrated Geographic Encoding and Referencing

DIS, *see* Demographic information system
Distance deterrence, in gravity model, 197–199
Distribution, 23
  and composition, 41–42, 53
  demand, 231, 252–254
  long-run, 173–174
  measuring, 24–37
Diversity, 300
  measuring, 301–303, 305
  political redistricting and, 307–308
Dominican Republic, 82
Dot maps, 24–25, 26(fig.), 341, 343(fig.)
Doubling time, 62
Dow Jones, *American Demographics,* 22
DPW, *see* Destination population weighted model
Duration of residence, 114–115

East, 46, 213
ECESIS model, 227
Ecological fallacy, 53
Ecological models, 279
Ecology, *see* Factorial ecology
Econometric gravity models, 189
Economic gravity models, 102, 219
  Lowry's, 220–222
Economic participation, 43–44
Economics
  forecasting, 233–234
  and migration, 102, 189–190, 215, 220–222
Efficiency of migration, 95
Eggs in Bird (Olsson), 233
Egypt, 5
Eigenvalues, in factor analysis, 311, 312–313
Emigrants, 92
Employment, 10. *See also* Jobs
Employment-to-population ratio (EPR), 44
Entropy index, 302, 307
Entropy statistic, 105–106
Environmental Protection Agency, 156
EPR, *see* Employment-to-population ratio
Equal Employment Opportunity, 303
Estimates, estimation, 127
  evaluating, 146–151
  techniques for, 130–146
  types of, 128–130
Estonia, 7

412 Subject Index

Ethnic groups, ethnicity, 31
  contingency table analysis of, 50–53
  in factorial ecology, 315–316, 321(fig.)
  location quotients and, 48–50, 56
Ethnic status, 308
Europe, 5, 7, 74, 94
Evanston (Ill.)
  life table for, 76(table), 78, 79
  population estimation for, 137–140
Exponential rates of change, determining, 59–61
Exposure index, 305–306
Extrapolation, 130
  geometric and exponential, 158–160
  linear, 132–133, 157–158
  population projections and, 156–157
Extremal tendencies, 118
  calculating, 119–121

Factor analysis, 308, 311(table)
Factorial ecology, 279, 301, 308
  in Cleveland, 309–320, 321(fig.), 322(fig.)
Factor loadings, 311–312
  in factorial ecology, 313–316
Factor scores, in factorial ecology, 316–319
Faculty, planning projections for, 183–184
Families, 8, 112, 223, 296
  composition of, 284–285
  and households, 325–326
  life cycle of, 313, 319
  sizes of, 145–146
Family status, 308, 319
Feeney model
  and migration, 215–216
  and population projections, 216–218
Fertility rates, 5, 9, 64
  in cohort component models, 160, 163–164, 165, 166, 167
  and dependency ratios, 43, 367
  measuring, 81–86
  migration and, 11–12
Fertility schedules, 108
Fibonnacci's rabbits, 2–4
Field counts, 143
Film optical sensing device for input to computers (FOSDIC), 343
Finland, 74

Fixed-period data, 94–95
Florida, 113, 145
  population change accounting in, 68–70
Focus groups, 300
Forecasting, forecasts, 17. See also Modeling
  decision making and, 269–270
  demand and, 231–232
  methods of, 91–92, 189–190, 211, 259–260
  net migration rates and, 192, 194–196
  population and economic, 233–234, 369–370
  recreation demand and supply, 254–255
  urban travel, 232–245
FOSDIC, see Filmoptical sensing device for input to computers
"4/5th-1/5th adjustment," 131–132
France, 24, 341
Friction of distance value, 256, 258
Frostbelt-to-Sunbelt migration, 10, 11, 201, 213, 369
Fund for Tomorrow, 366
Funding, allocation of, 155–156

GDW, see General destination weighted model
General destination weighted (GDW) model, 225–227
General fertility rate (GFR), 82
Genesee County (N.Y.), 58–59
Geocoding, 25, 343–344
Geographical units
  definitions of, 374–377
  hierarchical relationships of, 377–378
Geographic association, 23
  measuring, 50–53
Geographic information systems (GIS), 24, 339–340
  capabilities of, 350–356
  data in, 345–349
  limits of, 356–357
  and spatial analysis routines, 357–358, 365–366
Geography, 19–20
Geography in America (Gaile and Willmott), 20, 91
Geometric rates of change, 58–59, 61
Geopositioning systems (GPS), 25
Germany, 7, 287, 341
GFR, see General fertility rate

Gini index, 31, 305, 306
GIS, see Geographic information systems
GISPlus, 346, 356
Global scale, 23–24
GMR, see Gross migraproduction rates
GPS, see Geopositioning systems
Grade progression ratios, and school enrollment projections, 177, 178–181
Grafton County (N.H.), 48, 49(fig.)
Gravity models, see also Economic gravity models
  constrained and unconstrained, 199–204
  distance deterrence and, 197–199
  doubly constrained, 204–206
  and intervening opportunities model, 208–211
  migration and, 97, 189, 196–197
  trip distribution and, 239–240
Great Britain, see United Kingdom
Green Valley (Ariz.), 48
Grit Magazine, marketing of, 301, 329–334
Gross migraproduction rates (GMR), 109, 110
Gross migration rates, 97–98, 100–101
Gross reproduction rates (GRR), 84
Group meters, 144
Group-quarters population, 142
Growth, 1
  dynamics of, 57, 62–64
  exponential and geometric, 159–160
  forecasting, 369–370
  population, 4–7, 364–365
  rates of change in, 58–60
GRR, see Gross reproduction rates

Hanover (N.H.), 48
Hansen accessibility, 39. See also Aggregate accessibility index
Headship rate method, see Householder rate method
Health care, 10
Health departments, 137–138
Heterogeneity, and mobility, 114–115
Hierarchical agglomerative methods, 322

Hispanics, 308
  contingency table analysis of, 50–53
  location quotients and, 48–50
  school attendance zone planning, 304–305
Hoover index, 28, 30, 31, 304. *See also* Index of concentration
Householder rate method, 285–287
Householders, 285
Household membership rate method, 287, 290
Households, 9, 142, 258, 315
  cluster analysis and, 321–322, 325–326, 328–329
  composition of, 276–279, 284–285
  growth of, 275–276
  microsimulation models of, 291–293
  models of, 279–280
  projections for, 285–291
  trip generation, 236–238
  types of, 294–296
  vacancy chains and, 280–284
Housing, 8, 9, 315
  censuses of, 21–22, 25
  demand and supply, 293–296
  needs for, 279–280
Housing markets, 281–282, 320
Housing units
  and population estimation, 142–146
  school enrollment projections and, 181–182
Human capital, 12, 220
Hungary, 7

Illinois, 332. *See also* Evanston
Immigration, 92, 213, 315
Income, 112, 222, 241, 282
  contingency table analysis and, 50–53
Income tax returns, 141, 148
Indexes
  accessibility, 37–41
  concentration, 28–29, 30, 31
  diversity, 302–303
  segregation measures, 304–307
Index of concentration, 28–29, 30, 31, 304
Index of dissimilarity, 304–305
India, 5, 71
Indiana, 332
Individuals, 275, 278
Infant mortality, 74–75, 79

INFOMARK, 346
Information systems, *see* Demographic information systems; Geographic information systems
In-migration, 92, 97, 116, 191, 205(fig.)
  in demographic accounting, 66, 70
  modeling, 190, 194
  and out-migration, 100–106
Interaction index, 302–303
Intercensal estimates, 128–129, 133–134
Internal Revenue Service (IRS), 95
  migration data of, 141, 148, 172(table), 217–218
International Encyclopedia of Population, 4
International Geographical Union, 20
Interpolation
  areal, 351–352, 353(tables)
  constant-rate, 133–134
  linear, 130–132
Intervening opportunities model, 189, 196
  and constrained gravity models, 208–211
  formula of, 206–208
Ireland, 25, 340, 342(fig.)
IRS, *see* Internal Revenue Service
Isoline maps, 27
Iterative algorithms, 35
Iterative clustering, 327–328

Japan, 74, 103–104
Jobs
  intervening opportunities model and, 208–211
  migration and, 101, 221
Journal of Marketing, 20
Journal of Marketing Research, 20
Journal of Regional Science, The, 20
Journal of the American Planning Association, 20

Kansas City ADI, 334
Kansas Geographic Bureau, 301

Labor force, 8
  migration of, 101, 220–222
Labor force participation rates (LFPR), 43–44
Labor markets, 101, 222(Fig.), 223, 281

Land use, 143, 279, 234–235
Latvia, 7
"Laws of Migration" (Ravenstein), 196
Leslie Matrix, 165(fig.), 168(table), 170(fig.)
LFPR, *see* Labor force participation rate
Life course, 280
Life cycle, 223
  in factorial ecology, 313, 314
  of households, 279–280
  stage in, 308, 319
Life expectancies, 5, 46(fig.)
Life insurance companies, 75
Life space prism, 24
Life-style, 300
Life-style clusters, 303
  and market segmentation, 328–334
Life tables, 76(table), 108, 110(table), 140
  constructing, 77–80
  interpreting, 75, 77
  survival ratios in, 80–81
Local scale, 2
  school reassignment and, 13–18
Location, residential, 280, 281(fig.)
Locational stress, 280
Location quotients, 48–50
Logarithms, 379
Logistic growth, 62–64
Lorenz curve, 28
  data for, 32–33(table)
  description of, 30–31
  function of, 29–30
Los Angeles, 308
Lowry's model, 220–222

Maine, 197, 201, 218
Manhattan, 301, 332(fig.)
MapInfo, 346
Maps, mapping
  choropleth, 27, 354–356
  density, 350, 352–354
  dot, 24–25, 26(fig.)
  history of, 340–341
Marketing, 9, 20, 300–301
Marketing research, 223–224
Market segmentation, 300
  life-style clustering and, 328–334
Markov chain analysis, 115
Markov model, 282
  migration and, 176–177, 196, 211, 215, 216–218, 226

Markov model, *continued*
  for organizational projections, 183–184
  for population redistribution, 170–174
Massachusetts, 197, 201, 218
Matrix algebra, 380–381
Mean center, 31. *See also* Population centroid
Mean point. *See also* Population centroid
Median age, 42
Median center, calculating, 35–37
Merrill Lynch, 366
Metropolitan areas, 11, 24, 75, 278
Metropolitanization, 28, 29
Michigan, 332
Microsimulation, of household models, 291–293
Middle Ages, 2–4
Midwest, 10, 11(table), 28, 96, 123, 213
Migrants, contracted and speculative, 221
Migrant selectivity, 111–112
Migrant stock, 101–102
Migraproduction rates, and migration expectancies, 109–111
Migration, 8, 64, 67, 141, 160, 278, 280. *See also* In-migration; Net migration; Out-migration
  accounting constraints on, 224–225
  amenities and, 222–223
  behavioral research on, 223–224
  causative matrix model and, 211–214
  composition of, 12–13, 47–48, 368–369
  definition and measurement of, 92–95
  and duration of residence, 114–115
  economics of, 220–222
  generalized destination weighted model, 225–227
  geographic patterns of, 115–124
  gravity models and, 196–206
  information and, 112–113
  interregional, 189, 215–219
  intervening opportunities model and, 206–211
  Markov model and, 170–174

measures of, 95–99
modeling and forecasting, 91–92, 189–190
patterns of, 10–12, 28
repeat and return, 113–114
scale of analysis and, 361–362
sex ratios and, 45–46
Migration flows, 115, 201, 211
  interregional, 171–172
  and population distribution, 173–174, 213
Migration expectancies, table construction and, 108–111
Migration schedules model, 107–108
Migration streams, 113, 193(fig.), 201
  composition of, 106–115, 196
Minorities, 31, 307–308
Mobility, 91, 114, 212, 220. *See also* Spatial mobility
  age and, 13, 107, 109, 368–369
  in factorial ecology, 313–314, 322(fig.)
  of households, 277–279
  migration patterns and, 10, 110, 111
Modal split model, 244
Models, modeling, 108, 244, 254
  causative matrix, 211–214
  cohort component, 160–166, 174–177
  destination population weighted, 218–219
  economic-demographic, 225–227
  economic gravity, 219–225
  Feeney, 215–218
  gravity, 197–206
  of household distribution, 279–280, 291–293
  intervening opportunities, 206–211
  Markov, 170–174, 282
  migration and, 91–92, 97, 189–197
  transportation, 233–234
  trip generation, 235–238
Modifiable areal unit problem, 362–363
Mortality rates, 5, 64, 86
  age-specific, 71–72, 73, 77–78
  cohort component models and, 160, 166
  migration and, 11–12, 109
  sex and, 45, 47
Movement, 24, 93–94
Multiple-listing service data, 145

National Bureau of Standards, 343
National Center for Health Statistics, 9, 74–75
Natural increase, 194
Nearest neighbor distance, 41
Negative exponential distribution, 27
Neighborhoods, 258, 308
  cluster analysis of, 330–331(table)
  contingency table analysis of, 50–53
  location quotients and, 48–50
  schools and, 1–2, 14–18
  trips, 236–238, 239–240
Neoclassical trade-off models, 279
Net migraproduction rates, 109
Net migration, 66, 191, 192, 218–219
  and cohort survival, 169–170
  and in- and out-migration, 101, 104
  modeling, 190, 194–196
  rates of, 98, 105, 176
Net reproduction rates (NRR), 85–86
New England, migration in, 197, 200–201, 202(table), 203, 218–219
New Hampshire, 48, 49(fig.), 197, 197, 201, 218
New Jersey, 46, 47(fig.)
New York (state)
  migration and, 113, 205(fig.)
  mortality rates in, 72–74
  rates of change in, 58–59
  recreation planning for, 245–255, 256(figs.), 257(table)
  school enrollment in, 177(table), 178–179, 180(table)
New York-to-Florida migration stream, 113
New Zealand, 341
Nonhierarchical, iterative partitioning methods, 322
Nonmetropolitan areas, 278. *See also* Rural areas
Northeast, 213
  migration and, 10, 11(table), 43, 96, 123, 222
NRR, *see* Net reproduction rates

Occupancy certificates, 143
Occupancy rates, 144–145

Office of Population Research (Princeton), 19
Ohio, 157–158, 301, 332. *See also* Cleveland Metropolitan Area
Operations research, 260
Opportunity costs, 266, 267
Opportunity model, *see* Intervening opportunities model
Organizations, 183–184
Origin constraints, 263
Origin zones, 236, 240, 241–242
Out-migration, 92, 97, 205(fig.), 191, 216
  demographic accounting and, 66, 67
  and in-migration, 100–106
  modeling, 190, 194
  principal components analysis, 116, 117(fig.)

PAA. *see* Population Association of America
Packing, 307–308
Palo Alto (Calif.), 322, 323(fig.), 324(table)
Papers in Regional Science: The Journal of the RSAI, 20
PCensus, 346
Penetration rates, 329
Percentage rates of change, 58
Period rates, 86
Persons of opposite sex sharing living quarters (POSSLQs), 277
Persons-per-housing-unit factor, 145–146
Peru, 5
Physiological density, 25
Pima County (Ariz.), 131–132, 133, 134
Place-specific capital, 102
Place ties, 102, 223
Planning, 12, 20, 270
  cohort component model and, 165–166
  housing, 9, 284
  land use, 234–235
  manpower, 183–184, 185
  and population projections, 155–156
  recreation facilities, 245–255, 256(figs.)
  school assignment, 260–269
  transportation, 232–233
Polarization reversals, 219
Politics, redistricting, 307–308
Polygons, 349(fig.), 350–351

POPTAC, *see* Population Technical Advisory Committee
Population and Statistical Reports (United Nations), 22
Population Association of America (PAA), 19, 373
Population Bulletin, The (newsletter), 22
Population Commission, 20
Population Index, 19
Population potential, 39. *See also* Aggregate accessibility index
Population Reference Bureau (PRB), 4, 5, 22, 25, 60, 71, 373
Population Statistics Unit (Ariz.), 150–151
Population Technical Advisory Committee (POPTAC), 150, 151
POSSLQs, *see* Persons of opposite sex sharing living quarters
Postcensal estimates, 128–129, 133–134, 148
PPH, *see* Persons-per-housing unit factor
PRB, *see* Population Reference Bureau
Principal components analysis, 115–117
PRIZM Lifestyle Clusters, 301
  case study of, 328–334
Probabilities, 64–65, 70
  survival, 161–162
Professional Geographer (journal), 20
Projections
  accounting-based, 176–177
  choice of scale and, 361–362
  cohort component models for, 160–176
  extrapolative methods for, 156–160
  and geographic information systems, 356–357
  household, 285–293
  models of, 216–218, 219(table)
  net migration rates and, 195–196
  planning and, 155–156
  school enrollment and, 177–182, 363–364, 366
Project Link, 227
Pro rata (proportional) adjustment, 149–150
Prussia, 340

PSG, *see* Population Speciality Group
Psychographics, 300
Pyramids
  age-sex composition in, 46–48
  U.S., 8, 10(fig.), 49(fig.)
Pythagorean formula, in median center computation, 35–36

Race, 75, 315–316, 321(fig.)
Rates, 64–65, 86. *See also by type*
Ratio-correlation, estimates using, 135–137
Recreation, planning for, 245–255, 256(figs.), 257(table), 260
Redistribution, Markov model for, 170–174
Redistricting, political, 307–308
Rees-Wilson Accounting Matrix, 68(fig.)
Regional science, 20
Regional Science Association International (RSAI), 20, 373
Regions, 191
  cohort component models and, 160–164, 166–167, 174–176
Registries, 94
Regression analysis, 53, 200, 212, 222
  population estimates, 134–136
  recreation planning, 249–250
Relative attractiveness model, 254
Reproduction rates, 84–85
Residence, 24
Residential mobility, 92
Retirement, 11, 48, 223–224
Return migration, 113–114
Rhode Island, 197
Rotated factor loading matrix, in factor analysis, 313–316
RSAI, *see* Regional Science Association International
Rural areas, 42–43, 82, 102, 194, 248, 249
Rural-to-urban migration, 46

SAS, 358
Scales, 2. *See also* Local scale; World scale choice of method and, 361–366
  geographic, 354–356
Schedules, 108
School assignment, transportation and, 260–269

416 Subject Index

School attendance zone, 304–305
School districts, 9
   reassignment in, 13–18
School enrollment, 1–2, 17
   estimating, 138–139
   grade progression ratios and, 177, 178–181
   housing-unit method and, 181–182
   projections for, 363–364, 366
SDR, *see* Standardized death rates
SEAs, *see* State Economic Areas
Seasonal movement, 92
Second Report of the Irish Railway Commissioners, 340, 341(fig.)
Seemingly unrelated regression, 225
Segregation, 31, 282, 301
   measures of, 303–307
   political redistricting and, 307–308
Sex ratio
   measuring, 44–46
   population projections and, 160, 164
   and population pyramids, 46–48
Similarity measures, 322–323
Simpson's paradox, 43–44
Site location, determining, 255–259
Snowbelt, 28. *See also* Frostbelt-to-Sunbelt migration
Social area analyses, 308
Social classes, and housing markets, 281–282
Social rank, 308
Social Security, 10, 141
Social structure, urban, 319–320
Socioeconomic status, 308, 313
South, migration to, 10, 11, 28, 96, 123, 222
South Africa, 82
Soviet Union, 47, 204
Spatial analysis routines, 357–358
Spatial mobility, 92–93
Spatial variation, and scale of analysis, 364–366
SPSS, *see* Statistical Package for the Social Sciences Standardization
   direct, 71–73
   indirect, 73–74
Standardized death rates (SDR), methods of adjustment in, 71–74

Standardized tape files (STF), 309
State Data Centers, 22
State Economic Areas (SEAs), 102, 104
Stationarity, 212, 260
Statistical Abstract of the United States, 22
Statistical Package for the Social Sciences (SPSS), 309, 311, 358
Stepping stone method, and TPLP, 264–269
STF, *see* Standardized tape files
Stream effectiveness, 99
Sunbelt, migration to, 10, 11, 28, 201, 213, 223, 224, 369
Supply
   forecasting, 259–260
   housing, 293–296
   recreation, 253(table), 254–255
Survival, in cohort component models, 139, 161–162, 164, 165, 169–170, 174, 175
Survival ratios, 80–81
Sweden, 5, 74
Symptomatic variables, 128
"Synthetic" estimates, 129–130
System effectiveness, 99

Target marketing, 300–301, 333–334
Temporary movement, 92
Texas, 191, 192(fig.), 193(fig.)
TFR, *see* Total fertility rate
"Theory of migration, A" (Lee), 112
Threshold accessibility index, 37–38
TIGER, *see* Topologically Integrated Geographic Encoding and Referencing
Time, 62, 93, 94–95, 148
Time geography, 24
Topologically Integrated Geographic Encoding and Referencing (TIGER), 340
   development of, 341–344
Total fertility rate (TFR), 83–84
Total variance, 311
Traffic assignment, 244–243
Traffic zones, trip distribution and, 238–244
Transition probabilities, 95–97, 214(table), 216
Transportation
   and land use planning, 234–235

   school assignment and, 260–269
Transportation Problem of Linear Programming (TPLP), 18
   example of, 261–264
   solving, 264–269
Travel, 92
   forecasting, 232–245
Trips
   distribution of, 238–244
   generation of, 235–238
Tucson
   contingency table analysis of, 50–53
   location quotients in, 48–50
   school reassignment in, 14–18
Tucson Metropolitan Statistical Area, 131–132

Ukraine, 7
United Kingdom, 71, 205, 357
United Nations, 22
United States, 7, 44, 47(table), 85, 94(table), 144, 220, 300
   age composition in, 12–13
   baby boom in, 8–10
   censuses in, 149–150
   family size in, 145–146
   fertility rates in, 83–84
   geographers in, 19, 20
   households in, 275–276, 287, 289(table)
   infant mortality rates in, 74–75
   logistic growth and, 63–64
   map database for, 343–344
   migration in, 10–12, 45–46, 93, 95, 96, 98(table), 102, 112, 115, 117, 123, 141, 171–172, 205, 213, 214(table), 221, 222
   mobility rates in, 107, 109
   mortality rates in, 72–74
   population centroid in, 34, 35(fig.)
   population data in, 21–22
U.S. Bureau of the Census, 7, 142, 308, 373
   estimation techniques used by, 130, 134, 137, 140, 141, 148, 149–150
   on households, 277, 285–286
   on migration, 93, 95, 96, 113, 192(fig.), 203, 227
   mortality information of, 72(table), 74(table)
   population data in, 21–22

U.S. Bureau of the Census, *continued*
  population dot maps of, 25, 26(fig.)
  population projections by, 182, 216
  TIGER system of, 340, 341–344
U.S. Department of Health and Human Services, 72(table), 74(table), 85
U.S. Department of the Interior, 245
U.S. Population Data Sheet (PRB), 22
U.S. Postal Service, 300
Universities, 183–184
Utility data, 144

Urban areas, 47(table), 82, 92, 194, 279, 300, 308
  aggregate accessibility indexes and, 38–39
  density of, 27–31
  dependency ratios in, 42–43
  migration patterns in, 11, 46
  social structure of, 319–320
  travel forecasting, 232–245
Urbanization, 308

Vacancy chains, in households, 280–284
Vacancy rates, 144–145
Varimax (variance maximization) rotation technique, 312–313

Vermont, 30, 32–33(table), 197
Voting Rights Act, 307

Weighting, 150–151, 218–219
West, migration to, 10, 11, 28, 123, 213, 222
West Germany, 71
World Population Data Sheet (PRB), 22
World scale, population change and, 2–8

Youth, dependency ratios for, 43, 56

Zero Population Growth, 6, 373
Zone Improvement Plan (ZIP) codes, 300

Printed and bound in Singapore by Kin Keong Printing Co. Pte. Ltd.